Semiconductor Quantum Wells Intermixing

Optoelectronic Properties of Semiconductors and Superlattices

A series edited by *M. O. Manasreh*, Dept. of Electrical and Computer Engineering, University of New Mexico, Albuquerque, USA

See the back of this book for titles in preparation in Optoelectronic Properties of Semiconductors and Superlattices.

Semiconductor Quantum Wells Intermixing

Edited by

E. Herbert Li

University of Hong Kong

CRC Press
Taylor & Francis Group
Boca Raton London New York

CRC Press is an imprint of the
Taylor & Francis Group, an **informa** business

First published 2000 by Gordon and Breach Science Publishers

Published 2018 by CRC Press
Taylor & Francis Group
6000 Broken Sound Parkway NW, Suite 300
Boca Raton, FL 33487-2742

© 2000 by Taylor & Francis Group, LLC
CRC Press is an imprint of Taylor & Francis Group, an Informa business

First issued in paperback 2019

No claim to original U.S. Government works

ISBN-13: 978-0-367-44747-2 (pbk)
ISBN-13: 978-90-5699-689-5 (hbk)

Visit the Taylor & Francis Web site at
http://www.taylorandfrancis.com

and the CRC Press Web site at
http://www.crcpress.com

British Library Cataloguing in Publication Data

Semiconductor quantum wells intermixing : material
  properties and optoelectronic applications. –
  (Optoelectronic properties of semiconductors and
  superlattices ; v. 8 – ISSN 1023-6619)
  1. Quantum wells 2. Semiconductors
  I. Li, E. Herbert
  537.6′226

To

my wife *Elaine*
my daughter *Victoria*

and

my teacher *Prof. Bernard L. Weiss*

CONTENTS

ABOUT THE SERIES

The series *Optoelectronic Properties of Semiconductors and Superlattices* provides a forum for the latest research in optoelectronic properties of semiconductor quantum wells, superlattices, and related materials. It features a balance between original theoretical and experimental research in basic-physics, device physics, novel materials and quantum structures, processing, and systems—bearing in mind the transformation of research into products and services related to dual-use applications. The following sub-fields, as well as others at the cutting edge of research in this field, will be addressed: long wavelength infrared detectors, photodetectors (MWIR–visible–UV), infrared sources, vertical cavity surface-emitting lasers, wide-band gap materials (including blue-green lasers and LEDs), narrow-band gap materials and structures, low-dimensional systems in semiconductors, strained quantum wells and superlattices, ultrahigh-speed optoelectronics, and novel materials and devices.

The main objective of this book series is to provide readers with a basic understanding of new developments in recent research on optoelectronic properties of semiconductor quantum wells and superlattices. The volumes in this series are written for advanced graduate students majoring in solid state physics, electrical engineering, and materials science and engineering, as well as researchers involved in the field of semiconductor materials, growth, processing, and devices.

INTRODUCTION

Optoelectronics begins, of course, before the demonstration of the semiconductor laser (GaAs and GaAsP in 1962), but the semiconductor laser gave optoelectronics its main impetus towards becoming a big and important activity. From the beginning and the problem of adequate carrier and photon confinement in a Zn-diffused III-V homojunction laser, which approximates a "weak" form of p-i-n[1], it was clear that the double heterojunction, a "strong" form of p-i-n (i.e., one providing good carrier and photon confinement), was required for semiconductor lasers. Although epitaxial III-V heterojunctions were grown as early as 1960[1], lattice-matched double heterojunctions (DHs) were first realized with the development of the AlGaAs-GaAs system in 1967. It was then only a question of time until the old idea of quantum size effects (QSEs, "the particle in a box") would be manifest with shrinking of layer dimensions or with replication of thin DH layers into superlattices (SLs). For example, QSEs in a semiconductor were known in John Bardeen's laboratory (Urbana) in the early and mid 1950s when Bardeen had Schrieffer work on the conductivity of a thin Ge inversion layer, a carrier-confining layer of quantum well (QW) thickness ("size"). Why then would it not be possible to observe QSEs with III-V heterostructures as they reached QW dimensions ($L_z \lesssim 500$ Å), which is more a question of technology than basic principle?

At first it was thought that only a crystal growth technology such as molecular beam epitaxy (MBE) could grow QWs and superlattices (SLs), which, indeed, was a myth that was easily disproved by the demonstration of liquid phase epitaxial (LPE) InGaAsP-InP p-n QW lasers. In fact, the first LPE QW diode lasers demonstrated in 1977 were followed by metalorganic chemical vapor deposition (MOCVD) AlGaAs-GaAs p-n QW lasers in 1978.[1] The semiconductor laser made the QW practical, and the quantum well heterostructure (QWH), including the SL, offered many new opportunities for study and exploitation. For example, what would be simpler to modify, i.e., "process" into a patterned form, than a thin-layer structure (QWH or SL) with only thin layers of material to remove, move, or rearrange? The possibility of layer rearrangement (intermixing), i.e., crystal-conserving layer metamorphosis via planar processing, was not necessarily obvious, nor a question that even was asked (1980), particularly in view of the great resistance (stability) of an AlGaAs-GaAs QWH or SL to ordinary thermal annealing in spite of the ease of Al-Ga substitution in this prototype lattice-matched III-V system. The now classic example of selective layer modification with conservation of crystal, the subject of this volume edited by E. Herbert Li, is layer intermixing, or as we called it at the time of its discovery, "impurity induced layer disordering (IILD)."[2]

This not-so-obvious method of planar QWH processing IILD involved some element of accident in its discovery, which occurred when we attempted to convert, by low temperature Zn diffusion ($n_{Zn} \sim 5 \times 10^{17}/cm^3$), an undoped AlAs-GaAs SL to a doped SL for a phonon experiment[3,4], and found that the SL was not stable "against" low temperature Zn diffusion.[2] A "red" AlAs-GaAs SL (substrate removed) appeared "yellow" after Zn diffusion because of IILD or Al-Ga intermixing. In other words, Zn diffusion into GaAs dopes the crystal, but Zn diffusion into an AlAs-GaAs superlattice (1980) changed the crystal. A layered crystal, an ordered structure, became disordered. A QWH that proved to be stable against ordinary thermal annealing, turned-out to be unstable to impurity diffusion (at much lower temperatures!), and a new piece of QW and SL science and technology emerged. It did not take too long to establish that at temperatures and anneal intervals well below those required for ordinary QWH layer interdiffusion, impurities such as first Zn[2], next Si[3,4], and then various defects[3,4] and vacancies[3,4] could be diffused selectively (patterned) and promote intermixing of Al-bearing higher-gap barriers with lower-gap QWs, thus creating wider gap bulk crystal.

In effect, IILD or layer intermixing of QWHs established a new basis to define and to form QW devices. In addition, it provided a powerful new method to study diffusion mechanics and mechanisms in III-V compounds. It is interesting that at the time of the discovery of IILD in 1980 — the striking change of a red-gap AlAs-GaAs SL ($L_B \sim 150$ Å, $L_x \sim 45$ Å) into yellow-gap bulk-crystal $Al_xGa_{1-x}As(x \sim 0.77)$ by low temperature Zn diffusion[2] — it was easy, because of the unique nature of IILD, to engage John Bardeen in an old interest of his, atom diffusion in solids.[4] Some of Bardeen's ideas were of help to us in seeing that any process, e.g., a two-site diffusion process (Zn diffusion) that made a column-III atom site available in the region of a difference in Al and Ga concentration, such as at an AlAs-GaAs heteroboundary, provided the basis for intermixing.[4]

Although IILD turned out to be of great practical and fundamental value to us, i.e., providing a powerful method of QW device processing and for study of diffusion itself in III-Vs, it would not have had much real importance if this work remained only in Urbana. In spite of a few skeptics, who at first were prone to believe we were observing a weakness of MOCVD crystals that did not occur in MBE crystals, bit by bit our layer disordering (intermixing) studies were confirmed by others and the work spread. Now we see with this volume how far layer intermixing (IILD) of quantum well heterostructures has progressed and how far it has gone beyond what some time ago could be covered in a mere review article (e.g., Ref. 3). This area of study and work has truly expanded around the globe, which is clear from the authors listed in the chapter contents. This says something for the importance of the subject, for its breadth and its depth, and its value.

Now that we know about layer disordering — its very existence as well as many ramifications — our knowledge of it is bound to expand as QW science and technology expands. E. Herbert Li and the co-authors of this volume have done a great service for the workers and students in this field. They have provided a comprehensive survey and examination of the field so that either the beginner or the expert will find much of value and a convenient point of departure for more study and exploitation of QW layer intermixing (IILD), which is certain to increase in scale with the constantly increasing growth of QW science and technology. It will be interesting to witness the further developments, and have as a reference, a source of perspective, *Semiconductor Quantum Wells Intermixing*

Nick Holonyak, Jr.
John Bardeen Chair Professor of Electrical
and Computer Engineering and Physics
Center for Advanced Study
University of Illinois at Urbana-Champaign

REFERENCES

1. For a recent review see, N. Holonyak, Jr., "The Semiconductor Laser: A 35 Year Perspective," *Proc. IEEE*, **85**, 1678–1693 (Nov 1997).
2. W. D. Laidig, N. Holonyak, Jr., M. D. Camras, K. Hess, J. J. Coleman, P. D. Dapkus, and J. Bardeen, "Disorder of an AlAs-GaAs Superlattice by impurity Diffusion," *Appl. Phys. Lett.* **38**, 776–778 (1981).
3. D. G. Deppe and N. Holonyak, Jr., "Atom Diffusion and Impurity Induced Layer Disordering in Quantum Well III-V Semiconductor Heterostructures," *J. Appl. Phys.* **64**, R93–R113 (1988).
4. N. Holonyak, Jr., "Impurity-Induced Layer Disordering of Quantum-Well Heterostructures: Discovery and Prospects", *IEEE J. Selected Topics in Quantum Electronics*, **4**, 584–594 (July/Aug, 1998).

CHAPTER 1

Theories of Band Structure and Optical Properties of Interdiffused Quantum Wells

K.S. CHAN[a] AND E.H. LI[b]

[a]Department of Physics and Materials Science, City University of Hong Kong, Tat Chee Avenue Hong Kong
[b]Department of Electrical and Electronic Engineering, University of Hong Kong, Pokfulam Road, Hong Kong

1 INTRODUCTION

In recent years, there have been a large number of studies [1–22] of the defect enhanced interdiffusion of quantum well structures with the aim to develop the technique for monolithic integration of optoelectronic devices. With this technique, it is possible to selectively modify the composition variation within a small region in a wafer by increasing the defect concentration. The details of

1

these techniques will be discussed in other chapters in this book. To develop the technique for photonic integration and device performance optimization, it is necessary to know how the optical and electronic properties of a square quantum well is modified when the confinement profile is changed into a non-square one. The objective of this chapter is to present in detail models and formulations needed for the calculation of the optical properties of non-square quantum wells as well as to elucidate the mechanism by which interdiffusion modifies these properties. This chapter serves as a theoretical basis for understanding the performance of interdiffused non-square quantum wells in optoelectronic applications, some of which will be discussed in chapter 15 in this book.

This chapter is organized in the following way. In Section 2, the models for interdiffusion are presented and the effects of interdiffusion on composition variation within a quantum well structure are discussed. In Section 3, the envelop function formalism and the k.p approach for the determination of the conduction and valence band structures are described. The effects of interdiffusion on the band structures are described in detail in the same section. These results are important for the understanding of the optical properties of non-square quantum wells. The formalism for calculating the interband optical transition is presented in Section 4, which is the theoretical foundation for the determination of the optical absorption and optical gain of non-square quantum wells discussed in Sections 5 and 7 respectively. The quantum confined Stark effect, which is an important effect for the operation of quantum well modulators, is discussed in Section 6.

2   DIFFUSION MODELS

Interdiffusion of atoms across the heterointerface alters the composition profile across the QW structure [1–21]. Mathematical models are needed to describe and calculate the changes in the composition profile as a function of the duration of interdiffusion. In this section we present models used to calculate the composition profile of interdiffused non-square quantum wells. In this chapter, we adopt the notation $A_w B_{1-w} C_v D_{1-v}$ to denote the chemical formula of a III-V semiconductor material, where A and B represent the Group III atoms, and C and D represent the Group V atoms. w and v represent the mole fractions of A and C atoms respectively. After interdiffusion the mole fractions of Group III and V atoms are functions of position along the direction of crystal growth (the z-direction) and are denoted by $\tilde{w}(z)$ and $\tilde{v}(z)$

Interdiffusion in AlGaAs/GaAs and strained InGaAs/GaAs QW structures [1–5,7–9] both results in the diffusion of group-III atoms only, i.e. Al, In and Ga atoms, as there is no As concentration gradient across the interface. The diffusion of group III atoms across the quantum well structures is usually described by the Fick's Second law in the direction of crystal growth

$$\frac{\partial C}{\partial t} = D \frac{\partial^2 C}{\partial z^2} \tag{1.1}$$

where C denotes the conentration of group III atoms and D is the diffusion coefficient. The diffusion coefficients of all the Group III atoms are usually assumed to be identical, isotropic and independent of their respective concentrations and the crystal lattice. The concentrations are usually assumed to be continuous across the interface. The interdiffusion process is characterized by a diffusion length L_d, which is defined as $L_d = \sqrt{(Dt)}$, where D is the diffusion coefficient and t is the annealing time of thermal processing. Consider a single AlGaAs/GaAs quantum well with the as-grown barrier material $Al_w Ga_{1-w} As$, the compositional profile of Al in AlGaAs/GaAs QW after interdiffusion is given by:

$$\tilde{w}(z) = w \left\{ 1 - \frac{1}{2} \left[\operatorname{erf}\left(\frac{L_z + 2z}{4L_d}\right) + \operatorname{erf}\left(\frac{L_z - 2z}{4L_d}\right) \right] \right\} \tag{1.2}$$

where z is the coordinate along the crystal growth direction and L_z is the as-grown well width. In an InGaAs/GaAs quantum well, the well material is $In_{w_{In}} Ga_{1-w_{In}} As$. The In mole fraction across the InGaAs/GaAs QW structure after interdiffusion is given by:

$$\tilde{w}_{In}(z) = \frac{w_{In}}{2} \left\{ \left[\operatorname{erf}\left(\frac{L_z + 2z}{4L_d}\right) + \operatorname{erf}\left(\frac{L_z - 2z}{4L_d}\right) \right] \right\}$$

where w_{In} is the as-grown In mole fraction and the QW is centered at $z = 0$. In Figure 1, we show the composition profiles and the conduction band confinement potentials of an $Al_{0.3}Ga_{0.7}As$/GaAs and $In_{0.2}Ga_{0.8}As$/GaAs quantum well structures with various degrees of interdiffusion. The conduction band confinement potentials have an error function profile, which has been confirmed by the experimental study of Chang and Koma [18].

The interdiffusion process is more complicated in the case of InGaAs/InP quantum wells, in which atoms of both the group III and group V sublattices can interdiffuse. There is no a priori reasons to believe that the group III and group V atoms interdiffused with identical rates. In fact, various experimental studies have shown that the interdiffusion of each sublattice can be controlled by the enhancement process. Experiments have demonstrated that, by controlling the processing steps, either only group III atoms interdiffuse [22–25] or as in [26,27] only group V atoms interdiffuse. It is therefore more appropiate to assume that the group III and group V atoms interdiffuse with rates independent of each other, each with their own diffusion length L_d^{III} and L_d^{V}. The relative rate of interdiffusion of these two sublattices depends on the details of interdiffusion process and is still an open question which requires further studies.

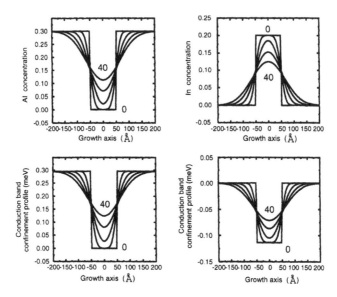

Fig. 1. The composition profile and the conduction band confinement potential of interdiffused (a) $Al_{0.3}Ga_{0.7}As$/GaAs and (b) $In_{0.2}Ga_{0.8}As$/GaAs quantum wells with an as-grown well width of 100 Å for $L_d = 0, 10, 20, 30, 40$ Å.

Currently $In_{0.53}Ga_{0.47}As$/InP QW structures are being investigated for the development of a variety of optoelectronic devices for telecommunication applications such as modulators, detectors, waveguides, and lasers for operations in the 1.3 to 1.55 μm wavelength region [28–30]. Here we consider an InGaAs/InP quantum well structure consisting a layer of $In_wGa_{1-w}As$ with thickness L_z sandwiched between two thick layers of InP. The composition of In after interdiffusion is given by

$$\tilde{w}(z) = 1 - \frac{(1-w)}{2}\left[\text{erf}\left(\frac{L_z + 2z}{4L_d^{III}}\right) + \text{erf}\left(\frac{L_z - 2z}{4L_d^{III}}\right)\right] \quad (1.3)$$

The As mole fraction after interdiffusion is

$$\tilde{v}(z) = \left[\text{erf}\left(\frac{L_z + 2z}{4L_d^V}\right) + \text{erf}\left(\frac{L_z - 2z}{4L_d^V}\right)\right] \quad (1.4)$$

The dependence of $\tilde{w}$ and $\tilde{v}$ on z are determined independently by the two diffusion lengths L_d^{III} and L_d^V. The composition profiles and conduction band confinement profile of (i) Group III diffusion only ($L_d^V = 0$ Å) and (ii) diffusion of both Group III and V atoms with identical diffusion rates

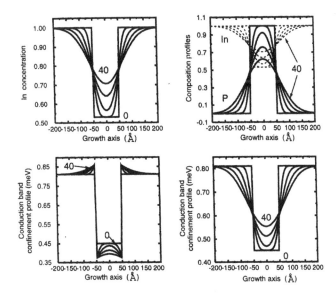

Fig. 2. The composition profile and conduction band confinement potential of interdiffused $In_{0.53}Ga_{0.47}As/InP$ QW with (a) group III interdiffusion only and (b) group III and V interdiffusion with identical diffusion rates for $L_d = 0, 10, 20, 30, 40$ Å.

are shown for comparison in Figure 2 with $w = 0.53$. The well layer $In_{0.53}Ga_{0.47}As$ is lattice-matched to the InP layer and the lattice-match condition is maintained in process (i) as the group III and V atoms are diffused with identical rates. In process (ii), the lattice-match condition is not maintained as the In mole fraction is deviated from 0.53 after interdiffusion resulting in a build-up of lattice strains. The confinement profile is no longer an error function profile as the lattice strain causes a shift in the band-edge energy. A detailed discussion of the strain induced potential can be found in Chapter 5 and Section 3.3.

Although the diffusion model described by Eq. (1.1) is widely accepted and confirmed in AlGaAs/GaAs QW structures experimentally in various studies [18,31], there are experimental evidences showing that the diffusion profile in an InGaAs/InP quantum well structure is not an error function profile as in an AlGaAs/GaAs QW. Nakashima et al. [32] have found the accumulation of Group V atoms near the interface between the InGaAs and InP layers using X-ray analysis. Fuji et al. [33] reported that the composition distribution is discontinuous at an interface. Mukai et al. [34] have studied in detail the energy shift of the subband levels using photoluminescence, which shows that the diffusion processes in the strained InGaAs/InP QW structures are different from those in the AlGaAs/GaAs quantum wells.

Mukai et al. [34] have developed a detailed model of interdiffusion of group V atoms in an InGaAsP QW structure to take into account the effects of the lattice distortion on the interdiffusion rate. In this model the group V atoms diffuse with different rates in the barrier and well lattices and the lattice distortion in a strained QW structure affects the diffusion process in such a way that concentrations of the constituent atoms are not continuous across the interface after interdiffusion. The concentrations across the interface are related to each other by a proportional constant called the distribution ratio, which is found experimentally to be a constant during the diffusion process. In Mukai's diffusion model, the interdiffusion process is described by a set of linear diffusion equations

$$\frac{\partial C_i(z, t)}{\partial t} = D_i \frac{\partial^2 C_i(z, t)}{\partial z^2} \tag{1.5}$$

where $i = b$ for the barrier region and is valid only for $t > 0$, $|z| > L$ (L is the well width with the origin defined at the centre of the well) or $i = w$ for the well region and valid only for $t > 0$, $|z| < L$; C_i are the concentrations of diffusion species in different layers; D_i are the diffusion coefficients of group-V atoms in different layers. The discontinuity of the concentrations across the interface and the continuity of diffusion fluxes give the following boundary conditions

$$C_b(z, t) = kC_w(z, t), \qquad z = \pm L \tag{1.6}$$

and

$$D_w \frac{\partial C_w(z, t)}{\partial z} = D_b \frac{\partial C_b(z, t)}{\partial z}, \qquad z = \pm L \tag{1.7}$$

where k is the interfacial distribution ratio of concentrations. Mukai have fitted the model results to the photoluminescence results and obtained empirical values for the distribution ratio for both AlGaAs and InGaAsP QWs. The distribution ratio of AlGaAs/GaAs interface is equal to 1 while the values for InGaAs/InP interface are between 1 and 40 depending on the temperature and the chemical composition. In the $In_{0.53}Ga_{0.47}As/InP$ QW, the distribution ratio is between 11 and 40, while in $In_{0.53}Ga_{0.47}As/In_{0.7}Ga_{0.3}As_{0.61}P_{0.39}$ the distribution ratio is between 1 and 2. This indicates that the distribution ratio is larger for a larger lattice distortion in $In_{0.53}Ga_{0.47}As/InP$ QW. The set of differential equations have been solved numerically using the finite difference method as in [35]. In Figure 3, we show the composition profiles of group V atoms obtained using the Mukai's model. It has to point out here that the large distribution ratio $k (k \cong 30)$ limits the phosphorous composition in the well layer as well as the shift in confinement energy. The large diffusion constant of group V atoms in the well layer gives rise to a quite uniform distribution of phosphorous atoms in the well layer.

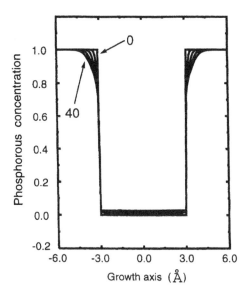

Fig. 3. The composition profile of phosphorous atoms in the $In_{0.53}Ga_{0.47}As/InP$ QW after interdiffusion to various degree obtained using the model described in Mukai et al. [34] for $L_d = 0, 10, 20, 30, 40$ Å

3   BAND STRUCTURE OF INTERDIFFUSED QUANTUM WELLS

3.1   Envelop Function Approximation

The optical properties of a non-square interdiffused quantum well depend on the band structure which can be found by using some existing models, such as the envelop function approximation [36–38], pseudopotential method [39–41], the tight-binding model [42,43] and the effective bond-orbital model [44]. For a non-square diffused quantum well, the chemical composition changes gradually in a spatial region with a thickness depending on the diffusion length within the whole quantum well structure. When L_d is large, we have to consider a thick layer of material with non-uniform composition in our models. This renders the use of computationally intensive approaches, such as the tight-binding model, inconvenient as it is not possible to use the known solutions for a bulk material with uniform composition to simplify the calculation [45]. To avoid these difficulties, we have to use the envelop function approximation which is computationally simple and accurate enough for describing phenomena occurring on a length scale large compared with the unit cell of the crystal. In the study of the optical properties of non-square quantum wells for optoelectronic applications, we are mainly interested in motion of carriers with energies close to the bandedge and wavevectors near to the Brillouin zone centre. The carrier wavefunction varies on a length scale in agreement with the approximation made in the envelop function approach. Apart from being computationally simple, the envelop function has an intuitive clarity which makes it easy to understand the relation between the optical properties and the non-square composition profiles.

In a crystal, the wavefunction of an electron with an energy E in the conduction or valence bands can be found by solving the Schroedinger equation [46]

$$H\psi = \left[\frac{P^2}{2m_0} + V(r)\right]\psi = E\psi \qquad (1.9)$$

where P is the momentum operator, r is the position operator, m_0 is the free electron mass, ψ is the electron wavefunction and $V(r)$ is the potential of the crystal lattice seen by the electron. Since $V(r)$ has the periodicity of the crystal lattice, the solution to the Schroedinger equation has the following form of a Bloch function [47]

$$\psi = e^{ik,r}u(k,r) \qquad (1.10)$$

where k is the wavevector of the electron and u is a periodic function with the periodicity of the crystal lattice. Function u is a function of the wavevector k and is rapidly changing on the length scale of a unit cell. However, if we consider wavevectors in the neighbourhood of the centre of the Brillouin zone ($k = 0$), the function u can be regarded as weakly dependent on k and approximated by $u(0, r)$. An electron wavefunction can be approximately written as the product of the slowly varying envelop function $F(r)$ and the rapidly varying periodic part [36–38]

$$\psi = F(r)u(0, r) \qquad (1.11)$$

$u(0, r)$ depends on the crystal structure of the semiconductor and therefore should be a function of the chemical composition in a non-square interdiffused quantum well. However, in the envelop function approximation, this dependence is assumed to be very weak and ignored. The changes in the chemical composition can only have effects on the slowly varying envelop function which usually satisfies a differential equation with a form similar to the Schroedinger equation. This modelling scheme is generally referred as the envelop function approach [36–38]. The differential equation governing the envelop function is usually derived from the Schroedinger equation using the k.p approach [47,48]. This approximation scheme is usually sufficient to describe the electron states in III-V compound semiconductor interdiffused quantum well devices as the range of energy considered is close to the conduction and valence band edges.

The conduction band in an interdiffused quantum well has only the spin degeneracy and is parabolic when it is well separated in energy from the other bands; the envelop function can be calculated by solving the Schroedinger-like equation

$$-\frac{\hbar^2}{2}\nabla\frac{1}{m(r)}\nabla F(r) + U_c(r)F(r) = EF(r) \qquad (1.12)$$

where $m(r)$ is the position-dependent effective mass of the electron in the band considered. In an interdiffused non-square quantum well structure, the chemical composition is a function of the coordinate along the growth direction z and is a constant within the $x-y$ plan. The motion in the z-direction can be separated from the motion in the $x-y$ plan. The envelop function for the z-direction can be obtained by solving the Schroedinger-like equation shown below

$$\frac{-\hbar^2}{2}\frac{d}{dz}\frac{1}{m_{c\perp}(\tilde{w}(z), \tilde{v}(z))}\frac{d\varphi_C(z)}{dz} + U_C(z)\varphi_C(z) = E\varphi_C(z) \qquad (1.13)$$

where $m_{C\perp}(\tilde{w}(z), \tilde{v}(z))$ is the effective mass of the conduction band, which depends on the position along the crystal growth direction z through the position dependent composition $\tilde{w}(z)$ and $\tilde{v}(z)$ of the Group III and Group V atoms in an interdiffused quantum well. $U_C(\tilde{w}(z), \tilde{v}(z))$ is the position-dependent potential energy due to the changes in the conduction band edge caused by the composition variation. The Hamiltonian for motion in the $x-y$ plan is the Hamiltonian for a free electron and the envelop function perpendicular to the growth direction is the free electron wavefunction $e^{ik.r}$. The differential equation for the z-direction can be easily solved using the finite-difference approximation with appropiate boundary conditions. Usually we consider the quantum well contained in a large box along the z-direction with the boundary condition $\psi = 0$ at the boundaries of the box.

In a strain-free III-V semiconductor (like AlGaAs) quantum well, we should consider two valence bands, namely, the heavy and light hole bands in order to determine the optical properties of a non-square quantum well as they are close in energy for device characteristics. In fact, in a bulk material and at the Brillouin zone centre, the heavy and light hole bands have the same energy. The valence band structure in an interdiffused quantum well is non-parabolic, because of the coupling between the heavy and light holes, which will be discussed in details in the Section 3.2. Nevertheless, if we are interested in their properties at the Brillouin zone centre where the heavy and light hole subbands are not coupled, their envelop functions can be found by a differential equation obtained from Eq. (1.12) by replacing U_c with U_i and m_c with m_i ($i = h$ for heavy hole and $i = l$ for light hole).

3.1.1 Conduction and valence subbands of an interdiffused quantum wells

Interdiffusion changes the square shape of an as-grown quantum well into a non-square profile which does not have a well-defined well width. It is difficult to define a single well width for a non-square quantum well with a depth-dependent well width. Various schemes proposed to define a single well width have been considered and discussed in details in Li et al. [49] To facilitate the discussion of a non-square well shape here, it is convenient to define an energy-dependent effective well-width in the following way. Consider an energy E in the conduction band, we first solve the equation $E = U_c(z_1) = U_c(z_2)$ for z_1 and z_2. The effective well width at energy E is then defined as $|z_1 - z_2|$. To understand the effects of interdiffusion on the well shape, we consider the potential of an as-grown square well structure and two non-square wells with different degrees of interdiffusion in Figure 4. In the figure, we also show the cross-over point which is defined as the intersection point of the potential energy of an interdiffused non-square well and the as-grown square well. In terms of the effective well width, a non-square quantum well has a larger or smaller effective well width in comparison with the width of the as-grown square well depending on the energy considered. For energy levels below the cross-over point, the effective well width is smaller than the as-grown values, while for energies above the cross-over point the effective well widths are larger than that of the square well. When the degree of interdiffusion increases the cross-over point gradually moves towards the bottom of the non-square well. If we consider an energy below the cross over point with a fixed energy separation from the cross-over point, the effective well width decreases with the increase in the degree of interdiffusion. For an energy above the cross-over point with a fixed energy separation from the cross-over point, the effective well-width increases with the decrease in the degree of interdiffusion.

With the concept of the cross-over point, it is easy to understand the effects of interdiffusion on the energies of the electron and hole subbands. In Figure 5, we show the electron and hole subband levels as a function of L_d for $Al_{0.3}Ga_{0.7}As/GaAs$ quantum well with an as-grown well width of 100 Å. We see that for low lying levels (C1, C2, HH1, HH2, HH3, LH1, LH2), when L_d is between 0 and 20 Å, the subband energy increases with L_d. This is due to the fact that the effective well width is reduced with the increase in L_d when L_d is between 0 and 20 Å. When L_d is greater than 20 Å, the effective well width starts to increase again for subbands either close to or above the cross-over point. For subbands with energies below the cross-over point, the subband energies and the cross-over point energy are moving towards each other and the effective well width is then increased to the value of the as-grown well width. For those subbands close to the top of the well (C3, HH4, LH3), the subband energies start to decrease when $L_d > 0$ Å, as the effective well width for these levels increases with L_d.

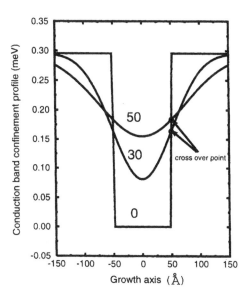

Fig. 4.   The conduction band confinement potentials of interdiffused $Al_{0.3}Ga_{0.7}As$/GaAs QW with an as-grown well width $= 100$ Å and $L_d = 0, 30, 50$ Å.

It is important to point out that in Figure 5 the increases in energies for C2, HH2, HH3 are larger than those increases for C1 and HH1, causing increase in the separation between the first subband and subbands lying above. This phenomenon can be explained by considering the probability distribution of the carriers in the well and the change in the confinement potential due to interdiffusion. If we consider the confinement potential in Figure 1, we notice that when L_d increases, the change in the confinement potential is negative outside the well and positive inside the well. The positive change in the confinement potential is largest when the position is near to the interface and is the smallest in the centre of the well (see Figure 6). For the first and second subbands in the quantum well, the probability distributions of which are shown schematically in Figure 7, the highest probabilities are respectively at the centre of the well and a distance of $L_z/4$ from the centre. So carriers occupying the second subband experience a larger change in energy as the change in the energy is proportional to product of the probability and the change of the confinement potential, i.e., $\Delta E \propto |\psi|^2 \Delta U$. For subbands near to the top of the well (C3, HH4, LH3), the wavefunctions penetrate quite deeply into the barrier ($|z| > L/2$) and therefore are affected strongly by the decrease in the confinement potential in.the barrier layer.

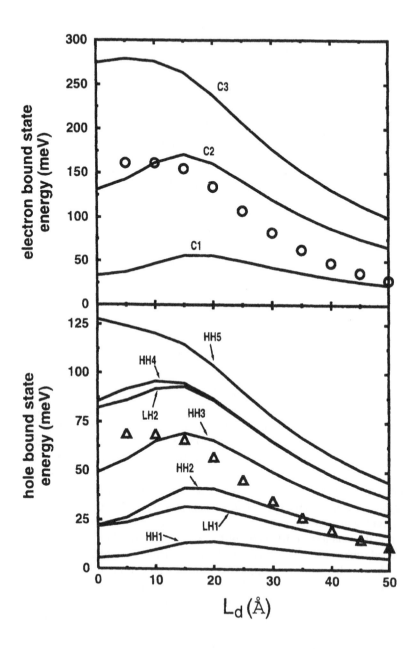

Fig. 5. The conduction and valence subband levels as a function of L_d for an $Al_{0.3}Ga_{0.7}As$/GaAs quantum well with an as-grown well width=100 Å. The energies of the cross-over points for conduction band (circle) and valence band (triangle) are also shown.

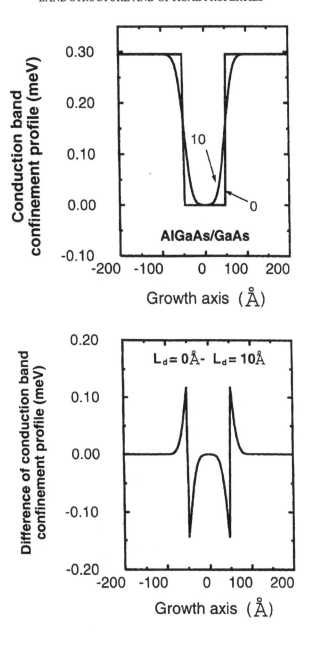

Fig. 6. The difference in energy between the confinement potentials of an interdiffused Al$_{0.3}$Ga$_{0.7}$As/GaAs quantum well with $L_d = 10$ Å and an as-grown square quantum well ($L_d = 0$ Å).

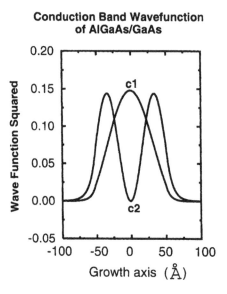

Fig. 7. The probabilities density of an electron in the first ($c1$) and second ($c2$) conduction subbands.

3.2   Valence Band Structure

The valence band structure in an interdiffused quantum well is more complicated as there is a four fold degeneracy (including spin degeneracy) at the top of the valence band. The periodic part of the Bloch function (not including the spin degeneracy) at the top of the valence band has the symmetry of a p type wavefunction which is three fold degenerate. Combining with the spin, there are six valence bands (the heavy hole band, the light hole band and spin-orbit split-off band) just below the conduction band. The spin-orbit split-off band is splitted from the heavy hole and light hole bands by the spin-orbit interaction. The heavy and light hole valence bands remain degenerate at the centre of the Brillouin zone ($k = 0$) in a bulk material but splitted in an interdiffused quantum well. If the separation in energy of the spin-orbit split-off band is far from the heavy and light hole bands as in AlGaAs quantum well, then the envelop function can be obtained by solving the equation

$$[H_V + U_V]F(r) = E F(r) \tag{1.14}$$

where H_V is the Luttinger-Kohn Hamiltonian [50] and U_V is the confinement potential given by

$$
H_V = \begin{bmatrix} P+Q & -S & R & 0 \\ -S & P-Q & 0 & R \\ R^+ & 0 & P-Q & S \\ 0 & R^+ & S^+ & P+Q \end{bmatrix} \tag{1.15}
$$

$$
U_V = \begin{bmatrix} U_{HH}(\tilde{v}, \tilde{w}) & 0 & 0 & 0 \\ 0 & U_{LH}(\tilde{v}, \tilde{w}) & 0 & 0 \\ 0 & 0 & U_{LH}(\tilde{v}, \tilde{w}) & 0 \\ 0 & 0 & 0 & U_{HH}(\tilde{v}, \tilde{w}) \end{bmatrix} \tag{1.16}
$$

$$
P = \frac{\hbar^2}{2m_0} \left(-\gamma_1(\tilde{w}, \tilde{v}) \left(\frac{\partial^2}{\partial x^2} + \frac{\partial^2}{\partial y^2} \right) - \frac{\partial}{\partial z} \gamma_1(\tilde{w}, \tilde{v}) \frac{\partial}{\partial z} \right) \tag{1.17}
$$

$$
Q = \frac{\hbar^2}{2m_0} \left(-\gamma_2(\tilde{w}, \tilde{v}) \left(\frac{\partial^2}{\partial x^2} + \frac{\partial^2}{\partial y^2} \right) + 2 \frac{\partial}{\partial z} \gamma_2(\tilde{w}, \tilde{v}) \frac{\partial}{\partial z} \right) \tag{1.18}
$$

$$
R = \frac{\hbar^2}{2m_0} \left(\sqrt{3}\gamma_2(\tilde{w}, \tilde{v}) \left(\frac{\partial^2}{\partial x^2} - \frac{\partial^2}{\partial y^2} \right) - i2\sqrt{3} \frac{\partial}{\partial z} \gamma_3(\tilde{w}, \tilde{v}) \frac{\partial}{\partial z} \right) \tag{1.19}
$$

$$
S = \frac{\hbar^2}{2m_0} \sqrt{3} \left(-\frac{\partial}{\partial x} + i \frac{\partial}{\partial y} \right) \left(\gamma_3(\tilde{w}, \tilde{v}) \frac{\partial}{\partial z} + \frac{\partial}{\partial z} \gamma_3(\tilde{w}, \tilde{v}) \right) \tag{1.20}
$$

$\gamma_1(\tilde{w}, \tilde{v})$, $\gamma_2(\tilde{w}, \tilde{v})$ and $\gamma_3(\tilde{w}, \tilde{v})$ are Luttinger-Kohn parameters depending on position z through compositions $\tilde{w}$ and $\tilde{v}$. In some cases, the warping of the valence band in the x–y plan is neglected and γ_2 and γ_3 are assumed to be identical to simplify the calculation. $F(r)$ is a column vector of envelop functions of heavy and light hole bands which has the general form

$$
\begin{bmatrix} F_{s=3/2}(r) \\ F_{s=1/2}(r) \\ F_{s=-1/2}(r) \\ F_{s=-3/2}(r) \end{bmatrix} \tag{1.21}
$$

where $s = \pm 3/2$ and $s = \pm 1/2$ denote the heavy hole and the light hole bands respectively. The Luttinger-Kohn parameters γ_i are dependent on z because of the change in the compositions in an interdiffused quantum well, which will make the calculation of the subband structure complicated. To simplify the Hamiltonian, we make the following assumptions:

1. In the term proportional to $\frac{\partial}{\partial x^2} + \frac{\partial}{\partial y^2}$ and describing motions on the x–y plan, γ_1 and γ_2 are assumed to be independent of z and are approximated by the values at the centre of the well, $\gamma_1^m = \gamma_1(w_m, v_m)$ and $\gamma_2^m = \gamma_2(w_m, v_m)$, where w_m and v_m are compositions at the centre of the well.

2. For the off-diagonal terms R and S, γ_2 and γ_3 are assumed to have the values at the centre of the well; $\gamma_2^m = \gamma_2(w_m, v_m)$ and $\gamma_3^m = \gamma_3(w_m, v_m)$.

These assumptions are justified on the ground that the wavefunctions are localized in the well and there is a large probability of finding the electron in a central region near the centre of the well. More complicated approximation schemes can be developed to take into account the fact that for some subbands the highest probability of finding the hole is not at the centre of the well but with a distance from the well centre (for example $L_z/4$ for the second subband). However, the present assumptions suffice for the study of optical properties for device applications. Taking the x–y plan wavefunctions to be the travelling wavefunction, the terms P, Q, R and S in the Luttinger-Kohn Hamitonian can be simplified as

$$P = \frac{\hbar^2}{2m_0} \left(\gamma_1^m(k_x^2 + k_y^2) - \frac{\partial}{\partial z}\gamma_1(\tilde{w}, \tilde{v})\frac{\partial}{\partial z} \right) \qquad (1.22)$$

$$Q = \frac{\hbar^2}{2m_0} \left(\gamma_2^m(k_x^2 + k_y^2) + 2\frac{\partial}{\partial z}\gamma_2(\tilde{w}, \tilde{v})\frac{\partial}{\partial z} \right) \qquad (1.23)$$

$$R = \frac{\hbar^2}{2m_0} \left(\sqrt{3}\gamma_2^m(-k_x^2 + k_y^2) - i2\sqrt{3}\gamma_3^m\frac{\partial^2}{\partial z^2} \right) \qquad (1.24)$$

$$S = \frac{\hbar^2}{m_0} \sqrt{3}(-ik_x - k_y)\gamma_3^m\frac{\partial^2}{\partial z^2} \qquad (1.25)$$

The Luttinger-Kohn Hamilitonian is actually a set of coupled linear differential equations for the envelop functions which can also be solved using the finite-difference approximation. Another approach which is less complicated is the effective Hamiltonian approach developed in [51]. One advantage of the effective Hamiltonian approach is the reduction of the computer memory for storing the envelop function for the calculation of the optical properties of the quantum well. The details of the effective Hamiltonian approach are included in the appendix.

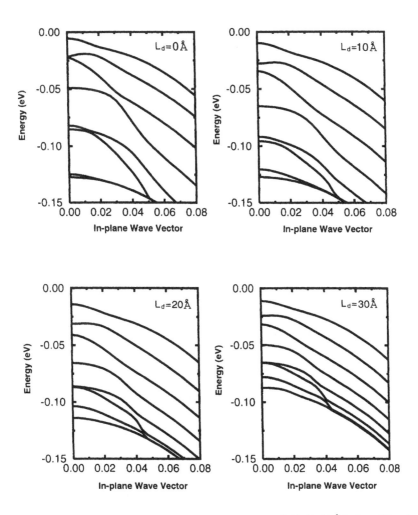

Fig. 8. Valence subband dispersions in the $x-y$ plan for $L_d = 0, 10, 20, 30$ Å for interdiffused $Al_{0.3}Ga_{0.7}As/GaAs$ quantum wells with an as-grown well width = 100 Å.

3.2.1 Valence band structure of diffused AlGaAs/GaAs quantum wells

In Figure 8, the valence band structures of $Al_{0.3}Ga_{0.7}As/GaAs$ quantum wells with an as-grown well-width of 100 Å are shown for various degrees of interdiffusion. The valence band structures are important properties of the quantum well structure as they determine the valence band density of states which is an important factor in determining the distribution of holes among the subbands. In an interdiffused quantum well laser, the distribution of the carriers among the subbands determines the magnitude of the gain spectrum. In the band structure the subbands arranged in a descending order in energy are correspondingly HH1, LH1, HH2, HH3, LH2, HH4 and HH5. Accordingly to the Luttinger-Kohn theory, the light hole subbands are coupled to the heavy hole subbands and the effect of coupling is strong if the heavy and light hole subbands are close in energy. Take the example of the as-grown square quantum well, LH1 and HH2 are very close in energy and the shape of these two dispersion curves are determined by the coupling between these two subbands at a region near to $k = 0$. The heavy and light hole coupling splits the two levels and as a result the LH1 dispersion curve bends up-ward and has a negative effective mass because of this coupling. At $k = 0.02$, the HH1 subband and LH1 subband are close in energy and the dispersion curves of HH1 and LH1 are affected by the coupling between HH1 and LH1. The result of this interaction is the bending down of the LH1 dispersion curve and the decrease in curvature of the HH1 subband at $k = 0.02$. Strong coupling between the HH2 and LH2 subbands occurs at $k = 0.03$ for $L_d = 0$ Å. The curvatures of HH2 and LH2 are respectively decreased and increased because of this coupling.

When the degree of interdiffusion increases, the relative energy positions of the heavy and light hole subbands change as shown in Figure 5 and discussed in Section 3.1; the hole subband dispersions are changed as a result. For $L_d = 10$ Å, the LH1 subband has a smaller negative effective mass at $k = 0$ than that in the as-grown square well because the HH2 shifts away from the LH1. The mixing of the HH2 and LH2 is reduced when Ld changes from 0 to 10 Å as the energy separation is increased. We are always interested in the effective masses of the HH1 subbands as they are important factors in deteremining the opitcal properties of the quantum well lasers and modulators. An increase in the effective mass of HH1 leads to an increase in the exciton binding energy and the optical absorption. From Figure 8, we notice that the HH1 effective mass does not have a strong dependence on L_d, as the energy separation between HH1 and LH1 is more or less independent of L_d (which is clear shown in Figure 5).

3.3   Interdiffused Strained Quantum Wells

A strained quantum well forms when the well material has a lattice constant different from that of the substrate. The strained quantum wells have been investigated intensely [52–57] because of the possibility to control the band structure by controlling the strain. In Chapter 5, we have shown that we can further tailor the strain profile by controlling the interdiffusion process. It is interesting to study the effects of interdiffusion on the band structure of strained quantum wells.

When a well material with a lattice constant $a(w, v)$ is grown on a substrate with a lattice constant a_0, the strain in the $x-y$ plan is

$$\varepsilon_{//} = \varepsilon_{xx} = \varepsilon_{yy} = \frac{a(w, v) - a_0}{a(w, v)} \tag{1.26}$$

In an interdiffused quantum well the composition is a function of the coordinate in the direction of crystal growth; the strain $\varepsilon_{//}$ is then position dependent which is given by

$$\varepsilon_{//} = \frac{a(\tilde{w}(z), \tilde{v}(z)) - a_0}{a(\tilde{w}(z), \tilde{v}(z))} \tag{1.27}$$

Since there is no stress along the growth direction, the well layer expands along the z-direction and the strain along the growth direction is

$$\varepsilon_{\perp}(\tilde{w}, \tilde{v}) = -\frac{2\sigma}{1 - \sigma}\varepsilon_{//} = -\frac{2C_{12}(\tilde{w}, \tilde{v})}{C_{11}(\tilde{w}, \tilde{v})}\varepsilon_{//}(\tilde{w}, \tilde{v}) \tag{1.28}$$

where σ is the Poison ratio, C_{11} and C_{12} are the elastic stiffness tensor. The strain can be separated into two components: hydrostatic and axial. The hydrostatic strain is just the ratio of the change in volume to the original volume and the axial strain has two components which are given by

$$\varepsilon_{vol} = \varepsilon_{xx} + \varepsilon_{yy} + \varepsilon_{zz}$$

$$\varepsilon_{ax} = \begin{cases} \varepsilon_{xx} - \varepsilon_{zz} \\ \varepsilon_{yy} - \varepsilon_{zz} \end{cases} \tag{1.29}$$

The hydrostatic strain does not change the symmetry of the crystal lattice and therefore shifts the band edges of the conduction band and the valence band in energy. The heavy hole and light hole bands are shifted by the hydrostatic strain with the same energy and there is no relative splitting of the heavy and light hole bands. On the contrary, the axial strain lowers the symmetry of the crystal lattice and hence splits the heavy and light holes with respect to each other. The shift in the band edge energy is given by

$$E_C(k = 0) = a_c(\tilde{w}, \tilde{v})[\varepsilon_{xx}(\tilde{w}, \tilde{v}) + \varepsilon_{yy}(\tilde{w}, \tilde{v}) + \varepsilon_{zz}(\tilde{w}, \tilde{v})] \tag{1.30}$$

$$\begin{aligned} E_{HH}(k = 0) = a_V(\tilde{w}, \tilde{v})[\varepsilon_{xx}(\tilde{w}, \tilde{v}) + \varepsilon_{yy}(\tilde{w}, \tilde{v}) + \varepsilon_{zz}(\tilde{w}, \tilde{v})) \\ + \frac{b}{2}(\varepsilon_{xx}(\tilde{w}, \tilde{v}) + \varepsilon_{yy}(\tilde{w}, \tilde{v}) - 2\varepsilon_{zz}(\tilde{w}, \tilde{v})] \end{aligned} \tag{1.31}$$

$$\begin{aligned} E_{LH}(k = 0) = a_V(\tilde{w}, \tilde{v})[\varepsilon_{xx}(\tilde{w}, \tilde{v}) + \varepsilon_{yy}(\tilde{w}, \tilde{v}) + \varepsilon_{zz}(\tilde{w}, \tilde{v})) \\ - \frac{b}{2}(\varepsilon_{xx}(\tilde{w}, \tilde{v}) + \varepsilon_{yy}(\tilde{w}, \tilde{v}) - 2\varepsilon_{zz}(\tilde{w}, \tilde{v})] \end{aligned} \tag{1.32}$$

where a_C is the conduction band deformation potential, a_V is the valence band deformation potential and b is the shear deformation potential. The strains ε are functions of the z coordinate through $\tilde{w}$ and $\tilde{v}$ in a diffused quantum well. The axial strain not only splits the heavy hole and light hole bands but also couples the spin-orbit split-off band to the light hole. The coupling between the light hole and the split-off bands is equal to

$$\sqrt{2}\frac{b(\tilde{w}, \tilde{v})}{2}[\varepsilon_{xx}(\tilde{w}, \tilde{v}) + \varepsilon_{yy}(\tilde{w}, \tilde{v}) - 2\varepsilon_{zz}(\tilde{w}, \tilde{v})] \qquad (1.33)$$

In most cases of practical interests, the split-off band can be ignored in the calculation of band structure as they are separated in energy by about hundreds of meV. On the contrary, in the calculation of strain effects, this coupling to the split-off band is not negligible as it can be in the range of tens of meV depending on the strain. The coupling is independent of the wavevector k and therefore causes a shift in the light hole band edge. The compositon (position) dependent shift in band-edge energy with the coupling to the split-off band taken into account using perturbation theory [46] is equal to

$$E_{LH}(k = 0) = a_V(\varepsilon_{xx} + \varepsilon_{yy} + \varepsilon_{zz}) - \frac{b}{2}(\varepsilon_{xx} + \varepsilon_{yy} - 2\varepsilon_{zz}) - \frac{\Delta_{so}}{2} +$$
$$\sqrt{\Delta_{so}^2 - 2b\Delta_{so}(\varepsilon_{xx} + \varepsilon_{yy} - 2\varepsilon_{zz}) + 9b^2(\varepsilon_{xx} + \varepsilon_{yy} - 2\varepsilon_{zz})^2}$$
$$(1.34)$$

By substituting appropiate z-dependent values, the band edge of the LH band can be easily obtained.

3.3.1   Interdiffused InGaAs/GaAs quantum wells

The distribution of strain and the confinement potential profiles of interdiffused quantum wells are shown in Figure 9 with an as-grown well width = 100 Å. The strain and the confinement potentials for electrons and heavy holes have error function profiles. For the light hole, the potential due to the quantum well layer is not a confining well potential but a barrier potential, which is the result of the shear strain splitting of the heavy and light holes. As a result, the light hole subband are not bound by the quantum well. In Figure 10, we show the valence subband dispersion curves for $L_d = 0, 10, 20, 30$ Å. The topmost three dispersion curves are for the HH subbands. We notice that the HH1 and HH2 subbands are parabolic like at wavevectors near to zero as a result of the fact that the HH subbands are well separated in energy from the LH subbands. When the wavevector is increased to about 0.04, the dispersion curves show deviations from the parabolic shape and a change in slope. The curvatures of HH2 and HH3

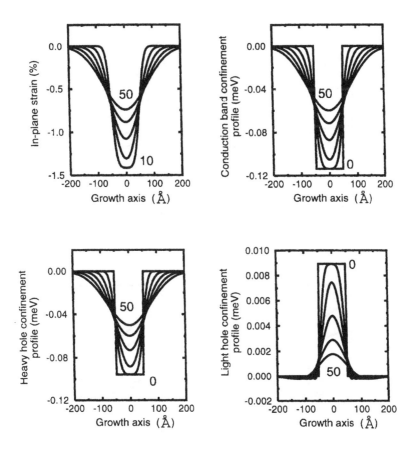

Fig. 9. Strain profile, conduction and valence band confinement potentials of interdiffused InGaAs/GaAs quantum wells ($L_d = 0, 10, 20, 30, 40, 50$ Å).

are obviously smaller than that of the HH1 as they are closer in energy to the LH bands and therefore bended by the heavy and light hole coupling. The curvatures of the HH subbands decrease (effective masses increase) as interdiffusion reduces the energy separation of HH and LH subbands and hence increases the HH and LH coupling.

3.4 Approximation of the Error Function Confinement Potential by a Hyperbolic Function

It is useful to use hyperbolic functions to model the error function confinement potential profiles of an interdiffused AlGaAs/GaAs quantum well, as the eigenstates of a hyperbolic function potential can be obtained analytically.

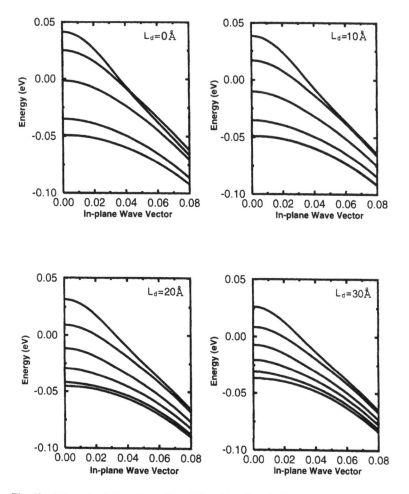

Fig. 10. Valence band dispersions of interdiffused $In_{0.2}Ga_{0.8}As/GaAs$ quantum wells with an as-grown well-width = 100 Å, for $L_d = 0, 10, 20, 30$ Å.

Li and Weiss [58] have obtained analytical solutions to the Schrodinger equation with a hyperbolic quantum well potential and studied the TE and TM optical absorption coefficients. The hyperbolic quantum well potential is defined here by

$$U_r(z) = U_{\text{eff},r}\left[1 - \frac{1}{\cosh^2(\beta z)}\right] \qquad (1.35)$$

where $U_{\text{eff},r} = \Delta\tilde{E}_r$ is the effective barrier depth after interdiffusion, and β

is determined by fitting $U_r(z)$ to the $U_{\text{COP},r}$, the cross over point potential at the cross over point defined in the following equation

$$U_{\text{COP},r} = Q_r \left[\tilde{E}_g \left(z = \frac{L_z}{2} \right) - \tilde{E}_g(z = 0) \right] \tag{1.36}$$

where Q_r is the band offset splitting ratio, $\tilde{E}_g(z)$ is the bandgap at position z. The origin of the z coordinate is at the centre of the as-grown square well with a well width of L_z. As a result, the positions of the barriers are at $z = -L_z/2$ and $z = L_z/2$. β is related to the cross over point potential by the following equation

$$\beta = \frac{2}{L_z} \left[\cosh^{-1} \sqrt{\frac{1}{1 - \hat{U}_{\text{COP}}}} \right] \tag{1.37}$$

where $\hat{U}_{\text{COP}} = U_{\text{COP}}/U_{\text{eff}}$ is the potential of the cross over point normalized to the well depth and the subscript r is removed from the U's since the Q_r's are canceled in the present unstrained quantum well case. However, for a strained quantum well structure, such as in InGaAs/GaAs quantum well structure, the ratio of $U_{\text{COP}}/U_{\text{eff}}$ should be dependent upon r. The electron confinement potentials of hyperbolic quantum well and error function quantum well are compared in Figure 11 for different values of L_d and it demonstrates a strong similarity of the profile.

The hyperbolic quantum well subband edge energies can be calculated, in the envelop function approximation using a constant effective mass approximation, by the 1-D Schrodinger-like equation which is written as follows:

$$-\frac{\hbar^2}{2m_r^*} \frac{d^2\chi_{rl}}{dz^2} + U_r(z)\chi_{rl}(z) = E_{rl}\chi_{rl}(z) \tag{1.38}$$

where $l = 1, 2, \ldots$ are the QW subband levels for either the electrons or holes, respectively, $m_r^* = m_r^*(z = 0)$ is the carrier effective mass in the z-direction. E_{rl} is the subband level, and the origin of the potential energy is taken to be the bottom of the hyperbolic quantum well. The boundary condition of the wavefunction is taken to be zero at infinity. The details of the calculation can be found in Li and Weiss [58].

It is useful to compare the approximate hyperbolic quantum well to the exact results of the error function quantum well in order to demonstrate the accuracy and applicability of the hyperbolic quantum well model. A comparison of subband levels for the error function quantum well and the hyperbolic quantum well is made in Table 1. The error function quantum well model for $L_z = 100$ Å and aluminium mole fraction = 0.3 is solved by a finite difference method using a variable effective mass. The confinement profiles for the hyperbolic quantum well and the error function quantum well models

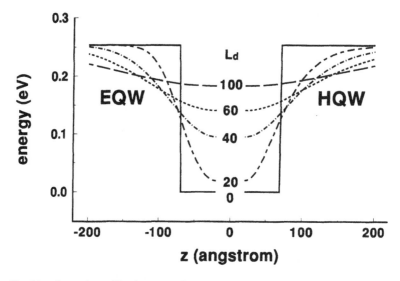

Fig. 11. Comparison of the electron confinement potentials of error function quantum well and hyperbolic quantum well. The as-grown square well has a well-width = 100 Å and aluminium mole fraction = 0.3.

are very similar for 20 Å < L_d < 100 Å, with the exception for larger values of L_d, where the hyperbolic quantum well is wider outside the well region. However, within the well region it is a good approximation since the two parameters U_{eff} and β are used to normalize (fit) the hyperbolic quantum well depth and to match the cross over point potential values with those of the error function quantum well, respectively. All the subband energy levels for both the hyperbolic quantum well and the error function quantum well are very similar for the cases of L_d = 20 Å and 40 Å, as shown in Table 1. The average difference in the corresponding subband energy levels is 3.17 meV (average error of 3.9%) in the case of L_d = 20 Å, and 1.21 meV (1.6%) in the case of L_d = 40 Å. The larger errors occur for the extremely small L_d (≤ 20 Å); this is because of the hyperbolic quantum well confinement profile fails to converge to a square as $L_d \to 0$.

4   OPTICAL ABSORPTION

One of the device applications of interdiffused quantum wells is optical modulators. In this section, we discuss the theory for the determination of the interband optical absorption of interdiffused quantum wells. The theory presented here also serves as theoretical background for calculating the optical gain of interdiffused quantum well lasers.

TABLE 1
Subband Energy (meV)

	Error Function QW		Hyperbolic QW		$|\Delta E|$		%	
L_d	20 Å	40 Å	20 Å	40 Å	20 Å	40 Å	20 Å	40 Å
C1	52.13	27.83	54.77	28.48	2.64	0.65	5.1	2.3
C2	146.87	78.59	147.18	78.58	0.31	0.01	0.2	0.0
C3	214.47	116.88	205.36	114.94	9.11	1.94	4.2	1.7
C4	–	141.23	–	137.55	–	3.68	–	2.6
HH1	19.04	10.6	21.11	10.85	2.07	0.25	10.9	2.4
HH2	56.21	30.85	59.98	31.19	3.77	0.34	6.7	1.1
HH3	89.47	48.71	92.1	48.82	2.63	0.11	2.9	0.2
HH4	117.64	64.18	117.47	63.47	0.17	0.44	0.1	0.7
HH5	139.48	77.19	136.09	75.95	3.39	1.24	2.4	1.6
HH6	152.39	87.54	147.96	85.45	4.43	2.09	2.9	2.4
HH7	–	94.82	–	92.25	–	2.57	–	2.7

4.1 Interband Optical Absorption

According to the Fermi Golden rule [46], the transition rate of an electron from the valence band to the conduction band is given as

$$W_{abs} = \sum_{i,f} \frac{2\pi}{\hbar} |\langle f|H_I|i\rangle|^2 \delta(E_f - E_i - \hbar\omega)(P_i - P_f) \qquad (1.39)$$

where H_I is the operator for the interaction between the electron and the electromagnetic interaction; P_i and P_f denote the probability of occupation of states $|i\rangle$ and $|f\rangle$. The states $|i\rangle$ and $|f\rangle$ are the initial valence and the final conduction band states given by

$$|i\rangle = e^{i\vec{k}.r_{//}}\phi_v(z)u_v(\vec{r}_{//})/\sqrt{A} \qquad (1.40)$$

$$|f\rangle = e^{i\vec{k}.r_{//}}\psi_c(z)u_c(\vec{r}_{//})/\sqrt{A} \qquad (1.41)$$

The envelop functions of the valence and conduction band states are obtained using the theory described in Section 3.1. $u_v(\vec{r}_{//})$ and $u_c(\vec{r}_{//})$ are the periodic part of the Bloch function, which can be written as the product of a orbital part and a spin part. We assume here for a interdiffused quantum well the periodic part of the Bloch function are not dependent on the coordinate z. For conduction band, there are two distinct $|u_c\rangle$ denoted by $|\frac{1}{2}, \pm\frac{1}{2}\rangle$ which, when expressed as the product of a spin and an orbital part, are as follows

$$\left|\frac{1}{2}, \frac{1}{2}\right\rangle = |S \uparrow\rangle, \left|\frac{1}{2}, \frac{1}{2}\right\rangle = |S \downarrow\rangle \qquad (1.42)$$

where S denotes the orbital part of the wavefunction which has a S-orbital symmetry. $\uparrow$ denotes the spin part of the wavefunction (spin $= +1/2$). The corresponding periodic part of the Bloch functions of heavy and light holes, $u_{HH}(\vec{r})$ and $u_{LH}(\vec{r})$, have the symmetry properties of angular momentum wavefunction $J = \frac{3}{2}$. $u_{HH}(\vec{r})$ has the symmetry of $M_J = \pm\frac{3}{2}$, which can be expressed as

$$u_{HH}(\vec{r}) = \left|\frac{3}{2},\frac{3}{2}\right\rangle = -\frac{1}{\sqrt{2}}(|X\uparrow\rangle + i|Y\uparrow\rangle)$$

and

$$\left|\frac{3}{2},-\frac{3}{2}\right\rangle = \frac{1}{\sqrt{2}}(|X\downarrow\rangle - i|Y\downarrow\rangle).$$

$u_{LH}(\vec{r})$ has the symmetry of $M_J = \pm\frac{1}{2}$, which can be written as

$$u_{LH}(\vec{r}) = \left|\frac{3}{2},-\frac{1}{2}\right\rangle = \frac{1}{\sqrt{6}}(|X\uparrow\rangle - i|Y\uparrow\rangle) + \frac{2}{\sqrt{3}}|Z\downarrow\rangle$$

and

$$\left|\frac{3}{2},\frac{1}{2}\right\rangle = \frac{1}{\sqrt{6}}(|X\downarrow\rangle - i|Y\downarrow\rangle) + \frac{2}{\sqrt{3}}|Z\uparrow\rangle \quad (1.43)$$

The matrix element $\langle f|H_I|i\rangle$ is expressed as

$$\langle f|H_I|i\rangle = \int d^3\vec{r}\, e^{-i\vec{k}.\vec{r}_{//}}\psi_c^*(z)u_c^*(\vec{r})H_I(\vec{r})e^{i\vec{k}.\vec{r}_{//}}\phi_v(z)u_v(\vec{r})/A \quad (1.44)$$

Since the envelop functions are slowly varying, the effect of the momentum operator on the envelop function is negligible, that is

$$\vec{p}\,e^{i\vec{k}.\vec{r}_{//}}\phi_v(z)u_v(\vec{r}) \cong e^{i\vec{k}.\vec{r}_{//}}\phi_v(z)\vec{p}\,u_v(\vec{r}) \quad (1.45)$$

and the integral can be simplified accordingly as

$$\langle f|H_I|i\rangle = \int dz\psi_c^*(z)\phi_v(z)\langle u_c|\vec{p}|u_v\rangle\frac{eA_0}{2m_0} \quad (1.46)$$

A_0 is the vector potential of the electromagnetic radiation related to the power intensity $\bar{S}$ of the photon flux by [61]

$$\bar{S} = \frac{n_r c\varepsilon_0\omega^2 A_0^2}{2} = N\hbar\omega \quad (1.47)$$

where N is the number flux of photon.

The absorption coefficient $\alpha(\omega)$ is given by

$$
\alpha(\omega) = \frac{W_{abs}\hbar\omega}{S} = \frac{\pi e^2}{nc\varepsilon_0 m_0\omega} \sum_{s_v,s_c,n,m} \frac{1}{L_z} \int_0^\infty \frac{kdk}{2\pi} \int_0^{2\pi} \frac{d\theta}{2\pi}
$$

$$
\times \left| \left\langle \psi_{3/2,s_v}^n \middle| \phi_{1/2,s_c}^m \right\rangle \left\langle \frac{3}{2}, s_v \middle| \vec{p} \middle| \frac{1}{2}, s_c \right\rangle \right|^2
$$

$$
\times \delta(E_m^c - E_n^v - \hbar\omega)(P_n^v - P_m^c)
$$

(1.48)

The matrix elements of the momentum operator between the periodic functions of the Bloch functions are usually treated as empirical parameters determined from experimental measurements and are related to the electron effective mass according to the Kane's model [47,48]. To determine the transition rates between the electron subbands and the hole subbands we have to know the following matrix elements

$$
\left\langle \frac{3}{2}, s_v \middle| \vec{p} \middle| \frac{1}{2}, s_c \right\rangle s_v = \pm\frac{3}{2}, \pm\frac{1}{2}; \, s_c = \pm\frac{1}{2}
$$

(1.49)

Owing to the symmetry of the crystal, the non-zero matrix elements are

$$
M_p = \langle S|p_x|X\rangle = \langle S\uparrow|p_x|X\uparrow\rangle = \langle S\uparrow|P_y|Y\uparrow\rangle = \langle S\uparrow|p_z|Z\uparrow\rangle
$$
$$
= \langle S\downarrow|p_x|X\downarrow\rangle = \langle S\downarrow|p_y|Y\downarrow\rangle = \langle S\downarrow|p_z|Z\downarrow\rangle
$$

(1.50)

The matrix element M_p is directly related to the electron effective mass in the Kane's k.p model [47,48] of the band structure as follows

$$
M_p^2 = \left(\frac{m_0}{2m_e(\tilde{v}, \tilde{w})} - 1 \right) \frac{m_0 E_G(\tilde{v}, \tilde{w})[E_G(\tilde{v}, \tilde{w}) + \Delta(\tilde{v}, \tilde{w})]}{E_G(\tilde{v}, \tilde{w}) + \frac{2}{3}\Delta(\tilde{v}, \tilde{w})}
$$

(1.51)

In an interdiffused quantum well the chemical composition is a slowly varying function of z coordinate and, as a result, the matrix element M_P is not uniquely defined. Nevertheless, as the approximation used in Section 3.1, we use the composition of the well centre in the expression to calculate the matrix element $|\langle X|p_x|S\rangle|^2$, which is a good approximation as the carrier has a large probability near to the well centre.

To illustrate how to calculate the matrix elements we determine the following expression for TE polarization

$$
\left\langle \frac{3}{2}, \pm\frac{3}{2} \middle| p_x \middle| \frac{1}{2}, \pm\frac{1}{2} \right\rangle = \mp\frac{1}{\sqrt{2}} \langle X|p_x|S\rangle
$$

(1.52)

and

$$
\left\langle \frac{3}{2}, \pm\frac{3}{2} \middle| p_y \middle| \frac{1}{2}, \pm\frac{1}{2} \right\rangle = \pm\frac{i}{\sqrt{2}} \langle X|p_x|S\rangle
$$

(1.53)

and for TM polarization

$$\left\langle \frac{1}{2}, \pm\frac{1}{2} \middle| p_z \middle| \frac{1}{2}, \pm\frac{1}{2} \right\rangle = \frac{2}{\sqrt{6}} \langle X|p_x|S\rangle \qquad (1.54)$$

According to the theory for valence band structure discussed in Section 3.2, there is a mixing of the heavy and light hole envelop functions, which depends on the wavevector. If we ignore the warping of the valence band (that is using the axial approximation) and integrate the matrix element with respect to the direction of the wavevector in the x–y plan, we can obtain the matrix elements for transition to the conduction band states $|1/2, \pm1/2\rangle$ as follows

$$\frac{1}{2\pi} \int_0^{2\pi} d\theta |\langle \psi_v|\phi_c\rangle \langle u_v|_{PTE}|1/2, 1/2\rangle|^2 \qquad (1.55)$$
$$= |\langle X|p_x|S\rangle|^2 \left[\frac{1}{2} \langle \psi_{3/2,3/2}|\phi_{1/2,1/2}\rangle^2 + \frac{1}{6} \langle \psi_{3/2,-1/2}|\phi_{1/2,1/2}\rangle^2 \right]$$

$$\frac{1}{2\pi} \int_0^{2\pi} d\theta |\langle \psi_v|\phi_c\rangle \langle u_v|_{PTE}|1/2, -1/2\rangle|^2 \qquad (1.56)$$
$$= |\langle X|p_x|S\rangle|^2 \left[\frac{1}{2} \langle \psi_{3/2,-3/2}|\phi_{1/2,-1/2}\rangle^2 + \frac{1}{6} \langle \psi_{3/2,1/2}|\phi_{1/2,-1/2}\rangle^2 \right]$$

$$\frac{1}{2\pi} \int_0^{2\pi} d\theta |\langle \psi_v|\phi_c\rangle \langle u_v|_{PTM}|1/2, 1/2\rangle|^2$$
$$= |\langle X|p_x|S\rangle|^2 \left[\frac{2}{3} \langle \psi_{3/2,1/2}|\phi_{1/2,1/2}\rangle^2 \right] \qquad (1.57)$$

$$\frac{1}{2\pi} \int_0^{2\pi} d\theta |\langle \psi_v|\phi_c\rangle \langle u_v|_{PTM}|1/2, -1/2\rangle|^2$$
$$= |\langle X|p_x|S\rangle|^2 \left[\frac{2}{3} \langle \psi_{3/2,-1/2}|\phi_{1/2,-1/2}\rangle^2 \right] \qquad (1.58)$$

where $\langle \psi|\phi\rangle$ represents the overlap integral of function $\psi(z)$ and $\phi(z)$ with respect to the z-coordinates. Substituting Eqs. (1.55), (1.56), (1.57), or (1.58) in Eq. (1.48) and integrating with respect to the magnitude of the wavevector k, we obtain the expression for the interband optical absorption (for free particle transition).

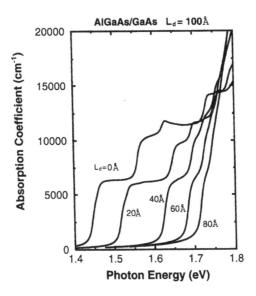

Fig. 12. Interband optical absorption coefficients of interdiffused AlGaAs/GaAs quantum wells for $L_d = 0, 20, 40, 60, 80$ Å. The Coulomb interaction between electron and hole is ignored.

In Figure 12, we show the interband optical absorption coefficients for free carriers in interdiffused $Al_{0.3}Ga_{0.7}As$/GaAs quantum wells with an as-grown well width of 100 Å. The absorption edge is blue-shifted by interdiffusion from about 1.45 eV to about 1.62 eV when L_d changes from 0 Å to 40 Å. The step-like feature of a two-dimensional system at the absorption edge is very clear when $L_d < 40$ Å, which implies that the quantum confinement of carriers is still important and the carrier motion is mainly confined to the $x-y$ plan with this extent of interdiffusion. For $L_d = 60$ and 80 Å, the optical absorption edge looks very much like the edge of a bulk material. The step height of the two dimensional absorption coefficients does not depend strongly on the value of L_d, although there is a slight decrease when L_d increases from 0 to 20 Å.

4.2 Excitons in Interdiffused Quantum Wells

Exciton is the bound state, like the ground state of a hydrogen atom, formed when an electron-hole pair is created by the absorption of a photon. When an electron in the valence band is excited to the conduction band after the absorption of a photon with an energy larger than the band-gap, it is attracted by the Coulomb force to the hole left behind in the valence band. The result of this attractive interaction is the formation of a bound state below the band-gap of the semiconductor. Optically the exciton bound state can appear as a strong peak in the absorption spectrum or the photoluminescence spectrum [60]. In the bulk material the bound state energy is small (about 4 meV for GaAs) [61,62] and the excitonic state can only be observed at very low temperatures. At room temperature, the life-time of the exciton in the bulk material is very short due to scattering with phonons and it is not possible to observe the excitonic peak in the optical spectrum.

In a non-square quantum well structure, when the interdiffusion length L_d is not very large in comparison to the as-grown well width, the carrier motions are confined along the direction of crystal growth by the confining potential to a region with size comparable to the quantum well width. As a result, the Coulomb attraction force between the electron and hole are increased by the reduction of electron-hole average separation along the z-direction as the Coulomb energy is proportion to $1/\sqrt{(z_e - z_h)^2 + (x_e - x_h)^2 + (y_e - y_h)^2}$. In a typical interdiffused quantum well structure with an as-grown width of 100 Å and $L_d < 40$ Å, the separation between the electron and hole (about 100 Å) is approximately half of the average electron-hole separation in a bulk exciton (the bulk exciton has a radius of about 300 Å). The increase in the Coulomb attraction leads to an increase in the binding energy of the exciton and the motion of the electron-hole pair in the exciton is confined to the 2-dimensional plan perpendicular to the growth direction. The binding energy of a two-dimensional exciton is exactly four times the value of a three-dimensional exciton [63]. An exciton in an interdiffused quantum well has a dimensionality between two and three and a binding energy between the bulk value and the two-dimensional value.

Excitons play a very important role in the application of interdiffused quantum well structures in optoelectronics. Owing to the large binding energies, interdiffused quantum well excitons can be observed at room temperatures [64]. In the room temperature absorption spectrum there is a large peak due to the absorption of light by the exciton ground state. Strong electro-optic effects can then be obtained from changing the absorption coefficient of the exciton by applying an electric field to modify the electronic states of the exciton [64–66]. The strength of the electro-optic effect depends on the height of the exciton absorption peak which in turn depends on the binding of the electron-hole pair by the Coulomb force and the carrier confinement potential [67–70].

According to the envelop function approximation discussed in Section 3.1 and assuming parabolic bands, the envelop function of an electron-hole pair without the effects of any confinement potentials is described by the

Schroedinger-like equation.

$$\left[E_g - \frac{\hbar^2 \nabla_e^2}{2m_e} - \frac{\hbar^2 \nabla_h^2}{2m_h} + \frac{e^2}{4\pi\varepsilon|\vec{r}_e - \vec{r}_h|} \right] \Psi(\vec{r}_e, \vec{r}_h) = E\Psi(\vec{r}_e, \vec{r}_h) \quad (1.59)$$

where ε is the material dielectric constant and Ψ is the exciton envelop function. In an interdiffused quantum well, it is necessary to add the confining potential and consider the anisotropy of the material properties. For an interdiffused quantum well, Eq. (1.59) should be rewritten as

$$\left[\begin{array}{l} E_g - \dfrac{\hbar^2}{2}\dfrac{\partial}{\partial z_e}\dfrac{1}{m_e^{\perp}}\dfrac{\partial}{\partial z_e} - \dfrac{\hbar^2}{2}\dfrac{\partial}{\partial z_h}\dfrac{1}{m_h^{\perp}}\dfrac{\partial}{\partial z_h} \\[2mm] - \dfrac{\hbar^2\nabla_{//,e}^2}{2m_e^{//}} - \dfrac{\hbar^2\nabla_{//,h}^2}{2m_h^{//}} + \dfrac{e^2}{4\pi\varepsilon|\vec{r}_e - \vec{r}_h|} \\[2mm] + U_c(\tilde{w}, \tilde{v}) + U_h(\tilde{w}, \tilde{v}) \end{array} \right] \Psi(\vec{r}_e, \vec{r}_h) = E\Psi(\vec{r}_e, \vec{r}_h)$$

$$(1.60)$$

where $//$ denotes directions in the x–y plan and $\perp$ denotes the direction along the crystal growth. When there is a quanum well confinement potential, the material properties along z-direction are different from those along the x and y directions and it is necessary to distinguish them in the equation. The effective masses $m^{//}$ in Eq. (1.60) should be a function of the composition distributions $\tilde{w}$ and $\tilde{v}$. To simplify the calculation, we use the effective values of the effective masses, by replacing $\tilde{w}$ and $\tilde{v}$ with w_m and v_m.

When the energy separation between subbands is larger than the binding energy of the exciton, the Coulomb interaction is not strong enough to affect the carrier motion along the z-direction which is mainly determined by the confinement potential. The dependence of the envelop function on z can be approximated by the envelop function obtained by solving the envelop function equations in Section 3.1. The envelop function $\Psi(\vec{r}_e, \vec{r}_h)$ can be approximated by

$$\Psi(\vec{r}_e, \vec{r}_h) \cong \phi_c^n(\tilde{w}, \tilde{v}, z_e)\psi_h^m(\tilde{w}, \tilde{v}, z_h)\chi(\vec{\rho}_e, \vec{\rho}_h) \quad (1.61)$$

where $\vec{\rho}$ is a vector in the x–y plan. Eq. (1.60) becomes

$$[E_g + E_{en}(\tilde{w}, \tilde{v}) + E_{en}(\tilde{w}, \tilde{v}) - \frac{\hbar^2\nabla_{//,e}^2}{2m_e^{//}} - \frac{\hbar^2\nabla_{//,h}^2}{2m_h^{//}}$$

$$+ V(\vec{\rho}_e - \vec{\rho}_h)]\Psi(\vec{r}_e, \vec{r}_h) = E\Psi(\vec{r}_e, \vec{r}_h) \quad (1.62)$$

$$V(\vec{\rho}_e - \vec{\rho}_h) = \int dz_e \int dz_h \frac{|\phi_e^n(z_e)\psi_h^m(z_h)|^2 e^2}{4\pi\varepsilon\sqrt{(z_e - z_h)^2 + |\vec{\rho}_e - \vec{\rho}_h|^2}} \quad (1.63)$$

By converting the coordinate to the centre of mass coordinate $\vec{\rho}_{CM} = (m_e\vec{\rho} + m_h\vec{\rho}_h)/(m_e + m_h)$ and relative coordinates $\vec{\rho} = \vec{\rho}_\varepsilon - \vec{\rho}_h$, we can eliminate the centre of mass motion as it is described by the free particle wavefunction $e^{i\vec{k}_{cm}\vec{\rho}_{cm}}$.

$$[E_g + E_{en}(\tilde{w}, \tilde{v}) + E_{hm}(\tilde{w}, \tilde{v}) - \frac{\hbar^2}{2\mu(w_m, v_m)}\left(\frac{\partial^2}{\partial x^2} + \frac{\partial^2}{\partial y^2}\right) \\ +U(\vec{\rho})]\Psi_l(\vec{\rho}) = E_l\Psi_l(\vec{\rho}) \qquad (1.64)$$

The eigenvalue equation (1.64) has eigenvalues E_l and eigenfunctions Ψ_l labelled by $l = 0, 1, 2, \ldots$. The optical absorption coefficient due to an excitonic state E_l is given by

$$\alpha_l(\omega) = \frac{\pi e^2}{n_r c\varepsilon_0 m_o^2\omega}\frac{2}{L_z}|\Psi_l(0)|^2|\langle\phi_c^n|\Psi_v^m\rangle|^2|\langle u_v|\vec{p}|u_c\rangle|^2 \qquad (1.65)$$

where $\Psi_l(0)$ is the value of the exciton wavefunction at the origin of the relative coordinate. The factor 2 in the expression is for the spin degeneracy. $\langle\phi_c^n|\Psi_v^m\rangle$ is the overlap integral of the appropiate envelop functions. The matrix element $|\langle u_c|\vec{p}|u_v\rangle|^2$ are given as follows

$$\text{TE mode}: |\langle u_c|\vec{p}|u_v\rangle|^2 = \begin{cases} M_p^2/2 & \text{Heavy hole} \\ M_p^2/6 & \text{Light hole} \end{cases} \qquad (1.66)$$

$$\text{TM mode } |\langle u_c|\vec{p}|u_v\rangle|^2 = \begin{cases} 0 & \text{Heavy hole} \\ 2M_p^2/3 & \text{Light hole} \end{cases} \qquad (1.67)$$

The total absorption coefficient is equal to $\sum_l \alpha_l(\omega)$. The eigenstates can be separated into two groups: (a) bound states with energies below the subband absorption edge (b) unbound states with energies above the absorption edge. For the unbound states the optical absorption coefficients can be written as

$$\sum_{l\in\text{unbound}} \alpha_l(\omega) = \alpha_{\text{free}}(\omega)|\phi_{E=\hbar\omega}(0)|^2 \qquad (1.68)$$

where α_{free} is the optical absorption coefficient due to the free electron-hole transition obtained in Section 4.1. $|\psi_{E=\hbar\omega}(0)|^2$ is usually referred as the Sommerfeld factor. It is quite a task to find the Sommerfeld factor for a quantum well and the Sommerfeld factor can be approximated by that of an exact 2-dimensional exciton. [63]

$$\alpha_{\text{unbound}}(\omega) = \frac{2}{1 + \exp\left(\frac{-2\pi}{\sqrt{E/Ry}}\right)}\alpha_{\text{free}}(\omega) \qquad (1.69)$$

For the bound states, the largest contribution is from the lowest ground state, so we can ignore the contribution of the high lying states. The total absorption can be approximately calculated by the following equation

$$\alpha_{\text{total}}(\omega) = \alpha_{l=0}(\omega) + \alpha_{\text{unbound}}(\omega) \tag{1.70}$$

It is difficult to calculate an exact solution to the exciton Hamiltonian either analytically or numerically. To find the binding energy, the approach usually adopted is the variational method, in which the exact envelop function of the exciton is approximated by a trial function. The physical dimension and shape of the trial wavefunction is determined by one or more variation parameters, which are varied to minimize the total energy calculated using the trial function.

In Figure 13, we show the exciton binding energies of $Al_{0.3}Ga_{0.7}As/GaAs$ and $In_{0.2}Ga_{0.8}As/GaAs$ QWs with an as-grown well width equal to 100 Å as a function of L_d. For both heavy and light holes, the exciton binding energies first increase with L_d until L_d reaches approximately 20 Å and the binding energies start to decrease. The explanation of this trend is that when L_d is smaller than 20 Å the energy levels are below the crossover point and the effective well width is smaller than the as-grown well width. As a result, the quantum confinement is larger than in the as-grown square quantum well, which leads to a larger binding energy. When L_d is larger than 20 Å, the effective well width seen by the carriers starts to increase with L_d and the quantum confinement by the well potential, as well as the binding energy, starts to decrease. It has to point out here that in AlGaAs/GaAs quantum wells, there are two competing forces determining the exciton binding energy when $L_d > 20$ Å: the mole fraction of Al in the well layer and the quantum confinement of the carriers along the growth direction. When L_d increases, the mole fraction of Al in the well centre increases resulting in increase in binding energy. This effects is opposite in sign to the effect of quantum confinement.

When the binding energy of the exciton is increased the probability of finding the electron and hole at the same place is increased (i.e. $|\Psi(0)|^2$ is increased) as the radius of the exciton is reduced. The optical absorption coefficient of the excition peak is increased as a result. This is clearly demonstrated in the absorption coefficients shown in Figure 14. The peak absorption coefficient of the exciton increases by about 10% when L_d is increased from 0 to 20 Å. The absorption peak starts to decrease when L_d is larger than 20 Å, which agrees well with the trend of the binding energy.

In the theory discussed above, the valence band structure is parabolic as the mixing of heavy and light holes is ignored. In section 3.2, we have shown that the hole subbands are highly non-parabolic because of valence band mixing, which can affect the exciton binding energy of an interdiffused

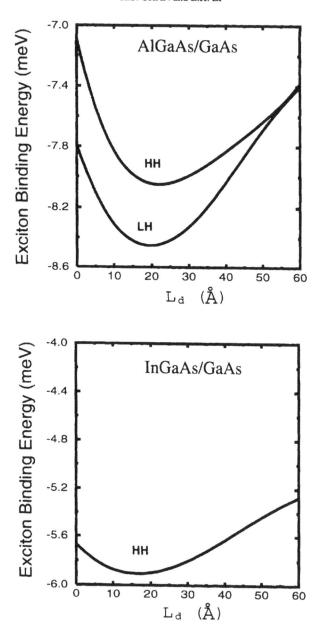

Fig. 13. Exciton binding energies of Interdiffused $Al_{0.3}Ga_{0.7}As$/GaAs and $In_{0.2}Ga_{0.8}As$/GaAs quantum wells as a function of Ld. The as-grown well width is 100 Å.

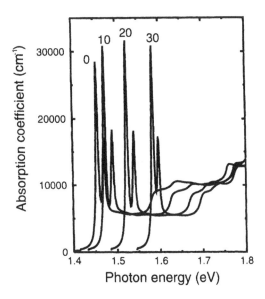

Fig. 14.   Interband optical absorption coefficients of interdiffused $Al_{0.3}Ga_{0.7}As$/GaAs quantum wells for $L_d = 0, 10, 20, 30$ Å. The Coulumb interaction between electron and hole is included.

quantum well. Meney [71] have calculated the exciton binding energy of a diffused quantum well including the valence band mixing effects. When the coupling of heavy and light holes are included, the exciton envelop function is no longer a scalar and should be represented by a spinor, with four components corresponding to the four heavy and light hole states. The exciton Hamiltonian is a 4×4 matrix. The heavy and light hole mixing couples the 1s exciton of the lowest subbands to the excitons with non-s symmetry of higher lying subbands. It is not possible to observe the non-s exciton as $|\Psi(0)|^2$ [51] is zero when the exciton does not have the s symmetry. An important effect of the mixing of non-s exciton is the reduction of optical absorption coefficient due to the mixing of s and non-s excitons. However, owing to the valence band mixing effects, the heavy hole subband has a larger effective mass than that in the parabolic band approximating. A larger hole effective mass implies a larger exciton reduced mass and a larger exciton binding energy as well as a larger absorption coefficient. The optical absorption coefficients are determined by the relative strengths of these two competing factors. Although, the valence band mixing effects have been included in Meney's study, no detailed comparison of two models have been made. Nevertheless, an estimate can be made from the comparison of these two models for square

wells in Bauer and Ando [72] in which the valence band mixing effects is
found to increase the optical absorption by about 15%.

5   QUANTUM CONFINED STARK EFFECTS IN INTERDIFFUSED QUANTUM WELLS

The electro-optic effect in interdiffused quantum wells has attracted wide
attention [64,67–69,73] because of its applicability to various optoelectronic
devices. The quantum-confined Stark effect is most notable, since strong
exciton absorption occcurs at room temperature in an interdiffused quantum
well [64]. When an electric field is applied to a diffused quantum well,
the subband energy levels as well as the energy position of the excitonic
absorption peak are shifted. The change in electroabsorption due to Stark
shift is enhanced because of the large optical absorption coefficients of the
exciton at the room termperatures. This is the main reason why quantum
wells, both square and non-square, are suitable for the realization of optical
switching devices such as the self-electro-optic effect devices (SEEDs). The
quantum confined Stark effects in interdiffused quantum well have been
studied experimental by Ghisoni et al. [73] and Ralston et al. [64], where
they both found the enhancement of quantum confined Stark effects and
small reduction of exciton absorption peak in interdiffused quantum well.

The magnitude of the Stark shift for a given applied field is dependent
upon the quantum well structure (i.e., the well width, the barrier height,
and the shape of the confinement profile). Some interdiffused non-square
quantum wells can provide an enhanced Stark shift because the energy
separation between the first and second subband can be smaller or greater in
an interdiffused quantum well than in a square well depending on the degree
of interdiffusion. In this section we present and discuss the theory and effects
of interdiffusion on the quantum confined Stark effects.

As an example for discussion, we consider an as-grown square Al-
GaAs/GaAs quantum well of width $L_z = 100$ Å, which is clad with 1000 Å
thick $Al_{0.3}Ga_{0.7}As$/GaAs barriers. The interdiffused quantum well profile is
modeled by an error function as discussed in Section 2. The subband energy
levels can be calculated using the following equation

$$\left[\frac{-\hbar^2}{2} \frac{\partial}{\partial z_i} \frac{1}{m_i(z_i)} \frac{\partial}{\partial z_i} + V_i(\tilde{w}, \tilde{v}, z_i) + p_i e \xi z_i \right] \phi_{i,n}(z_i) = E_{i,n} \phi_{i,n}(z_i)$$

$$(1.71)$$

where $i = e, h$ denote electron and hole respectively and, $p_e = -1$ and
$p_h = 1$. $E_{i,n}$ is the energy of the n-th level of the i carrier. This equation
is obtained using the envelop approximation discussed in Section 3 with
the addition of the interaction energy due to the external electric field. The

subband envelop functions obtained can then be used to calculate the exciton wavefunctions and hence the optical absorption coefficients as shown in Section 4. To determine the exciton wavefunction, the envelop function used in Eq. (1.61) should be replaced by the envelop function obtained from Eq. (1.71) given above. The envelop functions used in Eq. (1.65) for the optical absorption coefficients should also be replaced by envelop functions dependent on the applied electric field ξ.

The Stark shift energy ΔE for the transiton from the first conduction subband to the first heavy hole subband for various interdiffused quantum well structures are shown in Figure 15 up to an electric of 50 kV/cm. We have included the results for $L_d = 0$ Å, 20 Å, and 40 Å. We notice that the Stark shift energy first decreases with L_d increases from 0 to 20 Å. For the same electric field the Stark shift for $L_d = 20$ Å is smaller than the shift for $L_d = 0$ Å. However, the Stark shift for $L_d = 40$ Å is greater than that of $L_d = 0$ Å. At an electric field of 50 kV/cm, the Stark shift for $L_d = 40$ Å is about twice that of the as-grown square well, while that for $L_d = 20$ Å is about 60% of the as-grown square well.

The effects of interdiffusion on the quantum confined Stark effects can be understood in terms of the effects of interdifffusion on the energy separation between the first and second subband. According to the discussion in Section 3.1, when interdiffusion is first started, the energy separation between the first and second subband is increased with L_d. For a quantum well with width of 100 Å, this increase in energy separation occurs before $L_d = 20$ Å. After L_d increases beyond this value, the energy separation between the first and second subband starts to decrease. For small electric fields, the Stark shift for carrier i, ΔE_i, can be expressed in terms of the applied electric field using the second order perturbation theory [46] as follows

$$\Delta E_i \cong -\frac{|\langle E_{i,2}|e\xi z_i|E_{i,1}\rangle|^2}{E_{i,2} - E_{i,1}} \qquad (1.72)$$

This expression predicts that the Stark shift has a parabolic dependence on the electric field, which is approximately correct for small electric fields as shown in Figure 15. In Eq. (1.72), the Stark shift is inversely proportional to the energy separation between the first and the second subbands. Therefore when L_d is smaller than approximately 20 Å, the energy separation $E_{i,2} - E_{i,1}$ is increased with L_d and, according to equation (1.72), the Stark shift is smaller than the as-grown square well. For $L_d > 20$ Å, $E_{i,1} - E_{i,2}$ is reduced by the increase in interdiffusion and the Stark shift is then increased with the degree of interdiffusion.

The TE mode absorption coefficient for the as-grown square quantum well and the diffused quantum well ($L_d = 0$, 20 and 40 Å) are shown in Figure 16. For the same change in electric field from 0 to 50 kV/cm, the Stark shifts are

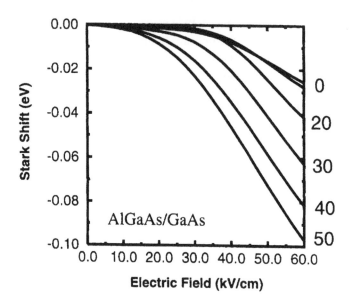

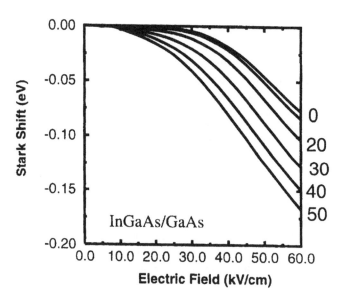

Fig. 15.   Stark shifts of interdiffused $Al_{0.3}Ga_{0.7}As/GaAs$ and $In_{0.2}Ga_{0.8}As/GaAs$ quantum wells for $L_d = 0, 10, 20, 30, 40, 50$ Å.

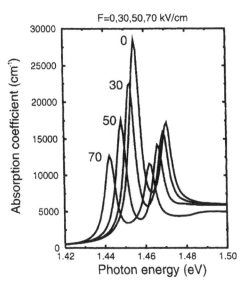

16(A)

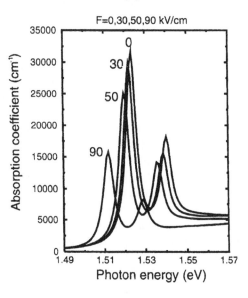

16(B)

F=0,10,20,30,40 kV/cm

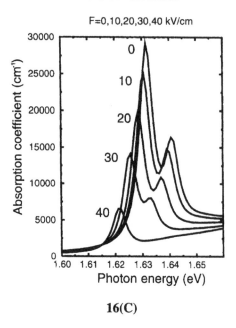

16(C)

Fig. 16. Effects of electric fields on the interband optical absorption coefficients of the interdiffused $Al_{0.3}Ga_{0.7}As/GaAs$ quantum wells for $L_d = 0, 20, 40$ Å.

approximately 150 meV and 100 meV for the as-grown square well and the interdiffused quantum well with $L_d = 20$ Å. The exciton peaks in both quantum well structures show significant reduction in magnitude when the electric field is increased. At a field of 50 kV/cm, the peak absorption is reduced from 32000 cm^{-1} to 25000 cm^{-1} while the change for the square well is correspondingly from 28000 to 17500 cm^{-1}. The reduction in the excitonic absorption peak is due to the reduction of the binding energy and the change in overlap integral of the z-direction envelop functions. When the subband level is Stark-shifted, the electron and the hole are pulled in opposite directions, resulting in an increase in the electron and hole separation and a decrease in the overlap integral and the exciton binding energy. When the Stark-shift is reduced by increasing L_d to 20 Å, the reduction in exciton peak due to the applied electric field is reduced. For $L_d = 40$ Å, the Stark shift is enhanced as shown in Figure 16. The Stark shift at a field of 30 kV/cm is comparable to the Stark shift for a square quantum well at a field of 50 kV/cm. However, the decrease in the exciton peak with the applied electric field is also enhanced with interdiffusion. We notice that the exciton peak for $L_d = 40$ Å decreases from 30000 cm^{-1} to about 14000 cm^{-1} when the field is increased from 0 to 30 kV/cm. For a square well, the decrease in exciton peak absorption is from 30000 cm^{-1} to 17000 cm^{-1} when the field changes from 0 to 50 kV/cm. This indicates that exciton binding in a non-square quantum well can be easily reduced by an external electric field.

6 OPTICAL GAIN OF INTERDIFFUSED QUANTUM WELLS

When the probability P_C is greater than the probability P_V in equation (1.48), the absorption coefficient becomes negative, which means that the optical intensity increases with distance travelled and there is a gain in power after travelling through the quantum well medium. The gain spectrum of an interdiffused quantum well is a sentive function of the subband energies and the electron occupation probabilities of the subbands, which depend on the degree of interdiffusion, as shown in equation (1.48). In Section 3.1, we show that the energy positions of the subbands in a non-square quantum well are not a linear function of the interdiffusion length as they are determined by the shape of the quantum well. It is obvious that the gain spectrum of an interdiffused quantum well does not depend linearly on the interdiffusion length. It is the purpose of this section that we examine the effects of interdiffusion on the gain, differential gain and linewidth enhancement factor of a quantum well.

In the past ten years, a large number of experimental studies [74–86] have been carried out on the application of interdiffusion techniques to the fabrication of quantum well lasers. In the early stage of this development, efforts were mainly concentrated on the application of interdiffusion techniques to disorder selected areas in the wafer and construct a waveguide for tranverse confinement of light in the laser structure. Zinc diffusion and silicon induced interdiffusion have been used to fabricate buried heterostructure lasers. [74–76,79] Among these early works, there were a few studies on tuning the operation wavelengths of lasers by using the interdiffusion techniques [81–85]. The concentration of efforts on using the technique to tune the wavelength of the laser occured in the past few years. O'Brien et al. [83] have used impurity-free interdiffusion technique to fabricate single mode quantum well ridge AlGaAs/GaAs lasers with wavelengths tuned by 11.7 nm. The threshold currents are found to remain unchanged and the differential quantum efficiency is lowered by a factor of 2 after interdiffusion. Poole et al. [84] have shifted the laser wavelengths by $50 \sim 60$ nm by ion-implanting P^+ into InGaAs/InP quantum wells and annealing afterwards. Ooi et al. [85] have studied the fabrication of multiple wavelength lasers in GaAs/AlGaAs structure using a one-step spattially controlled technique based on the technique of impurity-free interdiffusion. In this approach SrF_2 is used as diffusion barrier to prevent the out-diffusion of Ga atoms into the SiO_2 deposited on the wafer. The wavelength shift is controlled by changing the area covered by SrF_2.

In the application of the interdiffusion technique to fabricate buried heterostructure lasers, the optical properties of the quanutm well structure is modified by interdiffusion. The theory and the physics of the optical properties of interdiffused quantum well have been covered in Section 4 and will not be discussed here. To understand and interpret the experiments in which the laser wavelengths are shifted by interdiffusion, it is necessary to determine the gains of the interdiffused quantum wells. The gain of an interdiffused quantum well can be calculated using the expression of optical absorption with the appropiate occupation probabilities for electrons and holes. The electron and hole occupation probabilities are determined by the positions of the quasi Fermi levels (E_f^c and E_f^v) which are given by

$$N = \int_{-\infty}^{\infty} D_c(E) \frac{1}{\exp\left(\frac{E-E_f^c}{kT}\right)+1} dE = \int_{-\infty}^{\infty} D_v(E) \frac{1}{\exp\left(\frac{E-E_f^v}{kT}\right)+1} dE$$

(1.73)

where N is the number density of carriers injected into the quantum well and D_c and D_v are the density of states of the conduction and valence subbands. It is usually assumed that the numbers of holes and electrons are equal, although in reality there may be some deviation from this condition [88].

We are not only interested in the stimulated emission rates of an interdiffused quantum well, but also the spontaneous emission rates, as spontaneous emission constantly reduces the number of electron-hole pairs and change the non-equilibrium distribution into an equilibrium one. Photons from spontaneous emission are incoherent and therefore do not contribute to the laser light output. To maintain the condition for optical amplification in an interdiffused quantum well, a constant current is needed to replace carriers lost due to spontaneous emission. The threshold current is thus defined as the minimum current required to obtain positive optical gain for a specific laser structure. If a laser structure has a mirror loss α_M and an internal loss α_I. The condition of positive optical gain is $g(N) > \alpha_M + \alpha_I$, where $g(N)$ is the gain of the interdiffused quantum well which is a function of the injected carrier density N. The threshold carrier density is defined as the injected carrier density at which the condition of positive gain is satisfied. The threshold current density is defined as the required current for the threshold carrier density. To calculate the threshold current it is necessary to know the spontaneous emission rate as the injected current density is defined as the spontaneous emission rate times the charge of an electron. The spontaneous emission rate can be calculated from the expression for the gain by replacing the term $f_c - f_v$ by $f_c(1 - f_v)$. The total spontaneous emission rate is calculated by summing over all the polarization directions and averaging over all directions, which gives the following equation [89]

$$R_{\text{total}} = \frac{2}{3} R_{TE} + \frac{1}{3} R_{TM}$$

(1.74)

6.1   Fermi Levels of Interdiffused Quantum Well Lasers

When the quantum wells are interdiffused, the potential profiles of the quantum wells are modified to error function profiles according to expressions given in Section 2. The bottoms of the electron and hole confinement potentials are increased and the interband transition energies as well as the operation wavelengths are blue-shifted as a result of interdiffusion. To understand the effects of interdiffusion on characteristics such as gain and differential gain etc., it is sufficient to consider the subband energies as they have a very strong effect on the quasi Fermi levels. The subband edge energies of the $Al_{0.3}Ga_{0.7}As/GaAs$ interdiffused quantum wells considered here are shown in Figure 5 and discussed in detail in Section 3. At the early stage of interdiffusion ($L_d < 15$ Å) all the subband energies, except the top subbands (C3 and HH3 in Figure 5), increase with the increase in L_d. The increases in energies for C2, HH2, LH1 and LH2 in Figure 5 are larger than the increases in C1 and HH1, which leads to an increase in energy separation between the first subband and subbands lying above. When L_d is greater than 15 Å, the subband energies start to decrease gradually as the non-square quantum well potential starts to become shallow and the effective well width starts to increase.

The increase in the energy separation between the first subband and those lying above due to the decrease in the effective well width leads to an increase in the quasi Fermi energy (hereafter referred as Fermi energy) as more carriers occupy the first subband. The electron and hole Fermi energies measured from the bottom of the first subband are shown in Figure 17. For $Al_{0.3}Ga_{0.7}As/GaAs$ quantum wells, the electron and hole Fermi energies for various carrier densities shown in Figure 17 increase slightly with L_d when $L_d < 15$ Å and then gradually decrease with L_d for $L_d > 15$ Å. This trend can be explained by the changes in energy separation between subbands. The changes in Fermi energies discussed above will affect the optical gain through changing the conduction and valence band occupation probabilities, which will be discussed in detail in the following section.

6.2   Gain, Differential Gain and Linewidth Enhancement Spectra of Diffused Quantum Well Lasers

The gain, differential gain and linewidth enhancement factors for interdiffused $Al_{0.3}Ga_{0.7}As/GaAs$ quantum wells are shown in Figure 18. Only an injected carrier density of 5×10^{12} cm^{-2} is considered since the gain of the as grown square quantum well with this injected density exceeds the loss (about 1200 cm^{-1}) of a typical laser structures. To calculate the internal loss we take the following typical values for the optical confinement factor $\Gamma = 0.04$, laser cavity length $L = 300$ μm and mirror reflectivity $R = 0.32$,

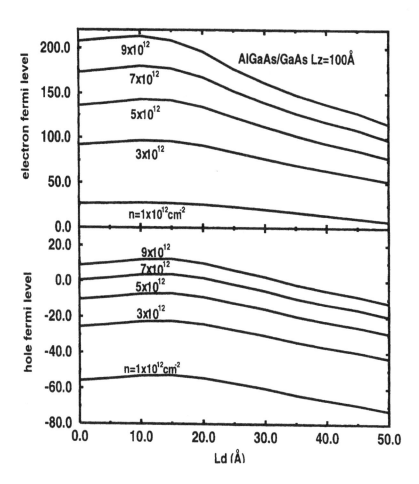

Fig. 17. The Fermi levels of injected carrriers in interdiffused $Al_{0.3}Ga_{0.7}As$/GaAs quantum wells is shown as a function L_d.

and non-radiative absorption $\alpha = 10\,cm^{-1}$. The total loss equals $48\,cm^{-1}$ and the minimum material gain needed to support lasing is $1200\,cm^{-1}$. When the interdiffusion length (L_d) increases, the gain, differential gain and linewidth enhancement factor spectra shift to shorter wavelengths, which is the result of the increase in transition energy between the hole and electron first subbands as the well composition changes. We notice that the gain peak first increases by 2–3% when L_d is increased to about 10 Å and then gradually decreases when L_d is further increased. The decrease in gain with L_d when $L_d > 20$ Å is quite rapid as the peak gain at $L_d = 20$ Å and 40 Å are respectively

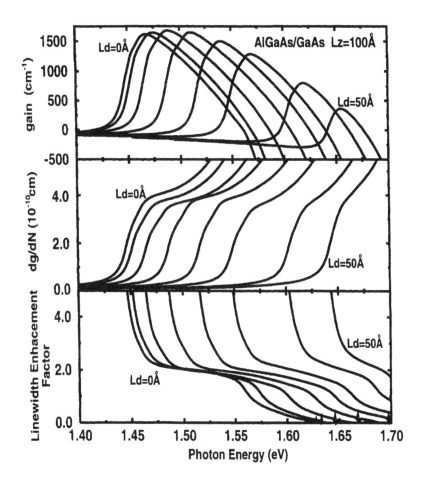

Fig. 18. The gain, differential gain and linewidth enhancement factor spectra of interdiffused $Al_{0.3}Ga_{0.7}As/GaAs$ quantum wells for $L_d = 0, 5, 10, 15, 20, 30, 40, 50$ Å.

95% and 50% of the peak gain of the as grown square quantum well. These changes can be explained by the behavior of the Fermi energy discussed in Section 6.1. The hole and electron Fermi energies for $Al_{0.3}Ga_{0.7}As/GaAs$ quantum wells first increase when $L_d < 10$ Å and then decrease as L_d is further increased beyond 10 Å. The increases in Fermi energies result in increases in the electron and hole occupation probabilities of the first subband and as a consequence increases the optical gain of the quantum well laser. This increase in gain gives a range of L_d in which one can tune the operation photon energy of the laser without any substantial deterioration of the gain and substantial increase in the injected current. Although the peak gain starts to decrease when $L_d > 10$ Å, we note that the peak gain at $L_d = 15$ Å is approximately the same as that of $L_d = 0$ Å. Therefore we can shift the operation energy by about 50 meV. A very similar trend is also observed in the numerical results for the gain of TM mode, which will not be discussed in detail here owing to the lack of space.

The differential gain is an important characteristic of interdiffused quantum well lasers as it is directly related to the modulation speed. The differential gain $\frac{dg}{dN}$ is calculated using the following expression

$$\frac{dg}{dN}(E, N) = \lim_{\Delta N \to 0} \frac{G(E, N + \Delta N) - G(E, N)}{\Delta N} \tag{1.75}$$

where E is the photon energy and N is the carrier density. The linewidth enhancement factor α is a key parameter that determines the performance of semiconductor laser both under cw operation and under high-frequency modulation. The factor α is the ratio of the change of the refractive index n with the carrier density N to the change in the optical gain g with the carrier density, which is expressed as

$$\alpha = -\frac{4\pi}{\lambda} \frac{\frac{dn}{dN}}{\frac{dg}{dN}} \tag{1.76}$$

where λ is the wavelength. $\frac{dn}{dN}$ in equation (17) can be obtained from $\frac{dg}{dN}$ by the Kramers-Kronig transformation.

As the degree of interdiffusion increases the differential gain and the linewidth enhancement factor spectra are blue-shifted as the gain spectra. However, in contrast with the gain spectra, it is interesting to note that the differential gain and the linewidth enhancement factor around the gain peak wavelengths are not strongly dependent on L_d as the gain spectra. In the differential gain (linewidth enhancement factor) spectra, there is a shoulder (elbow) around the transition energies of the gain peak due to the step in the joint density of states around the bandgap energy. The height of this shoulder (elbow) is approximately the largest differential gain (smallest linewidth enhancement factor) one can obtain from the material around the gain peak by tuning the operation wavelength. We notice that in both material systems the dependence of the height of the shoulder (elbow) on L_d is weak (about 10–20%) if the carrier density is kept constant. This implies that we can tune the operation wavelength of the laser by interdiffusion without any substantial deterioration in the relaxation oscillation frequency and chirping performance. To understand the effect of interdiffusion on the differential gain (we only need to consider the differential gain here as the linewidth

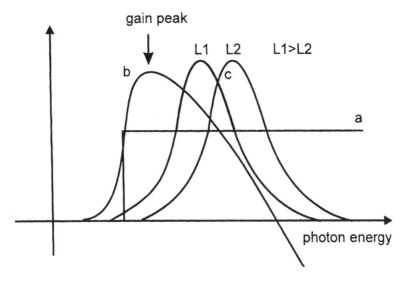

Fig. 19. The schematic diagram showing the relation between (a) the joint density of states, (b) the gain spectrum and (c) the function $f^c(1 - f^c)$. For (c) we show the curves for $L_d = L_1, L_2$.

enhancement factor is related to the differential gain by equation (1.76)) we consider the following approximate expression of the differential gain

$$\frac{dg}{dn} = D(E)|P|^2 \left[(f^c(1 - f^c)\frac{dE_F^C}{dn} - f^h(1 - f^h)\frac{dE_F^H}{dn} \right] \qquad (1.77)$$

where $D(E)$ and P^2 are respectively the joint density of states and the optical matrix element. The term $f^C(1 - f^C)\frac{dE_F^C}{dn}$ is greater than the term $f^h(1 - f^h)\frac{dE_F^H}{dn}$, which will be ignored in the following discussion, as the electron effective mass is smaller than the hole effective masses and $\frac{dE_F^C}{dn}$ is greater than $\frac{dE_F^H}{dn}$. To illustrate the relation between the gain, $D(E)$, $\frac{dg}{dn}$ and $f^C(1 - f^c)$, we show the dependence of these quantities schematically on the photon energy in Figure 19. The function $f^c(1 - f^c)$ has a peak at the Fermi energy. When L_d is increased the electron Fermi energy is decreased so the curve for the function $f^c(1 - f^c)$ shifts to the left in the schematic diagram and the value of the function $f^c(1 - f^c)$ at the gain peak wavelength increases with L_d. The factor $\frac{dE_F^C}{dn}$ decreases with the increase in L_d and so compensate the increase in $f^c(1 - f^c)$. As a result the differential gain is not strongly dependent on L_d.

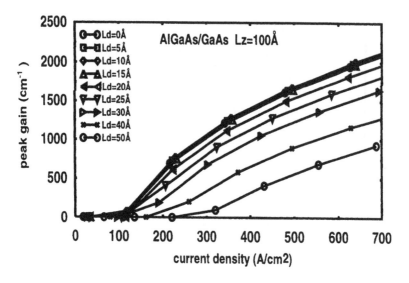

Fig. 20. The dependence of peak gain on the recombination current density of interdiffused quantum well lasers for various L_d.

6.3   Injected Current Density

In order to assess interdiffusion as a viable technique for shifting the operation wavelengths of lasers integrated monolithically, it is necessary to know how the injection current changes with the degree of interdiffusion. In Figure 20 we show the dependence of the peak gain on the injected current density due to spontaneous emission for L_d between 0 and 50 Å. For $0 < L_d < 15$ Å the peak gain of $Al_{0.3}Ga_{0.7}As$/GaAs QW has a small dependence on L_d and thus the curves in Figure 20 for these L_d cannot be resolved. The components of the injection current due to other mechanisms such as carrier leakage are ignored as the main focus here is in the material performance. The material peak gain shown in Figure 20 can be compared with the loss of 1200 cm^{-1} in a typical laser structure and determine the required injected current density. From Figure 20 the as-grown $Al_{0.3}Ga_{0.7}As$/GaAs quantum well requires a current density of 320 Acm^{-2} respectively to reach the lasing threshold. When L_d is increased from 0 to 50 Å, the current density should be increased to a value larger than 700 Acm^{-2} to keep the peak gain equal to the loss, which is more than two times the current requirement of an as-grown square well. For practical operation of interdiffused quantum well lasers, we have to limit the injected current after interdiffusion and therefore limit the degree of interdiffusion. Assuming that we can allow an increase in the injected current by 30% after interdiffusion and the peak gain is still 1200 cm^{-1}.

From Figure 20 the maximum L_d for $Al_{0.3}Ga_{0.7}As/GaAs$ quantum well that still satisfies the current and gain limits is 25 Å. From Figure 18 we find that with $L_d = 25$ Å the $Al_{0.3}Ga_{0.7}As/GaAs$ QW can be tuned approximately by about 100 meV, which is equivalent to 54 nm.

Acknowledgement

This work is partially supported by the City University of Hong Kong Strategic Research Grant and the Hong Kong Research Grant Council.

Referernces

1. L.B. Allard, G.C. Aers, S. Charbonneau, T.E. Jackman, R.L. Williams, I.M. Templeton, M. Buchanan, D. Stevanovic and F.J.D. Almeida, *J. Appl. Phys.*, **72**, 422 (1992).
2. L.B. Allard, G.C. Aers, P.G. Piva, Poole, M. Buchanan, I.M. Templeton, T.E. Jackman, S. Charbonneau, U. Akano, I.V. Mitchell, *Appl. Phys. Lett.*, **64**, 2412 (1994).
3. S.G. Ayling, J. Beauvais and J.H. Marsh, *Electron. Lett.*, **28**, 2240 (1992).
4. I.V. Bradley, W.P. Gillin, K.P. Homewood and R.P. Webb, *J. Appl. Phys.*, **73**, 1686 (1993).
5. I.V. Bradley, B.L. Weiss and J.S. Roberts, *Opt. Quantum Electron.*, **23**, S823 (1991).
6. S.P.J. Charbonneau, Poole, P.G. Piva, G.C. Aers, E.S. Koteles, M. Fallahi, J.-J. He, J.P. McCaffrey, M. Buchanan and M. Dion, *J. Applied Phy.*, **78**, 3697 (1995).
7. P. Chen and A.J. Stecki, *J. Appl. Phys.*, **77**, 5616 (1995).
8. J.J. Coleman, P.D. Dapkus, C.G. Kirkpatrick, M.D. Camras and N. Holonyak, *App. Phys. Lett.*, **40**, 904 (1982).
9. D.G. Deppe, L.J. Guido, N. Holonyak Jr. and K.C. Hsieh, *Appl. Phys. Lett.*, **49**, 510 (1986).
10. D.G. Deppe and N. Holonyak Jr., *J. Appl. Phys.*, **64**, R93 (1988).
11. B.B. Elenkrig, D.A. Thompson, J.G. Simmons, D.M. Bruce, Y. Si, J. Zhao, J.D. Evans and I.M. Templeton, *Appl. Phys. Lett.*, **65**, 1239 (1994).
12. A. Furuya, M. Makiuchi, O. Wada and T. Fujii, *IEEE Photon. Technol. Lett.*, **24**, 2448 (1988).
13. P. Gavrilovic, D.G. Deppe, K. Meehan, N. Holonyak and J.J. Coleman Jr., *App. Phys. Lett.*, **47**, 130 (1985).
14. L.J. Guido, N. Holonyak Jr., K.C. Hsieh and J.E. Baker, *Appl. Phys. Lett.*, **54**, 262 (1989).
15. Y. Hirayama, *Jpn. J. Appl. Phys.*, **28**, L162 (1989).
16. K.Y. Hsieh, Y.L. Hwang, J.H. Lee, R.M. Kolbas, *J. Electron. Mater.*, **19**, 1417 (1990).
17. S.R. Andrew, J.H. Marsh, M.C. Holland and A.H. Kean, *IEEE Photon. Technol. Lett.*, **4**, 426 (1992).
18. L.L. Chang and A. Koma, *Appl. Phys. Lett.*, **29**, 138 (1976).
19. F. Iikawa, P. Motisuke, F. Cerdeira, M.A. Sacilotti, R.A. Masut and A.P. Roth, *Superlattic. Microstruc.*, **5**, 273 (1989).
20. E.S. Koteles, B. Elman, C.A. Armiento and P. Melman, *Superlattic. Microstruc.*, **9**, 533 (1991).
21. H.H. Tan, J.S. Williams, C. Jagadish, P.T. Burke and M. Gal, *Appl. Phys. Lett.*, **68**, 2401 (1996).
22. A. Hamoudi, A. Ougazzaden, Ph. Krauz, E.V.K. Rao, M. Juhel and H. Thibierge, *Appl. Phys. Lett.* **66**, 718 (1995).
23. K. Nakashima, Y. Kawaguchi, Y. Kawamura and Y. Imamura, *Appl. Phys. Lett.*, **52**, 1383 (1988).
24. G.J. van Gurp, W.M. van de Wijgert, Fontijn and P.J.A. Thijs, *J. Appl. Phys.*, **67**, 2919 (1990).
25. S.L. Wong, R.J. Nicholas, R.W. Martin, J. Thompson, A. Wood, A. Moseley and N. Carr, *J. Appl. Phys.*, **79**, 6826 (1996).
26. C. Francis, P. Boucaud, F.H. Julien, J.Y. Emery and L. Goldstein, *J. Appl. Phys.*, **78**, 1944 (1995).

27. C. Francis, F.H. Julien, J.Y. Emery, R. Simes and L. Goldstein, *J. Appl. Phys.*, **75**, 3607 (1994).
28. T. Miyazawa, H. Iwamura and M. Naganuma, *IEEE Photonics Techn. Lett.*, **3**, 421 (1991).
29. H.C. Neitzert, C. Cacciatore, D. Campi, C. Rigo, Coriasso and A. Stano, *IEEE Photonics Tech. Lett.*, **7** 875 (1995).
30. T.P. Lee, *Proceedings of the IEEE*, **79**, 253 (1991).
31. I. Gontijo, T. Krauss, J.H. Marsh and De La Rue, *IEEE Journal Quantum Electron.*, **30**, 1189 (1994).
32. K. Nakashima, Y. Kawaguchi, Y. Kawamura, H. Asahi and Y. Imamura, *Jpn. J. Appl. Phys.*, **26**, L1620 (1987).
33. T. Fuji, M. Sugawara, S. Yamazaki and K. Nakajima, *J. Cryst. Growth*, **105**, 348 (1990).
34. K. Mukai, M. Sugawara and S. Yamazaki, *Phys. Rev. B*, **50**, 2273 (1994).
35. E.H. Li, J. Micallef and W.C. Shui, *Mat. Res. Soc. Symp. Proc.*, **417**, 289 (1996).
36. G. Bastard, *Phys. Rev. B*, **24**, 5693 (1981).
37. G. Bastard, *Phys. Rev. B*, **25**, 7584 (1982).
38. G. Bastard, J.A. Brum and R. Ferreira, *Solid State Phys.*, **44**, 229 (1991).
39. M. Jaros, K.B. Wong and M. Gell, *Phys. Rev. B*, **31**, 1205 (1985).
40. D. Ninno, K.B. Wong, M.A. Gell and M. Jaros, *Phys. Rev. B*, **32**, 2700 (1985).
41. M.A. Gell, K.B. Wong, D. Ninno and M. Jaros, *J. Phys. C*, **19**, 3821 (1986).
42. J.N. Schulman and Y.C. Chang, *Phys, Rev. B*, **24**, 4445 (1981).
43. Y.C. Chang and J.N. Schulman, *Appl. Phys. Lett.*, **43**, 536 (1983).
44. Y.C. Chang, *Phys. Rev. B*, **37**, 8215 (1988).
45. J.N. Schulman and Y.C. Chang, *Phys. Rev. B*, **31**, 2056 (1985).
46. L.I. Schiff, 1968, *Quantum Mechanics*, Third edition (Tokyo, McGraw Hill).
47. E.O. Kane, *J. Phys. Chem. Solids*, **1**, 82 (1956).
48. E.O. Kane, *J. Phys. Chem. Solids*, **1**, 249 (1957).
49. E.H. Li, B.L. Weiss and K.S. Chan, *Phy. Rev. B*, **46**, 15181–15192 (1992).
50. J.M. Luttinger and W. Kohn, *Phys. Rev.*, **97**, 869 (1955).
51. K.S. Chan, *J. Phys. C: Solid State Phys.*, **19**, L125 (1986).
52. E. Yablonovitch and E.O. Kane, *J. Lightwave Technol.*, **6**, 1292 (1988).
53. G.C. Osbourn, *Phys. Rev. B*, **27**, 5126 (1983).
54. D. Ahn and S.L. Chuang, *IEEE J. Quantum Electron.*, **24**, 2400 (1988).
55. T.C. Chong and C.G. Fonstad, *IEEE J. Quantum Electron.*, **25**, 171 (1989).
56. E.P. O'Reilly and A.R. Adams, *IEEE J. Quantum Electron.*, **30**, 366 (1994).
57. S.W. Corzine, R.H. Yan and L.A. Coldren, *Appl. Phys. Lett.*, **57**, 2835 (1990).
58. E.H. Li and B.L. Weiss, *IEEE J. Quantum Electron.*, **29**, 311 (1993).
59. F. Bassani and G.P. Parravicini, 1975, *Electronic States and Optical Transitions in Solids*, (Oxford UK, Pergamon Press).
60. M.D. Sturge, *Phys. Rev.*, **127**, 768 (1962).
61. R.J. Elliot, *Phys. Rev.*, **108**, 1384 (1957).
62. R.J. Elliot, *Polarons and Excitons, Scottish Universities' Summer School*, edited by C.G. Kuper and Whitfield, (New York, Plenum), 269–293 (1962).
63. M. Shinada and S. Sugano, *J. Phys. Soc. Jpn.*, **21**, 1936 (1996).
64. J.D. Ralston, W.J. Schaff, D.P. Bour and L.F. Eastman, *Appl. Phys. Lett.*, **54**, 534 (1989).
65. D.A.B. Miller, D.S. Chemla, T.C. Damen, A.C. Gossard, W. Weigmann, T.H. Wood and C.A. Burrus, *Phys. Rev. Lett.*, **53**, 2173 (1984).
66. D.A.B. Miller, D.S. Chemla, T.C. Damen, A.C. Gossard, W. Weigmann, T.H. Wood and C.A. Burrus, *Phys. Rev. B*, **32**, 1043 (1985).
67. E.H. Li, K.S. Chan, B.L. Weiss and J. Micallef, *Appl. Phys. Lett.*, **63**, 533 (1993).
68. J. Micallef, E.H. Li and B.L. Weiss, *Appl. Phys. Lett.*, **67**, 2768 (1995).
69. W. Seidel and P. Voisin, *Semicond. Sci. Technol.*, **8**, 1885 (1993).
70. E.H. Li and B.L. Weiss, *Proc. SPIE*, **1675**, 98 (1992).
71. A.T. Meney, *J. Appl. Phys.*, **72**, 5729 (1992).
72. G.E.W. Bauer and T. Ando, *Phys. Rev. B*, **38**, 6015 (1988).
73. M. Ghisoni, P.J. Stevens, G. Parry and J.S. Roberts, *Opt. Quantum Electron.*, **23**, S915 (1991).

74. T. Fukuzawa, S. Semura, H. Saito, T. Ohta, Y. Uchida and H. Nakashima H., "GaAlAs buried multiquantum well lasers fabricated by diffusion-induced disordering", *Appl. Phys. Lett.*, **45**, 1–3 (1984).

75. P. Gavrilovic, K. Meehan, L.J. Guido, N. Holonyak Jr., V. Eu, M. Feng and R.D. Burnham, "Si-implanted and disordered strip-geometry $Al_xGa_{1-x}As$-GaAs quantum well lasers", *Appl. Phys. Lett.*, **47**, 903–905 (1985).

76. D.F. Welch, D.R. Scifres, P.S. Cross and W. Streifer, "Buried heterostructure lasers by silicon implanted, impurity induced disordering", *Appl. Phys. Lett.*, **51**, 1401–1403 (1987).

77. J. Werner, E. Kapon, N.G. Stoffel, E. Colas, S.A. Schwarz, C.L. Schwartz and N. Andreadakis, "Integrated external cavity GaAs/AlGaAs lasers using selective quantum well disordering", *Appl. Phys. Lett.*, **55**, 540–542 (1989).

78. H. Ribot, K.W. Lee, R.J. Simes, R.H. Yan and L.A. Coldren, "Disordering of GaAs/AlGaAs multiple quantum well structures by thermal annealing for monolithic integration of laser and phase modulator", *Appl. Phys. Lett.*, **55**, 672–674 (1989).

79. K. Meehan, P. Gavrilovic and N. Holonyak Jr., R.D. Burnham and R.L. Thornton, "Stripe-geometry $Al_xGa_{1-x}As$-GaAs quantum well heterostructure lasers defined by Si diffusion and disordering", *Appl. Phys. Lett.*, **46**, 75–77 (1985).

80. P. Gavrilovic, K. Meehan, J.E. Epler and N. Holonyak Jr., "Impurity-disordered, coupled-stripe $Al_xGa_{1-x}As$/GaAs quantum well laser", *Appl. Phys. Lett.*, **46**, 857–859 (1985).

81. K. Meehan, J.M. Brown, P. Gavrilovic, N. Holonyak Jr., R.D. Burnham, T.L. Paoli and W. Streifer, "Thermal-anneal wavelength modification of multiple-well p-n $Al_xGa_{1-x}As$-GaAs quantum-well lasers", *J. Appl. Phys.*, **55**, 2672–2675 (1984).

82. M.D. Camras, N. Holonyak Jr., R.D. Burnham, W. Streifer, D.R. Scifres, T.L. Paoli and C. Lindstrom, "Wavelength modification of $Al_xGa_{1-x}As$ qunatum well heterostructure lasers by layer interdiffusion", **54**, 5637–5641 (1983).

83. S. O'Brien, J.R. Shealy, F.A. Chambers and G. Devane, "Tunable (Al)GaAs lasers using impurity-free partial interdiffusion", *J. Appl. Phys.*, **71**, 1067–1069 (1992).

84. P.J. Poole, S. Charbonneau, M. Dion, G.C. Aers, M. Buchanan, R.D. Goldberg and I.V. Mitchell, "Demonstration of an ion-implanted wavelength-shifted quantum well lasers", *IEEE Photon. Technol. Lett.*, **8**, 16–18 (1996).

85. B.S. Ooi, S.G. Ayling, A.C. Bryce and J.H. Marsh, *IEEE Photon. Technol. Lett.*, **7**, 944 (1995).

86. Y. Nagai, K. Shigihara, W. Karkida, S. Kakimoto, M. Otsubo and K. Ikeda, "Characteristics of laser diodes with a partially intermixed GaAs-AlGaAs quantum well", *IEEE J. Quantum Electron.*, **31**, 1364–1370 (1995).

87. E.H. Li and K.S. Chan, "Laser gain and current density in a disordered AlGaAs/GaAs quantum well", *Electron. Lett.*, **29**, 1233–1234 (1993).

88. S. Seki and K. Yokoyama K., *IEEE Photon. Technol. Lett.*, **7**, 251 (1995).

89. A. Yariv, *Quantum Electronics, Third Edition (New York, John Wiley & Sons)*, 166 (1989).

APPENDIX

Effective Hamiltonian Approach for Calculation of the Valence Band Structure

The mixing of the light hole and heavy hole in the valence band structure, which have strong effects on the optical properties, is described by the Luttinger-Kohn Hamiltonian [22]. As a result of band-mixing the envelope function as well as the periodic part of the Bloch wavefunction are functions of $k_{//}$, the wavevector in the plane perpendicular to the direction of crystal growth. To find the valence subband structures and the envelope functions of a quantum well, it is necessary to diagonalize the Luttinger-Kohn Hamiltonian with the appropriate confinement potentials for heavy and light holes. In this study, we adopt the effective Hamiltonian approach described in [23] to calculate the valence subband structure. In this approach, the hole envelope functions $\Psi_{sl}(k_{//}, z)$ for spin components $s = \pm 3/2, \pm 1/2$ of the Luttinger-Kohn Hamiltonian at any finite $k_{//}$ is expressed as a linear combination of the envelope functions $\psi_{sl}(z)$ at $k_{//} = 0$ as follows

$$\Psi_s(k_{//}, z) = \sum_{l=1}^{M} d_{s,l}(k_{//}) \cdot \psi_{s,l}(z), s = -\frac{3}{2}, -\frac{1}{2}, \frac{1}{2}, \frac{3}{2} \quad (4)$$

where $\psi_{s,l}(z)$ are the zone-centre ($k_{//} = 0$) envelope functions of valence subbands obtained by solving equation (3) and $d_{s,l}(k_{//})$ are coefficients to be determined by the Rayleigh-Ritz variational method. $s = \pm 1/2$ and $s = \pm 3/2$ denote the light hole and the heavy hole subbands considered in equation (3). The accuracy of this approximation depends on M, the number of wavefunctions used in the linear expansion. In this study, a basis set of 40 envelop functions is used in the calculation and the results obtained are accurate within the energy range which is typical for QW laser operation. According to the Raleigh-Ritz method, the coefficients $d_{s,l}(k_{//})$ are coefficients of the eigenvectors of the following effective Hamiltonian

$$\begin{bmatrix} \tilde{E}_{3/2} + \tilde{S}_{//} & \tilde{C} & \tilde{B} & 0 \\ \tilde{C}^* & \tilde{E}_{-1/2} - \tilde{S}_{//} & 0 & \tilde{B}^T \\ \tilde{B}^* & 0 & \tilde{E}_{+1/2} - \tilde{S}_{//} & \tilde{C}^T \\ 0 & \tilde{B}^* & \tilde{C}^* & \tilde{E}_{-3/2} + \tilde{S}_{//} \end{bmatrix} \quad (5)$$

where $\tilde{E}_s$, $\tilde{B}$ and $\tilde{C}$ are M by M submatrices with matrix elements given by

$$C_{jj} = \left[\frac{3}{4}\right]^{1/2} \frac{\hbar^2}{m_o} \gamma_2 (k_x - ik_y)^2 \int_{-\infty}^{\infty} dz \psi_{-3/2,j}(z) \psi_{1/2,j}(z) \quad (6a)$$

$$B_{jj} = 3^{1/2} \frac{\hbar^2}{m_o} \gamma_2 (-k_x - ik_y) \int_{-\infty}^{\infty} dz \psi_{3/2,j}(z) \frac{\partial}{\partial z} \psi_{-1/2,j}(z) \quad (6b)$$

$$E_{\pm 3/2, jj'} = \delta_{jj'} E_{H,j} - \frac{\hbar^2}{2m_{//H}} k_{//}^2 \quad (6c)$$

$$E_{\pm 1/2, jj'} = \delta_{jj'} E_{L,j} - \frac{\hbar^2}{2m_{//L}} k_{//}^2 \quad (6d)$$

where γ_2 is the Luttinger parameter and $j, j' = 1, 2, \ldots N$. $E_{H,j}$ and $E_{L,j}$ denote the j-th subband energy of the heavy and light hole respectively. $S_{//}$ is the diagonal submatrix containing the potential due to strains in the crystal lattice, the expressions of which will be discussed in the following section.

CHAPTER 2

Interdiffusion Mechanisms in III-V Materials

W.P. GILLIN

Department of Physics, Queen Mary and Westfield College, University of London, London, E1 4NS, United Kingdom

1 INTRODUCTION

The thermal stability of semiconductor heterostructures and in particular their interdiffusion has been an issue since the first heterostructures were produced. While some early work on self-diffusion in semiconductors goes back to the 1960's with work on silicon and germanium [1]. Some of the first work on interdiffusion was by Chang and Koma [2] in 1976, on the GaAs/AlAs system and since then there has been a proliferation of papers looking at most of the material systems that have been grown. In addition to the wide range of systems studied, the effects of a wide range of perturbations on the diffusion have been studied. These perturbations have included the surface condition (i.e. arsenic overpressure [3–6] or encapsulant [3–8]), doping [9–12], and ion implantation of a large variety of elements, for example [13,14], and in order to study these various effects a variety of techniques have been used, including Auger profiling [2], SIMS [9], TEM [15] and photoluminescence [14,16]. However, despite this huge quantity of work there is still some uncertainty as to the exact mechanisms responsible for the interdiffusion. Indeed the key

53

parameters necessary to characterise diffusion in materials, the activation energy and prefactor are still somewhat uncertain, with activation energies of 0.32 [17] to 6.2 eV [4] being quoted together with prefactors which cover a range of some 21 orders of magnitude. Much of this variation in the measured activation energies can be attributed to the use of small data sets and the resulting experimental uncertainties in E_A cause the massive variations in measured D_o values.

In this chapter we will provide an overview the experimental techniques used to measure interdiffusion and highlight the factors necessary for obtaining reliable data on interdiffusion. We will then go on to provide a review of the literature regarding the diffusion mechanisms thought to be responsible for interdiffusion of III-V heterostructures and the effects that the surface conditions, doping and strain have on the interdiffusion.

2   MEASURING INTERDIFFUSION

A number of techniques have been used to measure interdiffusion in III-V heterostructures. They can however, be divided in to two broad methods; direct compositional mapping and indirect compositional determination.

Direct compositional mapping techniques are in principle by far the best techniques for measuring interdiffusion as they should provide a compositional profile which can be directly fitted with a diffusion model. These techniques should be able to determine firstly whether the diffusion process is Fickian, and then the diffusion coefficient for the process. Auger profiling, which was used by Chang and Koma [2], to profile GaAs/AlAs heterostructures, was probably the first technique used to measure interdiffusion in III-V materials. However, it is not widely used and this is partly due to the fact that relatively thick layers are required, rather than the quantum well and superlattice structures which are routinely grown. Secondary Ion Mass Spectrometry (SIMS) which has been used by a number of groups, has the additional advantage that it can be used to correlate the diffusion of impurities with interdiffusion. This technique has been used to monitor diffusion following ion implantation [19], and the diffusion in the presence of impurity or dopant atoms. One of the most widely quoted examples of this is the work of Mei et al. [9] who studied the effects of dopants such as silicon and tellurium on the diffusion of GaAs/AlGaAs superlattices. Other direct compositional mapping techniques which have been applied to the study of III-V interdiffusion are Rutherford Backscattering Spectroscopy (RBS) [20,21] and Transmission Electron Microscopy [15]. RBS, like Auger profiling, has a limited resolution and hence is not very good for accurate measurements of diffusion. However, it can be used is certain situations where other techniques are not applicable in order to compare diffusion between samples [20]. TEM has good resolution and is able in theory to measure a wide range of diffusion lengths. However, the application of TEM to interdiffusion requires excellent high resolution electron microscopy (HREM) combined with image analysis [15] in order to obtain compositional maps and this, coupled with the difficult sample preparation required, has limited its application.

All of these techniques do however suffer from problems. With techniques such as Auger profiling and SIMS for example, the profiling is achieved through sputtering the surface with an ion beam. This sputtering will cause some intermixing of the layers and care has to be taken in the collection and analysis of the data to minimise such effects. Another problem, particularly with SIMS is in the calibration of the SIMS yield with composition. Neither of these problems are a fundamental obstacle but they do require good experimental practice in order to obtain reliable data. A rather more important obstacle in using these techniques for accurate determination of diffusion constants is that, with the possible exception of RBS, they are destructive. This means that a sample can be annealed and its diffusion measured, but in order to follow the diffusion as a function of time a number of samples must be used each of which has been annealed at a given temperature for different times. Following the time dependence in this manner is very time consuming and in consequence has only been done by a few groups. However, a failure to look for time dependent effects can seriously affect the diffusion coefficients measured. As an example of this one can look at the literature on the effect of gallium implantation on III-V diffusion. As gallium can be used in focused ion beam systems, for direct writing onto surfaces with fine feature sizes, it attracted a lot of attention as a possible candidate to use for selective area enhanced diffusion [22]. Consequently a number of groups looked at the effect of gallium implantation of the interdiffusion of quantum wells [23,24]. A number of these observed that gallium implants produced an enhancement of the interdiffusion. When the time dependence of the interdiffusion following annealing was measured [14] it could be seen that the diffusion coefficient for intermixing was unchanged following gallium implantation but that the implantation damage caused a rapid transient interdiffusion, furthermore this damage related interdiffusion was not confined to gallium implants but was present following any implant, Figure 1.

The second range of techniques available for measuring interdiffusion are those which yield the composition profile indirectly. The two most common of these are optical techniques and X-ray diffraction [25]. Of the optical techniques photoluminescence is the most widely used [12,14,16] although absorption [26], reflectance [27], cathodoluminescence [28] and Raman spectroscopy [29] have all been used. With the exception of Raman spectroscopy, which looks at the change in the phonon energies with

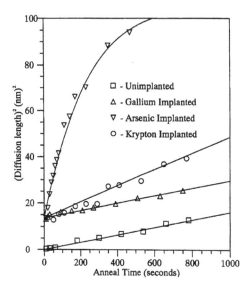

Fig. 1. Diffusion length squared against anneal time for samples implanted with 10^{13} cm^{-2} of either gallium, arsenic or krypton ions compared to an unimplanted sample. It can be seen that neither gallium or krypton have any effect upon the diffusion coefficient for intermixing, i.e. the gradient of the line. Whilst the arsenic implant produces an enhancement of the interdiffusion which decays with time. However, for all three implanted ions there is an initial fast process, which is over within the first anneal, which is equal for all species and is due to the diffusion of vacancies created by the implant.

interdiffusion, all of these techniques look for changes in the confined energy levels in quantum well systems and correlate these changes with a theoretical model for how the confined state should change with diffusion. The usual principle for the modelling is to assume that the diffusion is Fickian with a concentration-independent diffusion coefficient. The composition profile as a function of the diffusion length, L_D, can then be calculated using the error function solution for a thin layer in infinitely thick barriers,

$$C(z) = C_B + \left(\frac{C_W - C_B}{2} \right) \left[\mathrm{erf} \left(\frac{h-z}{L_D} \right) + \mathrm{erf} \left(\frac{h+z}{L_D} \right) \right] \quad (1)$$

where C_B is the concentration of the diffusing species in the barrier, C_W is the concentration of the diffusing species in the well, $2h$ is the well thickness and z is the depth with the well centre at $z = 0$.

Once the composition profile for a given diffusion length has been calculated, this can be used to calculate the conduction and valance band energy profiles and then the Schrödinger equation can be solved to determine the confined states and hence the photoluminescence transition energy,

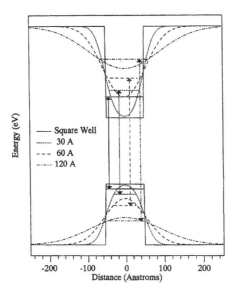

Fig. 2. A schematic diagram of the conduction and valance band potentials for various diffusion lengths and showing how the confined states change with annealing. The arrows indicate the photoluminescence transitions which can be monitored to measure diffusion.

Figure 2. In order to perform these calculations it is necessary to know the band-gap, band offset ratios and effective masses of the material system being studied as a function of the composition.

Using this method it is therefore possible to convert a shift in the photoluminescence peak following annealing to a diffusion length, see Figure 3. As the diffusion length, $L_D = 2\sqrt{Dt}$ where D is the diffusion coefficient and t is the anneal time, the photoluminescence peak shift can be converted directly in to a diffusion coefficient.

This method of performing a single anneal and calculating the diffusion coefficient would, however, have the same potential for error as the destructive methods and in addition it relies on the assumption that the diffusion is Fickian with a concentration independent D and has no means for checking the validity of this assumption. In order to overcome these deficiencies it is necessary to perform a number of anneals at a given temperature, on a single sample and measure the photoluminescence following each anneal. As the assumption used in the model is that the diffusion length is a square-root function of time it is possible to check the assumption by plotting L_D^2 against anneal time and looking for a linear dependence. As well as providing a consistency check on the experimental method this technique provides statistically more reliable data and allows for time dependent effects, such

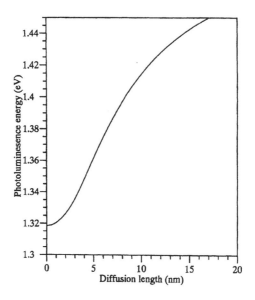

Fig. 3. A graph of photoluminescence peak shift against diffusion length calculated for a 10 nm $In_{0.2}Ga_{0.8}As$ well in GaAs barriers. Using this graph the shift in the photoluminescence peak can be converted in to a diffusion length.

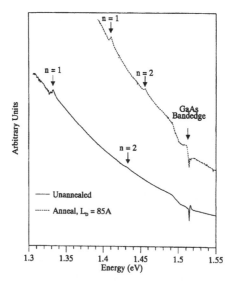

Fig. 4. Reflectance spectra for a 10 nm $In_{0.2}Ga_{0.8}As$ well in GaAs barriers before and after annealing at 1000°C. The positions of the $n = 1$ and $n = 2$ transitions are marked and agree with the theoretical predictions for a diffusion length of 8.5 nm.

as the post implant diffusion behaviour, to be observed. In order to provide further confirmation of the model used to interpret the data it is possible to measure the higher energy confined states and compare these with the same model. This has been done for the InGaAs/GaAs system using reflectance spectroscopy and has shown excellent agreement [27], Figure 4.

While this section has shown the range of techniques available for studying diffusion and their advantages and disadvantages, they are only part of the story for obtaining good diffusion data. Probably the key stage in a diffusion measurement and one which rarely receives much attention is the annealing. In order to measure accurate activation energies for diffusion processes it is necessary to measure diffusion coefficients over as wide a temperature range as possible. All of the techniques mentioned above, with the possible exception of TEM, would struggle to measure diffusion lengths over even two orders of magnitude and this would correspond to four order of magnitude in diffusion coefficient. In order to obtain a wide range of diffusion coefficients it is necessary to have anneal times over as wide a temperature range as possible. This is most commonly achieved by using rapid thermal annealing (RTA) systems for anneals in the range of seconds to tens of minutes and conventional furnace systems for longer anneal times, up to weeks [30]! The use of RTA systems in particular can cause serious problems for temperature calibration depending on the design of the system. It is not unknown for RTA systems to have $>50°C$ errors and for these errors to be a strong function of the annealing temperature. Thus even if the measurement technique was perfect serious errors in the values for activation energy and prefactor can occur.

3 INTERDIFFUSION OF GaAs BASED MATERIALS

The interdiffusion of heterostructures, like self-diffusion, must proceed through the movement of point defects in the crystal. The interdiffusion rate must therefore be a function of the diffusivity of those point defects and their concentration. If, for example, we consider the case of group III vacancy controlled interdiffusion of the group III sublattice, then we can express the interdiffusion coefficient, D_I, as,

$$D_I = f D_V [V_{III}] \tag{2}$$

where f is the correlation factor which is related to the crystal structure [31], but is of order unity, D_V is the vacancy diffusivity, and $[V_{III}]$ is the concentration of group III vacancies. The vacancy diffusivity term, D_V, is an Arrhenius expression of the form

$$D_V = a_0^2 \nu \exp\left(\frac{-E_D}{kT}\right) \tag{3}$$

where E_D is the activation energy for diffusion for the vacancy, a_0 is the jump distance and ν is the jump frequency.

The concentration term $[V_{III}]$ in equation 2 is normally assumed to be thermally activated with vacancies and interstitials being created through Frenkel formation. Thus the group III vacancy concentration is given by an Arrhenius expression,

$$[V_{III}] = C \exp \left(\frac{-E_F}{kT} \right) \tag{4}$$

where E_F is the formation energy of the point defect and C is a pre-exponential term which is related to the change in entropy of the system due to the introduction of the vacancies.

Thus in the Arrhenius expression for interdiffusion the measured activation energy is the sum of the formation energy, E_F, and the diffusion energy, E_D. Similarly the diffusion prefactor, D_0, is given by the expression

$$D_0 = f a_0^2 \nu C \tag{5}$$

Of the terms in equation 5 f, a_0 and n are fixed, and thus the entropy term is often used to explain observed variations in D_0 between experiments. However, as we mentioned in the introduction there is a 21 order of magnitude variation in the measured values of D_0 and it is not physically reasonable for all this to be attributable to entropy considerations between experiments. This would thus seem to suggests that either the simple theory is very wrong or that most of the variation in prefactors measured in the literature may be due to experimental error. In the following section we will take the GaAs/AlGaAs system as an example and review the activation energy data available in the literature.

4 THE ACTIVATION ENERGY FOR INTERDIFFUSION

One of the first reported studies of the interdiffusion of a III-V material was on an AlAs/GaAs heterostructure by Chang and Koma in 1976. Since then this material has become probably the most widely studied of all the III-V systems. However, despite the enormous amount of work on measuring the diffusion in this system the activation energy for the interdiffusion is still uncertain with values in the literature ranging from 0.32 eV to 6.2 eV. Many of these reported differences in E_A are used by authors as evidence for a change in the diffusion mechanism. In this section we will compare the data available in the literature on activation energies in the most common III-V materials and show how the data is consistent with a single activation energy and hence a single mechanism.

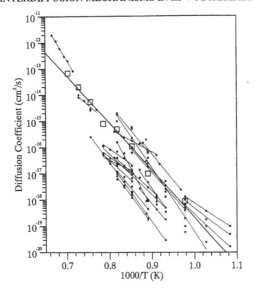

Fig. 5. An Arrhenius plot of our GaAs/AlGaAs interdiffusion data (squares) and much of the data form the literature (dots) the solid line is the least squares fit to our data, and the light lines are the least squares fits to all the other data sets. They literature data comes from [3,6,8,12,14,17,32–37,39,40].

In Figure 5 we show diffusion coefficients for GaAs/AlGaAs interdiffusion collected from the literature and plotted on an Arrhenius diagram (small dots) [3,6,8,12,14,17,32–37,39,40]. This graph includes data collected under a variety of annealing conditions, doping, etc. Also shown on this diagram are the various published fits to each of these data sets which give the wide range of activation energies quoted above. In plotting the data we have taken all the data to be perfect with no errors in either the measured diffusion coefficient or in the measurement temperature and then used a least squares fit to that data to obtain the activation energy with its corresponding uncertainty. These values are given in Table 1. The first thing that can be seen from Table 1 are the large experimental uncertainties present in the quoted activation energies. These large uncertainties are solely a result of small data sets taken over a limited temperature range. For example Guido et al. [3] have calculated activation energies for samples annealed at only four temperatures over a 75°C temperature range. While this may produce data which can be well fitted by a straight line. The value that is obtained is not very reliable. The squares shown on Figure 5 are our own results, $E_A = 3.6 \pm 0.2$ eV, obtained using photoluminescence over a temperature range of 750° to 1150°C [41] and even over this temperature range, and ignoring any experimental errors, we get an uncertainty of 6% of the measured value. One of the most striking things about Figure 5 is the clustering of data at temperatures between 800°C

TABLE 1

Activation energies and prefactors with their associated experimental uncertainty calculated from the experimental data in the literature using a least squares fitting routine.

Reference	$E_A/(eV)$	$D_0/(cm^2/s)$
41	3.6 ± 0.2	$10^{-0.7\pm0.7}$
38	5.6 ± 0.1	$10^{7.0\pm0.5}$
3	3.8 ± 0.1	$10^{-1.0\pm0.8}$
3	4.5 ± 0.5	$10^{2.9\pm2.5}$
4	6.2 ± 0.2	$10^{9.9\pm0.9}$
4	4.5 ± 0.1	$10^{1.5\pm0.5}$
4	4.0 ± 0.3	$10^{-0.4\pm1.4}$
32	3.0 ± 0.2	$10^{-4.6\pm0.9}$
32	2.4 ± 0.5	$10^{-6.8\pm2.3}$
32	2.3 ± 0.3	$10^{-6.8\pm1.3}$
32	1.5 ± 0.5	$10^{-10.4\pm1.9}$
9	3.6 ± 0.3	$10^{-0.5\pm1.5}$
10	3.1 ± 0.4	$10^{-4.0\pm1.8}$
39	4.5 ± 0.1	$10^{1.5\pm1.8}$
39	4.8 ± 0.2	$10^{5.0\pm1.0}$
36	2.8 ± 0.5	$10^{-3.7\pm2.0}$
37	4.0 ± 0.5	$10^{1.8\pm2.1}$
35	4.0 ± 0.1	$10^{1.8\pm0.2}$
18	2.9 ± 0.2	$10^{-4.4\pm1.0}$
18	1.7 ± 0.1	$10^{-8.2\pm0.5}$
17	0.3 ± 0.01	$10^{-11.6\pm0.1}$

and 1000°C. This is due to the relatively small range of diffusion coefficients which can easily be measured.

It is clear from Figure 5 that a large amount of the data is clustered around the line we have drawn through our own data. If we fit all of the data in this cluster we calculate an activation energy of 3.6 ± 0.1 eV. However, there is also a clear second group of data points which lie significantly below our line. When we make a least squares fit to this data we calculate an activation energy of 3.5 ± 0.2 eV. It thus appears that unless one makes an arbitrary choice of a small subset of the data it is difficult to argue for anything other than an activation energy for GaAs/AlGaAs interdiffusion of ~ 3.5 eV. Given that the data is consistent with a single activation energy process. We have fitted every data point with a line with activation energy 3.5 eV and determined what the corresponding prefactor would be. This data is presented in Figure 6 as a histogram where the height of each column is the number of data points giving a D_0 value in a given range. From this diagram it can be seen that the data falls into two distributions, one centred on a D_0 value of about 0.2 cm²/s and a second centred around a D_0 of 0.007 cm²/s.

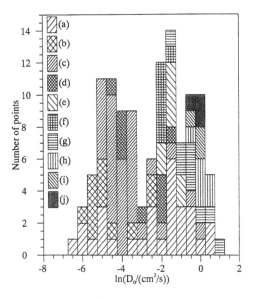

Fig. 6. A histogram of the $\ln(D_0/(\text{cm}^2/\text{s}))$ values for each of the data points plotted in Figure 5 having been fitted by our activation energy. The data points marked (a) are taken from [6,34], (b) are from [4], (c) from [18,32,40] and are all from one laboratory, (d) from [9,10], (e) this work, (f) from [36], (g) from [33], (h) from [38], (i) from [35] and (j) from [37].

As we suggest that there is a single diffusion mechanism describing all the data in the literature, we need now to account for the observed variations in D_0. It has been observed that in careful experiments in InGaAs/GaAs interdiffusion [12,14,16] there can be a factor of two variation in D_0 between two nominally identical wafers which were grown sequentially in the same MBE reactor, capped at the same time with silicon nitride and annealed together, see Figure 7. At the time this was attributed to differences in substrate materials. This factor alone could be enough to account for much of the random scatter seen in the data. If this were coupled with the experiment variations one would expect to see between different groups such as differences in furnace calibration, differences in the treatment of ramp times, surface passivation, etc. then the spread in the data is not unexpected. What is more interesting is the presence of the two clusters in the data. Hsieh et al. [4] demonstrated that annealing in a Ga rich environment resulted in a lower interdiffusion coefficient than an As rich environment. Their small temperature range produced an apparent change in E_A from 4 ± 0.3 to 6.2 ± 0.2 eV. Olmsted et al. [5,6,34] performed a more detailed study in 1993 of the effect of annealing in both Ga rich and As rich conditions. Their original data showed that under both annealing conditions they get essentially the same activation energy but with D_0 values approximately two

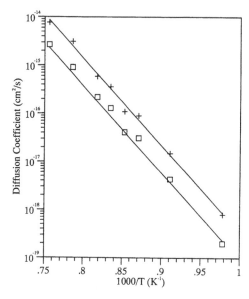

Fig. 7. An Arrhenius plot of diffusion coefficient for two InGaAs samples. The samples were grown sequentially in an MBE reactor, capped side by side in a PECVD reactor with silicon nitride and annealed together. Whilst the experimental scatter around the least squares line is clearly visible it should be noted that the difference in diffusion coefficient between the two samples is almost always constant at approximately a factor of four.

orders of magnitude different. Again with the Ga rich annealing producing the lower diffusion coefficients. Another paper which presents results of similar experiments was that of You et al. [18]. Their results also show no significant change in E_A with annealing ambient, although in contradiction to the results of Hsieh et al. [4] and Olmsted et al. [5,6,34] they seem to find that the Ga rich anneals give higher diffusion coefficients. This work was performed in collaboration with Holonyak from University of Illinois at Urbana-Champaign. The groups of Major et al. [7] and Guido and Holonyak [40] are responsible for most of the other data which lies in the second cluster consisting of approximately half of the data. This work is significant in that it contains a large amount of data obtained on samples which have been annealed with a silicon dioxide cap as well as some results with silicon nitride caps and some with As overpressure annealing. The results of Major et al. [7], like many other workers, show that an SiO_2 encapsulant enhances interdiffusion compared with Si_3N_4. Their results however, have diffusion coefficients which are nearly two orders of magnitude below those obtained under similar conditions by other groups (e.g. Ralston et al. [37]). The reasons for this large discrepancy is not clear. However, before we discuss a possible mechanism to explain the observed distributions in D_0 values shown in

Figure 6 we will discuss the various arsenic overpressure experiments in the literature.

5 ARSENIC OVERPRESSURE ANNEALING

According to the simple diffusion model discussed earlier, if the diffusion is directly related to the thermodynamic equilibrium concentration of point defects, anything which can effect this equilibrium concentration will directly effect the diffusivity. One obvious perturbation will be the surface condition during annealing. As we stated earlier under equilibrium conditions we would expect Frenkel defects to occur. Just considering the group III sublattice in GaAs we would have the following reaction occurring

$$0 \Longleftrightarrow I_{Ga} + V_{Ga} \quad \text{and} \quad [I_{Ga}][V_{Ga}] = k_1 \tag{6}$$

where k_1 is a temperature dependant constant. Thus at any temperature there would be an equilibrium concentration of both vacancies and interstitials. Interstitials are suggested [41] to be faster diffusers than vacancies and any gallium interstitial which reaches the surface under an arsenic rich annealing ambient could react with an arsenic atom in the gas phase effectively locking the interstitial at the surface. This would in effect act as a sink for gallium interstitials thus increasing the concentration of vacancies deeper in the crystal. Similarly annealing under an arsenic poor ambient would be expected to result in arsenic evaporating from the surface leaving a gallium rich surface. This excess gallium could then diffuse into the crystal increasing the gallium interstitial concentration. This reaction can be written as,

$$0 \Longleftrightarrow \frac{1}{4} As_4(\text{vapour}) + I_{Ga} \quad \text{and} \quad [I_{ga}] = k_{2_1} P_{As_4}^{-1/4} \tag{7}$$

By substituting equation 7 into equation 6 we get

$$[V_{Ga}] = \frac{k_1}{k_2} P_{As_4}^{1/4} \tag{8}$$

Thus it can be seen that if the diffusion is controlled by either vacancies or interstitials and the defect concentrations are in thermal equilibrium then one would expect to see a dependence of the interdiffusion coeffcient on the arsenic overpressure during annealing. Experiments to look for these effects have been performed by a number of groups. Guido et al. [3] for example compared the interdiffusion of an GaAs/AlGaAs single quantum well after annealing with an encapsulant of either silicon nitride or silicon dioxide, or an arsenic overpressure. The arsenic overpressure anneals were performed

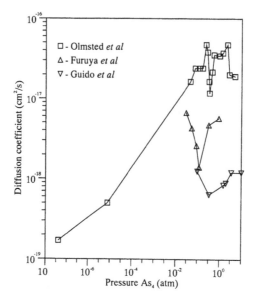

Fig. 8. Diffusion coefficient against arsenic overpressure taken from three papers. The data of Olmsted et al. shows a clear reduction in diffusion coefficient when the arsenic overpressure is reduced. Apparent dips in diffusivity for high overpressures are probably the result of experimental scatter.

with arsenic partial pressures between 10^{-1} and 10 atmospheres. The arsenic overpressure results showed some variation in diffusion coefficient with arsenic overpressure with an apparent minimum diffusion at a partial pressure of approximately 1 atmosphere of arsenic, Figure 8. The increasing diffusion with increased arsenic pressure is consistent with the group III vacancies being responsible for the diffusion and is hence in qualitative agreement with the thermodynamic equilibrium model. The observed increase in diffusion with arsenic pressures below 1 atm was attributed, in the original paper [3], to a possible role for group V vacancies on the group III interdiffusion. Deppe and Holonyak [42] in a review the following year interpreted the rise in diffusion at low arsenic pressure as evidence of group III interstitials becoming the dominant diffusion mechanism. In that paper they fitted the results to a model which appeared to fit the transition from vacancy to interstitial diffusion. However, the model they used was qualitative and did not provide convincing evidence of the change in mechanism. It should be noted that the changes in diffusion coefficient observed by Guido et al. [3] were only at most a factor of two. Considering the activation energy for the diffusion process, this change in diffusion could be caused be a temperature variation in annealing between samples of approximately 10°C. In addition,

Bradley et al. [14] demonstrated that even in carefully controlled experiments changes of a factor of two in diffusion coefficient were easily seen between two nominally identical samples.

Rather more convincing evidence for an effect of arsenic overpressure on diffusion coefficient was published by Olmsted et al. [5], Figure 8. They also looked at the effect of arsenic pressure during annealing on GaAs/AlGaAs interdiffusion, although they looked at an eight order of magnitude variation in arsenic pressure from 10^{-7} atm to 10 atm. Their work also showed the presence of a small dip at 1 atm but much more importantly it showed a two order of magnitude drop in diffusivity from an arsenic pressure of 1 atm to a pressure of 10^{-7} atm. This drop is strong evidence for the diffusion being controlled by vacancies on the group III sublattice over the entire arsenic overpressure range.

A third paper which reports results of GaAs/AlGaAs interdiffusion as a function of arsenic overpressure was published by Furuya et al. [43]. This work covers the arsenic pressure range from $\sim 10^{-2}$ atm to 1 atm and once again it shows an apparent dip in diffusivity although in this paper it occurs at $\sim 10^{-1}$ atm. Given Olmsted et al.'s results at very low arsenic pressures it is not reasonable to attribute these dips to a change in mechanism from a vacancy to an interstitial controlled process. Olmsted in their paper tentatively suggested some other point defect, such as the group V vacancy, might play a role and be responsible for these dips. However, given the strong evidence for vacancies dominating diffusion this seems unlikely. It is possible that given the amount of scatter seen in the Olmsted's diffusivity data at high arsenic pressures the apparent dip may be purely an effect of this scatter. This conclusion is not wholly satisfactory and more, careful experiments need to be performed before any firm conclusions can be drawn.

Thus it can be seen that whilst the arsenic overpressure results give strong evidence that group III vacancies are responsible for group III diffusion they do not provide any evidence that the vacancy concentrations are at thermal equilibrium values. It is thus important to measure the vacancy concentrations present in GaAs and in epilayers grown on GaAs as a function of temperature. This data would show whether the vacancy concentration is a function of temperature, as would be expected if it were in thermal equilibrium, and if there is a measureable change in concentration it would allow us to measure the activation energy for vacancy formation directly.

Relatively little data is available on the concentration of point defects in GaAs. Dannefaer et al. [44] measured the vacancy concentrations in commercial semi-insulating GaAs at temperatures up to 600°C, using positron lifetime spectroscopy, and measured vacancy concentrations of the order of $10^{17}/cm^3$. More recently Khreis et al. [45] measured both the vacancy concentrations and their diffusivity, over a temperature range

800°C–1000°C, in a single experiment. Their results showed that the background concentration of vacancies present in epilayers grown on GaAs was not a function of temperature but was rather constant at a value of $\sim 2 \times 10^{-17}/cm^3$, this value being effectively identical to that measured by Dannefaer et al. [44] at 600°C. This data is very strong evidence that the vacancy concentrations in GaAs are fixed at a 'grown-in' value, most probably during growth. In their paper Khreis et al. [45] measured the activation energy for diffusion of these vacancies and obtained a value of 3.4 ± 0.3 eV which, within experimental error, is the same as the value for interdiffusion. It thus appears that the interdiffusion of the group III sublattice in GaAs based materials is controlled by the concentration of vacancies incorporated in to the material during growth.

Given that the interdiffusion is not dependent upon thermal equilibrium defect concentrations it is necessary to explain the arsenic overpressure results. It was found by Olmsted et al. [5] that the diffusion coefficient for intermixing of GaAs/AlGaAs at ~ 850°C was of the order of $2 \times 10^{-17} cm^2/s$ for arsenic overpressures greater than $\sim 10^{-2}$ atm, which is comparable with most of the values in the literature, but reduces by two orders of magnitude as the arsenic overpressure is reduced to $\sim 10^{-7}$ atm. If this value is due to the grown in vacancy concentration then the two order of magnitude drop in diffusivity equates to a two order of magnitude drop in vacancy concentration during annealing. The very low arsenic overpressure would result in a large loss of arsenic from the surface of the sample, this would result in excess gallium in the surface region. It is not implausible that this excess gallium would exist as gallium interstitials which could annihilate some of the group III vacancies through equation 6.

We are now in a position to understand the distribution of D_0 values seen in Figure 6. As it appears that there is only one diffusion mechanism responsible for interdiffusion in GaAs/AlGaAs and as the prefactor for that process is directly proportional to a non-equilibrium concentration of vacancies present in the material. It would appear that the two distributions observed in Figure 6 are two different vacancy concentrations in the samples used. Part of the data in the lower distribution is from samples which were annealed in an extremely arsenic poor environment which we have suggested might be expected to reduce the group III concentrations in those samples whereas all of the data in the higher D_0 distribution are from samples annealed either under large arsenic overpressures or under dielectric encapsulation.

Whilst the arsenic overpressure data suggests that group III vacancies are responsible for group III interdiffusion there has been some discussion regarding their exact diffusion path, i.e. whether they diffuse via both group V and group III sites (nearest neighbour hopping, nnh) or via second nearest neighbour hopping (2nnh) solely on the group III sites. The first of these mechanisms would result in the creation of large quantities of antisite defects, unless the diffusion proceeded through some kind of ring mechanism [46] and it was impossible for any diffusing atoms to stop moving until they had completed their movement around the ring. The latter mechanism of 2nnh overcomes these difficulties but requires group III atoms to move around a group V atom in order for the vacancy to diffuse. Given the open lattice present in the III-V semiconductors this scenario is much more reasonable than the ring mechanisms needed for nnh.

6 DIELECTRIC ENCAPSULANT ANNEALING

In addition to arsenic overpressures various dielectric encapsulants have been used to stop the evaporation of arsenic from the surface during annealing and several papers have investigated the effects of different encapsulants on the interdiffsuion of heterostructures. The two most commonly used dielectric encapsulants are silicon nitride and silicon dioxide. In 1989 Kuzuhara et al. [47] studied the out diffusion of gallium from GaAs surfaces into silicon oxy-nitride films as a function of their composition. This study showed that silicon dioxide films caused significant gallium outdiffusion whilst silicon nitride films produced very little. This gallium outdiffusion is significant for interdiffusion studies as it is equivalent to the injection of gallium vacancies which mediate diffusion. Guido et al. [3] compared the interdiffusion of a GaAs/AlGaAs quantum well under both silicon nitride and silicon dioxide. As would be expected from the results of Kuzuhara et al., the samples annealed under silicon dioxide showed greater diffusion than those annealed under silicon nitride. The difference between the two being approximately an order of magnitude. It is interesting to note that in the same work the authors looked at the effects of arsenic overpressure and it can be seen from their data that the silicon nitride capped samples had slightly lower diffusion coefficients than the arsenic overpressure anneal samples. In this paper Guido et al. measured the activation energy under the three different annealing conditions and claimed that the dielectric encapsulated samples showed activation energies of ~3.8 eV whereas the arsenic overpressure annealed samples had an activation energy of 4.5 eV. This change in activation energy was described as 'significant'. If it were significantly outside the experimental uncertainty it would be physically very significant as it would be indicative of a change in the diffusion mechanism. However, when we take the authors raw data and perform a least squares fit to get the activation energy and its associated uncertainty, whilst we get the same activation energies as they report, we also find that they have an uncertainty of ±0.1 and 0.5 eV respectively. As this the one sigma value it is statistically likely that the two values are the same. As mentioned earlier, given the limited range of temperatures over which data has been collected (800°C–875°C) these activation energies are even less reliable than their statistical uncertainty would suggest. Thus it is probable that there is no real difference in activation energy between these samples, i.e. the diffusion mechanism is unchanged, and the differences in diffusivity are due to differences in the prefactor. This result is difficult to understand if the diffusion was mediated by equilibrium concentrations of vacancies.

The enhancement of diffusivity for samples annealed under silicon dioxide has been observed by a number of other groups and the effect has been used as a means of selectively intermixing regions of a wafer [48–50]. In our own work we have found that even for silicon nitride the deposition conditions are very important. We routinely measure the refractive index of the PECVD nitrides we deposit and we have found that nitrides with a refractive index of 2.1 produce reproducible low diffusion coefficients but silicon nitride layers with a refractive index < 0.1 below this can enhance the diffusivity by at least an order of magnitude. Similar effects may be present in the literature data comparing the performance of different encapsulants. Beauvais et al. [51] for example compared silicon nitride and silicon dioxide as encapsulants. Their results seemed to show no difference in the diffusivity for heterostructures with either silicon nitride or silicon dioxide caps. This result, in contradiction to other work in the area, may be due to the deposition conditions under which the nitride was grown.

In addition to the common encapsulants, (i.e. silicon nitride and silicon dioxide) there have been a few alternative materials which have been used in diffusion experiments. Beauvais et al. [51], for example, used strontium fluoride and compared this with both silicon dioxide and silicon nitride. While these experiments showed that the SrF caused less diffusion than either their silicon dioxide or silicon nitride it was still comparable to that produced by good quality silicon nitride. Allen et al. [7] compared SiO_2 and Si_3N_4 with tungsten nitride (WN_x) as an encapsulant. In their work they found that the WN_x film produced little intermixing of a superlattice during annealing at 900°C. In contrast the superlattice capped with SiO_2 was fully mixed to a depth of 2.5 μm and that under a Si_3N_4 film was fully mixed to a depth of 1.6 μm. They state that this is proof that the interdiffusion coefficient is higher under SiO_2 than Si_3N_4. This statement is not strictly true as what it shows is that both their SiO_2 and Si_3N_4 layers caused vacancies to be injected from the surface, although their SiO_2 film produced more vacancies. Again given the demonstration by other workers that Si_3N_4 can result in no vacancy injection, under certain growth conditions, this is probably a demonstration that all Si_3N_4 is not equal. What these results do show however is that WN_x is certainly an excellent means of suppressing vacancy injection, but whether it is better than good Si_3N_4 films is not clear.

7   THE FERMI LEVEL EFFECT

It was an early discovery in the study of III-V interdiffusion, that doping could enhance the interdiffusion rate, consistent with early work on self-diffusion in bulk Ge [1]. Laidig et al. [52] showed in 1981 that GaAs/AlAs heterostructures were unstable against zinc diffusing from a surface source and this opened up the possibility of selective intermixing to produce waveguides, lasers, etc. Following this initial discovery there was a lot of work with a variety of dopants introduced from surface sources, by ion implantation or introduced into the wafer during growth.

In 1985 Kawabe et al. [53] used SIMS to study the effect of doping on the intermixing of GaAs/Al$_x$Ga$_{1-x}$As superlattices which were co-doped with both silicon and beryllium. Their results showed that for layers with a silicon concentration of 7×10^{18} cm^{-3}, if the beryllium concentration was greater than that of the silicon then there was no observable intermixing of the superlattice after annealing at 750°C for 2 hr. However, without the beryllium being present in concentrations in excess of the silicon concentration the superlattice intermixes completely. In 1987 Mei et al. [9] also used SIMS to study the effect of silicon doping, which was incorporated during growth, on the subsequent interdiffusion of GaAs/Al$_x$Ga$_{1-x}$As superlattices in greater detail. Their results showed a clear dependence of the interdiffusion coefficient on the incorporated silicon concentration. Mei et al. [9] noted that the role of silicon in enhancing the interdiffusion is not in lowering the activation energy but in increasing the hopping probability, Figure 9. In order to explain this enhancement they proposed that the enhanced diffusion was due to the diffusion of silicon pairs which generate divacancies which are cause interdiffusion. Tan and Gösele [54] reinterpreted this data and showed that it appeared to be in agreement with their predictions for the interdiffusion being mediated by triply charged gallium vacancies and thus depending upon the position of the Fermi level. The Fermi level effect, as it has become known, relies on the assumption that the point defects controlling the diffusion are charged and that their concentrations are in thermal equilibrium. One of the main predictions of the Fermi level effect is that the measured diffusivity in n-type material should vary as a function of the carrier concentration n such that,

$$D_{Ga}(n) = D_{Ga}(n_i)(n/n_i)^p \tag{9}$$

where n_i is the intrinsic carrier concentration and p is the charge state of the mediating vacancy. From the original data of Mei et al. [9], where they measured the interdiffusion of an GaAs/AlGa superlattice as a function of silicon doping at levels between 2×10^{17}/cm^3 and 5×10^{18}/cm^3, Tan and Gösele showed that the data could be fitted with the Fermi level model where the diffusion mediating defect was a triply charged vacancy,

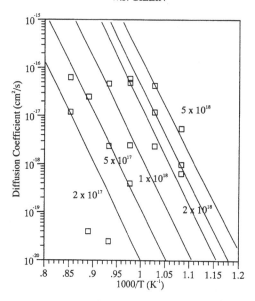

Fig. 9. An Arrhenius plot of interdiffusion coefficient at several silicon concentrations taken from Mei et al. [9]. The solid lines are their own fits showing that a single activation energy is responsible with a change in the prefactor causing the changes in diffusivity.

V_{III}^{3-}. This appeared to be a great success for the model and it has since been widely used as a qualitative explanation for many observations of enhanced diffusion in the presence of doping. Under more rigorous analysis it has been somewhat less successful. Following their work on silicon doping of GaAs/AlAs superlattices, Mei et al. [10] repeated their work, using tellurium as a dopant in place of silicon. From the Femi level effect this change of dopant should have no effect upon the diffusion behavior. However, rather than the cubic dependence of diffusivity on carrier concentration observed following the earlier experiments they found only a linear dependence. This difference was qualitatively explained by Tan and Gösele, by claiming that perhaps all of the tellurium present in the material was not electrically active and that this was the cause of the discrepancy.

It has been observed by a number of authors that [12,21,34,52,55], with the exception of Mei et al.'s samples, no enhancement in the interdiffusion is observed unless the doping concentration is greater than $\sim 10^{18}$ Si/cm^3. Again, this is in contradiction to the results one would expect from a Fermi level effect being responsible for the enhanced interdiffusion. One example of this is the work of Gillin et al. [12] who measured the interdiffusion in both GaAs/AlGaAs and InGaAs/GaAs structures that had been doped during growth with either silicon or beryllium at a constant level from a depth of 100 nm below the quantum well to the surface, 100 nm above the quantum well. They looked at samples with beryllium concentrations of 10^{17}, 10^{18} and 2.5×10^{19} Be/cm^3 and with silicon concentrations of 10^{17}, 10^{18} and 10^{19} Si/cm^3. These results showed that there was no measurable change in diffusion in any of the beryllium doped samples or in the silicon doped samples with 10^{18} Si/cm^3 or less. The sample doped with 10^{19} Si/cm^3 did however show a serious degradation of its optical quality and the photoluminescence spectra showed peaks which indicated the presence of defects believed to be related to group III vacancies.

In another attempt to quantify the role of any Fermi level effect on diffusion Seshadri et al. [56] studied p-i-n and n-i-p structures where the Fermi level is varying throughout the region where three quantum wells were placed. They found only small variations in the diffusion coefficient with the position of the quantum well within their structure and concluded that the Fermi level was playing no role in the interdiffusion of their samples.

A more recent paper by Jarfi et al. [21] showed that in the earlier work of Gillin et al. [12] the sample which was doped with silicon at 10^{19} Si/cm^3, did undergo some transient enhanced diffusion during it's first anneal but with following anneals the sample did not undergo any further enhanced diffusion. They attributed this to some of the silicon, which was originally on the group III sublattice, (i.e. electrically active donor sites), moving off these sites and creating group III vacancies. This process is related to the amphoteric behaviour of silicon in GaAs [57]. They suggest that this process may explain the behaviour of the samples of Mei et al. In Mei's samples there was a doping staircase with concentration plateaus of (I) 2×10^{17}, (II) 5×10^{17}, (III) 1×10^{18}, (IV) 2×10^{18}, (V) 5×10^{18}, and (VI) 2×10^{18}. From Jarfi et al. and from the work of Reynolds and Geva [55] it would be expected that layer V would act as a source of vacancies and hence experience a large amount of intermixing.

Thus if layer V is acting as a source of vacancies one can model their subsequent diffusion as a function of time using the standard solution for diffusion of a finite plane,

$$N_V(x, t) = N_0 \left\{ \mathrm{erf}\left(\frac{h/2 - x}{2\sqrt{D_v t}}\right) + \mathrm{erf}\left(\frac{h/2 + x}{2\sqrt{D_v t}}\right) \right\} \qquad (10)$$

where $N_V(x, t)$ is the concentration of vacancies as a function of depth and time, N_0 is the initial concentration of vacancies, h is the thickness of the initial vacancy layer, x is the distance in the growth direction (with $x = 0$ at the centre of the initial vacancy distribution), D_v is the diffusion coefficient for the vacancies and t is time. As the diffusion coefficient for intermixing is proportional to the vacancy concentration, equation 2, we can numerically integrate equation 10 as a function of time at different depths. Using this we can calculate a term proportional to the effective interdiffusion length in

W.P. GILLIN

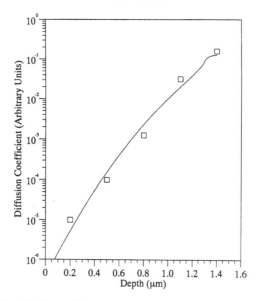

Fig. 10. The interdiffusion coefficients at 700°C as a function of depth taken from the paper of Mei et al. [9]. The solid line is the theoretical fit of Jafri et al. [21] calculated by assuming that only the 5×10^{18} Si/cm^3 is a source of vacancies.

each of the layers in Mei et al.'s samples and thus the effective diffusion coefficient they would have measured in their experiment. Figure 10 shows the diffusion coefficient data presented by Mei et al. for their 700°C anneals plotted as a function of depth and the results of this analysis. It can be seen that there is an excellent fit to the data for a diffusion length for the vacancies of 6.8 nm. This corresponds to a diffusion coefficient for the vacancies of $\sim 6 \times 10^{-16}$ cm^2/s at 700°C.

Mei et al. also observed that the measured intermixing caused by the silicon doping saturates for silicon concentrations greater than $\sim 10^{19}$ Si/cm^3 [9]. Indeed later work by them showed that the intermixing is inhibited by very high silicon concentrations [58]. In this work they have shown that these highly doped layers have dislocation loops in them and it is possible that these loops may act as an effective sink for vacancies removing them before they can cause intermixing. Alternatively, as they gave no electrical data for the silicon doped layers it is possible that whilst $>10^{19}$ Si/cm^3 was incorporated in the layers, not all of these were electrically active (i.e. on group III sites) and hence their subsequent behaviour on annealing may not create group III vacancies.

Jafri et al. then went on to see if their model could also explain the tellurium doped data collected by Mei et al. [10]. The first comment to note from Mei et al.'s tellurium data is that they measured the samples after both 30 minute and 3 hour anneals. If the diffusion mechanism in these samples were similar to the one we have described, at a given temperature one would expect to see an apparent diffusion coefficient that would decrease with time. This decrease would be more apparent at higher anneal temperatures where all the vacancies from this process would have diffused evenly throughout the layer during the early stages of the anneal, and this can be seen in their data. It should also be noted that the diffusion coefficient measured for the 2×10^{18} Te/cm^3 layer is two orders of magnitude less than that for silicon doping, and this indicates that there are correspondingly fewer vacancies being created.

Unfortunately there is much less data available on the annealing behaviour and site location of tellurium compared to silicon. However, the fact that tellurium sits preferentially on a group V site makes any mechanisms for the formation of group III vacancies speculative. However, as stated above, the group III vacancy concentrations required to obtain the diffusion coefficients they measured are quite small and one possibility is that at the relatively high doping concentrations used for their study there may have been some tellurium sitting on group III sites which was able to change site with annealing. This suggestion, whilst speculative, should be compared with the fact that in order for Mei et al.'s tellurium data to be fitted to the Fermi level model Tan and Gösele [59] had to assume that only $\sim 10\%$ of the tellurium was electrically active.

It is also interesting to speculate on possible causes for the silicon and beryllium data presented by Kawabe et al. [53]. They found that superlattice layers doped with either 6 or 7×10^{18} Si/cm^3 intermixed completely except where they were co-doped with 10^{19} cm^{-3} of beryllium. In their work it can be seen that even in the Be doped layers there is some intermixing near to the interfaces of the Be doped layer as well as deeper into the structure where there is no doping at all. The intermixing in the regions where there is no doping is consistent with the idea of vacancies diffusing out of the doped layer. However, the reduction in intermixing that can be observed with high Be doping may be due to a trapping mechanism for the vacancies. One possible such mechanism would be interstitial beryllium atoms reacting with the vacancies to form an electrically active beryllium acceptors. The use of co-implants of a dopant and a lattice constituent which sits on the opposite sublattice to enhance electrical activity has been shown in the literature and presumably operates through a similar mechanism.

Thus it can be seen that while there is some evidence for a Fermi level effect in III-V material it is only qualitative, and the only quantitative results in the literature which can be explained by the Fermi level effect can be explained equally well by other theories.

The Fermi level effect does however, in theory, appear quite convincing and it is often argued that it could play some role in diffusion in an ideal sample. This is because the physics upon which it is based are quite straightforward.

However, it should be remembered that the whole theory is based upon the assumption that the point defect concentrations responsible for intermixing are at their thermodynamic equilibrium concentrations and that the condition of the system remaining in equilibrium remains true during annealing. This core assumption of vacancies being at equilibrium, with more being generated through Frenkel pair formation as the temperature is increased, has recently been shown not to be true [45] and that the diffusion is instead controlled by a constant concentration of vacancies which are probably frozen in during the growth of the GaAs substrate. Thus without there being equilibrium vacancy concentrations the Fermi level effect cannot operate and the small changes occasionally observed in interdiffusion with doping must be caused by some other process.

8 STRAINED MATERIAL SYSTEMS

The majority of the work discussed in this chapter has been performed on unstrained GaAs/AlGaAs samples. However, there are a variety of III-V systems available many of which are strained and some of which appear to be more susceptible to diffusion than others. After GaAs/AlGaAs probably the most widely studied system is the strained InGaAs/GaAs.

The presence of strain in heterostructures has been argued to be another factor which may strongly affect interdiffusion rates. Iyer and LeGoues [60], for example, reported that a strained Si-SiGe interface could diffuse at 550°C at a rate which without strain would require temperatures in excess of 1000°C. They did not consider this result surprising as the effect of strain-enhanced diffusion had long been known in metals [61,62]. Kuan et al. [63] later reported significant interdiffusion in the Si-SiGe system at a temperature as low as 400°C, which, if due to the intrinsic strain in the structure, would make silicon-germanium pseudomorphic structures technologically very difficult to grow and process.

The first reported effects of strain on the interdiffusion of III-V systems were reported by Temkin et al. [64] in the InGaAs – InP system. They used both TEM and photoluminescence to investigate quantum wells before and after annealing and they suggest that following annealing their quantum wells remain square rather than the error functions one would expect for Fick's law diffusion. In order to explain these effects they suggest that the compositional driving force for diffusion is opposed by the increase in the lattice strain that diffusion would introduce, and that this resistance to strain adjusts the interdiffusion coefficient to produce a roughly lattice matched quantum well. While this is one possible explanation it is also possible that the miscibility gap which exists in the phase diagram for the InGaAsP system could also play a role. Cohen [65] showed that for a quantum well made from certain material combinations within the InGaAsP system, such as lattice matched InGaAs/InP, the miscibility gap could be expected to retard the interdiffusion of the layers.

Fujii et al. [66] have also studied the InGaAs-InP system, and have used photoluminescence to determine interdiffusion between 500°C and 640°C. They suggested that the diffusion rate is determined by the diffusion of the group V species across the heterointerface and that this is much lower than the diffusion rate in either the InGaAs or the InP. This they attribute to the diffusion at the interface creating local strain and this strain energy they suggest is sufficient to retard the diffusion. In this work they quote an estimated strain energy of about 1 eV per atom which they calculate from elastic constants. This strain energy is of the same order as the activation energy for interdiffusion that they determined, and they therefore suggest that strain can have a significant effect upon the diffusion through the exponential term in the Arrhenius equation. However, their strain energy is nearly three orders of magnitude greater than the correct value, which for a biaxially compressed pseudomorphic layer of strain, e_0, has an energy given by

$$E = \int_v \sum_{i,j=1}^{3} \frac{1}{2} \varepsilon_{ij} \sigma_{ij} dV = \frac{2Y}{1+v} \varepsilon_0^2 V \tag{11}$$

where Y is the Young's modulus, and which for a strain of $\varepsilon_0 = 0.01$, and using GaAs elastic constants gives an elastic energy of approximately 1.73×10^7 Jm^{-3}, or about 2.5 meV per atom.

Antonelli and Bernholc [67] in a theoretical paper calculate the change in formation energy of a vacancy in the presence of hydrostatic and biaxial strain. They calculated that while hydrostatic pressure would have some effect upon the formation energy of vacancies, pure biaxial strain would have a negligible effect. These results have been used by workers in SiGe as evidence to support a strain dependence of the interdiffusion. However, as the effect only acts upon the formation of energy of vacancies and, for III-V materials at least, the vacancy concentrations are not at equilibrium concentrations, the formation of vacancies is not an important issue in the interdiffusion and this mechanism should not effect interdiffusion.

Despite the results quoted above, other authors do not find a significant effect of strain on interdiffusion. Holländer et al. [68,69] also studied the Si-SiGe system, but they found that strain had only a minor effect on interdiffusion compared with concentration. While Gillin and co-workers have studied the InGaAs-GaAs system in detail and found that if there is any strain dependent term in the diffusion then is at a level too small to be reliably detectable given the accuracy of the data which can be collected [12,14,16,70]. Ryu et al. [71] have reanalysed some of the data of Gillin et al. and show that they can get excellent fits to it by applying a strain dependent term in to their diffusion equation. This term effectively gives a free fitting parameter which allows for the quality of the fit to their experimental data. What they do not show are error bars for the data, which were not present in the original paper, and hence whether the fits they obtain are statistically much better than the use of a strain independent diffusion model is impossible to judge. At the present time therefore the jury is still out as to whether strain does play a role in the diffusion of semiconductors, and it is clear that much better quality data will need to be produced before the matter can be fully settled.

In addition to the strained InGaAs/GaAs system, Gillin et al. [72–74] have also studied three strained InGaAsP based systems grown on InP. The first of these was $In_{0.66}Ga_{0.33}As/In_{0.66}Ga_{0.33}As_{0.7}P_{0.3}$. In this system the group III sublattice has no concentration change to drive diffusion and hence the photoluminescence measures solely the diffusion of the group V sublattice. These experiments were performed on samples grown on both sulphur and tin doped substrates in order to look for the substrate dependence of interdiffusion observed by Glew et al. [75]. The samples showed no effect of strain upon the diffusion and no effect of the substrate type. There did however, appear to be a low activation energy process starting to occur at temperatures below 700°C. Later experiments at temperatures down to 500°C [30], show that this low activation energy process does not exist and it serves to highlight the dangers of believing activation energies based on small data sets. In addition to measuring group V interdiffusion, Rao et al. [73] measured the group III interdiffusion in the InGaAsP system using samples of $In_{0.66}Ga_{0.34}As_{0.75}P_{0.25}/In_{0.79}Ga_{0.21}As_{0.75}P_{0.25}$ and $In_{0.67}Ga_{0.33}As/In_{0.53}Ga_{0.47}As$. In both of these structures there is no step in the group V sublattice and hence the photoluminescence only measures the group III interdiffusion. Again both these materials were grown on both sulphur and tin doped substrates to look for substrate effects. Like the group V diffusion there appeared to be no strain effect visible in these materials. What is interesting is that when the group III and group V diffusion are plotted on a single Arrhenius plot, Figure 11, it is clear that there is no difference in the activation energy or prefactor for diffusion of the two sublattices and the measured activation energy for both group III and group V interdiffusion in InGaAsP is 3.4 ± 0.1 eV, which is effectively identical to that measured in GaAs/AlGaAs [41] and InGaAs/GaAs [14]. As the activation energy in the GaAs based materials has been shown to be solely the diffusion energy for vacancies [45], and as the bonding in GaAs and InP based materials are effectively identical, it is reasonable to assume that the interdiffusion in InP based materials is also governed by non-equilibrium vacancy concentrations. Given the values measured for GaAs based materials this allows us to use the prefactor for InP based interdiffusion to estimate the vacancy concentrations in InP. From Figure 11 the prefactor for InP based interdiffusion can be

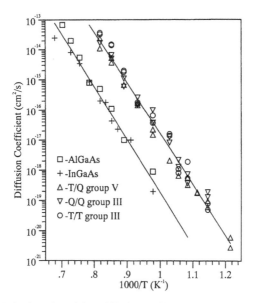

Fig. 11. An Arrhenius plot of interdiffusion coefficient for three InP based materials, $In_{0.66}Ga_{0.33}As/In_{0.66}Ga_{0.33}As_{0.7}P_{0.3}$ (T/Q), $In_{0.66}Ga_{0.34}As_{0.75}P_{0.25}/In_{0.79}Ga_{0.21}As_{0.75}P_{0.25}$ (Q/Q) and $In_{0.67}Ga_{0.33}As/In_{0.53}Ga_{0.47}As$ (T/T) and two GaAs based materials $In_{0.2}Ga_{0.8}As/$ GaAs and $GaAs/Al_{0.2}Ga_{0.8}As$. It can be clearly seen that for the InP based materials all the data can be described by a single activation energy although it includes both group III and group V sublattice diffusion. Furthermore the activation energy for the InP based material is identical to that for the two GaAs based materials.

seen to be an order of magnitude greater than that for GaAs based materials. This means that the vacancy concentrations on both sublattices are equal and they are also an order of magnitude higher than the group III vacancy concentration in GaAs, $\sim 10^{17}$ cm^{-3}. As this concentration of vacancies, $\sim 10^{18}$ cm^{-3}, is probably frozen in to the InP during growth it allows for the possibility of improving the thermal stability of lasers grown on InP through improvements in the substrate quality. Whether this vacancy concentration can be economically reduced is a different issue, but the possibility exists.

The fact that the group III and group V diffusion coefficients are identical are in contradiction to the results of some qualitative results based on the relative shifts of laser wavelengths during overgrowth for devices with either group III or group V compositional steps [76]. This apparent qualitative difference can be attribute to the much smaller band offsets available when only the group III sublattice is varied between well and barrier and this smaller band offset results in the smaller photoluminescence shifts with diffusion. This is highlighted in the photoluminescence shift diffusion for the three systems used by Gillin et al. [72] and Rao et al. [73], Figure 12. The results are also in contradiction to some studies of lattice matched InGaAs/InP.

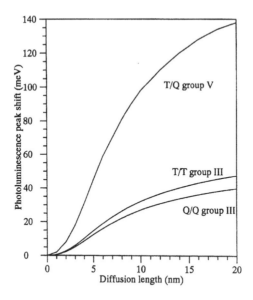

Fig. 12. The theoretical photoluminescence peak shift as a function of diffusion length for the three structures used in references 72 and 73. It can be seen that for an identical amount of diffusion the T/Q group V structure will have a much larger shift in photoluminescence peak position.

This contradiction may well be due to the miscibility gap mentioned earlier and the effect that it may be expected to have upon diffusion as discussed by Cohen [65].

9   CONCLUSIONS

In this chapter we have compared the various methods used to measure interdiffusion in III-V materials and discussed the experimental conditions necessary to obtain accurate and reliable measurements of both diffusion coefficients and activation energies for diffusion. We go on to look at the various measurements of activation energy made for the GaAs/AlGaAs system and show that all the data appears to be consistent with a single activation energy of ~3.5 eV. We then look at the arsenic overpressure annealing results which suggest that group III interdiffusion is controlled solely by vacancies diffusing on their own sublattice and with recent results which show that these vacancies are not thermally generated, but probably frozen in to the GaAs when the wafer is manufactured. Some comparison is given of the various capping technologies available and the effects that they have upon diffusion.

Finally some consideration is given to the two main perturbations which are thought to play a significant role in interdiffusion, doping and strain. We show that doping under technologically useful conditions appears to have little effect upon the thermal stability of III-V heterostructures and that the Fermi Level appears not to have a role in interdiffusion. Similarly, strain in III-V materials does not appear to have a theoretical basis for effecting diffusion in III-V materials and with present experimental techniques there is no hard evidence for strain effects.

References

1. A. Seeger and K.P. Chik, *Phys. Stat. Sol.*, **29**, 455 (1968).
2. L.L. Chang and A. Koma, *Appl. Phys. Lett.*, **29**(3), 138 (1976).
3. L.J. Guido, N. Holonyak, Jr., K.C. Hsieh, R.W. Kaliski, W.E. Plano, R.D. Burnham, R.L. Thornton, J.E. Epler and T.L. Paoli, *J. Appl. Phys.*, **61**(4), 1372 (1987).
4. K.Y. Hsieh, Y.C. Lo, J.H. Lee and R.M. Kolbas, *Inst. Phys. Conf. Ser.*, No. 96, 393 (1988).
5. B.L. Olmsted and S.N. Houde-Walter, *Appl. Phys. Lett.*, **60**(3), 368 (1992).
6. B.L. Olmsted and S.N. Houde-Walter, *Appl. Phys. Lett.*, **63**(4), 533, 1993.
7. E.L. Allen, C.J. Pass, M.D. Deal, J.D. Plummer and V.F.K. Chia, *Appl. Phys. Lett.*, **56**(25), 3252 (1991).
8. E.V.K. Rao, A. Hamoudi, Ph. Krauz, M. Juhel and H. Thibierge, *Appl. Phys. Lett.*, **66**(4), 472 (1995).
9. P. Mei, H.W. Yoon, T. Venkatesan, S.A. Schwarz and J.P. Harbison, *Appl. Phys. Lett.*, **50**(25), 1823 (1987).
10. P. Mei, C.L. Schwartz T. Venkatesan, and E. Colas, *J. Appl. Phys.*, **65**(5), 2165 (1989).
11. C.L. Reynolds, Jr. And M. Geva, *Appl. Phys. Lett.*, **61**(2), 165 (1992).
12. W.P. Gillin, I.V. Bradley, L.K. Howard, R. Gwilliam and K.P. Homewood, *J. Appl. Phys.*, **73**(11), 7715 (1993).
13. T. Venkatesan, S.A. Schwarz, D.M. Hwang, R. Bhat and H.W. Yoon, *Nucl. Inst. Methods*, **B19/20**, 777 (1987).
14. I.V. Bradley, W.P. Gillin, K.P. Homewood and R.P. Webb, *J. Appl. Phys.*, **73**(4), 1686 (1993).
15. A. Ourmazd. Defect and Diffusion Forum, **95-98**, 917 (1993).
16. W.P. Gillin, D.J. Dunstan, K.P. Homewood, L.K. Homewood and B.J. Sealy, *J. Appl. Phys.*, **73**(8), 3782 (1993).
17. I. Lahiri, D.D. Nolte, J.C.P. Chang, J.M. Woodall and M.R. Melloch, *Appl. Phys. Lett.*, **67**(9), 1244 (1995).
18. H.M. You, T.Y. Tan, U.M. Gÿ94sele, S.T. Lee, G.E. Höfler, K.C. Hsieh and N. Holonyak Jr., *J. Appl. Phys.*, **74**(4), 2450 (1993).
19. T. Humer-Hager, R. Treichler, P. Wurzinger, H. Tews and P. Zwicknagl, *J. Appl. Phys.*, **66**(1), 181 (1989).
20. A. Kozanecki, W.P. Gillin and B.J. Sealy, *Appl. Phys. Lett.*, **64**(1), 40 (1994).
21. Z.H. Jafri and W.P. Gillin, *J. Appl. Phys.*, **81**(5), 2179 (1997).
22. Y. Hirayama, Y. Suzuki, S. Tarucha and H. Okamoto, *Jpn. J. Appl. Phys.*, **24**(7), L516 (1985).
23. C. Vieu, M. Schneider, D. Mailly, R. Planel, H. Launois, J.Y. Marin and B. Descouts, *J. Appl. Phys.*, **70**(3), 1444 (1991).
24. L.B. Allard, G.C. Aers, S. Charbonneau, T.E. Jackman, R.L. Williams, D. Stevanovic and F.J.D. Almeida, *J. Appl. Phys.*, **72**(2), 422, 1992.
25. I. Karla, J.H.C. Hogg, W.E. Hagston, J. Fatah and D. Shaw, *J. Appl. Phys.*, **79**(4), 1898, 1996.
26. G.P. Kothiyal and P. Bhattacharya, *J. Appl. Phys.*, **63**(8), 2760, 1988.

27. W.P. Gillin, H. Peyre, J. Camassel, K.P. Homewood, I.V. Bradley and R. Grey, *Journal de Physique IV*, **3**, 291, 1993.
28. J. Cibert, P.M. Petroff, D.J. Werder, S.J. Pearton, A.C. Gossard and J.H. English, *Appl. Phys. Lett.*, **49**(4), 223, 1986.
29. K. Dettmer, W. Freiman, M. Levy, Yu. L. Khait and R. Beserman, *Appl. Phys. Lett.*, **66**(18), 2376, 1995.
30. S.S. Rao, PhD Thesis University of Surrey, 1995
31. P.G. Shewmon, Diffusion in Solids, McGraw Hill, 1963.
32. J.S. Major, Jr., F.A. Kish, T.A. Richard, A.R. Sugg, J.E. Baker and N. Holonyak, Jr., *J. Appl. Phys.*, **68**(12), 6199, 1990.
33. A. Ourmazd, Y. Kim and M. Bode, *Mat. Res. Soc. Symp. Proc.*, Vol. 163, 639, 1990.
34. B.L. Olmsted and S.N. Houde-Walter, *Appl. Phys. Lett.*, **63**(8), 1131 (1993).
35. J.S. Tsang, C.P. Lee, S.H. Lee, K.L. Tsai and H.R. Chen, *J. Appl. Phys.*, **77**(9), 4302 (1995).
36. H.D. Palfrey, M. Brown and A.F.W. Willoughby, *J. Electrochem. Soc.*, **128**(10), 2224 (1981).
37. J.D. Ralston, S. O'Brien, G.W. Wicks and L.F. Eastman, *Appl. Phys. Lett.*, **52**(18), 1511 (1988).
38. B. Goldstein, Physical Review, **121**(5), 1305 (1961).
39. B.L. Olmsted and S.N. Houde-Walter, *Appl. Phys. Lett.*, **63**(4), 530 (1993).
40. L.J. Guido and N. Holonyak, Jr., *Mat. Res. Symp. Proc.*, Vol 163, 697 (1990).
41. S.F. Wee, M.K. Chai, K.P. Homewood and W.P. Gillin, *J. Appl. Phys.*, (In Press)
42. D.G. Deepe and N. Holonyak Jr., *J. Appl. Phys.*, **64**(12), R93 (1988).
43. A. Furuya, M. Makiuchi, O. Wada, T. Fujii and H. Nobuhara, *Jpn. J. Appl. Phys.*, **26**, L926 (1987).
44. S.. Dannefaer, P. Mascher and D. Kerr, *J. Appl. Phys.*, **69**(7), 4080 (1991).
45. O.M. Khreis, W.P. Gillin and K.P. Homewood, *Phys. Rev. B*, **55**(24), (1997).
46. J.A. Van Vechten, *J. Appl. Phys.*, **53**, 7082 (1992).
47. M. Kuzuhara, T. Nozaki, and T. Kamejima, *J. Appl. Phys.*, **66**(12), 5833 (1989).
48. J.Y. Chi, X. Wen, E.S. Koteles and B. Elman, *Appl. Phys. Lett.*, **55**(9), 855 (1989).
49. M. Ghisoni, R. Murray, A.W. Rivers, M. Pate, G. Hill, K. Woodbridge and G. Parry, *Semicond. Sci. Technol.*, **8**, 1791 (1993).
50. B.S. Ooi, A.C. Bryce, J.H. Marsh and J.S. Roberts, *Semicond. Sci. Technol.*, **12**, 121 (1997).
51. J. Beauvais, J.H. Marsh, A.H. Kean, A.C. Bryce, C. Button, *Electron. Letts.*, **28**(17), 1670 (1992).
52. W.D. Laidig, N. Holonyak Jr., M.D. Camras, K. Hess, J.J. Coleman, P.D. Dapkus and J. Bardeen, *Appl. Phys. Lett.*, **38**(10), 776 (1981).
53. M. Kawabe, N. Shimizu, F. Hasegawa and Y. Nannichi, *Appl. Phys. Lett.*, **46**, 849 (1985).
54. T.Y. Tan and U. Gösele, *J. Appl. Phys.*, **61**, 1841 (1987).
55. C.L. Reynolds Jr., and M. Geva, *Appl. Phys. Lett.*, **61**(2), 165 (1992).
56. S. Seshadri, L.J. Guido and P. Mitev, *Appl. Phys. Lett.*, **67**(4), 497 (1995).
57. J.P. de Souza, D.K. Sadana, and H.J. Hovel, *Mat. Res. Soc. Symp. Proc.*, Vol. 144, (1989).
58. P. Mei, S.A. Schwarz, T. Vankatesan, C.L. Schwartz, E. Colas, *J. Appl. Phys.*, **65**(5), 2165, (1989).
59. T.Y. Tan and U. Gösele, *Mat. Res. Soc. Symp. Proc.* Vol. 144, 221 (1989).
60. S.S. Iyer and LeGoues, *J. Appl. Phys.*, **65**, 4693, (1989).
61. J.W. Cahn, *Trans. Metall. Soc. AIME*, **242**, 166, (1968).
62. B.M. Clemens, *Phys. Rev. B*, **33**, 7615, (1986).
63. T.S. Kuan and S.S. Iyer, *Appl. Phys. Lett.*, **59**(18), 2242, (1991).
64. H. Temkin, S.N.G. Chu, M.B. Panish and R.A. Logan, *Appl. Phys. Lett.*, **50**, 956, (1987).
65. R.M. Cohen, *J. Appl. Phys.*, **73**(10), 4903 (1993).
66. T. Fujii, M. Sugawara, S. Yamazaki and K. Nakajima, *J. Crystal Growth*, **105**, 348, (1990).
67. A. Antonelli and J. Bernholc, *Mat. Res. Soc. Symp. Proc.*, **163**, 523 (1990).
68. B. Holländer, S. Mantl, B. Stritzker and R. Butz, *Nucl. Inst. Meth. B*, **59/60**, 994, (1991).
69. B. Holländer, R. Butz and S. Mantl, *Phys. Rev. B*, **46**(11), 6975, (1992).
70. W.P. Gillin and D.J. Dunstan, *Phys. Rev. B*, **50**(11), 7495 (1994).
71. S.W. Ryu, I. Kim, B.D. Choe, W.G. Jeong, *Appl. Phys. Lett.*, **67**(10), 1417 (1995).
72. W.P. Gillin, S.S. Rao, I.V. Bradley, K.P. Homewood, A.D. Smith and A.T.R. Briggs, *Appl. Phys. Lett.*, **63**(6), 797 (1993).

73. S.S. Rao, W.P. Gillin and K.P. Homewood, *Phys. Rev. B*, **50**(11), 8071 (1994).
74. W.P. Gillin, S.D. Perrin and K.P. Homewood, *J. Appl. Phys.*, **77**(4), 1463 (1995).
75. R.W. Glew, A.T.R. Briggs, P.D. Greene and E.M. Allen, *Proceedings of the Fourth International Conference on InP and Related Materials*, Newport (IEEE, New York, 1992), p. 234.
76. S.D. Perrin, C.P. Seltzer and P.C. Spudens, *J. Electron. Mater.*, **27**, 81 (1994).

CHAPTER 3

Interdiffusion in Lattice-Matched Quantum Wells and Self-Formed Quantum Dots Composed of III-V Semiconductors

KOHKI MUKAI

Fujitsu Laboratories Ltd., Optical Semiconductor Devices Laboratory, 10-1 Morinosato-Wakamiya, Atsugi 243-0197, Japan

1 INTRODUCTION

The interdiffusion process in nanostructures of III-V compound semi-conductors cannot be measured directly due to the small size, although interdiffusion is a significant phenomenon in electronic and optoelectronic device applications. The profiles of quantum wells after interdiffusion are not necessarily the same as those measured directly in wide heterostructures. Since the quantum well layers are thin, interdiffusion penetrates alternate barrier layers, which complicates the interdiffusion profiles. In addition,

85

interdiffusion causes lattice strain in III-V compound semiconductors, and the lattice strain, in turn affects the interdiffusioon process. The strain in heterostructures wider than the critical thickness generates dislocations and defects. The profiles of a quantum wire and a quantum dot after interdiffusion are more complicated.

The interdiffusion in GaAs/AlGaAs quantum wells has been extensively studied because of its importance in practical applications and the simplicity of the analysis. Group-III atoms are the only possible interdiffusion species, and no lattice distortion occurs due to the interdiffusion. It is known that the interdiffusion in GaAs/AlGaAs quantum wells begins above 800°C, and is more thermally stable than in InGaAs/InP quantum wells. The influence of implanting active impurities [1–7] and lattice defects [8–11] on interdiffusion was studied using transmission electron microscopy or by measuring the quantum energy shifts. The interdiffused compositional profile of these materials was first reported by Chang and Koma [12] for 150-nm-wide GaAs/AlAs double heterostructures measured by Auger electron spectroscopy (Figure 1). The profiles were error-functional and symmetrical. The calculation includes two assumptions: 1) the interdiffusion coefficient is constant in both the GaAs and AlAs layers, and 2) the interface is smooth. Considering that the strain in interdiffused AlGaAs layers is negligible, several other researchers used Chang's profiles to explain how quantum energy shifts in GaAs/AlGaAs quantum wells relate to interdiffusion [11,13,14]. These researchers did not, however, directly examine or further confirm the interdiffusion profile of the quantum wells.

After GaAs/AlGaAs quantum wells, the next most studied structure is InGaAsP/InP quantum wells, which are the main components of 1-μm-region optoelectronic devices. This system has two possible interdiffusion species, group-III atoms and group-V atoms. The interdiffusion of group-III atoms and that of group-V atoms must be considered separately, since the atoms do not mix because of their polarities. Group-V atoms reportedly interdiffuse more easily than group-III atoms [15]. Arsenic and phosphorus atoms begin to interdiffuse at about 500°C [16]. The influence of impurities and defects on interdiffusion properties has been studied [17–23], but the interdiffusion profile in InGaAsP/InP quantum wells was difficult to understand. The interdiffusion profile cannot be estimated by direct observation of wide heterostructures since the lattice constant of InGaAsP materials depends on the composition, and the induced lattice strain affects the interdiffusion process. Some researchers have pointed out the difference between interdiffusion profiles of $In_{0.53}Ga_{0.47}As/InP$ quantum wells and those of GaAs/AlGaAs systems using indirect methods. Nakashima et al. [15] used X-ray analysis and the Fleming model [24] to show that arsenic composition in an $In_{0.53}Ga_{0.47}As$ well layer decreases uniformly and that the interdiffused arsenic atoms accumulate near the interface on the InP barrier layer side.

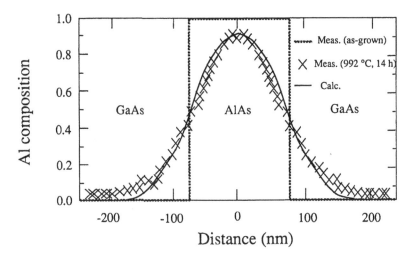

Fig. 1. Interdiffusion profile in GaAs/AlAs systems measured by Chang and Koma using Auger electron spectroscopy [12]. Dashed lines indicate the initial Al composition, x's show the data after 14 hours of annealing, and solid lines indicates the Al composition calculated using the simple error function.

Fujii et al. [16] reported that there must be a large compositional discontinuity of group-V atoms at the interface, even after interdiffusion, by roughly estimating the dependence of quantum energy shifts on well-layer width.

While quantum wells are now used in a wide range of device applications, the fabrication of quantum wires and dots is still a subject of research. Using self-formation phenomena under highly strained epitaxial growth, uniform high-quality quantum dots can now be grown [25,26]. Operation of quantum-dot lasers have been achieved using the self-formed quantum dots [27–29]. The expected excellent performances of dot lasers have, however, not yet been obtained. This may be attributed to fabrication problems such as nonuniformity of dots, low crystal quality, and small total dots' volume. Unavoidable interdiffusion during the growth of the laser structure may be also one of the reasons. An InGaP or AlGaAs cladding layer should be grown at above 600°C to obtain good optical quality though the growth temperature of dots is about 500°C. Interdiffusion changes the quantum energy levels and degrades the crystal quality [30]. In addition, we must note that the reduced performances may be related to the bottleneck effect where the carrier relaxation between intrabands is hindered due to the delta-functionlike state density of the quantum dots. The carrier relaxation and recombination process must be related to the quantum conefinement potential influenced by the interdiffusion.

The purpose of this chapter is to present our study on the interdiffusion process in III-V compound quantum wells and quantum dots. In Section 2, we show a formula which comprehensively describes interdiffusion profiles in quantum wells, and evaluate the interdiffusion process of group-V atoms in lattice-matched InGaAsP/InP and that of group-III atoms in GaAs/AlGaAs quantum wells using the formula [31,32]. We derive our formula by analytically solving diffusion equations for a stacked three-layer system, taking into account the different interdiffusion coefficients between layers and the compositional discontinuity of diffused species at interfaces. We relate our formula to the dependence of quantum energy shifts on annealing time and annealing temperature for various wide-well layers. We show that the dependence of quantum energy shifts is very different between InGaAsP/InP and GaAs/AlGaAs quantum wells, and that our formula explains the difference. Then, we demonstrate how to enhance the thermal stability of InGaAs/InP quantum wells and discuss the interdiffusion mechanism [33]. We examine the growth-condition dependence of interdiffusion coefficients using our formula. In Section 3, we discuss the effects of interdiffusion on energy and carrier relaxation process in sublevels of self-formed InGaAs/GaAs quantum dots. We first show the growth and the optical evaluation of quantum dots self-formed during alternate supply of precursors [25,34–36]. Then, we describe the quantum energy shift of the ground and the second sublevels as a function of annealing temperature, indicating the characteristics that are not observable in quantum wells. Considering that the interdiffusion caused the change of relative emission intensity of the sublevels, we prove the existence of the phonon bottleneck for carrier relaxation [37-39]. The phonon bottleneck phenomenon is peculiar to quantum dots, and the influence of interdiffusion on the bottleneck effect on the carrier process is presented by measuring time-resolved photoluminescence.

2 INTERDIFFUSION IN LATTICE-MATCHED QUANTUM WELLS

2.1 Theory

2.1.1 Model of interdiffusion profile

We solved the linear diffusion equations for a stacked three-layer structure. In our mathematical model, the outside layer extends infinitely in the $-x$ and $+x$ directions (Figure 2). We defined the origin of the x axis as the center of the middle layer. There are two boundaries, at $x = -L$ and $x = L$. First, we solved linear diffusion equations assuming a momentary plane source

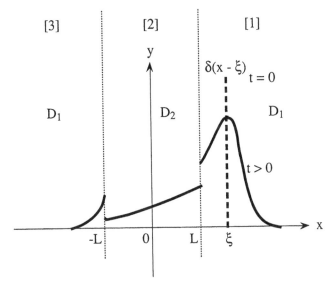

Fig. 2. Schematic of our model. We solved diffusion equations as a three-body system. We imposed three conditions in solving the diffusion equations: 1) the interdiffusion coefficients differ depending on the structural region, 2) the composition of diffused species is discontinuous at the interface, 3) diffusion from one barrier layer penetrates to the next barrier layer.

at $x = \xi$. We then integrated the solution, using an initial composition profile, to fit the actual structure. In the actual structure, diffusion species in the two outside layers can mix via the middle layer, which is equivalent to a single quantum well. To derive a solution, we made two generalizing assumptions [40]. First, we assumed that the diffusion coefficients of the outside layers are the same, but different from that of the middle layer. Second, we assumed a certain distribution ratio for diffused species at the interface [41], so the composition of diffused species can be discontinuous there. We made two more assumptions to eliminate difficulties in solving the diffusion equations: 1) both the interdiffusion coefficient in each layer and the distribution ratio at the interface are constant during interdiffusion, 2) we ignored the Smigelskas-Kirkendall effect, where unbalanced interdiffusion velocity causes the interface to move. We will show later that practical results are consistent with these assumptions. In addition, when the thicknesses of layers are less than 3 nm, a gradient correction term should be considered [42]. In this work, we neglected the correction term.

We derived our formula as shown below. The three linear diffusion equations are

$$\frac{\partial C_1(x,t)}{\partial t} = D_1 \frac{\partial^2 C_1(x,t)}{\partial x^2} \qquad t \geq 0, x \geq L, \qquad (2\text{-}1)$$

$$\frac{\partial C_2(x, t)}{\partial t} = D_2 \frac{\partial^2 C_2(x, t)}{\partial x^2} \qquad t \geq 0, |x| < L, \tag{2-2}$$

and

$$\frac{\partial C_3(x, t)}{\partial t} = D_1 \frac{\partial^2 C_3(x, t)}{\partial x^2} \qquad t \geq 0, x \leq -L. \tag{2-3}$$

The boundary conditions are

$$C_1(x, t) = kC_2(x, t) \qquad \text{for } x \to L, \tag{2-4}$$

$$C_3(x, t) = kC_2(x, t) \qquad \text{for } x \to -L, \tag{2-5}$$

$$D_1 \frac{\partial C_1(x, t)}{\partial x} = D_2 \frac{\partial C_2(x, t)}{\partial x} \qquad \text{for } x \to L, \tag{2-6}$$

$$D_1 \frac{\partial C_3(x, t)}{\partial x} = D_2 \frac{\partial C_2(x, t)}{\partial x} \qquad \text{for } x \to -L, \tag{2-7}$$

and the initial conditions are

$$C_1(x, t) = \delta(x - \xi), \quad C_2(x, t) = C_3(x, t) = 0 \qquad \text{for } t \to 0. \tag{2-8}$$

Here, $C_i (i = 1, 3)$ is the concentration of diffusion species in the outside layers and C_2 is the concentration in the middle layer. $D_i (i = 1, 2)$ is the diffusion coefficient in the outside layers and the middle layer, respectively. The interfacial distribution ratio of concentration is $k = C_i/C_2 (i = 1, 3)$. Equations (2-4) and (2-5) model the discontinuous concentration at interfaces, while Eqs. (2-6) and (2-7) express flux continuity. $\delta(x - \xi)$ is a delta function.

We solved these diffusion equations. We rearranged Eqs. (2-4)–(2-7) to give the conventional solution. C_2 can be expressed by a Maclaurin expression using the value at $x = L$:

$$C_2(x', t) = C_2(x', t)\Big|_0 + x' \frac{\partial C_2(x', t)}{\partial x'}\Big|_0$$
$$+ \frac{x'^2}{2!} \frac{\partial^2 C_2(x', t)}{\partial x'^2}\Big|_0 + \cdots = \sum_{n=0}^{\infty} \frac{x'^n}{n!} \frac{\partial^n C_2(x', t)}{\partial x'^n}\Big|_0, \tag{2-9}$$

where $x' = x - L$. From Eqs. (2-1) and (2-2), we get

$$\frac{\partial^n C_1(x', t)}{\partial t^n} = D_1^n \frac{\partial^{2n} C_1(x', t)}{\partial x'^{2n}}, \tag{2-10}$$

$$\frac{\partial^n C_2(x', t)}{\partial t^n} = D_2^n \frac{\partial^{2n} C_2(x', t)}{\partial x'^{2n}}, \tag{2-11}$$

We differentiate Eqs. (2-4) and (2-6) by t, substitute them into Eq. (2-10) and get

$$D_1^n \frac{\partial^{2n} C_1(x',t)}{\partial x'^{2n}}\bigg|_0 = kD_2^n \frac{\partial^{2n} C_2(x',t)}{\partial x'^{2n}}\bigg|_0, \tag{2-12}$$

$$D_1^{n+1} \frac{\partial^{2n+1} C_1(x',t)}{\partial x'^{2n+1}}\bigg|_0 = kD_2^{n+1} \frac{\partial^{2n+1} C_2(x',t)}{\partial x'^{2n+1}}\bigg|_0, \tag{2-13}$$

We substitute Eqs. (2-10)–(2-13) into Eq. (2-9):

$$C_2(x',t) = \frac{1}{2}\left(\frac{1}{k}+a\right)\sum_{n=0}^{\infty}\frac{1}{n!}(ax')^n \frac{\partial^n C_1(x',t)}{\partial x'^n}\bigg|_0$$

$$+ \frac{1}{2}\left(\frac{1}{k}-a\right)\sum_{n=0}^{\infty}\frac{1}{n!}(-ax')^n \frac{\partial^n C_1(x',t)}{\partial x'^n}\bigg|_0, \tag{2-14}$$

where $a = \sqrt{D_1/D_2}$. We then get the new condition

$$C_2(x-L,t) = \frac{1}{2}\left(\frac{1}{k}+a\right)C_1(a(x-L),t)+\frac{1}{2}\left(\frac{1}{k}-a\right)C_1(-a(x-L),t), \tag{2-15}$$

Using the same solution as above, we get

$$C_1(x-L,t) = \frac{1}{2}\left(k+\frac{1}{a}\right)C_2(\frac{1}{a}(x-L),t)+\frac{1}{2}\left(k-\frac{1}{a}\right)C_2(-\frac{1}{a}(x-L),t), \tag{2-16}$$

At $x = -L$, in the same way as for $C_1(x,t)$ and $C_2(x,t)$ above, we get

$$C_2(x+L,t) = \frac{1}{2}\left(\frac{1}{k}+a\right)C_3(a(x+L),t)+\frac{1}{2}\left(\frac{1}{k}-a\right)C_3(-a(x+L),t), \tag{2-17}$$

and

$$C_3(x+L,t) = \frac{1}{2}\left(k+\frac{1}{a}\right)C_2(\frac{1}{a}(x+L),t)+\frac{1}{2}\left(k-\frac{1}{a}\right)C_2(-\frac{1}{a}(x+L),t), \tag{2-18}$$

Note that Eqs. (2-15)–(2-18) are sufficient conditions for Eqs. (2-4)–(2-7).

The solution must satisfy diffusion equations (2-1)–(2-3), boundary conditions (2-15)–(2-18), and the initial condition (2-8). If we consider only two layers separated by the boundary at $x = L$, the solutions are [40]

(i) $x \geq L$

$$C_1(x,t) = \frac{1}{2\sqrt{\pi D_1 t}}\left\{\exp\left(-\frac{(x-\xi)^2}{4D_1 t}\right)\right.$$

$$\left. -\frac{1-ak}{1+ak}\exp\left(-\frac{(x-2L+\xi)^2}{4D_1 t}\right)\right\}, \tag{2-19}$$

(ii) $|x| < L$

$$C_2(x, t) = \frac{1}{2\sqrt{\pi D_1 t}} \frac{2a}{1 + ak} \exp\left(-\frac{\left(x - L + \frac{L}{a} - \frac{\xi}{a}\right)^2}{4D_2 t}\right), \quad (2\text{-}20)$$

These are normalized. Considering the boundary at $x = -L$ [Eqs. (2-17), (2-18)] and the initial condition [Eq. (2-8)], Eq. (2-20) must be transformed to

(ii) $|x| < L$

$$C_2(x, t) = \frac{1}{2\sqrt{\pi D_1 t}} \left\{ \frac{2a}{1 + ak} \exp\left(-\frac{\left(x - L + \frac{L}{a} - \frac{\xi}{a}\right)^2}{4D_2 t}\right) \right.$$
$$\left. + \frac{2a(1 - ak)}{(1 + ak)^2} \exp\left(-\frac{\left(x + 3L - \frac{L}{a} + \frac{\xi}{a}\right)^2}{4D_2 t}\right) \right\}$$

$$(2\text{-}21)$$

and then,

(iii) $x \leq L$

$$C_3(x, t) = \frac{1}{2\sqrt{\pi D_1 t}} \frac{4ak}{(1 + ak)^2} \exp\left(-\frac{(x + 2L - 2aL - \xi)^2}{4D_1 t}\right).$$

$$(2\text{-}22)$$

Due to the added second term of Eq. (2-21), Eq. (2-19) must be transformed to

(i) $x \geq L$

$$C_1(x, t) = \frac{1}{2\sqrt{\pi D_1 t}} \left\{ \exp\left(-\frac{(x - \xi)^2}{4D_1 t}\right) \right.$$
$$- \frac{1 - ak}{1 + ak} \exp\left(-\frac{(x - 2L + \xi)^2}{4D_1 t}\right)$$
$$+ \frac{1 - ak}{1 + ak} \exp\left(-\frac{(x - 2L + 4aL + \xi)^2}{4D_1 t}\right)$$
$$\left. - \left(\frac{1 - ak}{1 + ak}\right)^2 \exp\left(-\frac{(x - 4aL - \xi)^2}{4D_1 t}\right) \right\}.$$

$$(2\text{-}23)$$

Equation (2-23) satisfies the diffusion equations and the boundary conditions, but not the initial conditions. The last term of Eq. (2-23) shows that there is a diffusion source at $x = 4aL + x$, which contradicts the single source in Eq. (2-8). In order to cancel this last term, we introduce a negative source at $x = 4aL + x$, which has the same absolute value as the last term of Eq. (2-23). This imaginary source needs additional terms in regions $x \geq L$, $|x| < L$, and $x \leq -L$ to satisfy the boundary conditions (2-15)–(2-18). This, however, also produces a diffusion source in $x \geq L$. We finally found that solutions which satisfy all conditions are infinite series.

(i) $x \geq L$

$$
C_1(x, t) = \frac{1}{2\sqrt{\pi D_1 t}} \left[\exp\left\{ -\frac{(x - \xi)^2}{4D_1 t} \right\} - \frac{1 - ak}{1 + ak} \exp\left\{ -\frac{(x - 2L + \xi)^2}{4D_1 t} \right\} \right.
$$
$$
\left. + \frac{4ak}{(1 + ak)^2} \sum_{n=1}^{\infty} \left(\frac{1 - ak}{1 + ak} \right)^{2n-1} \exp\left\{ -\frac{(x - 2L + 4aL + \xi)^2}{4D_1 t} \right\} \right]
$$

(2-24)

(ii) $|x| < L$

$$
C_2(x, t) = \frac{1}{2\sqrt{\pi D_1 t}} \frac{2a}{1 + ak} \sum_{n=1}^{\infty} \left(\frac{1 - ak}{1 + ak} \right)^{2(n-1)} \times
$$
$$
\left(\exp\left[-\frac{\left\{ x + \frac{L}{a} - (4n - 3)L - \frac{\xi}{a} \right\}^2}{4D_2 t} \right] \right.
$$

(2-25)

$$
\left. + \frac{1 - ak}{1 + ak} \exp\left[-\frac{\left\{ x - \frac{L}{a} - (4n - 1)L + \frac{\xi}{a} \right\}^2}{4D_2 t} \right] \right)
$$

(iii) $x \leq -L$

$$
C_3(x, t) = \frac{1}{2\sqrt{\pi D_1 t}} \frac{4ak}{(1 + ak)^2} \sum_{n=1}^{\infty} \left(\frac{1 - ak}{1 + ak} \right)^{2(n-1)} \times
$$

(2-26)

$$
\exp\left[-\frac{\{ x + 2L - (4n - 2)aL - \xi \}^2}{4D_1 t} \right]
$$

We next integrated Eqs. (2-24)–(2-26) using the initial composition profile. We used the initial conditions

$$
F(x, 0) = F_0 \qquad |x| \geq L, \tag{2-27}
$$

$$
F(x, 0) = 0 \qquad |x| < L \tag{2-28}
$$

and calculated
(i) $x \geq L$

$$F(x,t) = F_0 \left[\int_L^\infty C_1(x,t)d\xi + \int_{-\infty}^{-L} C_3(-x,t)d\xi \right], \qquad (2\text{-}29)$$

(ii) $|x| < L$

$$F(x,t) = F_0 \left[\int_L^\infty C_2(x,t)d\xi + \int_{-\infty}^{-L} C_2(-x,t)d\xi \right], \qquad (2\text{-}30)$$

(iii) $x \leq -L$

$$F(x,t) = F_0 \left[\int_L^\infty C_3(x,t)d\xi + \int_{-\infty}^{-L} C_1(-x,t)d\xi \right], \qquad (2\text{-}31)$$

We then derived the following formulae:
(i) $|x| \geq L$

$$F(x,t) = \frac{F_0}{2} \left(1 + \mathrm{erf}\left(\frac{|x|-L}{2\sqrt{D_1 t}}\right) - \frac{1-ak}{1+ak}\left\{1 - \mathrm{erf}\left(\frac{|x|-L}{2\sqrt{D_1 t}}\right)\right\} \right.$$

$$+ \frac{4ak}{(1+ak)^2} \sum_{n=1}^\infty \left(\frac{1-ak}{1+ak}\right)^{2(n-1)} \left[1 - \mathrm{erf}\left(\frac{|x|-L+(4n-2)aL}{2\sqrt{D,t}}\right)\right.$$

$$\left.\left. + \frac{1-ak}{1+ak}\left\{1 - \mathrm{erf}\left(\frac{|x|-L+4naL}{2\sqrt{D_1 t}}\right)\right\}\right]\right),$$
$$\tag{2-32}$$

(ii) $|x| < L$

$$F(x,t) = \frac{F_0 a}{1+ak} \sum_{n=1}^\infty \left(\frac{1-ak}{1+ak}\right)^{2(n-1)} \times$$

$$\left[2 + \mathrm{erf}\left(\frac{x-(4n-3)L}{2\sqrt{D_2 t}}\right) - \mathrm{erf}\left(\frac{x+(4n-3)L}{2\sqrt{D_2 t}}\right)\right.$$

$$+ \frac{1-ak}{1+ak}\left\{2 - \mathrm{erf}\left(\frac{x+(4n-1)L}{2\sqrt{D_2 t}}\right)\right.$$

$$\left.\left. + \mathrm{erf}\left(\frac{x-(4n-1)L}{2\sqrt{D_2 t}}\right)\right\}\right]. \qquad (2\text{-}33)$$

Equations (2-32) and (2-33) generally give the concentrations of diffused species; these equations describe the interdiffusion profile when $F(x,t)$ is regarded as the composition of layers.

In a practical analysis of interdiffusion in quantum wells, we determine three unknown material-specific parameters, D_1 (in the barrier layer), D_2 (in the well layer) and k, in the formula.

2.1.2   Calculation of quantum energy shift

We determined the three unknown parameters, two interdiffusion coefficients, and the ratio of interfacial discontinuity, by fitting quantum energy shifts calculated with our formula to measured ones. We used a numerical method to calculate quantum energies because the effective mass equation [43] cannot be solved analytically in these cases.

The effective mass equation is

$$\left(-\frac{\hbar^2}{2m_i}\frac{\partial^2}{\partial x^2} + V_i(x) + S_i(x)\right)\varphi_i(x) = E_i\varphi_i(x), \quad i = e, hh, lh, \tag{2-34}$$

where $V_i(x)$ is the potential energy, $S_i(x)$ is the strain energy, E_i is the eigenenergy, and $\varphi_i(x)$ is the envelope wave function. Subscripts i = 'e', 'hh', and 'lh' mean 'electron', 'heavy hole', and 'light hole', respectively. The potential energy $V_i(x)$ is calculated as a function of x in our formula. The strain energy $S_i(x)$ is

$$S_c(x) = 2a_c\frac{C_{11} - C_{12}}{C_{11}}\varepsilon(x) \tag{2-35}$$

and

$$S_v(x) = \left\{2a_v\frac{C_{11} - C_{12}}{C_{11}} \pm b\frac{C_{11} + 2C_{12}}{C_{11}}\right\}\varepsilon(x), \tag{2-36}$$

where a_c and a_v are hydrostatic deformation potentials in the conduction and the valence band, respectively, and b is the shear deformation potential [44–46]. C_{11} and C_{12} are elastic stiffness. $\varepsilon(x)$ is misfit strain. Upper and lower signs correspond to light-hole band and heavy-hole band, respectively.

We used the Runge-Kutta method suggested by Sakurai [47] to solve the effective mass equations. We computed the envelope function numerically, with the boundary conditions that both $\varphi_i(x)$ and $1/m_i \cdot \partial\varphi_i(x)/\partial x$ are continuous at the interface, and we determined eigenenergies for both the conduction band and the valence band. We calculated total energy shifts as the sum of the energy shift in the two bands.

We used the following parameters to calculate the quantum energy levels in $In_{1-x}Ga_xAs_yP_{1-y}/InP$ quantum wells: effective mass $m_e = 0.041m_0$, $m_{hh} = 0.50m_0$, and $m_{lh} = 0.052m_0$ in an $In_{0.53}Ga_{0.47}As$ well layer, $m_e = 0.08m_0$ and $m_{hh} = 0.56m_0$ in an InP barrier layer, $m_e = 0.044m_0$ and $m_{hh} = 0.49m_0$ in an $In_{0.70}Ga_{0.30}As_{0.61}P_{0.39}$ barrier layer, elastic stiffness

$C_{11} = 1.016 \times 10^{11}$ dyn/cm^2 and $C_{12} = 0.509 \times 10^{11}$ dyn/cm^2 [48], and distribution ratio of conduction band offset $\Delta E_c = 0.4 \cdot \Delta E_g$ [49]. We used the hydrostatic deformation potentials $a_c = a_v = -3.94$ eV and the shear deformation potential $b = -1.75$ eV [50], that is, assumed that the hydrostatic deformation potential was distributed evenly between the conduction and valence band, and neglected the compositional dependence of these three deformation potentials. Since the compositional dependence of the energy gap at 4.2 K has not been reported to our knowledge, we used the compositional dependence at 295 K [51]:

$$E_g^{295 \ K}(x, y) = 1.35 + 0.672x - 1.091y + 0.758x^2 + 0.101y^2$$
$$+ 0.111xy - 0.580x^2y - 0.159xy^2 + 0.268x^2y^2.$$

$$(2\text{-}37)$$

We believe this is a good approximation when calculating shifts of quantum energies since band offset does not depend on temperature.

We used the following material parameters when calculating quantum energy levels in GaAs/Al$_x$Ga$_{1-x}$As quantum wells: effective mass $m_e = 0.067m_0$ and $m_{hh} = 0.42m_0$ in a GaAs well layer, $m_e = 0.088m_0$ and $m_{hh} = 0.53m_0$ in an Al$_{0.25}$Ga$_{0.75}$As barrier layer, and distribution ratio of conduction band offset $\Delta E_c = 0.65 \cdot \Delta E_g$ [14]. Strain energy is negligible in this material. For the same reason as with an In$_{1-x}$Ga$_x$As$_y$P$_{1-y}$ system, we used the compositional dependence of the energy gap of Al$_x$Ga$_{1-x}$As at 300 K [52],

$$E_g^{300 \ K}(x) = 1.424 + 1.247x, \qquad (2\text{-}38)$$

and calculated shifts of quantum energies.

2.2 Interdiffusion Process in Quantum Wells

2.2.1 InGaAsP/InP quantum wells

We grew experimental samples by metalorganic vapor-phase epitaxy (MOVPE). Samples were undoped In$_{0.53}$Ga$_{0.47}$As/InP and In$_{0.53}$Ga$_{0.47}$As/In$_{0.70}$Ga$_{0.30}$As$_{0.61}$P$_{0.39}$ single quantum wells (SQWs) on (001)-InP substrates grown at 570°C. Each sample had a cap layer, quantum well layers, and a buffer layer on the substrate. Quantum well layers were composed of four well layers having different thicknesses to give us enough data to find the three unknown material specific parameters. Well layers of In$_{0.53}$Ga$_{0.47}$As/InP SQWs were 20, 10, 7.5, and 5 nm wide, separated by 50-nm-wide barrier layers. Well layers of In$_{0.53}$Ga$_{0.47}$As/In$_{0.70}$Ga$_{0.30}$As$_{0.61}$P$_{0.39}$ SQWs were 20, 15, 10, and 5 nm wide, separated by 30-nm-wide barrier layers. Cap and buffer layers were 200 nm wide in all samples. We controlled the width and the composition via the growth conditions, which were checked by X-ray diffraction and transmission electron microscopy (TEM).

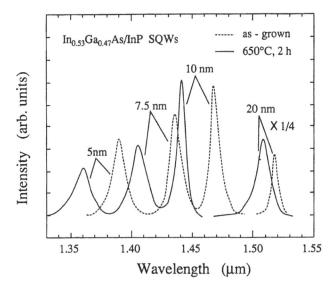

Fig. 3. Photoluminescence spectra of as-grown and annealed samples of $In_{0.53}Ga_{0.47}As/InP$ quantum wells 5, 7.5, 10, and 20 nm wide at 4.2 K. Annealing was performed at 650°C for 2 hours.

We annealed samples in a reactor tube of a liquid-phase epitaxy system. During the annealing, we placed an InP plate on $In_xGa_{1-x}As_yP_{1-y}$ samples and introduced pure H_2 gas. We believe the semiconductor plates raise the phosphorous vapor pressure during annealing.

We measured photoluminescence (PL) spectra of these samples at 4.2 K before and after annealing by immersing samples in liquid helium. We excited luminescence with the 647.1 nm line of a Kr^+ ion laser and detected luminescence using a PbS detector. Figure 3 is an example of the measured photoluminescence spectra of $In_{0.53}Ga_{0.47}As/InP$ SQWs. The four peaks correspond to four quantum wells of different widths, and all of them shifted to a shorter wavelength after annealing. These peak shifts are caused by changes of quantum energy levels, which are sensitive to compositional profile.

We compared the measured and calculated dependences of quantum energy shifts on well layer width for various annealing times in $In_{0.53}Ga_{0.47}As/InP$ (Figure 4). Solid lines indicate shifts calculated using our formula. The measured shifts closely agree. The energy shift increased gradually in 20-nm-wide well layers, and in narrower well layers, the shift increased rapidly up to 2 hours and then became saturated. Our formula explained each of these characteristics using a set of parameters, that is interdiffusion

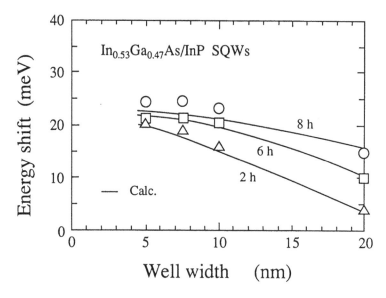

Fig. 4. Quantum energy shifts as a function of well width: Ina$_{0.53}$Ga$_{0.47}$As/InP quantum wells after annealing at 650°C for 2, 6, and 8 hours. Symbols show measured results and solid lines indicate calculated results.

coefficients in each layer and the interfacial distribution ratio. This result is consistent with our fundamental assumption that the three parameters can be regarded as constant during interdiffusion. Note that this assumption for the interdiffusion coefficient is equivalent to assuming that the coefficient depends negligibly on the composition of group-V atoms. Our result suggests that the assumption is valid, at least under our experimental conditions.

We measured the dependence of quantum energy shifts on well layer width at various annealing temperatures for In$_{0.53}$Ga$_{0.47}$As/InP quantum wells (Figure 5) and In$_{0.53}$Ga$_{0.47}$As/In$_{0.70}$Ga$_{0.30}$As$_{0.61}$P$_{0.39}$ quantum wells (Figure 6). The solid lines indicate the calculated shifts and agree well with the measured shifts. Table 1 lists calculated interdiffusion coefficients and distribution ratios for In$_{1-x}$Ga$_x$As$_y$P$_{1-y}$/InP quantum wells. We found that interdiffusion coefficients in In$_{0.53}$Ga$_{0.47}$As layers were common to quantum wells having different barrier layers. The distribution ratio, k, decreased as annealing temperature increased in In$_{0.53}$Ga$_{0.47}$As/InP quantum wells.

We made Arrhenius plots of interdiffusion coefficients (Figure 7). The solid lines are for an Arrhenius expression with a single activation energy. We found that activation energy was 0.5 eV in an In$_{0.53}$Ga$_{0.47}$As layer, 2.0 eV in an In$_{0.70}$Ga$_{0.30}$As$_{0.61}$P$_{0.39}$ layer, and 8.4 eV in an InP layer. The activation energy in InP was larger than the energy of self-diffusion of phosphorus atoms

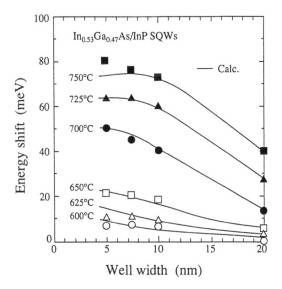

Fig. 5. Quantum energy shifts as a function of well width: $In_{0.53}Ga_{0.47}As/InP$ quantum wells after annealing at between 600 and 750°C for 2 hours. Symbols show measured results and solid lines indicate calculated results.

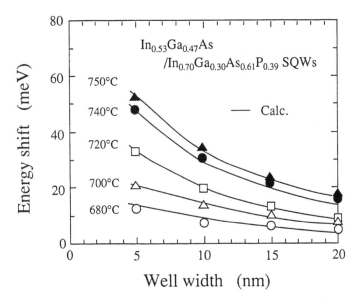

Fig. 6. Quantum energy shifts as a function of well width: $In_{0.53}Ga_{0.47}As/In_{0.70}Ga_{0.30}As_{0.61}P_{0.39}$ quantum wells after annealing at between 680 and 750°C for 2 hours. Symbols show measured results and solid lines indicate calculated results.

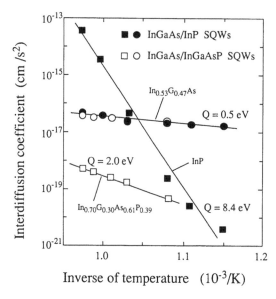

Fig. 7.   Interdiffusion coefficients in $In_xGa_{1-x}As_yP_{1-y}$ quantum wells as a function of the inverse of annealing temperature.

in InP, which has been measured as 5.65 eV [53]. We suggest that one reason for this discrepancy is lattice distortion which adds excess energy to the activation energy. We cannot compare activation energies in $In_{0.53}Ga_{0.47}As$ and $In_{0.70}Ga_{0.30}As_{0.61}P_{0.39}$ layers with other values due to the lack of reported data.

We concluded the following concerning the characteristics of the interdiffusion of group-V atoms in $In_{1-x}Ga_xAs_yP_{1-y}$ materials.

We assume that interfacial compositional discontinuity is related to lattice distortion, which is first induced at the interface during interdiffusion. If there is no interfacial discontinuity, the greater the difference in the initial composition of group-V atoms between well and barrier layers, the greater the lattice distortion produced by interdiffusion will be. In Table 1, distribution ratios of $In_{0.53}Ga_{0.47}As/InP$ quantum wells were larger than those of $In_{0.53}Ga_{0.47}As/In_{0.70}Ga_{0.30}As_{0.61}P_{0.39}$ quantum wells. The ratio in GaAs/AlGaAs quantum wells is regarded to be one (see Figure 1 and will be shown in detail in Table 2). This shows that the distribution ratio was larger at the interface where the greater lattice distortion was expected. We can also see from Table 1 that the distribution ratio decreased as annealing temperature increased. These results support the assumption that compositional discontinuity starts by a thermodynamic potential barrier

TABLE 1

Interdiffusion coefficients and distribution ratios of $In_{0.53}Ga_{0.47}As/InP$ and $In_{0.53}Ga_{0.47}As/In_{0.70}Ga_{0.30}As_{0.61}P_{0.39}$ quantum wells.

T (°C)	$In_{0.53}Ga_{0.47}As/InP$			$In_{0.53}Ga_{0.47}As/In_{0.70}Ga_{0.30}As_{0.61}P_{0.39}$		
	D_{InGaAs} (cm²/s)	D_{InP} (cm²/s)	k	D_{InGaAs} (cm²/s)	$D_{InGaAsP}$ (cm²/s)	k
600	1.7 E-17	4.3 E-21	40			
625	2.0 E-17	2.0 E-20	35			
650	2.1 E-17	2.1 E-19	30	2.6 E-17	5.1 E-20	2
700	2.8 E-17	5.6 E-17	17	2.8 E-17	1.4 E-19	2
720				3.5 E-17	2.3 E-19	1
725	3.8 E-17	3.8 E-15	13			
740				3.5 E-17	4.3 E-19	1
750	4.7 E-17	4.7 E-14	11	3.9 E-17	5.5 E-19	1

related to lattice distortion. Yu et al. reported that lattice mismatch was avoided when both group-III atoms and group-V atoms interdiffused [54]. They speculated that this was because interdiffusion which produces lattice distortion requires excess energy. Considering that the strain energy will follow the compositional profiles in layers, the ratio k is expected to vary versus time. We suppose that the annealing time in our experiment is too short for the effect on the ratio k to be observed in this work.

Even though we neglected the Smigelskas-Kirkendall effect, our model accurately explained quantum energy shifts due to interdiffusion. This suggests that the Smigelskas-Kirkendall effect was negligible in our samples. We assume that this result is also related to the lattice distortion which adds excess energy to the interdiffusion activation energy. When group-III atoms do not move, as in our experiment, the motion of the interface of group-V atoms must produce a large lattice distortion between the stable interface of group-III atoms and the moved interface of group-V atoms.

2.2.2   GaAs/AlGaAs quantum wells

We grew experimental samples by MOVPE. Samples were undoped GaAs/$Al_{0.25}Ga_{0.75}As$ SQWs on (001)-GaAs substrates grown at 720°C. Each sample had a cap layer, SQWs, and a buffer layer on the substrate. Well layers of GaAs/$Al_{0.25}Ga_{0.75}As$ SQWs were 20, 15, 10, and 5 nm wide, separated by 50-nm-wide barrier layers. Cap and buffer layers were 200 nm wide. We annealed samples in a reactor tube of a liquid-phase epitaxy system. During annealing, we placed a GaAs plate on $Al_xGa_{1-x}As$ samples and introduced pure H_2 gas.

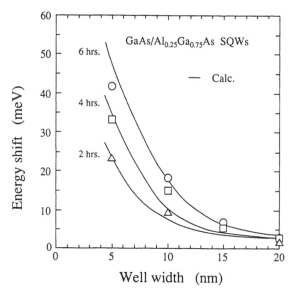

Fig. 8. Quantum energy shifts as a function of well width: GaAs/Al$_{0.25}$Ga$_{0.75}$As quantum wells after annealing at 825°C for 2, 4, and 6 hours. Symbols show measured results and solid lines indicate calculated results.

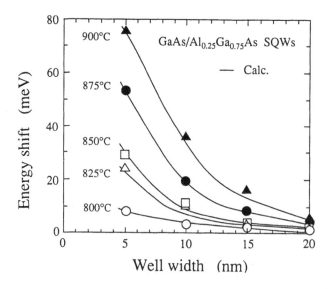

Fig. 9. Quantum energy shifts as a function of well width: GaAs/Al$_{0.25}$Ga$_{0.75}$As quantum wells after annealing at between 800 and 900°C for 2 hours. Symbols show measured results and solid lines indicate calculated results.

TABLE 2

Interdiffusion coefficients and distribution ratios of GaAs/Al$_{0.25}$Ga$_{0.75}$As quantum wells. Interdiffusion coefficients of GaAs and Al$_{0.25}$Ga$_{0.75}$As layers were equal.

GaAs/Al$_{0.25}$Ga$_{0.75}$As T (°C)	D (cm^2/s)	k
800	2.2 E-19	1
825	6.9 E-19	1
850	1.1 E-18	1
875	2.3 E-18	1
900	4.0 E-18	1

The dependence of quantum energy shifts on the well width in GaAs/ AlGaAs SQWs differed greatly from that in InGaAsP/InP SQWs. Figure 8 shows the dependence of quantum energy shifts on well-layer width for various annealing times. The energy in the 20-nm-wide well layer was almost constant, and in narrower well layers, energy shift increased and did not saturate. Our formula explained each of these characteristics. Figure 9 shows the quantum energy shifts for various annealing temperatures. We see the same tendency of energy shifts as in Figure 8, and this tendency differs greatly from that in In$_{0.53}$Ga$_{0.47}$As/InP quantum wells (see Figure 5).

Table 2 lists calculated interdiffusion coefficients and distribution ratios for GaAs/Al$_{0.25}$Ga$_{0.75}$As quantum wells. The parameters in Tables 1 and 2 give an indication of why the tendency in Figure 6 looks a blend of those in Figure 5 and Figure 7. Interdiffusion coefficients of In$_{0.53}$Ga$_{0.47}$As/In$_{0.70}$ Ga$_{0.30}$As$_{0.61}$P$_{0.39}$ quantum wells are different between layers, which is a characteristic of In$_{0.53}$Ga$_{0.47}$As/InP, and interfacial discontinuity is small, which is a characteristic of GaAs/Al$_{0.25}$Ga$_{0.75}$As. Also, the parameters in Table 2 suggest that Chang's model is applicable to GaAs/Al$_x$Ga$_{1-x}$As quantum wells, i.e., interdiffusion coefficients were common to both layers and distribution ratios were 1 at all the annealing temperatures we adopted for GaAs/Al$_{0.25}$Ga$_{0.75}$As quantum wells.

We compared the interdiffusion profiles of In$_{0.53}$Ga$_{0.47}$As/InP and GaAs/Al$_{0.25}$Ga$_{0.75}$As quantum wells. Figure 10(a) explains the results found by Nakashima et al. (arsenic composition in an In$_{0.53}$Ga$_{0.47}$As well layer decreased uniformly and the interdiffused arsenic atoms stagnated near the interface on an InP barrier layer side) [15] and Fujii et al. (a large compositional discontinuity of group-V atoms existed at the interface even after interdiffusion) [16]. These profiles also help us to understand the characteristics of energy shift in two types of quantum well. In In$_{0.53}$Ga$_{0.47}$As/InP quantum wells, there is a large interfacial discontinuity, which starts as the distribution ratio. The distribution ratio limits the increase

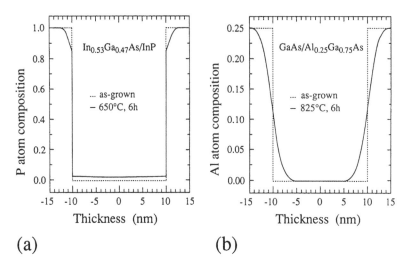

Fig. 10. Calculated composition profiles: (a) phosphorus composition profile in an In$_{0.53}$Ga$_{0.47}$As/InP quantum well 20 nm wide after annealing at 650°C for 6 hours, and (b) aluminum composition profile in a GaAs/Al$_{0.25}$Ga$_{0.75}$As quantum well 20 nm wide after annealing at 825°C for 6 hours.

of phosphorus composition in the well layer. Because of this compositional limitation, the shift of quantum energy saturates. With high velocity (see Table 1), interdiffusion easily advances to the center of the well layer. Quantum energies therefore shift even in a wide well layer. Distribution ratios were independent of well-layer width, so saturation and shift occur earlier in narrower well layers. With GaAs/Al$_{0.25}$Ga$_{0.75}$As quantum wells (Figure 10(b)), the distribution ratio was 1, so energy shifts did not saturate and increased gradually.

We made Arrhenius plots of interdiffusion coefficients in GaAs/Al$_x$Ga$_{1-x}$As quantum wells (Figure 11). The solid lines are for an Arrhenius expression with a single activation energy. We found an interdiffusion activation energy of 2.8 eV in GaAs/Al$_{0.25}$Ga$_{0.75}$As quantum wells. We superimposed previously reported data in Figure 11. We believe that the discrepancy between our value and others is caused by a difference in crystal quality. Chang and Koma found an interdiffusion activation energy of 4.1 eV in Al$_{0.25}$Ga$_{0.75}$As [12]. They speculated that the disagreement between their activation energy and the Ga vacancy, which was 2.1 eV [55], was due to As vacancies via the formation of divacancies [56]. Our activation energy was closer to the Ga vacancy than Chang amd Koma's, so it is possible that our samples had fewer As vacancies than their sample. This would also explain why our interdiffusion coefficients were lower than Chang ans Koma's.

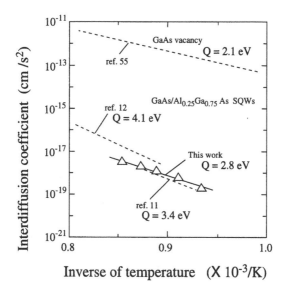

Fig. 11. Interdiffusion coefficients in GaAs/Al$_{0.25}$Ga$_{0.75}$As quantum wells as a function of the inverse of annealing temperature. Results of previous research are shown for reference.

Guido et al. studied the influence of vacancy concentration on interdiffusion [11]. The smallest activation energy among their samples was 3.4 eV. Guido et al.'s results are very similar to ours, though their activation energy is different.

2.3 Growth-condition Dependence

We studied the influence of the growth temperature and the buffer-layer thickness on the set of parameters (i.e., interdiffusion coefficients and interfacial distribution ratio) of In$_{0.53}$Ga$_{0.47}$As/InP quantum wells. We grew three samples of undoped In$_{0.53}$Ga$_{0.47}$As/InP single quantum wells on (100)-oriented InP substrates using MOVPE. Each sample consisted of a 100 nm InP cap layer, SQWs, and an InP buffer layer. The SQWs consisted of 20, 10, 7.5, and 5 nm In$_{0.53}$Ga$_{0.47}$As well layers separated by 50 nm InP barrier layers. The growth temperatures and the buffer layer thicknesses varied among the samples. Sample I was grown at 615°C with a 1 μm buffer layer. Sample II was grown at 570°C with a 1 μm buffer layer. Sample III was grown at 615°C with a 0.1 μm buffer layer. After etching away the epitaxial layers to expose the buffer layers, we observed etch pits which originated from substrate dislocations. Etch pit density (EPD) was about 12,000/cm^2 on a 0.1 μm buffer layer and about 6,000/cm^2 on a 1 μm buffer layer. EPD of

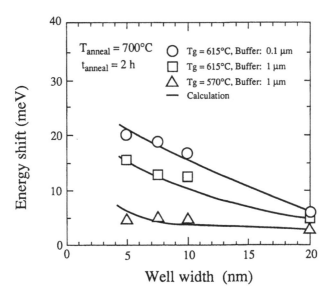

Fig. 12. Quantum energy shift in InGaAs/InP quantum well grown under various conditions. Annealing was performed at 700°C for 2 hours.

the substrate was below 5,000/cm². We annealed samples at between 680 and 740°C for 2 hours using the reactor tube of a liquid-phase epitaxy system. We put an InP plate on the samples while flowing pure H_2 gas over the samples to protect them from oxidation as before.

We measured the 4.2 K PL spectra before and after annealing. In the measurements, samples were immersed in liquid helium and were excited with the 641.7 nm line of a Kr^+ laser. Luminescence was measured with a PbS detector. Figure 12 shows the measured and calculated energy shifts as a function of well width after annealing at 700°C. The lower the growth temperature and the thicker the buffer layer, the more stable the peak energy. This tendency was consistent for annealing between 680 and 740°C. The results suggest that the interdiffusion parameters varied among these samples. The calculated energy shifts agreed very well with measured shifts, and we determined each sample's interdiffusion coefficients and interfacial distribution ratios for temperatures, between 680 and 740°C.

Figure 13 shows the interdiffusion coefficients of samples I, II, and III as functions of temperature. In the InP layer, the interdiffusion coefficients for the sample grown at 615°C (sample I) were about 20 times that for the sample grown at 570°C (sample II). The diffusion coefficients for the sample with a 0.1 μm buffer (sample III) were also about 20 times that for the sample with

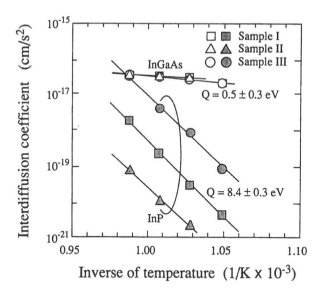

Fig. 13. Interdiffusion coefficient in each layer of $In_{0.53}Ga_{0.43}As/InP$ quantum wells as a function of temperature. Open symbols represent interdiffusion coefficients in the well layers and solid marks show coefficients in the barrier layers.

a 1 μm buffer (sample I). All samples had a common interdiffusion activation energy of 8.4 eV. In $In_{0.53}Ga_{0.47}As$ layers, the interdiffusion coefficients were almost independent of growth temperature and buffer layer thickness, and they showed an activation energy of 0.5 eV.

Table 3 shows the interfacial phosphorous composition ratios k. The ratios were constant for a given temperature. This means that the ratios were independent of growth temperature and buffer layer thickness.

The interdiffusion activation energy must be affected by the interdiffusion mechanism. The activation energies were the same in layers having the same composition under all our experimental conditions (even the values

TABLE 3

Interfacial phosphorous composition ratios of samples at various annealing temperatures. The error is about 10%. The sample R value for 680°C was not termined because it showed no energy shift.

Sample	680	700	720	740
I	100	35	25	20
II	–	35	25	20
III	100	35	30	20

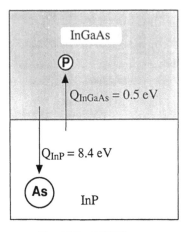

Defile: Defect-dependent diffusion
is divided into three types.

1. Vacancy-related diffusion:

    5.6 eV (P in InP), 3.9 eV (In in InP)

2. Dislocation-related diffusion:

    3.7 eV (P in Si), 2.8 eV (In in Ge)

3. Interstitial-site-related diffusion:

    0.51 eV (Li in Ge), 0.38 eV (Au in Si)

T = 600 ~ 750°C

Fig. 14. Defect-dependent diffusion is divided into three types and the obtained diffusion coefficient in each layer suggests the diffusion mechanism.

in Figure 13 are equal to those in Figure 9). We infer that the difference in the activation energy between layers indicates the difference in the interdiffusion mechanism. In general, defect-dependent diffusion can be categorized based on the effective defects: vacancy, dislocation, and interstitial (Figure 14). The diffusion activation energies are specified according to the defect types. The activation energy of vacancy or dislocation diffusion is several electron volts and the energy of interstitial diffusion is under one electron volt [57]. Based on these findings, we conclude that interdiffusion in the InP layer occurs due to vacancy and/or dislocation defects, and that the interdiffusion in the $In_{0.53}Ga_{0.47}As$ layer is related to interstitial phenomena. The growth-condition dependence of the interdiffusion coefficient supports this idea if we assume the following. The higher the temperature is, the more vacancies are generated. If this trend is retained after quenching, vacancy-assisted interdiffusion will be enhanced in the high-temperature-grown sample. Remember that EPD was twice as large in the 0.1 μm buffer layer than in the 1 μm buffer layer. If the number of dislocation paths follows the EPD trends, dislocation-assisted interdiffusion will be enhanced in layers on a high-EPD buffer layer. Measurements for the InP layer matched these trends. Note that we cannot distinguish these two mechanism since the number of vacancies and of dislocation paths may correlate to each other. On the other hand, it is reasonable that interdiffusion in the $In_{0.53}Ga_{0.47}As$ layer is not affected by

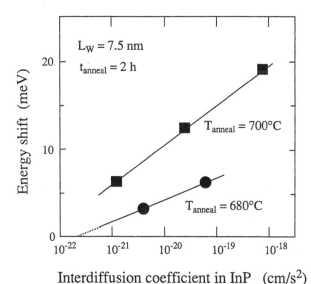

Fig. 15. Quantum energy shift in a single 7.5-nm-wide quantum well as a function of the interdiffusion coefficient in the InP layer, after annealing at 680 and 700°C for 2 hours.

the growth temperature or the buffer-layer thickness if the interdiffusion is mainly due to the interstitial effects. We consider that the interdiffusion in the $In_{0.53}Ga_{0.47}As$ layer resembles the Zn diffusion in GaAs [58]. We expect that comparatively small phosphorous atoms can enter an interstitial path in the $In_{0.53}Ga_{0.47}As$ layer and extend the effective interdiffusion length.

Quantum energy shifts are presented in Figure 15 as a function of the interdiffusion coefficient in the InP layer for 7.5-nm $In_{0.53}Ga_{0.47}As$ well layers after annealing at 680 and 700°C. We attained no energy shift from the 1 μm buffer layer sample grown at 570°C and annealed at 680°C. This region is represented by a dashed line because we extrapolated D_b for the case when no energy shift was observed. We can see that the thermal stability of our samples was enhanced by improving the crystal quality of the barrier (InP) layer, which was realized by reducing the growth temperature and making the buffer layer thicker.

In this section, it is demonstrated that the thermal stability is related to the crystal quality and that the lattice strain plays an important role in the interdiffusion process. Our method enables a basic understanding of the interdiffusion process in III-V semiconductor quantum wells, that will aid in their application to practical devices.

3 INTERDFFFUSION IN SELF-FORMED QUANTUM DOTS

3.1 Growth of Self-fomed InGaAs/GaAs Quantum Dots

Three-dimensional quantum confinement of electrons, holes, and excitons in semiconductor quantum dots is predicted to produce new physical phenomena and significantly improve optoelectronic devices such as laser, optical modulator, and some nonlinear devices [59,62]. Primarily, artificial techniques such as high-resolution patterning and regrowth have been studied [63,64]; These techniques have advantages in the controllability of positioning, but suffer from poor interface quality, low numerical density, and a nonuniform dot size due to their complex nanofabrication processes. Self-organization of microcrystals was a breakthrough in the fabrication of quantum dots, and will realize experimentally the predictions made on quantum dots in the near future. Using three-dimensional (islanding) growth under the Stranski-Krastanov mode, Ge/Si dots [65], $Si_x Ge_{1-x}$/Si dots [66], $In_x Ga_{1-x}$As/GaAs dots [26], InAs/GaAs dots [67], and $In_x Ga_{1-x}$N/GaN dots [68] were produced by MOVPE, molecular beam epitaxy (MBE), and chemical vapor deposition (CVD). These self-organized dots have uniform size smaller than the exciton Bohr radius, high spontaneous emission efficiencies, and high numerical densities, all of which are not possible in artificially fabricated structures.

Although the dots grown by the Stranski-Krastanov mode (SK dots) are the most widely used, we have grown another type of highly uniform self-organized $In_{0.5}Ga_{0.5}$As/GaAs quantum dots during alternate supply of precursors, using the atomic layer epitaxy technique [25,34]. (We have called this type of dots 'ALE-grown dots' or simply 'ALE dots'. We hereafter call them 'ALS dots' after 'ALternate Supply'.) The ALS dots emit at the wavelength of 1.3 μm at room temperature, which is favorable for optical telecommunications systems and the wavelength is longer than that of ordinary SK dots. The TEM images reveal the different structures between the two types of dots (Figure 16). Cross-sectional images indicate that the ALS dots are spherical and are buried in the quantum-well layer. The thickness of the quantum well layer is discernibly greater than the wetting layer of the SK dots. Plane-view images also indicate the difference of the structure.

To grow the ALS quantum dots, we use a low-pressure MOVPE apparatus with a chimney reactor [69]. Our growth system is designed for pulse jet epitaxy, in which source gases are supplied in a fast pulsed stream. For the gas handling system, we use a fast switching manifold with a pressure-balanced vent-and-run configuration. Source materials were trimethylindium-dimethylethylamine adduct (TMIDMEA), trimethylgallium (TMG_a) and arsme (AsH_3). A growth temperature of 460°C yields distinct self-limiting growth for both InAs ALE by TMG_a. We perform

Cross-sectional TEM   Plan-view TEM

Sphere (ALS growth)

Island (S-K growth)

MBE & MOVPE

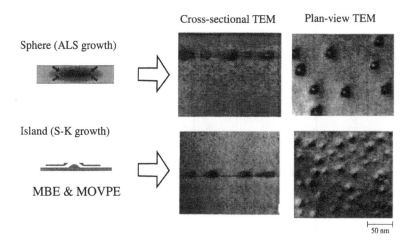

50 nm

Fig. 16. Two types of self-formed InGaAs/GaAs quantum dots. Cross-sectional and plan-view TEM images are indicated.

InAs/GaAs alternate supply separated by H_2 purging pulses based on the ALE conditions, typically on a 300-nm-wide GaAs buffer layer on a (001)-GaAs substrate. Instead of a short-periodic layer, a quantum-well layer involving many quantum dots (a quantum-dot layer) was self-formed. A 30-nm-wide GaAs layer is grown on the quantam-dot layer. No dislocation around the dots was observed in cross-sectional TEM images. We also observed the lattice image by high-resolution TEM in a cross-sectional sample, and again did not find any dislocations or defects. We evaluated the spatial composition distribution of sample cross sections with energy dispersive X-ray microanalysis (EDX). We obtained the proved indium composition of 0.5 at the center of the dots (i.e. $In_{0.5}Ga_{0.5}As$) and that of about 0.1 in the quantum well layer surrounding the dots (i.e. $In_{0.1}Ga_{0.9}As$).

We confirmed that the self-formed microcrystals were actually quantum dots, base on the following four criteria. I) The size is smaller than the excitdn Bohr radius which is about 20 nm in an InGaAsP compound semiconductor; TEM images showed that the microcrystals satisfy this criterion. II) Spectra width is independent of temperature since carriers cannot distribute thermally in a quantum dot where the state density is delta-functionlike. We proved the temperature independence of the full width at half-maximum (FWHM) for the ALS dots (Figure 17). The FWHM was about 30 meV for the ALS dots, and 80–100 meV for the SK dots. Considering that the observable spectral broadening must be determined by the nonuniformity of dot size and composition, the results suggest a high uniformity of the ALS dots.

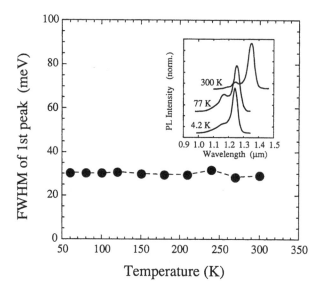

Fig. 17. FWHM of PL spectra of the ALS dots as a function of temperature. PL spectra at 300, 77, and 4.2 K are shown in the inset.

III) The radiative lifetime should be independent of temperature in a quantum dot where the thermal carrier distribution among sublevels is limited. This characteristic was observed, and the detailed discussion will appear in section 3.3. IV) Energy should be insensitive to magnetic fields. The occurrence of diamagnetic energy shift is caused by the localization of excitons. In quantum dots where excitons are laterally confined, in addition to in the growth direction, the energy shift under magnetic field perpendicular to the sample surface is smaller than in the quantum well. The stronger the lateral quantum confinement is, the smaller the diamagnetic shift is. Figure 18 shows the diamagnetic shifts of the ground level of ALS dots [35]. The shifts of 1.3 μm-emission InGaAsP/InP quantum wells are superimposed for a reference. The dot size was controlled by the number of cycles of alternate supply, and we see that the diamagnetic shifts were smaller than that of quantum wells and that the diamagnetic shift decreased as the number of cycles decreased.

We assume that the unique structure of the ALS dots indicates a self-formation process different from that of the SK dots. The surface of the ALS-dot layer is flat owing to the quantum-well layer, and the ALS dots do not have a wetting layer. They are also unique in the proportion between their height and diameter. The height of SK dots is generally under 1/4 of

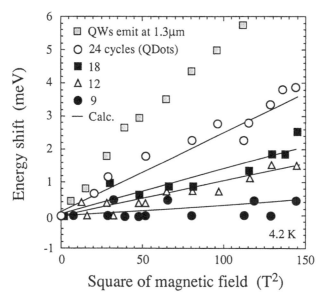

Fig. 18. Diamagnetic energy shifts as a function of the square of the magnetic field. The shifts of InGaAsP/InP quantum well emitting at the wavelength of 1.3 μm are shown for reference.

the in-plane diameter, while the height of ALS dots is 1/2 of the diameter. We believe that the dots were formed as compositional nonuniformities during a two-dimensional growth (Figure 19). The flat surface can be explained only by assuming two-dimensional growth. ALE is better for layer-by-layer growth than MOVPE or MBE. Our growth temperature (460°C) and growth rate (1 monolayer per 10 s) were significantly lower than those of other techniques. The smaller the growth rate and the growth temperature are, the longer the surface-migration length is; the long surface-migration length enhances two-dimensional growth [70]. Additionally, the group-III precursor was provided only for single-layer coverage during one cycle of atomic layer epitaxy. Therefore, when islands were constructed during the supply of group-III atoms, all the islands can deform to a two-dimensional structure during the supply of group-V atom [71]. Since an $(InAa)_1/(GaAs)_1$ bulk monolayer superlattice is intrinsically unstable against phase segregation [72], atoms may form 3D microcrystals which reduce the enthalpy of the system.

3.2   Energy Shift of Discrete Sublevels by Interdiffusion

In the ALS dots, we observed multiple discrete sublevels of self-organized quantum dots for the first time [25]. Figure 20 shows the excitation power

Model 1:

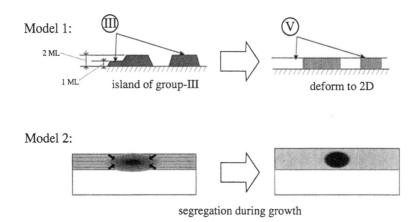

2 ML

1 ML   island of group-III   deform to 2D

Model 2:

segregation during growth

Fig. 19. Schematics of growth mechanism of the ALS dots. We assume the mechanism to be related to compositional nonuniformity during two-dimensional growth.

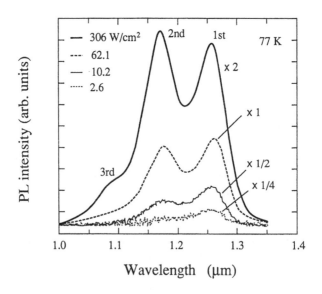

Fig. 20. The excitation power dependence of PL spectra in the dots grown with 9 cycles.

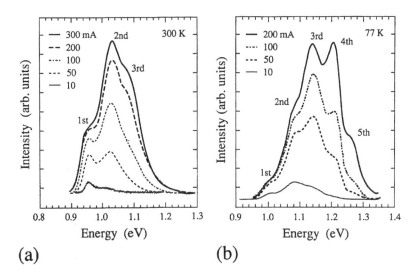

Fig. 21. Measured EL spectra of the quantum dots at (a) 300 K and (b) 77 K.

dependence of PL spectra in the dots grown with 9 ALE cycles [36]. Two main peaks were observed in this figure, and as the excitation power increased, the relative intensity of the second peak increased. At the highest excitation, we observed the third peak. Electron carrier injection realizes higher carrier density than in PL measurement. Figure 21 shows the EL spectra of dots at 300 K and 77 K. At both temperatures, emission from the levels appeared first in the low-energy position, and then gradually moved upward as the current increased. Three peaks appeared at 300 K as the injected current was increased, while five peaks appeared at 77 K. Figure 22 shows the photoluminescence excitation (PLE) spectra [38]. We can see two peaks in the PLE spectra at the energy where peaks appeared in PL spectra, indicating that these peaks correspond to the sublevels in the dots. Note that the peak intensity of the PLE spectra does not suggest the state density of each sublevel since the intensity depends on the carrier relaxation rate among sublevels.

Figure 23 shows the emission energy of the ground and the second level as a function of alternate supply cycle [35]. As the number of cycles increased, sublevel energies and the energy interval decreased. The diameter of the dots observed by TEM increased as the number of supply cycles increased. The results suggest that the energy of sublevels was determined by the quantum size effect. The dot size was changed by post-growth annealing as shown in the following.

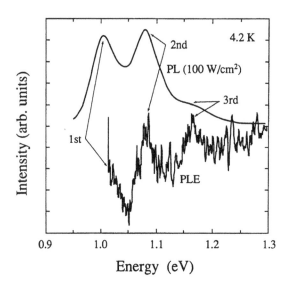

Fig. 22.   PLE and PL spectra at 4.2 K. Discrete sublevels up to the third level were proved.

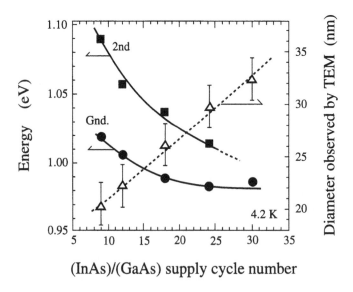

Fig. 23.   Energy of the ground and the second level as a function of number of cycles. Dot's diameter observed by TEM is also indicated.

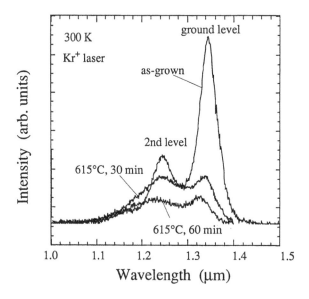

Fig. 24. PL spectra of quantum dots as-grown and after annealing at 615°C for 30 and 60 minutes.

We measured PL spectra before and after annealing and compared the energy of sublevels [39]. We prepared the InGaAs dots with 18 cycles of $(TMIDMEA)/(TMG_a)/(AsH_3)$ alternate supply. Samples were annealed at 500 to 680°C for 30 and 60 minutes in the reactor tube of a liquid-phase epitaxy system. During the annealing, we flowed pure N_2 gas to protect samples from oxidation, and placed a GaAs plate on the samples to prevent arsenic desorption by raising the arsenic pressure Figure 24 shows the PL spectra before and after annealing at 615°C for 30 and 60 minutes. We see energies of both the ground and the second levels shifted upon annealing. The integrated PL intensity decreased simultaneously, indicating that a nonradiative channel was introduced in the samples by the annealing [30]. It is noteworthy that the peak intensity of the ground level weakened much more than that of the second level.

We plotted the energies of the ground and the second levels as a function of annealing temperature (Figure 25). We see that the energy shift occurred above 600°C. The shift was larger in the second level up to 650°C, that is, energy separation of the levels increased upon annealing. We observed a single peak in the measurement at 680°C, and we attributed it to the second or higher excited sublevel, considering that the ground-level emission is weakened by annealing and that the energy shift is suppressed when the level energy gets close to the band gap of the GaAs cap layer.

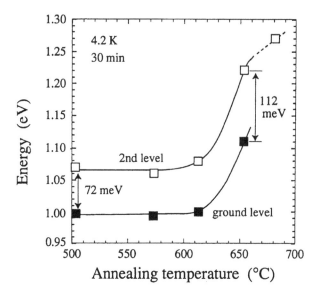

Fig. 25. Energy of the ground and the second level at 4.2 K as a function of annealing temperature.

Enhancement of level separation by annealing was not observed in quantum wells. Level separation decreases due to interdiffusion in a quantum well since the interdiffusion always weakens the one-dimensional quantum confinement. We suppose that quantum confinements in additional (in-plane) dimensions characterized the shift in quantum dots. Sallese et al. predicted the increase of level separation due to interdiffusion for a quantum wire system, considering the lateral potential structure [73]. The quantum wire in their model had a crescent-type cross-sectional profile. At the first stage of interdiffusion the lateral potential increases at a higher rate at the edge of the crescent due to its thinness, resulting in the enhancement of energy splitting. The magnitude of the enhancement depends on the initial energy split (i.e., the strength of the initial quantum confinement potential). We assume that a similar phenomenon occurred in our quantum dots (Figure 26). Remember that their dot structure was anisotropic. The cross section of our dots is oval, and quantum confinement may become severe due to the increase of lateral potential at the comparatively thin area of the dots.

To our regret, it is too difficult to analyze the energy shift quantitatively in the quantum dots. We do not know the exact initial confinement potential in the dot. The boundary around the dots must not be abrupt but gradual in composition. We do not know how the crystal lattice is distorted in the dot (biaxial or hydrostatic ?). Additionally, there is no formula that treats

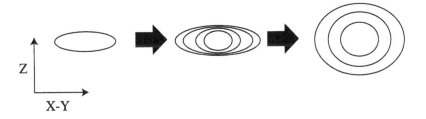

Fig. 26. Schematic of potential in anisotropic quantum-dot structure during interdiffusion. Quantum confinement can be severe due to the increase of lateral potential height. Motion of interface may occur simultaneously due to unbalanced diffusion coefficient.

intermixing in three dimensions in the anistropic structure. The quantitative analysis is a subject of future works.

3.3   Phonon Bottleneck in Anealed Quantum Dots.

In the following we demonstrate the influence of annealing on the carrier lifetimes at sublevels of quantum dots [39]. It is noteworthy in Figure 24 that the peak intensity of the ground level weakened by annealing much greater than that of the second level. We observed the same tendency at other annealing temperatures over 600°C where the peak energies shifted. These results suggest that the carrier population at the ground level decreased after annealing. Remember that in the EL spectra in Figure 21, a higher level emission appeared before the emission intensity of the lowest level reached the maximum value. These phenomena suggest the presence of the phonon bottleneck effect in this system.

The phonon bottleneck is a phenomenon occurring in the zero-dimensional semiconductor structure, the carrier relaxation between intrabands is hindered due to the delta-functionue state density [74]. The bottleneck effect was pointed out as being an obstruction to the predicted excellent performance of dot lasers where the discrete state density enables considerable reduction of the threshold current and temperature sensitivity [59]. The slow relaxation rate will degrade the laser performance. Before the self-formation technique was developed to produce semiconductor quantum dots, the phonon bottleneck had been a mere theoretical research subject. Benisty and co-workers predicted that the slow relaxation lifetime will enhance the nonradiative recombination process in the quantum dots (Figure 27) [75,76]. It was believed that the intraband

Energy

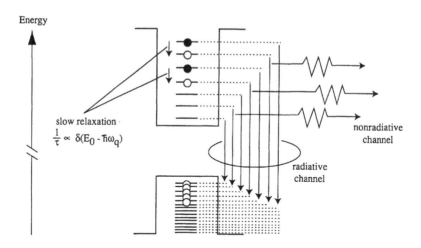

slow relaxation
$\frac{1}{\tau} \propto \delta(E_0 - \hbar\omega_q)$

nonradiative
channel

radiative
channel

Fig. 27. Schematic of the phonon bottleneck effect.

spacing must agree with the LO-phonon energy for the fast carrier relaxation, and the carrier relaxation rate was calculated as a function of intraband spacing considering electron-hole interaction [77,78], LA–LO phonon process [79], and phonon-electron mode coupling [80]. Reduction of PL intensity in III-V compound semiconductor dots prepared by lithography and regrowth was reported and used as evidence of the existence of the phonon bottleneck [81]. However defects introduced during the nanofabrication process require a more complex picture for PL intensity analysis. Lack of high-quality semiconductor microcrystals made it difficult to examine phonon bottleneck experimentally.

We measured the luminescence decay of sublevels in the ALS dots as a function of temperature using a time-resolved PL system [38,39]. Examples of the decay curves obtained are shown in Figures 28 and 29. We found that in the as-grown sample the higher the temperature was, the faster the decay progressed and that the lower the order of a level, the slower the progression of decay was. In samples annealed at 615°C for 60 minutes, we saw that the decay time of the ground level decreased as measurement temperature was raised. In the as-grown sample, the decay time of the ground level was independent of temperature.

We assumed that the decay curves can be analyzed using a double-exponential curve-fitting equation; $[\exp(-t/\tau_r) + \exp(-t/\tau_{0i})]$, where τ_r is the radiative recombination lifetime and τ_{0i} is the relaxation lifetime for the i-th level. The second term is neglected in the decay of the ground level. This assumption is based on the following observation. Since the energy separation

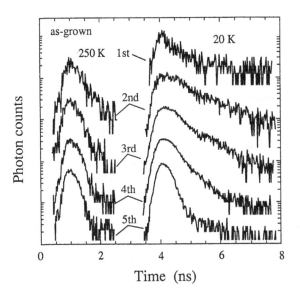

Fig. 28. Decay curves for the five sublevels at 250 and 20 K. Emission decay for the first level at 250 K was outside the detectable range of measurement.

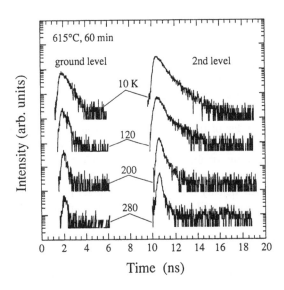

Fig. 29. Decay curves of the ground- and the second-level emission at various temperatures after annealing at 615°C for 60 minutes.

of sublevels in each dot is comparatively larger than the value of kT and the physical distance between the dots is too long (severaltens of manometers) for carriers to tunnel, carriers in one dot cannot move to another dot and each decay curve is the sum of decay in individual dots. The carrier relaxation rate in the individual dot strongly depends on the number of carriers in the dot. Carriers cannot relax when the lower levels are filled (case 1). Relaxation is assumed to be faster than recombination when the lower levels are empty (case 2) since the lowest-energy peak was dominant at low carrier density in PL at every temperature. Considering that each level in a dot is filled with a few carriers, the decay process between sublevels can be categorized into the above two cases approximately. After deconvoluting the decay curves with excitation pulse spectra using the double-exponential equation, we determined the lifetimes in the equation. We found that the two lifetimes in each level can be characterized as follows, especially for the as-grown sample. One is long (0.7–1.8 ns) and independent of the level order and temperature (Figure 30(a)). The other is short (10 ps–1 ns), varies among levels, and is temperature dependent (Figure 30(b)). We denote the former as recombination and the latter as relaxation, as described below [38].

The near-one-nanosecond lifetime and the temperature independence match well with predicted characteristics of the radiative recombination lifetime for quantum dots. We have proposed a theory on the spontaneous emission lifetime of excitons in quantum dots and predicted that as the in-plane diameter of an exciton decreases to several manometers, the lifetime will increase to a few nanosecond [82]. The spontaneous emission rate is not strongly temperature dependent since the excitons in the dots have discrete energy states. Experimentally, Wang et al. [83] found the decay time for ground-level spontaneous emission of InGaAs/GaAs quantum dots to be ~ 1 ns, which is in good agreement with our results.

The shorter lifetime can be well explained if we treat it as a relaxation lifetime. As the order of the level increased from the second to the fifth, lifetime decreased by one order of magnitude. Since relaxation rate can be expressed in terms of the sum of all possible transitions between the levels, the relaxation lifetime of a level should decrease as the state density increases. The state density is expected to be larger in higher levels [36]. In fact, we have observed in EL spectra that the intensity increases as the level order increases. We also saw in the experiment that lifetime decreased as the temperature was raised from 20 K to 300 K. This observation agrees with the increase in the number of phonons predicted by the Bose distribution function: $[\exp(\hbar\omega/kT) - 1]^{-1}$. The relaxation lifetimes are much longer than those in systems of higher dimensionality (10^{-14} s in GaAs bulk [84] and 10^{-13} s in a GaAs quantum well [85]), which suggests the existence of the phonon bottleneck.

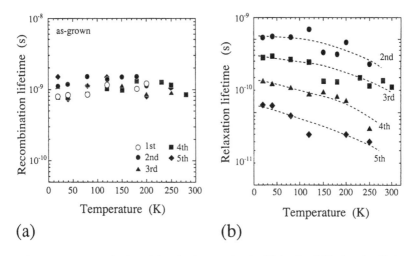

Fig. 30. (a) Recombination lifetime for the ground to the fifth level and (b) relaxation lifetime for the second to the fifth level in quantum dots as a function of temperature.

We simulated the EL specta using the simplifid model with the obtained lifetimes. We considered that there were five electron levels in the dots where radiative decay is allowed, assuming that the hole energy separation was within the thermal energy owing to the heavy mass. We also assumed that electron relaxation occurs toward the neighboring lower level and ignored the upward thermal excitation of electrons. We assumed that the relaxation rate in the i-th level (τ_1^{-1}) is proportional to the filled proportion of the $(i-1)$-th level (f_{i-1}; $0 \leq f_{i-1} \leq 1$); $\tau_i^{-1} = (1 - f_{i-1})\tau_{0i}^{-1}$, where τ_{0i} is the intrinsic relaxation lifetime in the i-th level. Assuming that the recombination lifetime, τ_r, is common to all levels, we solved the rate equations

$$\frac{f_5 N_5}{\tau_5} + \frac{f_5 N_5}{\tau_r} - G = 0 \tag{3-1}$$

$$\frac{f_i N_i}{\tau_i} + \frac{f_i N_i}{\tau_r} - \frac{f_{i+1} N_{i+1}}{\tau_{i+1}} = 0 \qquad (i = 2, 3, 4) \tag{3-2}$$

$$\frac{f_1 N_1}{\tau_r} - \frac{f_2 N_2}{\tau_2} = 0, \tag{3-3}$$

where G is the carrier generation rate, and N_i ($i = 1$ to 5) is the density of states in the i-th level and τ_{nr} is the nonradiative lifetime.

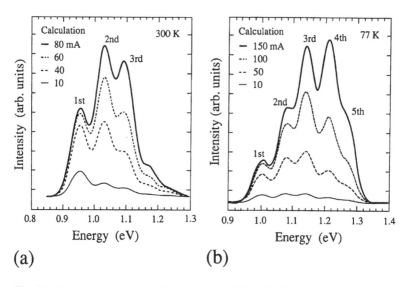

Fig. 31. Measured and simulated EL spectra for 300 K and 77 K.

Figure 31 shows the calculated EL spectra at 300 and 77 K. Results of the calculation correlate very well with measurements shown in Figure 21. In the measured spectra, we can see that high-energy peaks appeared earlier for 77 K than for 300 K as the current increased. For instance, when the emission intensity of the ground level reached half-maximum, the intensity of the second level was stronger than that of the ground level at 77 K (10 mA), but almost equal at 300 K (50 mA). In the calculation, we can see that the trend of peak intensities was explained well, and understood that the variation due to temperature was because the relaxation lifetime at 77 K was longer than at 300 K. We noted for 300 K that the calculated values of current are much smaller than the measured values. This discrepancy worsens when the injected current is large. We suspect that this discrepancy is attributable to the nonradiative recombination process outside the dots [37]. Carriers may have been trapped in defects in the quantum well surrounding the dots where the indium atoms are missing. High current injection increases local temperature and may promote the nonradiative recombination process. At 77 K, we see a good agreement between the calculated and measured current values, which suggests a rather negligible influence of the nonradiative process.

Obtained lifetimes of the annealed sample are shown in Figures 32(a) and 32(b) as a function of temperature. We see that the relaxation lifetimes decreased slightly after annealing. The recombination lifetime of the second level did not change much after annealing while the recombination lifetime

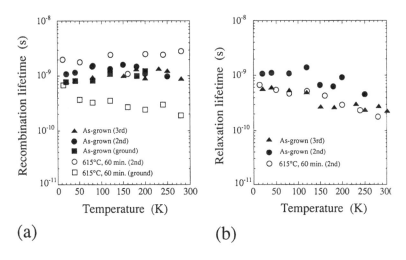

Fig. 32. (a) Recombination lifetimes and (b) relaxation lifetimes of sublevels before and after annealing at 615°C for 60 minutes.

at the ground level decreased drastically to a few hundred picoseconds and seemed to become temperature dependent after annealing. In a quantum dot where the state density is a series of delta functions, the recombination lifetime will be insensitive to temperature when the energy separation is larger than the thermal energy. The decrease in lifetimes at the ground level suggests that the nonradiative recombination channel affected the ground level. From the observed lifetimes, the nonradiative lifetime can be estimated to be a few hundred picoseconds. In a quantum well, carrier trapping by defect levels is believed to occur from the bottom of the energy band. As one possibility, we consider that the similar phenomenon occurred in quantum dots and that the nonradiative channel only affected the ground level. As another possibility, we consider the existence of the phonon bottleneck for the carrier trapping process. If carrier trapping requires phonon scattering the second level energy may not satisfy the conditions related to the phonon energy. To enable further discussion, additional research on the origin of the nonradiative level and the details of its trapping process is required.

Using the measured lifetimes, we can estimate the carrier recombination process in the ground and the second levels of quantum dots (inset in Figure 33). The relative emission intensity of the ground and the second level is determined by the equilibrium between the radiative recombination process ($r1, r2$), the relaxation process ($x2$), and the nonradiative recombination process ($n1, n2$). In the as-grown sample, the nonradiative lifetimes τ_{n1}

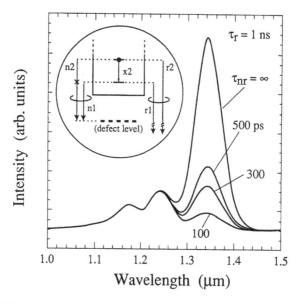

Fig. 33. PL spectra calculated with various values of nonradiative lifetime using the measured relaxation and the recombination lifetime. Schematic of electron recombination process in the ground and the second level is shown in the inset.

and τ_{n2} were negligible. After annealing, the recombination lifetime at the ground level decreased to $(\tau_{r1}^{-1} + \tau_{n1}^{-1})^{-1}$ while the nonradiative lifetime τ_{n2} was still negligible. The one-sided reduction of the recombination lifetime at the ground level caused a significant decrease in the relative emission intensity of the ground level. Another view is that this phenomenon was a consequence of the phonon bottlenerk for the carrier relaxation. If the relaxation lifetime τ_{x2} is negligible, the second-level emission will decrease drastically when the nonradiative channel is generated inside the dot.

Figure 33 shows the calculated PL spectra with several values of nonradiative lifetime. In the calculation, we used the same model as for EL spectra. Assuming that the nonradiative recombination occurred only at the ground level, we solved the same rate equations for five levels, replacing Eq. (3-3) by

$$\frac{f_1 N_1}{\tau_r} + \frac{f_1 N_1}{\tau_{nr}} - \frac{f_2 N_2}{\tau_2} = 0. \tag{3-4}$$

In Figure 33, we see that as the nonradiative lifetime decreases from infinity to a few hundred picoseconds, the emission intensity of the ground level decreases and becomes comparable to that of the second level. The value of a few hundred picoseconds agrees with the lifetime estimated in Figure 32(b). The intensity of higher sublevels is constant since the relaxation rate is not influenced by the nonradiative recombination, due to the low filled proportion at the ground level. In the measured spectra, emission intensity of higher sublevels was reduced by annealing. We suppose that the number of carriers injected into dots decreased in the actual annealed samples since the crystal quality outside the dots may also be degraded by annealing.

The phonon bottleneck rendered the carrier density at the ground level sensitive to the nonradiative channel generated by annealing. In the growth of a laser structure, InGaP or AlGaAs cladding layers sandwiching the dot region were grown at over 600°C in order to obtain good optical quality. The reported dot-laser performances are, therefore, expected to be degraded by the generation of nonradiative channels, although the relaxation rate itself may be improved by the thermal treatment. As a future research subject with the aim of practical application, we should study the nonradiative defects generated by thermal treatment to find out how to eliminate the defects. The structural dependence of the carrier relaxation rate is also a significant research area.

4 SUMMARY

We presented our study on interdiffusion in III-V compound semiconductor quantum wells and self-formed quantum dots. First, we presented a formula which comprehensively describes interdiffusion profiles of quantum wells. We derived the formula by solving diffusion equations, assuming that interdiffusion coefficients differ between layers and that there is interfacial discontinuity in the compositions of interdiffused species. The formula includes the penetration of interdiffusion between alternate layers, which is common in quantum wells. We applied the formula to the analysis of the dependence of quantum energy shift on annealing time, annealing temperature, and well layer width in lattice-matched $In_xGa_{1-x}As_yP_{1-y}$/InP and GaAs/Al_xGa_{1-x}As quantum wells. Our formula correctly predicted the dependence and clarified the differences between interdiffusion processes in these two materials. We then showed the dependence of interdiffusion parameters on growth conditions in $In_{0.53}Ga_{0.47}$As/InP quantum wells, investigating their interdiffusion mechanisms. We investigated the influence of the growth temperature and the buffer-layer thickness. From the observed dependence of interdiffusion activation energies and interdiffusion coefficients on the conditions, we inferred that the interdiffusion mechanism in the InP layer is related to vacancies and dislocations, and in $In_{0.53}Ga_{0.47}$As layer, to interstitial site propagation. Next, we reviewed the effects of thermal treatment on our self-formed $In_{0.5}Ga_{0.5}$As quantum dots. After showing our unique technique of growing quantum dots with an alternating supply of precursors, we show that the quantum energy split between the ground and second sublevel increased as annealing temperature was raised, which was not observable in quantum wells. We also showed that the nonradiative channel introduced by interdiffusion caused the degradation of the relative emission intensity of the ground level, taking into account the existence of the phonon bottleneck. The influence of interdiffusion on the carrier lifetimes was quantitatively presented by measuring time-resolved photoluminescence.

Acknowledgements

The author is grateful to many colleagues who have cooperated in various aspects of these researches in Fujitsu Laboratories Ltd. Thanks are due in particular to Takuya Fujii, Toshiyuki Tanahashi, Susumu Yamazaki, Niro Okazaki, Osamu Aoki, Nobuyuki Ohtsuka, Hajime Shoji, Hiroshi Ishikawa, and Shigenobu Yamakoshi. The stimulating discussions with Mitsuru Sugawara are also gratefully acknowledged. The author also thanks Sachiko Mukai for her encouragements on the writing of this chapter.

References

1. W.D. Laidig, N. Holonyak, Jr., M.D. Camras, K. Hess, J.J. Coleman, P.D. Dapkus, and J. Bardeen, *Appl. Phys. Lett.*, **38**, 776 (1981).
2. M. Kawabe, N. Matsuura, N. Shimizu, F. Hasegawa, and Y. Nannichi, *Jpn. J. Appl. Phys.*, **23**, L623 (1984).
3. K. Meehan, N. Holonyak, Jr., J.M. Brown, M.A. Nixon, and P. Gavrilovic, *Appl. Phys. Lett.*, **45**, 549 (1984).
4. E.V.K. Rao, H. Thibiergr, F. Brillouet, F. Alexandre, and R. Azoulay, *Appl. Phys. Lett.*, **46**, 867 (1985).
5. R.L. Thornton, R.D. Burnham, T.L. Paoli, N. Holonyak, Jr., and D.G. Deppe, *Appl. Phys. Lett.*, **48**, 7 (1986).
6. R.L. Thornton, R.D. Burnham, T.L. Paoli, N. Holonyak, Jr., and D.G. Deppe, *Appl. Phys. Lett.*, **47**, 1239 (1985).
7. T. Fukuzawa, S. Semura, H. Saito, T. Ohta, Y. Uchida, and H. Nakashima, *Appl. Phys. Lett.*, **45**, 1 (1984).
8. J. Cibert, P.M. Petroff, D.J. Werder, S.J. Pearton, A.C. Gossard, and J.H. English, *Appl. Phys. Lett.*, **49**, 223 (1986).
9. D.G. Deppe, L.J. Guido, N. Holonyak, Jr., K.C. Hsieh, R.D. Burnham, R.L. Thornton, and T.L. Paoli, *Appl. Phys. Lett.*, **49**, 510 (1986).
10. J. Cibert, P.M. Petroff, G.J. Dolan, S.J. Pearton, A.C. Gossard, and J.H. English, *Appl. Phys. Lett.*, **49**, 1275 (1986).
11. L.J. Guido, N. Holonyak, Jr., K.C. Hsieh, R.W. Kaliski, W.E. Plano, R.D. Burnham, R.L. Thornton, J.E. Epler, and T.L. Paoli, *J. Appl. Phys.*, **61**, 1372 (1987).
12. L.L. Chang and Koma, *Appl. Phys. Lett.*, **29**, 138 (1976).
13. M.D. Camras, T.L. Paoli, and C. Lindstrom, *J. Appl. Phys.*, **54**, 5637 (1983).
14. H. Leier, H. Rothfritz, and Forchel, *J. Appl. Phys.*, **95**, 277 (1989).
15. K. Nakashima, Y. Kawaguchi, Y. Kawamura, H. Asahi, and Y. Imamura, *Jpn. J. Appl. Phys.*, **26**, L1620 (1987).
16. T. Fujii, N. Sugawara, S. Yamazaki, and K. Nakajima, *J. Cryst. Growth*, *105*, 348 (1990).
17. B. Tell, B.C. Johnson, J.L. Zyskind, J.M. Brown, J.W. Sulhoff, K.F. Brown-Goebeler, B.I. Miller, and U. Koren, *Appl. Phys. Lett.*, **52**, 1428 (1988).
18. M. Razeghi, O. Acher, and F. Launay, *Semicond. Sci. Technol.*, **2**, 793 (1987).

19. K. Nakashima, Y. Kawaguchi, Y. Kawamura, and Y. Imamura, *Appl. Phys. Lett.*, **52**, 1383 (1988).
20. S.A. Schwarz, P. Mei, T. Venkatesan, R. Bhat, D.M. Hwang, C.L. Schwartz, M. Koza, L. Nazar, and B.J. Skromme, *Appl. Phys. Lett.*, **53**, 1051 (1988).
21. H. Sumida, H. Asahi, S.J. Yu, K. Asami, and S. Gonda, *Appl. Phys. Lett.*, **54**, 520 (1989).
22. H. Ribot, K.W. Lee, R.J. Simes, R.H. Yan, and L.A. Coldren, *Appl. Phys. Lett.*, **55**, 672 (1989).
23. E.S. Koteles, A.N.M.M. Choudhury, A. Levy, B. Elman, P. Melman, M.A. Koza, and R. Bhat, 'Materials Research Society Symposium Proceedings', 240, 171, (1991).
24. R.M. Flemming, D.B. McWhan, A.C. Gossard, W. Wiegmann, and R.A. Logan, *J. Appl. Phys.*, **51**, 357 (1980).
25. K. Mukai, N. Ohtsuka, M. Sugawara, and S. Yamazaki, *Jpn. J. Appl. Phys.*, **33**, L1710 (1994).
26. D. Leonard, M. Kishnamurthy, C.M. Reaves, S.P. Denbaars, and P.M. Petroff, *Appl. Phys. Lett.*, **63** (1993), 3203.
27. N. Kirstaedter, N.N. Ledentsov, M. Grundmann, D. Bimberg, V.M. Ustinov, S.S. Ruvimov, M.V. Maximov, P.S. Kop'ev, Zh. I. Alferov, U. Richter, P. Werner, U. Gössele, and J. Heydenreich, *Electron. Lett.*, **30**, 1416 (1994).
28. H. Shoji, Y. Nakata, K. Mukai, M. Sugiyama, M. Sugawara, N. Yokoyama, and H. Ishikawa, *Electron. Lett.*, **32**, 2023 (1996).
29. K. Mukai, Y. Nakata, H. Shoji, M. Sugawara, K. Ohtsubo, N. Yokoyama, and H. Ishikawa, *Electron. Lett.*, **34**, 1588 (1998).
30. R. Leon, Y. Kim, C. Jagadish, M. Gal, J. Zou, and D.J.H. Cockayne, *Appl. Phys. Lett.*, **69**, 1888 (1996).
31. K. Mukai, M. Sugawara, and S. Yamazaki, *J. Crystal Growth*, **115**, 433 (1991).
32. K. Mukai, M. Sugawara, and S. Yamazaki, *Phys. Rev. B*, **50**, 2273 (1994).
33. K. Mukai, M. Sugawara, and S. Yamazaki, *J. Crystal Growth*, **137**, 388 (1994).
34. K. Mukai, H. Shoji, N. Ohtsuka, and M. Sugawara, *Appl. Surface Science*, **112**, 102 (1997).
35. K. Mukai, M. Sugawara, and S. Yamazaki, *Jpn. J. Appl. Phys.*, **35**, L262 (1996).
36. K. Mukai, N. Ohtsuka, S. Hajime, and M. Sugawara, *Appl. Phys. Lett.*, **68**, 3013 (1996).
37. K. Mukai, M. Sugawara, and S. Yamazaki, *Appl. Phys. Lett.*, **70**, 2416 (1997).
38. K. Mukai, N. Ohtsuka, S. Hajime, and M. Sugawara, *Phys. Rev. B*, **54**, R5243 (1996).
39. K. Mukai, and M. Sugawara, *Jpn. J. Appl. Phys.*, **37**, 5451 (1998).
40. W. Jost, 'Diffusion' (Academic Press, New York, 1960) p. 68.
41. R. Smoluchowski, *Phys. Rev.*, **62**, 539 (1942).
42. H.E. Cook and J.E. Hilliard, *J. Appl. Phys.*, **40**, 2191 (1969).
43. H. Haken, 'Quantenfeldtheorie des Festkorpers' (B.G. Teubner, Stuttgart, 1973).
44. G.E. Pikus and G.L. Bir, *Sov. Phys. Solid State*, **1**, 136 (1959).
45. G.E. Pikus and G.L. Bir, *Sov. Phys. Solid State*, **1**, 1502 (1960).
46. A. Gavini and M. Cardona, *Phys. Rev. B*, **1**, 672 (1970).
47. K. Sakurai, 'Introduction to Quantum Mechanics by Personal Computer' (in Japanese, Shokabo, Tokyo, 1989).
48. M.C. Joncour, J.L. Benchimol, J. Burgeat, and M. Quillec, *J. Phys. (Paris)*, **C5**, 3 (1982).
49. S.R. Forest, P.H. Schmidt, R.B. Wilson, and M.L. Kaplan, *Appl. Phys. Lett.*, **45**, 1199 (1984).
50. C.P. Kuo, S.K. Vong, R.M. Cohen, and G.B. Stringfellow, *J. Appl. Phys.*, **57**, 5428 (1985).
51. Y. Suematsu, 'Semiconductor Laser and OEIC' (Ohm Inc., Tokyo, 1984) p. 70 (in Japanese).
52. H.C. Casey, Jr., *J. Appl. Phys.*, **49**, 3684 (1978).
53. B. Goldstein, *Phys. Rev.*, **121**, 1305 (1961).
54. S.J. Yu, H. Asahi, S. Emura, S. Gonda, and K. Nakashima, *J. Appl. Phys.*, **70**, 204 (1991).
55. S.Y. Chiang and G.L. Pearson, *J. Appl. Phys.*, **46**, 2986 (1975).
56. D.L. Kendal, in 'Semiconductors and Semimetals' (Academic Press, New York, 1968), Vol. **4**, p. 163.
57. For example, D. Shaw, 'Atomic Diffusion in Semiconductors' (Plenum Press, London and New York, 1973).
58. C.H. Ting and G.L. Pearson, *J. Electrochem. Soc.*, **118**, 1454 (1971).
59. Y. Arakawa and H. Sakaki, *Appl. Phys. Lett.*, **40**, 939 (1982).

60. D.S. Chemla and D.A.B. Miller, *Opt. Lett.*, **11**, 522 (1986).
61. L. Brus, *IEEE J. Quantum Electron*, **22**, 1909 (1986).
62. J.N. Randall, M.A. Reed, and G.A. Frazier, *J. Vac. Sci. & Technol.*, **B7**, 1398 (1989).
63. K. Kash, A. Scherer, J.M. Worlock, H.G. Craighead, and M.C. Tamargo, *Appl. Phys. Lett.*, **49**, 1043 (1986).
64. T. Fukui, S. Ando, Y. Tokura, and T. Toriyama, *Appl. Phys. Lett.*, **58**, 2018 (1991).
65. D.J. Eablesham and M. Cerullo, *Phys. Rev. Lett.*, **64**, 1943 (1990).
66. R. Apetz, L. Vescan, A. Hartmann, C. Dieker, and H. Luth, *Appl. Phys. Lett.*, **66**, 445 (1995).
67. J.M. Moison, F. Houzay, F. Barthe, L. Leprince, E. Andre, and O. Vatel, *Appl. Phys. Lett.*, **64**, 196 (1994).
68. Y. Narukawa, Y. Kawakami, M. Funato, S. Fujita, S. Fujita, and S. Nakamura, *Appl. Phys. Lett.*, **70**, 981 (1997).
69. K. Mukai et al., *Semiconductors and Semimetals* (Academic Press, New York, 1999), Vol. 60, Chap. 3.
70. N. Grandjean and J. Massies, *J. Cryst. Growth*, **134**, 51 (1993).
71. J. Osaka, N. Inoue, Y. Mada, K. Yamada and K. Wada, *J. Cryst. Growth*, **99**, 120 (1990).
72. P. Boguslawski and A. Baldereschi, *Phys. Rev. B*, **39**, 8055 (1989).
73. J.M. Sallese, J.F. Carlin, M. Gaihanou, and P. Grunberg, *Appl. Phys. Lett.*, **67**, 2633 (1995).
74. U. Bockelmann and G. Bastard, *Phys. Rev. B*, **42**, 8947 (1990).
75. H. Benisty, C.M. Sotomayor-Torres, and C. Weisbuch, *Phys. Rev. B*, **44**, 10945 (1991).
76. H. Benisty, *Phys. Rev. B*, **51**, 13281 (1995).
77. U. Bockelmann and T. Egeler, *Phys. Rev. B*, **46**, 15574 (1992).
78. U. Bockelmann, *Phys. Rev. B*, **48**, 17637 (1993).
79. T. Inoshita and H. Sakaki, *Phys. Rev. B*, **46**, 7260 (1992).
80. H. Nakayama, and Y. Arakawa, 7th Int. Conf. Superlattices, Microstructures, and Microdevices, Banff, Canada (1994).
81. K. Brunner, U. Bockelmann, G. Abstreiter, M. Walther, G. Bohm, G. Trankle, and G. Weimann, *Phys. Rev. Lett.*, **69**, 3216 (1992).
82. M. Sugawara, *Phys. Rev. B*, **51**, 10743 (1995).
83. G. Wang, S. Fafard, D. Leonard, J.E. Bowers, J.L. Merz, and P.M. Petroff, *Appl. Phys. Lett.*, **64**, 2815 (1994).
84. M. Yamada, H. Ishiguro, and H. Nagato, *Jpn. J. Appl. Phys.*, **19**, 135 (1980).
85. M. Asada, *IEEE J. Quantum Electron.*, **25**, 2019 (1989).

CHAPTER 4

Interdiffusion in Strained Layer In$_x$Ga$_{1-x}$As/GaAs Heterostructures

FERNANDO IIKAWA

IFGW - Universidade Estadual de Campinas, CEP-13083.970, CP-6165,
Campinas-SP, Brazil
E-mail: iikawa@ifi.unicamp.br

1   INTRODUCTION

Strained layer In$_x$Ga$_{1-x}$As/GaAs systems have received a great deal of attention in the last few years due to their potential applications to electronic and opto-electronic semiconductor devices. The In$_x$Ga$_{1-x}$As alloy layer is biaxially compressed due to the larger lattice constant compared to the GaAs layer. Despite the fact that the alloy layer is under high strain (from 0 to 7%), high quality of epitaxial layers can be obtained (see [1] and references therein). The In$_x$Ga$_{1-x}$As alloy band gap is smaller than the GaAs one resulting in higher potential barrier between In$_x$Ga$_{1-x}$As and Al$_y$Ga$_{1-y}$As heterojunctions when compared to the GaAs/Al$_y$Ga$_{1-y}$As based systems. This change in the band structure becomes useful for applications in high speed digital and optical devices, such as high electron mobility transistors (HEMT) [2–5] and quantum well laser diodes [6–11]. The advance of crystal growth techniques enabled the fabrication of semiconductor devices with multi-layered materials. The fabrication steps of these devices involves unintentional annealing processes during impurity diffusion and electric contact metal deposition. In quantum well and superlattice based devices, the intermixing of atoms at the interface during the thermal annealing can change the band structure potential profile, affecting the device properties and their performance.

The interdiffusion effects have been intensively investigated in different heterostructure systems. As for self-diffusion processes in bulk semiconductors, it has been demonstrated that the layer intermixing in heterostructures also depends on the native point defects generated at certain experimental conditions during the heat treatment. Several experimental works showed the enhancement of the interdiffusion process in the presence of point defects. In the work reported by Laidig et al. [12] it was shown that the impurities also strongly affect the interdiffusion process in heterostructure systems. The introduction of impurities and the subsequent creation of point defects in III-V semiconductors are well understood and they are used in several experiments in order to identify the interdiffusion mechanisms. They were also recently used for applications in optical devices [13–19].

The interdiffusion effects in $Al_yGa_{1-y}As/GaAs$ lattice matched heterostructures have been intensively investigated [20,21]. The thermal annealing in this system yields the interdiffusion at the interface smoothing the Al (or Ga) concentration profile at the interface and consequently changing the energy band potential profile. In lattice mismatched heterostructures, such as $In_xGa_{1-x}As/GaAs$, Si/Ge, and others, the interdiffusion produces graded strain along the growth direction (corresponding to the alloy composition profile) and in some cases the strain is relieved by generating misfit dislocations. The strain relaxation occurs when the alloy layer thickness is close or larger than the critical thickness [22,23].

The interdiffusion coefficient (D_i) is a good parameter to characterize the degree of interdiffusion effect in semiconductor heterostructures. This parameter can be obtained from different experimental methods. The most common technique involves indirect optical measurements, such as photoluminescence [24–30], photoconductivity [31], transmission spectroscopy [29], photoreflectance [32] and Raman spectroscopy [33, 34]. The optical measurements provide important informations about the transition energies and the potential energy profile. Other methods such as secondary ion mass spectroscopy (SIMS) [35] and Auger-electron spectroscopy [36–38] were also employed to study the interdiffusion in semiconductor heterostructures. These techniques give direct measurements of the composition profile of a heterostructure.

In this chapter, thermal annealing effects in $In_xGa_{1-x}As/GaAs$ heterostructures are discussed taking into account interdiffusion and strain effects. In Section 2, a theoretical analysis related to the interdiffusion effect, the lattice stability and the calculation of the transition energies in layer intermixed quantum wells and superlattices is described. In section 3, we present the main experimental results and some theoretical models that were used to explain the interdiffusion mechanisms in $In_xGa_{1-x}As/GaAs$ systems. In Section 3, we also discuss the impurity induced disorder and strain effects on the interdiffusion processes. In Section 4, we present the application of interdiffusion effects in optoelectonic devices. Finally, our conclusions are given in Section 5.

2 THEORY

2.1 Band Structure

GaAs and $In_xGa_{1-x}As$ are direct gap materials. $In_xGa_{1-x}As$ ternary alloy can be seen as a mixture of GaAs and InAs binary semiconductors. A parabolic energy dispersion is a reasonable approximation for conduction and valence bands $In_xGa_{1-x}As/GaAs$ heterostructures. In the latter case, this is justified by the presence of the biaxial strain which diminishes the effects of the valence band mixing effects. The effects of the strain in the valence band will be discussed below. The ternary alloy energy gap as a function of In content for three different temperatures are:

$$E_g^o = 1.519 - 1.594x + 0.485x^2 \quad \text{(2 K) [39]} \qquad (2.1a)$$

$$E_g^0 = 1.508 - 1.470x + 0.375x^2 \quad \text{(77 K) [40]} \qquad (2.1b)$$

$$E_g^0 = 1.425 - 1.501x + 0.426x^2 \quad \text{(300 K) [41]} \qquad (2.1c)$$

These equations are used to analyze the results of optical measurements in interdiffusion studies. These experiments are usually performed at these temperatures. A detailed discussion concerning the compositional dependence of the Γ, X and L ternary alloy band energies can be found in Ref. [42]. All ternary alloy parameters, excepting the energy gaps, are obtained from a linear combination of the binaries GaAs and InAs materials. The parameter values of GaAs and InAs used in this chapter are given in Table 1.

2.2 Strain Dependent Band Structure

The lattice constant of bulk $In_xGa_{1-x}As$, for $x > 0$, is larger than the GaAs one. The pseudomorphic $In_xGa_{1-x}As$ layer is grown on GaAs substrate

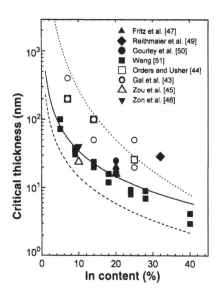

Fig. 1. Critical layer thickness of $In_xGa_{1-x}As/GaAs$ heterostructures as a function of In content. The symbols represent the experimental data: open symbols for single layer and the solid ones for quantum wells. The solid and dashed curves correspond to the calculations using Matthews and Blakeslee's model for quantum wells and single heterostructures, respectively. The dotted curve was obtained using People and Bean's model for single heterostructures.

generating a built-in biaxial elastic strain in the $In_xGa_{1-x}As$ layer when this layer thickness (L) is smaller than a critical layer thickness (h_c). The amount of strain varies from 0 (GaAs) to $\sim 7\%$ (InAs). If the ternary layer thickness exceeds h_c, plastic relaxation occurs with the generation of misfit dislocations at the interface [22,23]. The In content and the layer thickness must be well controlled to obtain pseudomorphic layers free of defects.

The critical thickness h_c depends on the lattice mismatch between the ternary alloy and the GaAs layers. There are several data of h_c data obtained experimentally using different methods [43–52]. Figure 1 shows the h_c data for $In_xGa_{1-x}As/GaAs$ heterostructures obtained by several authors. The open symbols correspond to the data for single heterostructure without a GaAs capping layer, and the solid ones correspond to quantum well and superlattice data. These data present a wide spread of the h_c values. This discrepancy was pointed out by Fritz [53] and was attributed to differences in the sensitivity of experimental techniques used in the respective experiments.

Several theoretical models were proposed to determine the h_c values [23,54,55]. The continuous curves shown in Figure 1 are calculated values for h_c using different models: dashed and solid curves are the Matthews and Blakeslee's model [23] for single heterostructures and quantum wells, respectively. The dotted curve represents the People and Bean's model [54] developed for single heterostructures. The correspondent equations for h_c are written as:

$$h_c^{MB} = \frac{b(1 - v\cos^2\theta)}{c\pi f(1 + v)\cos\theta_1} \left(\ln\left(\frac{h_c^{MB}}{b}\right) + 1 \right), \qquad (2.2.1a)$$

for Matthews and Blakeslee's model [23], where, $c = 4$ is for quantum wells and $c = 8$ for single layer heterostructures. For People and Bean's model [54]:

$$h_c^{PB} = \frac{b(1 - v)}{32\pi(1 + v)f^2} \left(\ln\left(\frac{h_c^{PB}}{b}\right) \right) \qquad (2.2.1b)$$

where,

$$b = a(x)/2^{1/2} \qquad (2.2.2a)$$
$$f = (a(x) - a_0)/a_0 \qquad (2.2.2b)$$
$$v = C_{12}/(C_{11} + C_{12}) \qquad (2.2.2c)$$

$a(x)$ and a_0 are the lattice constants of the ternary alloy (x is the In fraction) and GaAs, respectively. b is the magnitude of the Burgers vector, v is the Poisson ratio, C_{ij} are the elastic constants, θ is the angle between the dislocation line and its Burgers vector, and θ_1 is the angle between the slip direction and that direction in the film plane which is perpendicular to the line of intersection of the slip plane and the interface. The parameters b, f and C_{ij} of the ternary alloy depend on the In content. They can be obtained from linear interpolation of the data given in Table 1.

Since the $In_x Ga_{1-x} As$ ternary alloy built-in strain in heterostructures is not relaxed the value of the strain in the alloy layer is the same of the lattice mismatch. As a result, the band structure can be calculated using the elastic properties of the material.

The elastic properties of cubic materials under stress are described by the matrix equation:

$$
\begin{bmatrix}
\varepsilon_{11} \\
\varepsilon_{22} \\
\varepsilon_{33} \\
\varepsilon_{12} \\
\varepsilon_{23} \\
\varepsilon_{31}
\end{bmatrix}
=
\begin{bmatrix}
S_{11} & S_{12} & S_{12} & 0 & 0 & 0 \\
S_{12} & S_{11} & S_{12} & 0 & 0 & 0 \\
S_{12} & S_{12} & S_{11} & 0 & 0 & 0 \\
0 & 0 & 0 & S_{44} & 0 & 0 \\
0 & 0 & 0 & 0 & S_{44} & 0 \\
0 & 0 & 0 & 0 & 0 & S_{44}
\end{bmatrix}
\begin{bmatrix}
\sigma_{11} \\
\sigma_{22} \\
\sigma_{33} \\
\sigma_{12} \\
\sigma_{23} \\
\sigma_{31}
\end{bmatrix}
\qquad (2.2.3)
$$

TABLE 1
Parameters for bulk GaAs and InAs*.

Material	a (Å)	$C_{11} \times 10^{11}$ (dyn/cm^2)	$C_{12} \times 10^{11}$ (dyn/cm^2)	$\mathcal{A}$ (eV)	$\mathcal{B}$ (eV)	Δ_0 (eV)	m_e (m_0)	m_{hh} (m_0)	m_{lh} (m_0)
GaAs	5.6533	12.11	5.48	-8.93^a	-1.76^a	0.341	0.067	0.45	0.087
InAs	6.0583	8.33	4.53	-6	-1.8	0.38	0.024	0.41	0.026

*) Landoldt-Börnstein, *Numerical Data and Functional Relationships in Science and Technology*, edited by O. Madelung (Springer-Verlag, Berlin, 1982), New Series, Vol. III/17a, pp.218 and 297.
a) From M. Chandrasekahr and F.H. Pollak, *Phys. Rev. B*, *15*, 2127 (1977).
b) From P. Wickboldt, E. Anastassakis, R. Sauer and M. Cardona, *Phys. Rev. B*, **35**, 1362 (1987).
c) From F. Cerdeira, C.J. Buchenauer, F.H. Pollak and M. Cardona, *Phys. Rev. B*, **5**, 580 (1972).

where ε_{ij} and σ_{ij} are the components of the strain and stress tensors, respectively. The index 1, 2, 3 refers to the crystallographic axis parallel to (100), (010), and (001), respectively. The $S_{ij}s$ are tensor elements of the elastic compliance constant, which are related to the elastic constants C_{ij} by the following equations:

$$S_{11} = \frac{C_{11} + C_{12}}{(C_{11} - C_{12})(C_{11} + 2C_{12})} \qquad (2.2.4a)$$

$$S_{12} = \frac{C_{12}}{(C_{11} - C_{12})(C_{11} + 2C_{12})} \qquad (2.2.4b)$$

$$S_{44} = \frac{1}{C_{44}} \qquad (2.2.4c)$$

The systems grown along (001) direction produce a tetrahedral distortion where the biaxial stress components are $\sigma_{11} = \sigma_{22} = \sigma$ and the other components are zero. In this case, the strain tensor components can be written as:

$$\varepsilon_{11} = \varepsilon_{22} = (S_{11} + S_{12})\sigma = \varepsilon \qquad (2.2.5a)$$

$$\varepsilon_{33} = 2S_{12}\sigma = -\alpha\varepsilon = \varepsilon_\perp \qquad (2.2.5b)$$

where

$$\alpha = -\frac{2S_{12}}{(S_{11} + S_{12})} = \frac{2C_{12}}{C_{11}}, \qquad (2.2.5c)$$

and other components of strain tensor are zero. The value of the parameter α is ~ 1 in both GaAs and In$_x$Ga$_{1-x}$As materials (see Table 1). In this situation, $\varepsilon_\perp \sim -\varepsilon$.

The biaxial strain configuration can be analyzed as a combination of a hydrostatic compression (or tension) minus an uniaxial tension (or compression) along the growth direction. The influence of the strain in the band structure, in the framework of the effective mass approximation, is given by Birus-Pikus Hamiltonian [56]. The evaluation of this Hamiltonian for the conduction band and the valence band at $k = 0$ gives the energy gap as a function of the biaxial strain [56]. The break of the degeneracy of the valence band at $k = 0$ yields two different energy gaps [57]:

(i) for the heavy hole band,

$$E_g^{HH} = E_g^o + \delta E^{HH} \tag{2.2.6a}$$

where E_g^o is the energy gap of the unperturbed crystal and

$$\delta E^{HH} = [A(2 - \alpha) + B(1 + \alpha)]\varepsilon \tag{2.2.6b}$$

and

(ii) for the light hole band,

$$E_g^{LH} = E_g^o + \delta E^{LH} \tag{2.2.6c}$$

where,

$$\delta E^{LH} = [A(2 - \alpha) - B(1 + \alpha)]\varepsilon + 2[B(1 + \alpha)\varepsilon]^2/\Delta_o \tag{2.2.6d}$$

and,

$$\varepsilon = (a_0 - a(x))/a(x). \tag{2.2.6e}$$

The parameters A and B are the hydrostatic and the shear deformation potentials, respectively, and Δ_o is the spin-orbit splitting energy. The last term in Eq. 2.2.6d is much smaller than the first two terms, and it is usually neglected. The hydrostatic stress contribution (the first term of Eq. 2.2.6b and Eq. 2.2.6d) affects only the band gap energy. Other terms (the second term of Eq. 2.2.6.b and the second and third terms of Eq. 2.2.6.d) contribute to the valence band and it is responsible to the degeneracy break at $k = 0$, which splits heavy- and light-hole bands. The direction of band shift depends on the sign of ε, i.e., if it is a compressive ($\varepsilon < 0$) or a tensile ($\varepsilon > 0$) strain. Figure 2 shows schematically the energy dispersion in the parabolic band approximation for different ε values. For $\varepsilon < 0$, the band gap is related to the heavy hole band and it is larger than for unstrained crystal ($\varepsilon = 0$). For $\varepsilon > 0$, the band gap is smaller and is related to the light hole band. For a pseudomorphic $In_xGa_{1-x}As$ layer grown on GaAs substrate, $\varepsilon < 0$, the $In_xGa_{1-x}As$ layer is compressed and the energy gap is larger than the unperturbed layer one.

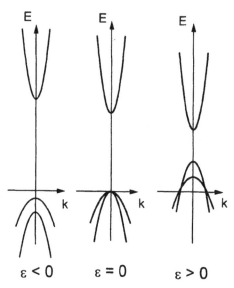

Fig. 2. Effects of biaxial strain on the energy dispersion of a direct gap material in the parabolic band approximation.

2.3   In/Ga Interdiffusion Effect

To study the interdiffusion effects in heterostructures we use the Fick's model. The evolution of the In (or Ga) composition profile at the interface with the thermal annealing time (t) is evaluated by solving the Fick's second law of diffusion, calculated along the growth direction (z-axis) [58]:

$$\frac{\partial}{\partial t}x(z, t) = D_i \frac{\partial^2}{\partial z^2}x(z, t) \qquad (2.3.1)$$

where, D_i is the interdiffusion coefficient and $x(z, t)$ is the In content at the z position relative to the center of the quantum well along the growth direction. The interdiffusion coefficient depends only on the annealing temperature. Before annealing the In content profile in the $In_x Ga_{1-x}As$/GaAs quantum well is considered to have a square-type shape: the GaAs barrier layers $x(|z| > L, t = 0) = 0$ and the ternary $In_x Ga_{1-x}As$ alloy layer with thickness $Lx(|z| < L, t = 0) = x_0$, where x_0 is the initial (as-grown) alloy In composition. The function $x(z, t = 0)$ is shown in Figure 3a. The solution of this type of equation can be obtained by using the method of the Laplace transformation [58]. Using the boundary condition given above and the continuity of $x(z, t)$ and $\lim_{x \to \pm\infty} x(z, t) = 0$ we obtain [57–59]

$$x(z, t) = \frac{x_0}{2}[\text{erf}(\xi_+) - \text{erf}(\xi_-)], \qquad (2.3.2a)$$

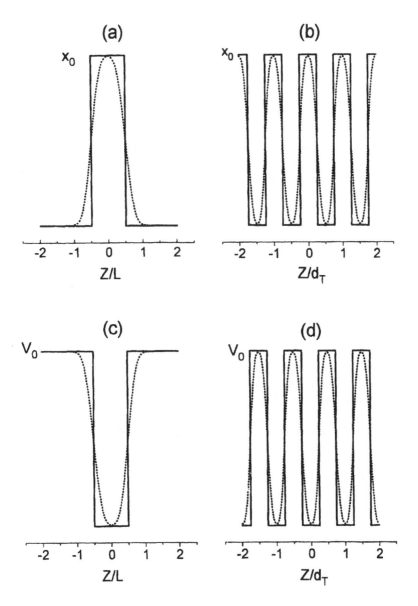

Fig. 3. Calculated In content and potential energy profiles along the growth direction-z before (solid line) and after (dotted line) thermal annealing: a) In content profile for single quantum well and b) for a superlattice. c) Potential energy profiles for a single quantum well and d) for a superlattice.

where,

$$\xi_\pm = \frac{|z| \pm \frac{L}{2}}{\sqrt{4D_i t}} \tag{2.3.2b}$$

and erf(y) is the error function given by:

$$\text{erf}(y) = \frac{2}{\sqrt{\pi}} \int_o^y \exp(-\eta^2) d\eta \tag{2.3.3}$$

Figure 3a shows a typical function x(z,t) before and after annealing at a given value of D_i. The square like shape of the In content profile is smoothed after annealing. The strength of the layer intermixing can be estimated from the diffusion length, which is defined as $L_D = (D_i t)^{1/2}$.

The interdiffusion coefficient D_i in $In_x Ga_{1-x}$As/GaAs systems exhibits an exponential behavior with the inverse of the annealing temperature (T), i e., an Arrhenius type equation [60]:

$$D_i(T) = D_o \exp\left[-\frac{E_a}{k_B T}\right], \tag{2.3.4}$$

where E_a is the activation energy, D_o is the pre-exponential factor and k_B is the Boltzmann constant.

We can extend the same analysis presented above to the study of the interdiffusion in superlattices. The periodicity of the alternated layers A and B, for example GaAs and $In_x Ga_{1-x}$As respectively, must be included in the calculation. In this case, $x(z, t)$ is a periodic function and has a period of $d_T = d_A + d_B$, where d_A and d_B are the thickness of the layers A and B, respectively, before thermal annealing. The initial condition, considering abrupt interfaces, is $x(d_T/2 > |z| > d_B/2, t = 0) = 0$ and $x(|z| < d_B/2, t = 0) = x_0$. The boundary condition in this case is $x(z, t) = x(z + nd_T, t)$, where n is an integer. To calculate the solution of Eq. 2.3.1 in a periodic system we can use the method of separation of variables. The solution is given by [27,29,57]:

$$x(z, t) = \frac{x_o d_B}{d_T} + \sum_{m=1}^{\infty} \frac{2x_o}{\pi m} \sin\left(\frac{k_m}{2} d_B\right) \cos(k_m z) \exp[-(k_m L_D)^2] \tag{2.3.5a}$$

where,

$$k_m = \frac{2\pi m}{d_T}. \tag{2.3.5b}$$

The convergence of the second term in the Eq. 2.3.5a is obtained for a maximum m around 10 [29,57]. Figure 3b shows the $x(z, t)$ for superlattices before and after annealing.

The In composition profiles described by Eq. 2.3.2 and Eq. 2.3.5 exhibit good agreement with experimental data for symmetric quantum wells and superlattices. There are other methods to calculate the composition profile such as the Green function model proposed by Homewood and Dunstan [61]. This model is more general and can be used in any distribution of alloy composition profile.

2.4   Transition Energy

The as-grown abrupt $In_x Ga_{1-x} As$/GaAs interface is smoothed after thermal annealing due to the interdiffusion of In and Ga atoms, as discussed above. The energy gap profile can be obtained by replacing the In concentration profile (given by the Eq. 2.3.2 for quantum wells and Eq. 2.3.5 for superlattices) into the equation of the composition dependent band gap energy evaluated in the Section 2.2 which includes the strain effect. An important parameter to be evaluated in the transition energy calculation is the band offset between GaAs and $In_x Ga_{1-x} As$ junctions. The band offset gives the discontinuity between the conduction and valence bands. This discontinuity determines the electron and hole confinement energies. The conduction band offset (Q_c) is defined as $Q_c = \Delta E_c / \Delta E_g$, where ΔE_c and ΔE_g are the conduction band and the band gap energy discontinuities between the GaAs and the $In_x Ga_{1-x} As$ layers, respectively. The values of Q_c were measured from optical experiments and are in the range of ~ 0.4 to ~ 0.8 [62–69]. The reported values are summarized in Figure 4. The parameter Q_c apparently varies smoothly with the In content. The dotted line shown in Figure 4 is a guide for the eyes in order to show the average behavior of experimental data, as suggested by Joyce et al. [67].

Some works determine Q_c using the fact that the band offset of the light hole band form either a type-I or a type-II interface [62–69]. The optical spectra related to each interface type give rise to distinct features. For the type-II interface, the light hole is localized in the GaAs layer, resulting in a decrease of the transition oscillator strength (indirect transition) compared to the type-I interface (direct transition). However, the Q_c values shown in Figure 4 yield small light hole confinement energy either in GaAs or $In_x Ga_{1-x} As$ layer, making the determination of Q_c values very imprecise. Moreover, the position of the light hole band depends strongly on the material parameter values, such as the deformation potentials and elastic constants. These parameter values are poorly known mainly for the InAs material, making the precise determination of Q_c even more difficult.

For a given Q_c value we can determine the potential energy profile of a quantum well or a superlattice for each value of D_i. The fitting parameter D_i, at a given annealing temperature and annealing time, is found from a direct comparison between calculated and experimental transition energies.

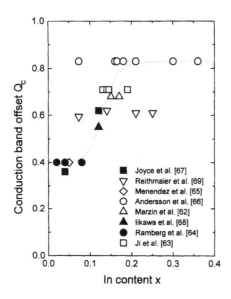

Fig. 4. Conduction band offset for $In_x Ga_{1-x} As/GaAs$ junctions as a function of the In content according to several authors.

The transition energies are calculated using a z-dependent effective mass Schrödinger equation given by [70,71]:

$$\left\{ -\frac{\hbar^2}{2} \frac{d}{dz} \left(\frac{1}{m_\alpha^*(z)} \frac{d}{dz} \right) + V_\alpha(z) \right\} \phi_{\alpha j}(z) = \lambda_\alpha \phi_{\alpha j}(z) \qquad (2.4.1)$$

where, α is the index for the electron (e) or heavy hole (h), $m_\alpha^*(z)$ is the effective mass obtained by the linear interpolation between GaAs and InAs data. $\lambda_{\alpha j}$ and $\Phi_{\alpha j}(z)$ are the energy and the envelope function at the subband j, respectively. The potential energies for electrons and holes are given by

$$V_e(z) = Q_c[E_g(z) - E_g(z = 0)] \qquad (2.4.2a)$$

$$V_h(z) = (1 - Q_c)[E_g(z) - E_g(z = 0)] \qquad (2.4.2b)$$

where, $E_g(z)$ is the energy gap, which depends on the In content profile (given by Eq. 2.1.1 and Eq. 2.2.6). The In content as a function of z is given by Eq. 2.3.2. The solution of the Eq. 2.4.1 is calculated numerically with wave function $\Phi_{\alpha j}(z) = 0$ when $z \to \pm\infty$ as boundary conditions.

Usually the In content in ternary alloys is smaller than $\sim 25\%$. For higher In contents the layer thickness is limited by the critical thickness, due to the strain relaxation effects. For small values of In composition the change of the effective mass is small. In this range, the effective mass variation between the GaAs and the ternary alloy with 25% of In content is 15% for electron and 2% for heavy hole. Since the effective masses of the alloy do not change significantly with In content an average effective mass, instead of $m^*(z)$, can be used. In this case, Eq. 2.4.1 can be simplified as [71]:

$$\left\{ -\frac{\hbar^2}{2} \frac{1}{\mu_\alpha^*} \frac{d^2}{dz^2} + V_\alpha(z) \right\} \phi_{\alpha j}(z) = \lambda_{\alpha j} \phi_{\alpha j}(z) \tag{2.4.4}$$

where,

$$\frac{1}{\mu_\alpha^*} = \int |\phi_{\alpha j}(z)|^2 \frac{1}{m_\alpha^*(z)} dz \tag{2.4.4b}$$

This approximation has been successfully applied to the layer intermixing in quantum wells [29,57,59]. The solution of the Eq. 2.4.4 is obtained numerically.

The Eq. 2.4.4 can also be applied to calculate the transition energies of periodic systems such as superlattices. The boundary condition is the periodicity of the envelope wavefunction, $\Phi_\alpha j(z) = \Phi_\alpha j(z + d_T)$. The solution of Eq. 2.4.4 in superlattices can be obtained by assuming the wavefunctions of the Eq. 2.4.4 for electrons or holes, at the Brillouin zone center along the growth direction z, as combination of sine or cosine functions:

$$\varphi_{\text{even}}(z) = \sum_{n=0} b_n \cos\left(\frac{2\pi n}{d_T} z\right) \tag{2.4.5a}$$

$$\varphi_{\text{odd}}(z) = \sum_{m=0} c_m \sin\left(\frac{2\pi m}{d_T} z\right) \tag{2.4.5b}$$

where φ_{even} and φ_{odd} are the wavefunctions for even ($j = 0, 2, 3, \ldots$) and odd states ($j = 1, 3, 5, \ldots$), respectively. Substituting equations (2.4.5) into (2.4.4) the following secular equation can be obtained:

(i) for even states:

$$\begin{bmatrix} J_{00}/2 & J_{01}/2 & J_{02}/2 & \cdots \\ J_{10} & R_1 + J_{11} & J_{12} & \cdots \\ J_{20} & J_{21} & R_2 + J_{22} & \cdots \\ \cdot & \cdot & \cdot & \end{bmatrix} \begin{bmatrix} b_0 \\ b_1 \\ b_3 \\ \cdot \end{bmatrix} = \lambda \begin{bmatrix} b_0 \\ b_1 \\ b_2 \\ \cdot \end{bmatrix} \tag{2.4.6a}$$

and ii) for odd states:

$$\begin{bmatrix} J_{00}'/2 & J_{01}'/2 & J_{02}'/2 & \cdots \\ J_{10}' & R_1 + J_{11}' & J_{12}' & \cdots \\ J_{20}' & J_{21}' & R_2 + J_{22}' & \cdots \\ \cdot & \cdot & \cdot & \end{bmatrix} \begin{bmatrix} c_0 \\ c_1 \\ c_3 \\ \cdot \end{bmatrix} = \lambda \begin{bmatrix} c_0 \\ c_1 \\ c_2 \\ \cdot \end{bmatrix} \tag{2.4.6b}$$

where,

$$J_{mn} = \frac{2}{d_T} \int_{-d_T/2}^{d_T/2} \cos\left(\frac{2\pi m}{d_T}z\right) V_\alpha(z) \cos\left(\frac{2\pi n}{d_T}z\right) dz \tag{2.4.6c}$$

$$J'_{mn} = \frac{2}{d_T} \int_{-d_T/2}^{d_T/2} \sin\left(\frac{2\pi m}{d_T}z\right) V_\alpha(z) \sin\left(\frac{2\pi n}{d_T}z\right) dz \tag{2.4.6d}$$

and

$$R_n = \frac{\hbar^2}{2m_\alpha^*} \left(\frac{2\pi n}{d_T}\right)^2 \tag{2.4.6e}$$

The solution of equations 2.4.6 is obtained numerically. The potential energy $V(z)$ in J_{mn} and J'_{mn} is also given by Eq. 2.4.2. The periodic function $V(z)$ is derived from Eq. 2.3.5 for $x(z, t)$. A fast convergence of the solution can be easily obtained for a matrix dimension equal 10 [29,57].

3   THERMALLY INDUCED EFFECTS

3.1   In/Ga Interdiffusion in Undoped Systems

The interdiffusion in $In_x Ga_{1-x}As/GaAs$ heterostructures occurs between the group III atoms, In and Ga. The interdiffusion process in this system is very similar to that observed in $Al_y Ga_{1-y}As/GaAs$. Most of the analysis used in $Al_y Ga_{1-y}As/GaAs$ systems can then be applied to $In_x Ga_{1-x}As/GaAs$ ones.

The degree of layer intermixing is described by the interdiffusion coefficient D_i and the activation energy E_a. Figure 5 shows the In/Ga interdiffusion coefficient in $In_x Ga_{1-x}As/GaAs$ heterostructures as a function of the inverse of the annealing temperature obtained by several authors using different experimental techniques [30,59,72,73–79]. A large spread of D_i values is clearly observed in this figure varying by almost two orders of magnitude at certain temperatures. Two distinct sets of data can be clearly seen in Figure 5 with an order of magnitude of difference between them. A great variation of activation energy ranging from ~ 1.5 eV to ~ 9 eV was obtained in these works. The main differences observed for D_i and E_a were attributed to the different experimental methods used for the annealing process. However, the understanding of the main mechanisms underlying the interdiffusion process remain unclear. Here, we present some possible mechanisms that may contribute to the interdiffusion process and the influence of the experimental environment conditions on the interdiffusion process.

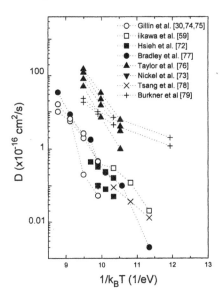

Fig. 5. In/Ga interdiffusion coefficient versus the inverse of the annealing temperature measured by several groups in $In_xGa_{1-x}As/GaAs$ heterostructures.

The interdiffusion process in heterostructures have exhibited a similar behavior to the self-diffusion of the host atom in bulk semiconductor materials. The self-diffusion can be described by the mechanism involving intrinsic defects such as vacancies, interstitials and anti-structures, or combinations of them [60]. The connection between intermixing and self-diffusion was discussed by Tan and Gösele [80] based on experimental results obtained for lattice matched $Al_yGa_{1-y}As/GaAs$ systems. Despite the similarity in several experimental results between self-diffusion in GaAs and interdiffusion in $Al_yGa_{1-y}As/GaAs$ and $In_xGa_{1-x}As/GaAs$ this connection is not fully understood.

The studies involving interdiffusion in $In_xGa_{1-x}As/GaAs$ systems reveal a strong influence of native defects such as vacancies and interstitials. The diffusion of these defects into the heterostructures strongly enhances the layer intermixing. The stoichiometric equilibrium of a III-V crystal at a given annealing temperature and atmosphere are responsible for the creation of native point defects. The concentration control of these defects depends on the experimental condition, as is described below.

A well known experimental condition to induce generation of native point defects in GaAs is the use of As or Ga atmosphere with different vapor pressure. At an As rich atmosphere conditions the most likely point defects

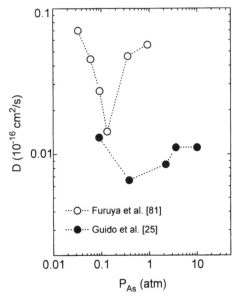

Fig. 6. Influence of the As vapor pressure to the Al/Ga interdiffusion coefficient obtained by Guido et al. [25] (solid circles) and Furuya et al. [81] (open circles) for $Al_yGa_{1-y}As/GaAs$ systems.

created close to the surface are: As interstitial (I_{As}), Ga vacancy (V_{Ga}) and As anti-site (As_{Ga}). At a Ga rich atmosphere the point defects are: Ga interstitial (I_{Ga}), As vacancy (V_{As}) and Ga anti-site (Ga_{As}). Several works involving self-diffusion and interdiffusion studies have been carried out in As and Ga atmospheres with different vapor pressures in order to study their influence on the diffusion process.

Guido et al. [25] and Furuya et al. [81] have independently shown that in $Al_yGa_{1-y}As/GaAs$ systems the interdiffusion coefficient varies with an arsenic vapor pressure. Figure 6 shows the Al/Ga interdiffusion coefficient as a function of As vapor pressure. They observed that the interdiffusion coefficient presents a minimum. At a low As vapor pressure, below the minimum of the curve in Figure 6, the diffusion of Ga interstitials from the surface is responsible for the increase of the interdiffusion coefficient. The interdiffusion mechanism involving Ga (Al) interstitials is the main layer intermixing process when compared to the mechanism involving other point defects created in that condition. At high pressures, the diffusion of Ga vacancies from the surface is dominant due to the As rich atmosphere condition which favors the creation of Ga vacancies. In this case, the main intermixing mechanism involves the Ga (Al) vacancies. These experiments show the competition between contributions of group III vacancies and interstitials to the interdiffusion coefficient. The minimum observed in interdiffusion coefficient as a function of As vapor pressure shown in Figure 6 can be explained by these mechanisms. The variation of the interdiffusion coefficient with the As vapor pressure suggests that the group III point defects, such as vacancies and interstitials, are candidates for the layer intermixing in $Al_y Ga_{1-y} As/GaAs$ systems. Other studies reveal the contribution of group III vacancies and interstitials to the Al/Ga interdiffusion process. Recently, Tan et al. [82] showed that the native point defect contribution to the Ga self-diffusion in GaAs is due to V_{Ga}^{3-} and I_{Ga}^{2+}. They compared the self-diffusion data with Al/Ga interdiffusion ones and concluded that both processes are nearly identical. Therefore, the Al/Ga interdiffusion mechanism also occurs via these native point defects. For further detail discussion on this subject see the review works of Deppe and Holonyak [20] and Harrison [21].

In the case of the interdiffusion in $In_x Ga_{1-x} As/GaAs$ systems the native point defects demonstrated to be an important element involved in intermixing mechanisms. Hsieh et al. [72] reported the effects of the As and Ga vapor overpressure on the interdiffusion in $In_x Ga_{1-x} As/GaAs$ systems. These authors used $In_x Ga_{1-x} As/GaAs$ single quantum wells with and without $Al_y Ga_{1-y} As$ cladding layer. They observed strong intermixing enhancement under As vapor overpressure for both types of samples and a slight variation under Ga overpressure condition. These results are shown in Figure 7. The enhancement of intermixing is attributed to the Ga vacancies diffused into the quantum well. The Ga vacancies were created at the surface and also via dislocations produced by the relaxation of the strain at a high As vapor pressure condition. The influence of the dislocations created during the strain relaxation is reported by Major et al. [19]. The interdiffusion coefficient for the sample with an $Al_y Ga_{1-y} As$ cladding layer presents larger values compared to those data measured in the sample grown without it, as it is clearly shown in Figure 7. This effect was attributed to the Al diffusion into the $In_x Ga_{1-x} As/GaAs$ interface or the movement of point defects due to the $Al_y Ga_{1-y} As$ layer.

Hsieh et al. also studied the influence of dislocations generated due to the strain relaxation in the interdiffusion process for $In_x Ga_{1-x} As/GaAs$ systems with different In content [72,83]. Two atmosphere conditions (As and Ga vapor overpressure) were used in order to induce or inhibit the creation of point defects. The In/Ga interdiffusion coefficients with different In content and As and Ga atmosphere conditions are shown in Figure 8. For an As overpressure condition the values of interdiffusion coefficients are larger than those obtained under Ga overpressure condition. The dependence of the interdiffusion coefficient with In content presents an opposite behavior for both atmosphere species. At an As overpressure the main mechanisms responsible for the increase of the layer intermixing with the In content

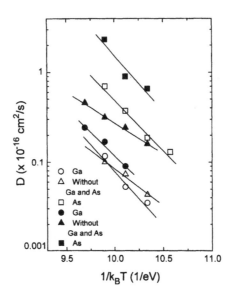

Fig. 7. Influence of As and Ga overpressure to the In/Ga interdiffusion coefficient obtained for pseudomorphic $In_{0.15}Ga_{0.85}As/GaAs$ (open symbols) and $In_{0.15}Ga_{0.85}As/GaAs/Al_{0.35}Ga_{0.65}As$ (solid symbols) quantum wells. After Hsieh et al. [72].

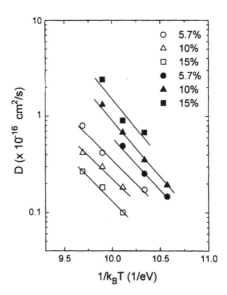

Fig. 8. Dependence of the In/Ga interdiffusion coefficient with In content for a pseudomorphic $In_xGa_{1-x}As/GaAs$ quantum well. The solid symbols are data measured under an As overpressure condition and the open symbols under a Ga overpressure condition [72, 83].

are the presence of Ga vacancies generated close to the surface and the dislocation created due to the strain relaxation for higher In contents. At a Ga overpressure condition the interdiffusion coefficient decreases with the increase of In content. This result can not be explained by the mechanism used above to interpret the As overpressure data. In this case another mechanism may be involved in intermixing for a Ga overpressure condition.

Usually $In_x Ga_{1-x} As$/GaAs heterostructures is capped with a GaAs layer. The GaAs surface at annealing conditions is unstable and it must be protected by a dielectric layer (silicon nitride or silicon oxide) or performed at an As atmosphere in order to avoid the degradation of the surface by As atom evaporation. The influence of the dielectric layer on the interdiffusion is well known and is discussed by Choi in chapter 1 of this book. We limit our discussion here to its application to $In_x Ga_{1-x} As$/GaAs systems.

The SiO_2 dielectric protection layer used to cover the samples during the thermal annealing process has been also used as a tool to induce the generation of Ga vacancies in the GaAs capped heterostructures. The SiO_2 deposited on GaAs has the capability to absorb the Ga atoms during the thermal annealing. That is a consequence of the higher Ga diffusion in SiO_2 [84]. This results in the generation of Ga vacancies at the $GaAs/SiO_2$ interface. The Si diffusion into GaAs layer is less efficient than the Ga diffusion into SiO_2 [85,86]. The dielectric Si_3N_4 is also used as a sample protector during the heat treatment process. The Si_3N_4 material is less efficient to create Ga vacancies when compared to SiO_2 one [85,87]. However, the former is normally used in several heterostructures to avoid the interdiffusion enhancement. Deppe et al. [86] reported the application of the SiO_2 cap layer in the Al/Ga interdiffusion study of a $GaAs/Al_y Ga_{1-y} As$ heterostructure. They showed that the intermixing is strongly enhanced during the annealing process when SiO_2 is used as the cap layer. They did not observe a significant variation in the interdiffusion coefficient when Si_3N_4 layer was used. Several works [17–19,25,31,33,79,88–91] showed the influence of the dielectric capping layer on the interdiffusion of GaAs based heterostructures. In $In_x Ga_{1-x} As$/GaAs systems [19,79,89], the SiO_2 capping layer also enhances the layer intermixing. Figure 9 shows the In/Ga interdiffusion coefficient for $In_{0.35} Ga_{0.65} As$/GaAs quantum wells using SiO_2 and GaAs capping layer obtained by Börkner et al. [79]. Using SiO_2 capping layer the interdiffusion coefficient values are almost an order of magnitude bigger compared to those using GaAs one. Samples with GaAs capping layer and different number of wells (4 and 9) exhibited essentially the same interdiffusion coefficient values (see Figure 9). The difference observed in the interdiffusion coefficient values for samples with SiO_2 and GaAs cap layers is attributed to the diffusion of the Ga vacancies created at the $GaAs/SiO_2$ interface. The presence of the Ga vacancies in the vicinity of the $In_x Ga_{1-x} As$/GaAs interfaces may contribute to a faster layer intermixing. This result shows the strong influence of the

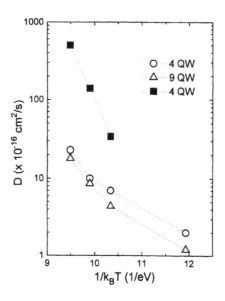

Fig. 9. In/Ga interdiffusion coefficient for $In_{0.35}Ga_{0.65}As/GaAs$ multi-quantum wells capped either with a GaAs (open symbols) or with a SiO_2 (solid symbols) layer. After Bürkner et al. [79].

group III vacancies on the interdiffusion mechanism in $In_xGa_{1-x}As/GaAs$ heterostructures.

Another indication of the influence of Ga vacancies on the interdiffusion process involves the use of a capping or buffer layer grown at low temperature GaAs (LT-GaAs) [78,83,92–96]. The GaAs related material grown at low temperatures, below the usual growth temperature ($\sim 550°C$), exhibits high resistivity properties. The high resistivity of LT-GaAs layers is obtained after heat treatment and it is attributed to the deep levels, due to the native point defects generated during the growth and after annealing [97, 98]. They obtained concentrations of defects as high as 10^{19} cm^{-3} for growth temperature of 200°C [98–101]. The LT-GaAs layer is usually grown under arsenic rich condition. The As rich atmosphere favors the formation of As antisite, As interstitial and Ga vacancy. The use of LT-GaAs in heterostructures is believed to be an efficient source of Ga vacancies during the thermal annealing process. The layer intermixing enhancement in $In_xGa_{1-x}As/GaAs$ [78] and $Al_yGa_{1-y}As/GaAs$ [83, 92–94] heterostructures was observed in the presence of a LT-GaAs layer. Figure 10 shows the interdiffusion coefficient obtained by Tsang et al. [78] for $In_{0.2}Ga_{0.8}As/GaAs$ quantum wells with and without a LT-GaAs capping layer. The interdiffusion coefficient for samples with a LT-GaAs layer increases by almost two orders

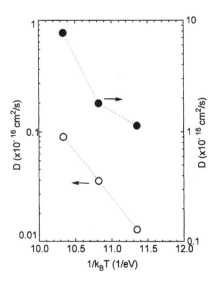

Fig. 10. In/Ga interdiffusion coefficient obtained for a pseudomorphic $In_xGa_{1-x}As/GaAs$ quantum well grown with (solid) and without (open) LT-GaAs layer. After Tsang et al. [78].

of magnitude. This enhancement is attributed to the diffusion of Ga vacancies from LT-GaAs into the quantum wells. These results is another indication that group III vacancies play an important role in the interdiffusion mechanism in $In_xGa_{1-x}As/GaAs$ heterostructures.

3.2   Impurity Induced Interdiffusion

Laidig et al. [12] reported the strong enhancement of layer intermixing in Zn diffused $Al_yGa_{1-y}As/GaAs$ quantum wells. Several other works also dedicated to the study of interdiffusion effects using different impurity species. An extensive discussion concerning impurity induced interdiffusion in III-V heterostructure systems was carried out by Deppe and Holonyak [20] and Harrison [21].

A theoretical model to explain the impurity induced disorder is based on the Fermi level position in the material band gap [102]. As discussed above, the interdiffusion mechanism in $In_xGa_{1-x}As/GaAs$ heterostructures involves native point defects of the group III. The density of native defect charge states depends on the Fermi energy of the crystal. The position of the Fermi level is related to the density of donor or acceptor impurities. There is a direct correlation between defect charge state concentration and the impurity concentration. Recently, Tan and Gösele [80] proposed that the interdiffusion in $Al_yGa_{1-y}As/GaAs$ heterostructures is due to the group III vacancies in n-type doped systems and group III interstitials in p-type doped ones. The authors also suggested that the triply negatively charged group III vacancies and the positively charged group III interstitials are the key elements contributing to the layer intermixing enhancement. The interpretation proposed by Tan and Gösele [80] is consistent with calculations of the absolute formation energy developed by Zhang and Northrup [103] for Ga self-diffusion in GaAs bulk layer. Zhang and Northrup showed that the Ga self-diffusion via V_{Ga}^{3-} is energetically favorable for n-type structure and via I_{Ga}^{3+} for p-type ones. The authors suggested that this model can also be applied to the interdiffusion in $Al_yGa_{1-y}As/GaAs$ heterostructures.

In $In_xGa_{1-x}As/GaAs$ systems, Laidig et al. [36] studied the Zn impurity induced interdiffusion process in $In_xGa_{1-x}As/GaAs$ superlattices. The layer intermixing observed in Zn diffused samples was faster than without Zn diffusion. The Zn atoms in GaAs can act either as an acceptor impurity substituting group III (Zn_{Ga}) or as a donorlike interstitials (Zn_i). The Zn diffusion mechanism in GaAs involves an interstitial-substitutional process [60, 104, 105]. Gösele and Morehead [106] proposed that the interstitial Zn moving into a group III lattice site creates group III interstitial through a "kick-out" mechanism described by the following reaction:

$$Zn_i^{++} \rightleftarrows Zn_{III}^- + I_{III}^+ + 2h^+$$

where Zn_i^{++}, Zn_{III}^- and I_{III}^+ represent the charge state of the Zn interstitial, the Zn substitutional and the group III interstitial, respectively, and "h" represents the free hole. In p-type GaAs material the group III interstitials are positively charged. A high concentration of the group III interstitials can be found in the region where Zn diffusion occurs due to the "kick out" mechanisms. This explains the high disorder observed in $In_xGa_{1-x}As/GaAs$ superlattice with Zn diffusion [36,107]. The layer intermixing enhancement is dominated by the mechanism via group III interstitial as suggested by Tan and Gösele [80] and Zhang and Northrup [103] for p-type $Al_yGa_{1-y}As/GaAs$ systems.

Hsieh et al. [72] also studied Zn induced interdiffusion in $In_{0.15}Ga_{0.85}As/GaAs$ quantum wells under different annealing conditions. The samples were exposed to a $ZnAs_2$ overpressure environment with and without Ga or As vapor overpressure. At heat treatments under a $ZnAs_2$ overpressure they observed partial intermixing of the quantum wells. The addition of Ga overpressure enhances the layer intermixing. When the Ga vapor overpressure was substituted by the As vapor overpressure the structure presented a more stable behavior during the heat treatment. They explained these results based on the dependence of the interdiffusion process on the Zn interstitial concentration. The latter can be controlled by changing the

As or Ga overpressure conditions. At an As overpressure condition the Ga vacancy concentration increases. In this situation most of interstitial Zn occupy the Ga vacancies becoming Zn substitutional. For small interstitial Zn, concentration the creation of group III interstitial, via "kick-out" mechanism discussed above, decreases. As a result, the interdiffusion process becomes less efficient. At a Ga overpressure condition an opposite behavior was observed. The Ga vacancy concentration is small and the interstitial Zn concentration is kept unchanged. In this case, the Zn diffusion occurs at a faster ratio resulting in a faster layer intermixing induced by the presence of group III interstitials.

Gillin et al. [75,108,109] reported the layer intermixing induced by Si and Be atoms in $In_x Ga_{1-x} As/GaAs$ systems and also by Si atoms in $Al_y Ga_{1-y} As/GaAs$ systems. The impurities were introduced in the samples during the growth process. The results are shown in Figure 11. They did not observe any significant layer intermixing enhancement in n-type samples for Si concentration lower than 10^{18} cm^{-3} (for both systems, $In_x Ga_{1-x} As/GaAs$ and $Al_y Ga_{1-y} As/GaAs$) and also in p-type samples for Be concentration lower than 10^{19} cm^{-3} (for $In_x Ga_{1-x} As/GaAs$) [75]. The interdiffusion coefficient variation with the impurity concentration obtained is considerably smaller than those observed by Mey et al. [110] in AlAs/GaAs superlattices. The interdiffusion coefficient estimated from the results obtained by Mey et al. [110] is expected to vary by almost two orders of magnitude for the same range of the Si concentration. However, Gillin et al. observed only a factor ~ 2 for Si concentration 10^{18} cm^{-3} (see Figure 11). Gillin et al. calculated the Fermi energy in $Al_{0.2} Ga_{0.8} As$ system and showed that the Fermi level for $N_{Si} = 10^{18}$ cm^{-3} is considerably above the expected intrinsic Fermi level (10^{16} cm^{-3}) at 1000°C. In this case, an enhancement of the interdiffusion process is expected to happen based on the model proposed by Tan and Gösele [80]. Gillin et al. [75] suggested that some additional effect may drive the interdiffusion process instead the Fermi level effect.

Other works involving impurity induced disorder in $In_x Ga_{1-x} As/GaAs$ are performed with ion implantation techniques. The ion implantation is widely used in device applications due to its high impurity penetration control. The damage created during the implantation [111–114] is recovered after annealing and the implanted impurities become active in the crystal resulting in a n- or p-type doped region, depending on the impurity species. Myers et al. [111–113] studied the implantation damage in $In_x Ga_{1-x} As/GaAs$ superlattices using different ion species: Si, Zn, N, and Ar. They observed that the implanted ions induce a stress in the crystal. The in-plane compressive stress at the saturation condition results in an average stress value of 5×10^9 dyn/cm^2 for Si, Zn, and N ions [111]. This value is about 4 times smaller than that compared to the lattice mismatch stress of 1.4% for a 20% of In content superlattices.

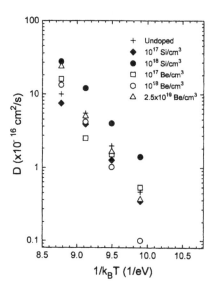

Fig. 11. In/Ga interdiffusion coefficient obtained for $In_{0.2}Ga_{0.8}As/GaAs$ quantum wells for several Si and Be concentration. After Gillin et al. [75].

Bradley et al. [77] reported results of the interdiffusion studies in $In_xGa_{1-x}As/GaAs$ implanted with As, Ga and Kr. The interdiffusion coefficients obtained in this study are shown in Figure 12. They observed that the implantation with Ga ions did not affect the layer intermixing when compared to an unimplanted sample. For As and Kr ion implanted samples a slight enhancement of the interdiffusion coefficients was observed (see Figure 12). They also observed that all the three ions affect the interdiffusion process at the initial stage ($t = 0$), i.e., the interdiffusion length is not zero at $t = 0$. This is attributed to implantation annealing effects. The results obtained for the As implanted sample can be described by three different stages: i) for t = 0 initial state, as discussed above, ii) an enhanced intermixing region and iii) a steady-state regimen. The second stage was attributed to the lattice accommodation due to the motion of As atoms into to group V lattice sites, giving rise to group III vacancies. The excess of group III vacancies is the main responsible for the disorder enhancement. In the steady-state region, the interdiffusion coefficients exhibited the same values as those obtained in Kr implanted samples. The activation energy obtained for all the three ions was the same: $E_a = 3.7$ eV.

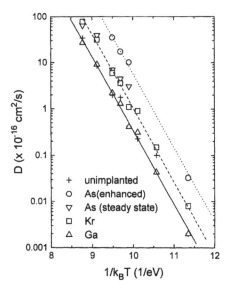

Fig. 12. In/Ga interdiffusion coefficient obtained for Ga, Kr and As ions implanted in $In_{0.2}Ga_{0.8}As/GaAs$ quantum wells. After Bradley et al. [77].

3.3   Strain Effects on the Layer Interdiffusion and the Lattice Relaxation

Several works were devoted to the study of the layer intermixing induced by the misfit strain in strained layer heterostructures such as, $In_xGa_{1-x}As/InP$ [26,115,116], SiGe/Si [117–120], and $In_xGa_{1-x}As/GaAs$ [116,121,122]. Recently, Cowern et al. [120] proposed an exponential dependence of the interdiffusion coefficient on the misfit strain to explain the layer intermixing in strained layer SiGe/Si systems. This study was based on the linear dependence of the native point defect (self-interstitial and vacancy) formation energy in Si with the strain. This leads to a linear dependence of Si diffusion activation energy with the strain. In this case, the diffusion coefficient will have an exponential dependence on the strain (or on the alloy composition). The Ge compositional profiles obtained by SIMS [120] agree quantitatively with the proposed model. The authors demonstrated that the composition profile calculated by the new model exhibited a better agreement with the experimental data when compared to Fick's model.

Recently, Ryu et al. [122] extended Cowern et al.'s model to the $In_xGa_{1-x}As/GaAs$ systems and compared with the experimental data measured from low temperature photoluminescence. They observe a good agreement between their model and the experimental data and they also demonstrate that Fick's model can not be confidently applied to this case. Ryu et al. [122] also applied their calculation to the previous data reported by Gillin et al. [74]. In this case, this calculations also give a good agreement. Gillin et al. [74] observed that their experimental results follows two stages: the first occurs at the annealing time $t = 0$ (similar to that observed in implanted samples [77]) and the second stage occurs at a steady-state regimen. The first stage is a fast diffusion due to the presence of defects introduced during the growth and the second one is the usual steady-state diffusion which obeys Fick's law. In this case, the interdiffusion length can be written as: $L_D^2 = 4Dt + (L_D^2)_0$, where $(L_D^2)_0$ is the initial interdiffusion ($t = 0$) and D is the interdiffusion coefficient at the steady-state. In a recent work, Gillin and Dunstan [121] showed that the strain does not significantly contribute to the interdiffusion process in $In_x Ga_{1-x} As/GaAs$ systems, that is due to the strain energy per atom is about three orders of magnitude lower than the diffusion activation energy.

The different interpretations of the interdiffusion process in strained layer systems show that it still is a topic that must be systematically studied. However, the contributions of these works are important to future investigations.

4   APPLICATION

The advance of the interdiffusion studies in semiconductor heterostructures resulted in an improvement on the performance of optical and electronic devices. The degree of layer intermixing in heterostructures can be controlled using an appropriate environment condition to produce a faster or stable layer intermixing. The interdiffusion control can be done using different methods: i) the environment atmosphere and its vapor pressure; ii) capping layer with a dielectric material; and iii) impurity by diffusion or ion implantation. Each of these methods and its effects were discussed in Section 3. The main applications involving the interdiffusion process were successfully obtained in optical devices such as quantum well lasers, waveguides and optical modulators. We present here some remarkable technological applications based on interdiffusion effects in $In_x Ga_{1-x} As/GaAs$ systems.

Few works have been dedicated to the technological applications of the interdiffusion effect in $In_x Ga_{1-x} As/GaAs$ heterostructures. They are related to the selective layer intermixing in laser diodes, emitting at ~ 1 μm, based on the impurity induced interdiffusion effects. The layer intermixed can be selectively controlled using stripe lasers [13–17,19] or coupled stripe laser arrays [18].

One of the first impurity induced disordered $In_x Ga_{1-x} As/GaAs$ heterostructure laser was reported by Yang et al. [13]. They used Si_3N_4 mask to protect the sample surface and ZnA_2 and arsenic vapor overpressure to induce layer intermixing. After heat treatment the layers in the region without dielectric was completely mixed and the protected region remained unchanged. The mixed layers produce different refractive index, compared to the unchanged active region of the laser, giving rise to a lateral optical waveguide. High quality lasers with reduced threshold currents and high efficiency were obtained using the layer intermixing properties.

Major et al. [17] used SiO_2 and Si_3N_4 as cap layers to select regions of the laser structure. As discussed in Section 3, during the heat treatment the region covered by SiO_2 produces faster group III atoms (In, Ga and Al) intermixing compared to the region protected by Si_3N_4. The laser exhibits threshold current as low as 7 mA, differential quantum efficiency of 41% per facet and output power of 20 mW/facet. Major et al. [18] also reported a high power coupled stripe laser arrays. They used the same layer intermixing method of a previous work [17]. In this case the interdiffusion condition was controlled in order to obtain a high coupling efficiency between the stripe lasers. The obtained threshold current was 75 mA, with a differential quantum efficiency of 26% per facet and an output power of 280 mW for both uncoated facets.

Zou et al. [14,15] also employed impurity induced disorder heterostructure stripe lasers to obtain very low threshold currents. A thin Si layer was selectively deposited on the structure in order to obtain complete layer intermixing. The wafer was capped with a SiO_2 layer. The region covered by the Si layer are completely mixed after heat treatment due to the Si diffusion. The stripe region, however, the intermixing process is less affected. The lasers obtained with this process exhibited threshold current of 2.2 mA with a differential quantum efficiency of 73% for both facets.

These examples show that $In_x Ga_{1-x}As/GaAs$ quantum well lasers can be successfully fabricated using impurity induced disordered heterostructures. The interdiffusion can be easily monitored by dielectric, annealing time, temperature, and impurities in laser structures. The interdiffusion process becomes also an additional parameter to control the emission wavelength, since the interdiffusion effects induces a blue shift.

5 CONCLUSION

The interdiffusion studies of lattice mismatched $In_x Ga_{1-x}As/GaAs$ systems contributed to a new subject: the strain effect. This is an additional parameter that can affect the physical properties, mainly the energy band gap structure of the system during the heat treatment. It also introduces defects (misfit dislocations) that may limit the device performance. These studies have also contributed to a better understanding of the diffusion phenomena. Several experimental works demonstrated that the interdiffusion mechanisms in $In_x Ga_{1-x} As/GaAs$ heterostructures involve two native point defects: group III vacancies and interstitials. The group III vacancies dominate the interdiffusion process for both intrinsic and n-type doped structures, and the group III interstitials for p-type ones. Although the theoretical aspects of the interdiffusion mechanisms are not fully understood, experimentally the layer intermixing control is well dominated. The layer intermixing in semiconductor heterostructures can be monitored by controlling the environment condition such as the capping layer, the atmosphere, the impurity diffusion, the ion implantation, or in combinations of them. These methods were successfully applied to optical device fabrication.

References

1. G.C. Osbourn, *IEEE Journal of Quantum Electronics*, **QE-22**, 1677 (1986).
2. T.E. Zipperian and T.J. Drummond, *Electronic Lett.*, **21**, 823 (1985).
3. J.J. Rosenberg, M. Benlamri, P.D. Kirchner, J.M. Woodall, and G.D. Petti, *IEEE Electron Device Lett.*, **21**, 491 (1985).
4. A.A. Ketterson, W.T. Masselink, W.F. Kopp, H. Morcoÿ87, and K.R. Gleason, *IEEE Transactions on Electron Devices*, **ED-33**, 564 (1986).
5. H. Morcoÿ87 and H. Unler, *Semiconductors and Semimetals*, vol. 24, volume editor R. Dingle, edited by R.K. Willardson and A.C. Beer (Academic Press, New York, 1987), p. 135.
6. M.J. Ludowise, W.T. Dietze, C.R. Lewis, M.D. Camras, N. Holonyak,Jr., B.K. Fuller, and M.A. Nixon, *Appl. Phys. Lett.*, **42**, 487 (1983).
7. W.D. Laidig, P.J. Caldwell, Y.F. Lin, and C.K. Pen, *Appl. Phys. Lett.*, **44**, 653 (1984).
8. J.N. Baillargeon, P.K. York, C.A. Zmudzinski, G.E. Fernández, K.J. Beernink, and J.J. Coleman, *Appl. Phys. Lett.*, **53**, 457 (1988).
9. P.K. York, K.J. Beernink, G.E. Fernández, and J.J. Coleman, *Appl. Phys. Lett.*, **54**, 499 (1989).
10. S.E. Fischer, D. Fekete, G.B. Feak, and J.M. Ballantyne, *Appl. Phys. Lett.*, **50**, 714 (1987).
11. D.P. Bour, D.B. Gilbert, L. Elbaum, and M.G. Harvey, *Appl. Phys. Lett.*, **53**, 2371 (1988).
12. W.D. Laidig, N. Holonyak,Jr., M.D. Camras, K. Hess, J.J. Coleman, P.D. Dapkus, and J. Bardeen, *Appl. Phys. Lett.*, **38**, 776 (1981).
13. Y.J. Yang, K.Y. Hsieh, and R.M. Kolbas, *Appl. Phys. Lett.*, **51**, 215 (1987).
14. W.X. Zou, J.L. Merz, R.J. Fu, and C.S. Hong, *Electronics Letters*, **27**, 1243 (1991).
15. W.X. Zou, K.-K. Law, J.L. Merz, R.J. Fu, and C.S. Hong, *Appl. Phys. Lett.*, **59**, 3375 (1991). 16 R.B. Bylsma, W.S. Hobson, J. Lopata, G.J. Zydzik, M. Geva, M.T. Asom, S.J. Pearton, P.M. Thomas, P.M. Bridenbaugh, M.A. Washington, D.D. Roccasecca, and D.P. Wilt, *J. Appl. Phys.*, **76**, 590 (1994).
17. J.S. Major, Jr., D.C. Hall, L.J. Guido, N. Holonyak,Jr., W. Stutius, P. Gravilovic, and J.E. Williams, *Appl. Phys. Lett.*, **54**, 913 (1989).
18. J.S. Major, Jr., D.C. Hall, L.J. Guido, N. Holonyak,Jr., P. Gravilovic, K. Meehan, J.E. Williams, and W. Stutius, *Appl. Phys. Lett.*, **55**, 271 (1989).
19. J.S. Major, Jr., L.J. Guido, N. Holonyak,Jr., K.C. Hsieh, E.J. Vesely, D.W. Nam, D.C. Hall and J.E. Baker, *J. of Electronic Materials*, **19**, 59 (1990).
20. D.G. Deppe and H. Holonyak, Jr., *J. Appl. Phys.*, **64**, R93 (1988).
21. I. Harrison, *J. of Materials Science: Material in Electronics*, **4**, 1 (1993).
22. J.H. van der Merwe, *J. Appl. Phys.*, **34**, 123 (1963).
23. J.W. Matthews and A.E. Blakeslee, *J. Crystal Growth*, **27**, 118 (1974).
24. T.E. Schlesinger and T. Kuech, *Appl. Phys. Lett.*, **49**, 519 (1986).
25. L.J. Guido, N. Holonjak Jr., K.C. Hsieh, R.W. Kaliski, W.E. Plano, R.D. Burnham, R.L. Thornton, J.E. Epler and T.L. Paoli, *J. Appl. Phys.*, **61**, 1372 (1987).
26. H. Temkin, S.N.G. Chu, M.B. Panish, and R.A. Logan, *Appl. Phys. Lett.*, **50**, 956 (1987).

27. J.-C. Lee, T.E. Schlesinger, and T.F. Kuech, *J. Vac. Technol.*, **B5**, 1187 (1987).
28. K. Kash, B. Tell, P. Grabbe, E.a. Dobsz, H.G. Craighead, and M.C. Tamargo, *J. Appl. Phys.*, **63**, 190 (1988).
29. F. Iikawa, P. Motisuke, F. Cerdeira, M.A. Sacilotti, R.A. Masut, and A.P. Roth, *Superlattices and Microstructures*, **5**, 273 (1989).
30. W. P. Gillin, Y.S. Tsang, N.J. Whitehead, K.P. Homewood, B.J. Sealy, M.T. Emeny, and C.R. Whitehouse, *Appl. Phys. Lett.*, **56**, 1116 (1990).
31. J.D. Ralston, S. O'Brien, G.W. Wicks, and L.F. Eastman, *Appl. Phys. Lett.*, **52**, 1511 (1988).
32. P.J. Hughes, E.H. Li, and B.L. Weiss, *J. Vac. Sci. Technol. B* **13**, 2276 (1995).
33. S. O'Brien, D.P. Bour, and J.R. Shealy, *Appl. Phys. Lett.*, **53**, 1859 (1988).
34. H. Peyre, F. Alsina, J. Camassel, J. Pascual, and R.W. Glew, *J. Appl. Phys.*, **73**, 3760 (1993).
35. T. Venkatesan, S.A. Schwarz, D.M. Hwang, R. Bhat, M. Koza, H.W. Yoon, P. Mei, Y. Arakawa, and A. Yariv, *Appl. Phys. Lett.*, **49**, 701 (1986).
36. W.D. Laidig, J.W. Lee, P.K. Simpson, and S.M. Beair, *J. Appl. Phys.*, **54**, 6382 (1983).
37. L.L. Chang and A. Koma, *Appl. Phys. Lett.*, **29**, 138 (1976).
38. E.V.K. Rao, P. Ossart, F. Alexandre, and H. Thibierge, *Appl. Phys. Lett.*, **50**, 588 (1987).
39. K.-H. Goetz, D. Bimberg, H. Jurgensen, J. Selders, A.V. Solomonov, G.F. Glinskii, and M. Razechi, *J. Appl. Phys*, **54**, 4543 (1983).
40. Y.-T. Leu, F.A. Thiel, H. Scheiber,Jr., J.J. Rubin, B.I. Miller, and K.J. Bachmann, *J. of Electronic Materials*, **8**, 663 (1979).
41. R.E. Nahory, M.A. Pollak, W.D. Johnston, and R.L. Barns, *Appl. Phys. Lett.*, **33**, 659 (1978).
42. W. Porod and D.K. Ferry, *Phys. Rev. B* **27**, 2587 (1983).
43. M. Gal, P.C. Taylor, B.F. Usher, and P.J. Orders, *J. Appl. Phys.*, **62**, 3898 (1987).
44. P.J. Orders and B.F. Usher, *Appl. Phys. Lett.*, **50**,980 (1987).
45. J. Zou, B.F. Usher, D.J. H. Cockayne, and R. Glaisher, *J. of Electronic Materials* **20**, 855 (1991).
46. J. Zou, D.J.H. Cockayne, B.F. Usher, *J. Appl. Phys.*, **73**, 619 (1993).
47. I.J. Fritz, S.T. Picraux, L.R. Dawson, T.J. Drummond, W.D. Laidig, and N.G. Anderson, *Appl. Phys. Lett.*, **46**, 967 (1985).
48. I.J. Fritz, P.L. Gourley, and L.R. Dawson, *Appl. Phys. Lett.*, **51**, 1004 (1987).
49. J.-P. Reithmaier, H. Cerva, and R. Lörsch, *Appl. Phys. Lett.*, **54**, 48 (1989).
50. P.L. Gourley, I.J. Fritz, and L.R. Dawson, *Appl. Phys. Lett.*, **52**, 377 (1988).
51. S.-L. Weng, *J. Appl. Phys.*, **66**, 2217 (1989).
52. R.G. Andersson, Z.G. Chen, V.D. Kulakovskii, A. Uddin, and J.T. Vallin, *Appl. Phys. Lett.*, **51**, 752 (1987).
53. I.J. Fritz, *Appl. Phys. Lett.*, **51**, 1080 (1987).
54. R. People and J.C. Bean, *Appl. Phys. Lett.*, **47**, 322 (1985); ibid *Appl. Phys. Lett.*, **49**, 229 (1986).
55. B.W. Dodson and J.Y. Tsao, *Appl. Phys. Lett.*, **51**, 1325 (1987).
56. F.H. Pollak and M. Cardona, 172, 816 (1968).
57. F. Iikawa, PhD Thesis in University of Campinas, 1988.
58. J. Crank, *The Mathematics of Diffusion*, (Clarendon Press, Oxford, 1975) second edition, cap. 1 and 2.
59. F. Iikawa, P. Motisuke, J.A. Brum, M.A. Sacilotti, A.P. Roth, and R.A. Masut, *J. Crystal Growth*, **93**, 336 (1988).
60. Don L. Kendall, *Semiconductor and Semimetals — Physics of III-V Compounds*, Vol. 4, edited by R.K. Willardson and A.C. Beer (Academic Press, New York 1968) p. 163.
61. K.P. Homewood and D.J. Dunstan, *J. Appl. Phys.*, **69**, 7581 (1991).
62. J.-Y. Marzin, M.N. Charasse, and B. Semage, *Phys. Rev. B*, **31**, 8298 (1985).
63. G. Ji, D. Huang, U.K. Reddy, T.S. Henderson, R. Houdré, and H. Morkoç, *J. Appl. Phys.*, **62**, 3366 (1987).
64. L.P. Ramberg, P.M. Enquist, Y.-K. Chen, F.E. Najjar, L.F. Eastman, E.A. Fitzgerald, and K.L. Kavanagh, *J. Appl. Phys.*, **61**, 1234 (1987).
65. J. Menéndez, A. Pinczuk, D.J. Werder, S.K. Sputz, R.C. Miller, D.L. Sivco, and A.Y. Cho, *Phys. Rev. B*, **36**, 8165 (1987).

66. R.G. Andersson, Z.G. Chen, V.D. Kulakovskii, A. Uddin, and J.T. Vallin, *Phys. Rev. B* **37**, 4032 (1988).
67. M.J. Joyce, M.J. Johnson, M. Gal, and B.F. Usher, *Phys. Rev. B* **38**, 10978, (1988).
68. F. Iikawa, F. Cerdeira, C. Vazquez-Lopez, P. Motisuke, M.A. Sacilotti, A.P. Roth, and R.A. Masut, *Phys. Rev. B*, **38**, 8473 (1988).
69. J.-P. Reithmaier, R. Höger, H. Riechert, P. Hiergeist, and G. Abstreiter, *Appl. Phys. Lett.*, **57**, 957 (1990).
70. G. Bastard, *Phys. Rev. B*, **24**, 5693 (1981) and ibid B **25**, 7584 (1982).
71. G. Bastard, in "Wave mechanics applied to semiconductor heterostructures", (Halsted Press, New York, 1988), cap. V, p. 155.
72. K.Y. Hsieh, Y.L. Hwang, J.H. Lee, and R.M. Kolbas, *J. of Elect. Materials*, **19**, 1417 (1990).
73. H. Nickel, R. Lösch, w. Schlapp, H. Leier, and A. Forchel, *Surf. Sci.*, **228**, 340 (1990).
74. W.P. Gillin, D.J. Dunstan, K.P. Homewood, L.K. Howard, and B.J. Sealy, *J. Appl. Phys.*, **73**, 3782 (1993).
75. W.P. Gillin, I.V. Bradley, L.K. Howard, R.Gwilliam, and K.P. Homewood, *J. Appl. Phys.*, **73**, 7715 (1993).
76. W.J. Taylor, N. Kuwata, I. Yoshida, T. Katsuyama, and H. Hayashi, *J. Appl. Phys.*, **73**, 8653 (1993).
77. I.V. Bradley, W.P. Gillin, K.P. Homewood, and R.P. Webb, *J. Appl. Phys.*, **73**, 1686 (1993).
78. J.S. Tsang, C.P. Lee, S.H. Lee, K.L. Tsai, C.M. Tsai, and J.C. Fan, *J. Appl. Phys.*, **79**, 664 (1996).
79. S. Bürkner, M. Baeumler, J. Wagner, E.C. Larkins, W. Rothemund, and J.D. Ralston, *J. Appl. Phys.*, **79**, 6818 (1996).
80. T.Y. Tan and U. Gösele, *J. Appl. Phys.*, **52**, 1240 (1988).
81. A. Furuya, O. Wada, A. Takamori and H. Hashimoto, *Jpn. J. Appl. Phys.*, **26**, L926 (1987).
82. T.Y. Tan, S.Yu and U. Gösele, *J. Appl. Phys.*, **70**, 4823 (1991).
83. R.M. Kolbas, Y.L. Hwang, T. Zhang, M. Prairie, K.Y. Hsieh, and U.K. Mishra, *Optical and Quantum Electronics*, **23**, S805 (1991).
84. S.Y. Chang and G.L. Pearson, *J. Appl. Phys.*, **46**, 2986 (1975).
85. B. Molnar, *J. Electrochem. Soc.*, **123**, 767 (1976).
86. D.G Deppe, L.J. Guido, N. Holonyak,Jr., K.C. Hsieh, R.D. Burnham, R.L. Thornton, and T.L. Paoli, *Appl. Phys. Lett.*, **49**, 510 (1986).
87. K.V. Vaidyanathan, M.J. Helix, D.J. Wolford, B.G. Streetman, R.J. Blattner, and C.A. Evans, Jr., *J. Electrochem. Soc.*, **124**, 1781 (1977).
88. J.Y. Chi, X. Wen, E.S. Koteles, and B. Elman, *Appl. Phys. Lett.*, **55**, 855 (1989).
89. R.W. Kaliski, D.W. Nam, D.G. Deppe, N. Holonyak,Jr., and K.C. Hsieh, *J. Appl. Phys.*, **62**, 998 (1987).
90. E.S. Koteles, B. Elman, P. Melman, J.Y. Chi, and C.A. Armiento, *Optical and Quantum Electronics*, **23**, S779 (1991).
91. K.J. Beernink, D. Sun, D.W. Treat, and B.P. Bour, *Appl. Phys. Lett.*, **66**, 3597 (1995).
92. Y. Hwang, K.Y.Hsieh, J.H. Lee, T. Zhang, U.K. Mishra, and R.M. Kolbas, Proc. of the 6th Conf. on Semi-Insulating III-V Materials, in Semi-insulating III-V Materials, Edited by A.G. Milnes and C.J. Miner, (Adam Hilger, Philadelphia, 1990) pag. 77.
93. C. Kielowski, A.R. Calawa, and Z. Liliental-Weber, *J. Appl. Phys.*, **80**, 156 (1996).
94. J.S. Tsang, C.P. Lee, S.H. Lee, K.L. Tsai, and H.R. Chen, *J. Appl. Phys.*, **77**, 4302 (1995).
95. J.C.P. Chang, J.M. Woodall, M.R. Melloch, I. Lahiri, D.D. Nolte, N.Y. Li, and C.W. Tu, *Appl. Phys. Lett.*, **67**, 3491 (1995).
96. I. Lahiri, D.D. Nolte, J.C.P. Chang, J.M. Woodall, and M.R. Melloch, *Appl. Phys. Lett.*, **67**, 1244 (1995).
97. M. Kaminska, E.R. Weber, Z. Liliental-Weber, R. Leon, and Z.U. Rek, *J. Vaccum Science Technology B* **7**, 710 (1989).
98. M. Kaminska, Z. Liliental-Weber, E.R. Weber, T. George, J.B. Kortright, F.W. Smith, B.-Y. Tsaur, and A.R. Calawa, *Appl. Phys. Lett.*, **54**, 1881 (1989).
99. D.C. Look, D.C. Walters, M.O. Manasreh, J.R. Sizelove, C.E. Stutz, and K.R. Evans, *Phys. Rev. B*, **42**, 3578 (1990).

100. D.C. Look, D.C. Walters, G.D. Robinson, J.R. Sizelove, M.G. Mier, and C.E. Stutz, *J. Appl. Phys.*, **74**, 306 (1993).

101. J. Störmer, W. Triftshäuser, N. Hozhabri, and K. Alavi, *Appl. Phys. Lett.*, **69**,1867 (1996).

102. T.Y. Tan and U. Gösele, *J. Appl. Phys.*, **61**, 1841 (1987).

103. S.B. Zhang and J.E. Northrup, *Phys. Rev. Lett.*, **67**, 2339 (1991).

104. R.L. Longini, *Solid State Electron.*, **5**, 127 (1962).

105. H.C. Casey, Jr., *Atomic Diffusion in Semiconductors*, edited by D. Shaw (Plenum, New York, 1973), Chap. 6, pp. 369.

106. U. Gösele and F. Morehead, *J. Appl. Phys.*, **52**, 4617 (1981).

107. M.T. Furtado and M.S.S. Loural, in *Defect and Diffusion Forum*, Vol. 127–128, 9 (1995).

108. W.P. Gillin, B.J. Sealy, K.P. Homewood, *Optical and Quantum Electronics*, **23**, S975 (1991).

109. W.P. Gillin, K.P. Homewood, L.K. Howard, and M.T. Emeny, *Superlattices and Microstrucutres*, **9**, 39 (1991).

110. P. Mei, H.W. Yoon, T. Venkatesan, S.A. Schwarz, and J.P. Harbison, *Appl. Phys. Lett.*, **50** 1823 (1987).

111. G.W. Arnold, S.T. Picraux, P.S. Peercy, D.R. Myers, and L.R. Dawson, *Appl. Phys. Lett.*, **45**, 382 (1984).

112. D.R. Myers, G.W. Arnold, C.R. Hills, L.R. Dawson, and B.L. Doyle, *Appl. Phys. Lett.*, **45**, 382 (1984).

113. D.R. Myers, G.W. Arnold, I.J. Fritz, L.R. Dawson, R.M. Biefeld, C.R. Hills, and B.L. Doyle, *J. of Electronic Materials*, **17**, 405 (1988).

114. P.G. Piva, P.J. Poole, M. Buchanan, G. Champion, I. Templeton, G.C. Aers, R. Williams, Z.R. Wasilewski, E.S. Koteles, and S. Charbonneau, *Appl. Phys. Lett.*, **65**, 621 (1994).

115. T. Fujii, M. Sugawara, S. Yamazaki, and K. Nakashima, *J. Crystal Growth*, **105**, 348 (1990).

116. S.N.G. Chu, R.A. Logan, and W.T. Tsang, *J. Appl. Phys.*, **79**, 1397 (1996).

117. S.S. Iyer and F.K. Legoues, *J. Appl. Phys.*, **65**, 4693 (1989).

118. T.S. Kuan and S.S. Iyer, *Appl. Phys. Lett.*, **59**, 2242 (1991).

119. J.-M. Baribeau, R. Pascual and S. Saimoto, *Appl. Phys. Lett.*, **57**, 1502 (1990).

120. N.E.B. Cowern, P.C. Zalm, P. van der Sluis, D.J. Gravestijn, and W.B. Boer, *Phys. Rev. Lett.*, **72**, 2585 (1994).

121. W.P. Gillin and D.J. Dunstan, *Phys. Rev. B*, **50**, 7495 (1994).

122. S.-W. Ryu, I. Kim, B.-D. Choe, and W. G. Jeong, *Appl. Phys. Lett.*, **67**, 1417 (1995).

CHAPTER 5

Strain in Interdiffused In$_{0.53}$Ga$_{0.47}$As/InP Quantum Wells

JOSEPH MICALLEF

Department of Microelectronics, University of Malta, Msida, Malta

1 INTRODUCTION

Developments in semiconductor growth technologies, such as metalorganic vapor phase epitaxy [1] and molecular beam epitaxy [2], have enabled fabrication of high-quality lattice-matched and strained quantum well (QW) structures which exhibit many unique material properties that cannot be realized in bulk semiconductors [3]. These properties have resulted in improved photonic devices [3,4] as well as in completely new types of devices [5,6]. For instance, the improvement in laser performance [3], such as lower threshold current density, operation at higher temperatures, modulation at higher frequencies and reduced spectral linewidths [4], is so dramatic that quantum well lasers are now available from a number of commercial sources. QW electroabsorption modulators [7] are possible as a result of the quantum-confined Stark effect (QCSE) [8] which describes the change in the absorption properties of QW structures when an electric field is applied across the structure. The shift in the exciton absorption edge with electric field is observable at room temperature for QW structures but not for bulk semiconductors. The novel self-electro-optic device (SEED) and related devices [5,9], which are proving of importance for large-density optical signal processing and computer applications, are also based on the quantum-confined Stark effect. Very large numbers of extremely small (~ 10 μm square) devices with low-power consumption can be fabricated permitting the realisation of highly parallel and complex optical architectures. For instance, a 64×32 array of symmetric SEED devices, each of which can be operated as a memory element or logic gate has been reported [6].

1.1 InGaAs/InP QUANTUM WELLS

Progress in epitaxial growth techniques using phosphorus has made it possible to obtain high-quality InGaAs(P)/InP QW structures [10,11]. In a conventional $In_x Ga_{1-x} As$/InP QW the active $In_x Ga_{1-x} As$ layer is confined between two InP barrier layers. $In_x Ga_{1-x} As$/InP QW structures are lattice-matched for an In composition $x = 0.53$. For $x < 0.53$, the $In_x Ga_{1-x} As$ layer is under tension, with the strain increasing as x decreases. For $x > 0.53$, the $In_x Ga_{1-x} As$ layer is under compressive stress, with the modulus of the strain increasing as x increases.

Lattice-matched $In_{0.53} Ga_{0.47} As$/InP QW structures have been investigated for a variety of photonic devices, such as lasers [12,13], detectors [14], optical waveguides [15,16], and QCSE modulators [17–19], for operation in the 1.3 μm to 1.55 μm wavelength range, where high-speed, long-distance optical communication systems operate as a result of the low loss and minimum dispersion characteristics exhibited by optical fibres in this wavelength range. There is also much interest in LEDs for operation around 1.3 μm where material dispersion in optical fibres goes through zero and where the wide linewidth of the LED imposes far less limitation on link length than intermodal dispersion within the fibre. Broad spectrum LEDs fabricated using QW intermixing have also been reported [20]. SEED devices using InGaAs/InP MQWs have also been demonstrated [9].

The possibility of having the In$_x$Ga$_{1-x}$As well layers under tension or under compression adds to the attraction of the In$_x$Ga$_{1-x}$As/InP material system. Strained InGaAs/InP QW lasers, for operation at optical communication wavelengths, grown under both compressive and tensile strain have been fabricated [13,21] with significantly improved performance compared to their lattice-matched counterparts. Strained InGaAs/InP QW lasers operating in the 1.8 μm range have also been demonstrated [22]. Electroabsorption enhancement in tensile strained QWs has been studied theoretically [23], and demonstarted experimentally [24], for InGaAs/InP QW structures.

The ability to grow InGaAs layers on InP substrates also provides a transparent substrate at these long wavelengths. This allows easy optical access to the active region without the need to etch away the substrate, which is advantageous also for photonic integration. It is also possible to fabricate devices such as junction field-effect transistors using In$_{0.53}$Ga$_{0.47}$As as the channel layer material, grown on an InP substrate [25], so that In$_{0.53}$Ga$_{0.47}$As/InP is being investigated not only for photonic integration but also for applications in optoelectronic integration [26].

1.2   Strain in Quantum Well Structures

Lattice-matched material systems, such as AlGaAs/GaAs and, to a lesser extent lattice-matched In$_{0.53}$Ga$_{0.47}$As/InP, have been intensively investigated due to the relative ease in the growth of high-quality QWs. However, the available lattice-matched QW material systems do not cover the whole potential wavelength range available for photonic devices, so that the study of strained QW structures has proceeded systematically despite the difficulties associated with the growth of coherently strained structures. In addition, strained QW structures provide freedom from the need for precise lattice-matching between the well and the barrier, thus widening the choice of compatible materials, and increasing the ability to control the optical and transport properties of QW structures. Moreover, the suggestion by Adams [27], and independently by Yablonovich and Kane [28], that reshaping of the band structure of a semiconductor by the introduction of compressive strain could lead to reduced threshold current and improved differential gain in semiconductor QW lasers, provided a considerable impetus to increased research in strained QW structures. As a result various strained-layer material systems, such as InGaAs/GaAs [29], strained In$_x$Ga$_{1-x}$As/InP [30], and

GaAsP/GaP [31], are being widely investigated for photonic and electronic applications. $In_x Ga_{1-x}As/GaAs$ QW structures enable applications such as waveguides, lasers, modulators and detectors operating in the 880–1100 nm wavelength range, while strained $In_x Ga_{1-x}As/InP$ is essential for 1.55 μm wavelength strained devices required in optical communications.

1.2.1 Strained-Layer Structures

When a film is grown epitaxially on a substrate with a different lattice constant, the lattice mismatch can be accommodated either by coherent strain, resulting in high-quality crystalline structures, or by other mechanisms, such as dislocation generation at the interface, metastable epitaxy (a combination of strain and misfit dislocations), bending of the epitaxial layer, and tilt of the lattices with respect to each other, in which cases poor crystalline quality results. In optoelectronic devices, it is important to realise conditions in which no misfit dislocations are present, so that the structure is coherently strained. The individual layers in such a strained-layer structure match up the lattice constants in the plane parallel to the interface by compression or expansion in this plane [31,32], Figure 1. A tetragonal deformation in the layers also takes place so that the layers are compressed or expanded in the direction perpendicular to the interfacial plane [33]. Thus the resultant strains in a coherently strained structure consist of a biaxial hydrostatic strain parallel to the interface and a uniaxial shear component parallel to the direction of epitaxial growth. The strain induced in a strained-layer structure depends on the composition of the substrate, as well as on that of any buffer layer present between the strained-layer structure and the substrate [34,35].

Theoretical models [36, 37] and experimental evidence [30, 38, 39] demonstrate that for a small enough lattice mismatch in a heterostructure (less than about 3%) [38], there is a critical layer thickness (CLT) below which dislocations are not generated because it is energetically more favourable to accommodate the strain energy by stretching the epitaxial layer than to create dislocations. The force-balance model developed by Matthews and Blakeslee [36] and the energy-balance model developed by People and Bean [37] have both gained experimental support despite predicting values of CLTs which differ by as much as an order of magnitude. Temkin et al [30] have demonstrated that the CLT of both tensile-strained and compressively-strained InGaAs/InP QW structures are limited by the Matthews and Blakeslee force-balance model. Calculated CLTs, using the the Matthews and Blakeslee model, for InGaAs/InP as a function of In composition are shown in Figure 2. The CLT is a function of the lattice mismatch and thus it is dependent on the composition of the layers in the structure. Starting from the lattice-matched x = 0.53, the CLT of $In_x Ga_{1-x}As$ layers in pseudomorphic $In_x Ga_{1-x}As/InP$ QW structures decreases both as

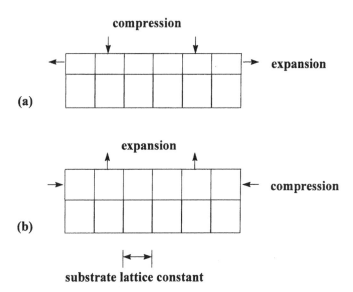

Fig. 1. Coherently strained epitaxial layer in (a) biaxial tension, (b) biaxial compression.

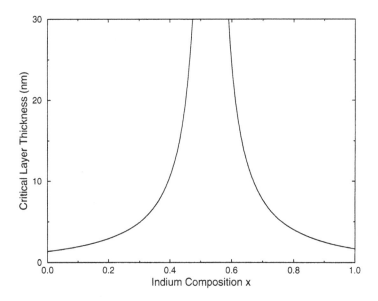

Fig. 2. Calculated critical layer thickness as a function of indium concentration for In$_x$Ga$_{1-x}$As on InP.

$x \rightarrow 0$, and as $x \rightarrow 1.0$. The growth of high-quality strained QW structures is therefore conditioned by the need to remain within the critical layer thickness limits for the specific material system.

1.2.2 Band Structure Considerations

The band structure of bulk III-V compound semiconductors is characterized by a conduction band with light effective mass, a valence band that is doubly degenerate at the Brillouin zone centre Γ, and a spin-orbit split-off valence band. The doubly degenerate bands are identified as the heavy hole (hh) and light hole (lh) bands due to their different effective masses. In QW structures carrier confinement potentials result and quantum size effects are introduced, leading to a modification of the effective bandgap and to the splitting of the hh and lh subbands due to the different effective masses, Figure 3.

The band structure is further modified in coherently strained QWs. The biaxial component of strain alters the bulk bandgap between the conduction band and both the degenerate valence bands and the spin-orbit split-off valence band. The uniaxial component of the strain lifts the degeneracy of the hh and lh bands at Γ. In the case of compressive strain the biaxial component causes an increase in the bulk bandgap energy, while the uniaxial component shifts the hh band towards the conduction band and the lh band away from the conduction band [34]. This valence band splitting is increased by the quantum size effects in the case of compressive strain, Figure 3(b). In the case of tensile strain, the biaxial component decreases the bulk band gap energy while the uniaxial component causes the hh band to shift away from the conduction band and the lh band to move toward the conduction band. The quantum size effects now reduce the valence band splitting, Figure 3(c). The uniaxial component also causes the lh band to couple with the spin-orbit split-off band [40].

The splitting between the hh and lh states alters the band-to-band mixing, so that strain also induces a change in the effective masses parallel to the interface. For compressively strained layers the hh states, which are at the top of the valence band, acquire a smaller in-plane effective mass, while the lh states exhibit a comparatively heavy in-plane effective mass [41]. In the case of tensile strain, the light holes are shifted to the top of the valence band, and have a relatively light in-plane effective mass. These effects are of fundamental importance in the application of strained layers to semiconductor QW lasers, since the differential gain in a laser is optimal for equal electron and hole effective masses.

The depths of the carrier confinement profiles shown in Figure 3 are strongly dependent on the value of band offsets. The bandgap offset ΔE_g in a semiconductor heterojunction defines the energy difference between the

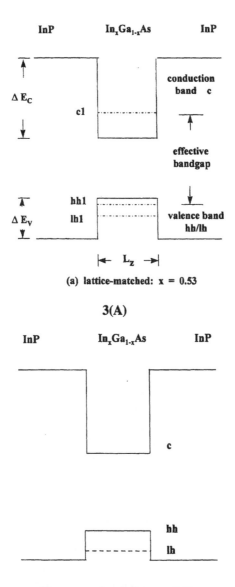

(a) lattice-matched: x = 0.53

3(A)

(b) compressive strain: x > 0.53

3(B)

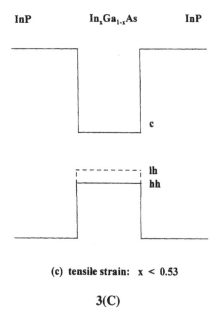

InP   In$_x$Ga$_{1-x}$As   InP

(c) tensile strain: x < 0.53

3(C)

Fig. 3.   Conduction-band and valence-band confinement profiles of In$_x$Ga$_{1-x}$As/InP quantum well structure for (a) $x = 0.53$, showing electron (c), heavy hole (hh) and light hole (lh) ground states, (b) $x > 0.53$ and (c) $x < 0.53$.

respective bandgaps (conduction-band minimum to valence-band maximum at Γ) of the two layers forming the heterostructure, and is the sum of the conduction-band discontinuity ΔE_c and the valence-band discontinuity ΔE_v, Figure 3(a). The conduction band offset Q_c is defined as $Q_c = \Delta E_c/\Delta E_g$, while the valence band offset Q_v is defined as $Q_v = \Delta E_v/\Delta E_g$. As for most III-V heterojunctions, a large discrepancy exists in reported band offset values, determined through different experimental techniques [42], for In$_{0.53}$Ga$_{0.47}$As/InP. Early determinations based on optical spectroscopy indicated a Q_c of about 0.65 [43] while measurements using capacitance-voltage data indicated a Q_c of about 0.40 [44]. More recent reports on Q_c measurements indicated a value of about 0.33 [45,46], while a value of 0.4 has been used to provide an adequate fit to exciton peak energies in this material system as a function of QW width [47]. The question of the band alignment at the InGaAs/InP heterojunction has thus not yet been resolved. Nevertheless, the band offsets constitute an important parameter affecting the optical properties of a QW structure.

The lowest, direct bandgap in a QW structure is the (interband) transition energy between the ground state in the conduction and hh valence band,

Figure 3(a). The principal transition energy is therefore dependent on the strain and quantization effects. By altering the QW width or the ternary alloy composition the conduction and valence band confinement profiles are altered. As a result, the position of the confined subbands is changed, and with it the principal interband transition energy, as well as higher order interband, transition energies, of the QW structure. Thus QW structures offer a wider range of principal transition energies than that possible by the use of bulk materials. Strained QW structures offer an extra degree of flexibility in tailoring the interband transition energies, and hence the optical properties, of a structure by making use of both strain and quantization effects.

2   QUANTUM WELL INTERMIXING

Intermixing of QW structures involves the interdiffusion of constituent atoms across the well-barrier interfaces, and is of particular interest since it offers the possibility of continuous, controlled modification of the material composition [48]. This change in composition alters the confinement profile and subband edge structure in the QW, resulting in a modified effective bandgap of the QW structure. The extent of the intermixing, and thus of the modification of the subband edge structure, is a function of the QW intermixing technique parameters, such as the nature and concentration of the impurity species present, and the process temperature and time. Moreover the disordering process can be localised to selected regions of the QW structure so that the optical properties of only the selected areas are modified by the disordering process. QW intermixing can thus provide a useful tool for bandgap engineering.

Quantum well intermixing of III-V semiconductor structures, often also referred to as QW disordering, is being actively investigated, with two principal objectives: obtaining high-performance device applications and achieving monolithic integration. QW intermixing of both lattice-matched and strained QWs has been carried out employing various techniques to study the potential of enhanced performance of various optical devices. Selective area disordering of an In$_{0.53}$Ga$_{0.47}$As/InP superlattice by argon implantation [49] has been used to modify the fundamental bandgap, and thus the refractive index, in localised areas, resulting in lateral photon confinement in optical waveguides. An impurity-free disordering process has been used to successfully fabricate optical channel waveguides which can selectively confine TE and TM modes in an In$_{0.53}$Ga$_{0.47}$As/InP disordered superlattice [50]. QW intermixing has been successfully used to obtain very low threshold currents in strained InGaAs/GaAs QW lasers [51,52]. A two-wavelength demultiplexing *p-i-n* photodetector using disordered multiple QW structures has been demonstrated [53]. High energy implantation has been used to selectively blue shift the emission wavelength of an InP-based QW laser structure while maintaining high crystal quality after the QW intermixing [54].

Investigation into QW intermixing as an effective tool for spatially selective bandgap tuning of QW structures is also driven by the requirement to make optoelectronic integrated circuits (OEICs) and photonic integrated circuits (PICs) [55]. Although a limited level of integration can be achieved by complex growth techniques such as selective area etch and regrowth, these processes suffer from low yields due to the complex processing entailed [56].

To integrate monolithically optical devices such as lasers, detectors, modulators and optical switches in a PIC, it is necessary to be able to achieve precise bandgap shifting compatible with the different bandgap requirements of the optical devices on the chip, as well as to produce low-loss waveguides as interconnects. For instance, for a PIC consisting of lasers, detectors, and waveguides on a single chip, the detector would operate with smaller bandgaps than the lasers, while the waveguides would benefit from larger bandgaps to minimize losses due to absorption. InP-based electroabsorption modulators with high on/off ratios, low drive voltages and short interaction lengths, employing the QCSE, are also of considerable interest for compact monolithic integration with other InP-based optoelectronic devices for communication and optical signal processing [57], since they offer higher speed operation than direct modulation of the laser source. Successful integration with lasers again requires bandgap shifted QCSE modulators.

Monolithic integration of a transparent dielectric waveguide into an active laser cavity by impurity-induced disordering [58] has been demonstrated. Selective disordering of a multiple QW structure has been used to integrate a laser and phase modulator, which required modification of the modulator section to make it efficient at the lasing wavelength [59]. Cap-annealing disordering has been used for the integration of lattice-matched InGaAs/InP lasers and low-loss waveguides [60], and of an InGaAsP/InP strained-layer distributed feedback laser and an external electroabsorption modulator [61]. Bandgap shifting of InGaAs(P)/InGaAsP MQW structures for integration of lasers and electroabsorption modulators has been demonstrated using SiO:P encapsulant [62], and laser intermixing [63]. QW intermixing is emerging as a powerful tool for fabricating PICs and optoelectronic integrated circuits.

2.1   Interdiffusion Techniques

Thermal annealing of III-V semiconductor heterostructures can result in the interdiffusion of constituent atoms across the well-barrier interface if the annealing is carried out at a suitably high temperature and for a sufficiently long time [48,64]. Detectable intermixing in InGaAs/GaAs QWs is reported to occur above 800°C, while intermixing in InGaAs/InP single QW and multiquantum well (MQW) structures is reported for temperatures greater than 600°C [65,66]. Detectable bandgap shifts as a result of compositional intermixing have been reported for as-grown InGaAs/InP MQW structures subjected to rapid thermal anneals above 700°C [67].

QW intermixing can occur at a lower temperature in the presence of point defects, such as vacancies and interstitials. The major processes [68] that have been employed to introduce these point defects and enhance QW intermixing are impurity-induced disordering (IID) [69], and impurity-free vacancy diffusion (IFVD) [70,71]. More recently, laser-induced annealing has also been investigated as a potential tool for QW intermixing [68].

2.1.1 Impurity-Induced Disordering

In impurity-induced disordering [69], surface diffusion or ion-implantation of an impurity species is used to enhance the interdiffusion process.

Zn diffusion has been widely investigated in both lattice-matched QW structures grown on GaAs and InP substrates [72–75], and strained-layer QW structures [76], and has been shown to greatly enhance the interdiffusion of group III atoms across the interface, thereby disordering the QW structure. Disordering of QWs has also been demonstrated for other impurity species, such as Si diffusion [39], and S diffusion [77]. Ion-implantation has also been employed to enhance the disordering of QW structures. Implantation-induced disordering of QW structures has been demonstrated for various species implants, including Zn [78], Si [79], and Ge [80] implants.

The high free-carrier absorption associated with the diffusion or implantation of electrically active impurities limits the use of dopants, such as Zn and Si, for QW intermixing in photonic devices. Consequently, attention has also been focused on the use of non-electrically active impurity implantation, such as B, F and Ne implants, [67,81,82] which does not increase free-carrier absorption.

Ion-beam intermixing has also been demonstrated as an alternative process for QW intermixing. In ion-beam mixing, implantation of a low mass ion is carried out at an elevated temperature which, however, is lower than the anneal temperatures used in IID or IFVD processes [83]. Implantation at the elevated temperature eliminates the need for a high-temperature anneal to induce compositional disordering. For instance, starting with a 50 period lattice-matched InGaAs/InP superlattice, 550°C oxygen ion-beam mixing has been used to simultaneously achieve compositional disordering and electrical isolation, and form guided-wave modulators [84].

2.1.2   Impurity-Free Vacancy Disordering

Impurity-free vacancy disordering also results in an enhanced interdiffusion process. SiO_2 caps are employed for both AlGaAs/GaAs and InGaAs/GaAs QW intermixing [71,85–87]. The SiO_2 capping layers are deposited on selected areas of the QW structure. The enhanced intermixing is attributed to Ga atoms outdiffusing into the encapsulating SiO_2, which has been shown to have a great affinity for Ga [88]. Group III vacancies are thus created at the surface, and these vacancies then diffuse through the structure during a rapid thermal annealing process, enhancing the intermixing [85]. The surface defects can alternatively be generated by shallow ion implantation and are then driven deeper into the QW structure during the rapid thermal anneal [86]. Again the presence of these vacancies at the well-barrier interfaces enhances the anneal driven interdiffusion of the group III atoms.

Si_3N_4 cap annealing has been used to promote IFVD of InGaAs/InP QW structures. This annealing partially intermixes the region under the cap but does not significantly alter the characteristics of other regions [60, 89]. Unfortunately, Si_3N_4 films usually contain a substantial amount of SiO_2 and are also usually highly strained, adversely affecting the disordering process beneath the cap. The use of SrF_2 caps has been investigated [90], resulting in enhanced discrimination between the areas where interdiffusion was required and those which were meant to remain as-grown.

Impurity-free vacancy disordering is of particular interest for various applications of different complexities, from waveguides to multiple wavelength laser arrays, since IFVD does not introduce additional carriers and avoids implantation-induced device damage. In applications where different wavelength shifts are required, such as multiple-wavelength laser structures, which are used for wavelength division multiplexing, laterally selective interdiffusion can be achieved by locally modifying the thickness, composition or porosity of the dielectric cap [91].

2.1.3   Laser-Induced Disordering

Laser-induced QW intermixing is also under active investigation. Photo-absorption-induced disordering (PAID) relies on the bandgap dependent absorption of the incident cw laser radiation within the active region of a multilayer structure. Heat is then generated by carrier cooling and nonradiative recombination, causing interdiffusion between the well and barrier materials to take place. The use of high-power pulsed lasers has also been studied. The disruption of the lattice due to rapid thermal expansion is thought to lead to an increase in the point defect density, thereby enhancing the QW intermixing. Laser-induced intermixing has been investigated and shown to be effective for intermixing of AlGaAs/GaAs [92,93] and InGaAs/InGaAsP QW structures [94,95].

2.2 Experimental Results of In$_{0.53}$Ga$_{0.47}$As/InP QW Intermixing

Disordering of lattice-matched AlGaAs/GaAs quantum well structures results in the interdiffusion of only group III atoms Al, Ga [69], as there is no As concentration gradient across the interface, while disordering of strained InGaAs/GaAs QW structures results in the interdiffusion of the group III atoms In, Ga [64]. In comparison, the disordering of lattice-matched In$_{0.53}$Ga$_{0.47}$As/InP QW structures is more complicated since interdiffusion can occur for both group III (In, Ga) and group V (As, P) atoms [39], resulting in a quaternary In$_x$Ga$_{1-x}$As$_y$P$_{1-y}$ layer. The process used for QW intermixing will determine the properties of the structure after interdiffusion. For instance, in impurity-induced disordering (IID), the impurity used to enhance the intermixing of the QW structure determines the interdiffusion process and can result in either a short wavelength [96] or a long wavelength shift [73] in the fundamental absorption edge. If the compositional interdiffusion in lattice-matched In$_{0.53}$Ga$_{0.47}$As/InP QWs takes place at a comparable rate on the two sublattices, then the lattice-matching condition is maintained. When interdiffusion is dominant on one of the sublattices a strained layer structure will result after interdiffusion [39].

Intermixing of In$_{0.53}$Ga$_{0.47}$As/InP QW structures by means of various impurity species, as well as by impurity-free vacancy diffusion has been widely investigated experimentally. The nature of the interdiffusion mechanisms on the two sublattices in InGaAs/InP is not yet fully established, and the role of strain in promoting or inhibiting the interdiffusion is still the subject of active debate and study [97,98]. Nevertheless, most reported experimental results have been interpreted in terms of the relative interdiffusion rates between the two sublattices, and can be grouped as follows:

(i) comparable interdiffusion rates on both group III and group V sublattices;

(ii) dominant group III interdiffusion;

(iii) dominant group V interdiffusion.

Lattice-matching of an In$_x$Ga$_{1-x}$As$_y$P$_{1-y}$ layer to InP occurs only when $y \approx 2.2(1 - x)$ [99]. Thus, in the first case, the intermixed QW structure remains lattice-matched, while in the other two cases QW intermixing results in a strained structure, where the well layer can be either in tension or in compression.

2.2.1   Comparable Interdiffusion Rates

Pape et al. [77] reported results of S diffusion in InGaAs/InP MQW structures. The initial S diffusion was carried out at 600°C for 20 min, with further annealing steps after the sample was covered with a SiN$_x$ cap. Using photoluminescence (PL) and optical absorption analysis, S diffusion was shown to induce interdiffusion on both group III and group V sublattices.

Julien et al. [100] annealed undoped InGaAs/InP QW structures after i) Ge and ii) S implantation. Both SQW and MQW structures were used for Ge implantation, with the SQW samples annealed at 650°C for 10 min, while the MQW samples were annealed at 600°C for 20 min. From PL and secondary ion mass spectroscopy (SIMS) profiles it was found that Ge implantation and annealing enhanced interdiffusion on both group III and group V sublattices with the same order of magnitude, indicating preservation of lattice-matched condition. Characterization of the S implanted MQW samples was carried out using PL, X-ray analysis and Auger electron spectroscopy (AES). The results were interpreted in terms of enhanced interdiffusion on both sublattices with the MQW structure again maintaining lattice-matching to InP.

Schwarz et al. [39] reported on Si diffusion at 700°C for 3 h in a InGaAs/InP (SL). SIMS profiles were interpreted in terms of comparable interdiffusion on the cation and anion sublattices within a narrow range of Si concentration.

Miyazawa et al. [71] reported compositional intermixing of InGaAs/InP MQW structures capped by a Si$_3$N$_4$ film. Repetitive (rapid) thermal annealing was carried out at 800°C at a rate of 30°C/s. Optical transmission measurements were used to study interdiffusion. The blue shifts of the spectra were interpreted as indicative of comparable interdiffusion of all atoms.

2.2.2   Dominant Group III Interdiffusion

Nakashima et al. [101] investigated Zn diffusion at 550°C for 1 h in InGaAs/InP MQW structures. SIMS and X-ray analysis were used to characterize the samples. Group III interdiffusion was found to dominate leading to strained layers.

Schwarz et al. [39] reported on Zn diffusion at 600°C for 1 h in a InGaAs/InP SL. SIMS characterization revealed substantial cation interdiffusion with little or no anion interdiffusion. TEM performed on the interdiffused samples showed that the resulting 3.1% lattice-mismatch was accommodated coherently by the lattice.

Julien et al. [100] reported results on Zn diffusion over the temperature range 400 to 700°C for time intervals up to 150 min, for both SQW and MQW samples. Low-temperature PL showed shifts to longer wavelengths which were again interpreted in terms of enhanced interdiffusion on group III sublattice only, resulting in a highly strained QW structure.

Pape et al. [73] reported on Zn diffusion, at 475°C and 600°C for 20 min each, in MQWs of InGaAs/InP. Using room-temperature PL and SIMS analysis it was found that Zn induced interdiffusion led to shifts to lower energies, attributable to interdiffusion on the group III sublattice.

Razeghi et al. [96] investigated Zn diffusion into QWs of InGaAsP/InP. PL results for a SQW subjected to Zn diffusion up to a temperature of 700°C showed a shift to longer wavelengths which saturated around 600°C, followed by a return to shorter wavelengths at 700°C. These results were interpreted in terms of enhanced interdiffusion on the group III sublattice, with the change of direction in the PL shift at 700°C suggesting a group V interdiffusion rate larger than group III interdiffusion rate at these temperatures. In contrast thermal annealing without Zn diffusion at 650°C did not result in any wavelength change

van Gurp et al. [102] reported on Zn-enhanced interdiffusion in InGaAsP/InP MQW structures. SIMS, AES, PL and X-ray diffraction were used for characterization. Significant interdiffusion of In and Ga was found while interdiffusion of As and P was negligible. PL showed that the interdiffusion started at temperatures above 420°C. In contrast no measurable interdiffusion occurred at 750°C in the absence of Zn.

2.2.3   Dominant Group V Interdiffusion

Temkin et al. [66] subjected SQW and MQW to brief anneals at temperatures in the 600 to 850°C range. Low-temperature PL and TEM were used for characterization. Changes were observed only for annealing temperatures above 700°C, when PL results showed shifts to higher energies. TEM for samples annealed near 800°C showed unexpectedly sharp well-barrier interfaces and approximately doubling of well widths. The observed PL shift was attributed to phosphorus diffusion into the well layer.

Nakashima et al. [103] carried out 1 h thermal anneals in N$_2$ atmosphere of MQW structures capped with SiO$_2$ films. Optical absorption experiments indicated that the structures remained stable below 650°C, while a blue shift of the excitonic absorption energy was observed for temperatures above 650°C. SIMS and X-ray diffraction analyses showed that the MQW structure remained basically intact and the intermixing process was attributed to a minor diffusion of group V phosphorus atom, resulting in a strained system.

Fujii et al. [104] studied interdiffusion in undoped InGaAs/InP QWs between 500 and 640°C for time intervals 1 to 3 h. Results of low-temperature PL spectra were interpreted in terms of group V species interdiffusion, with rapid interdiffusion in both the well and the barrier, but with a much lower interdiffusion rate across the heterointerface. This limiting rate of interdiffusion was attributed to strain effects.

Yu et al. [105] performed thermal annealing of InGaAs/InP SLs over the temperature range 700 to 850°C for 5 to 60 min, capped with Si_3N_4 film. Raman spectroscopy was used for characterization. It was concluded that the group III and group V atoms interdiffused maintaining the lattice-matched condition, except for the 700°C annealing. At this lower temperature group V interdiffusion was higher than group III interdiffusion.

Wan et al. [67] investigated interdiffusion in undoped InGaAs/InP MQW structures (lattice-matched and strained MQWs) after shallow, neutral ion (Ne^+ and F^+) implantation and rapid thermal annealing. PL results for anneals at 700°C showed a bandgap shift to higher energies, which saturated after 20 s annealing time, followed by a reduction in the blue shift for longer anneal times. These results were interpreted in terms of different interdiffusion rates on the two sublattices, with the group V interdiffusion initially dominating, with a subsequent increased contribution of group III interdiffusion.

2.3   Strain and Interdiffusion

The interdiffusion of the constituent atoms across the well-barrier interface leads to a transition from an as-grown square well compositional profile to a nonlinear graded profile, which can be described by an error function distribution [106]. This compositional modification in strained material systems, such as InGaAs/GaAs and strained InGaAs/InP, raises the question of the QW quality after interdiffusion. It has been demonstrated that the FWHM of PL spectra of thermally annealed InGaAs/GaAs QWs show little broadening for QWs within the critical thickness regime, indicating that no degradation in the quality of the strained QW has taken place [64]. In the case of Zn enhanced interdiffusion, some broadening of PL spectra occurred after disordering [76] but the spectra indicated that the QWs were still coherently strained. InGaAs/GaAs single QWs disordered by shallow As implantation and thermal annealing also showed little broadening of PL peaks [87], while an early study reported that strained layer superlattices demonstrate good structural integrity after ion implantation [107]. The issue of the crystal quality needs also to be addressed in the case of compositional interdiffusion in lattice-matched InGaAs/InP that gives rise to a strained structure after QW intermixing. Investigation into Zn diffusion induced disordering of $In_{0.53}Ga_{0.47}As/InP$ QWs, which results in a strained structure, also reported [39] that high-resolution transmission electron microscopy lattice images of the disordered QW structure showed no misfit dislocations or microtwins, indicating that the disordered structure is coherently strained. The intermixed structure of a wavelength-shifted QW laser, obtained with post-growth high-energy phosphorus implantation, showed no degradation in lasing threshold current, as compared to the as-grown QW structure, indicating high-quality material after the intermixing process [54]. With careful processing, coherently strained QW structures are possible even after disordering, so that QW intermixing in In$_x$Ga$_{1-x}$As/InP can be used to introduce strain, either tensile or compressive, in selected areas of the QW structure. It is thus useful to model the combined effects of compositional interdiffusion, and the strain that could result from this interdiffusion, on the subband edge structure of the intermixed In$_{0.53}$Ga$_{0.47}$As/InP QWs.

3 SUBBAND EDGE STRUCTURE

In order to be able to model the optical properties of the intermixed QW structure, it is necessary to determine the carrier confinement profiles and the subband-edge structure of the QW. Different mathematical approaches have been employed in arriving at the subband-edge structure of intermixed QWs. Almost all of them are based on the assumption that Fick's laws of diffusion apply to the species interdiffusion in the QW structure, and that the compositional profiles after interdiffusion are considered to follow an error function distribution.

3.1 Computational Considerations

Variational calculations have been used to obtain the electron and hh ground states in disordered AlGaAs/GaAs QWs [106]. The electron and hh subband states for disordered AlGaAs/GaAs QWs were obtained for the error function potential well using a particle transmission calculation [108]. The subband energies and wavefunctions for an error function distribution have also been calculated in the AlGaAs/GaAs material system taking into consideration a non-parabolic band model for the electron effective mass, and the valence subband mixing between the heavy and light holes [109]. In the strained InGaAs/GaAs material system, calculations of the ground state have been reported for disordered QWs, using a Green's function approach for the compositional profile and a variational technique to obtain the ground state [110]. By taking into account the effects of strain and disorder on the carrier confinement profile [111], the energy and wavefunctions for all the subband states in disordered InGaAs/GaAs single QWs have been determined [112] by solving numerically the one-electron Schrödinger equation using the envelope function scheme with an effective mass approximation. This model is described in detail in the next section and then applied to In$_{0.53}$Ga$_{0.47}$As/InP single QW intermixing.

In all the above approaches it is assumed that the interdiffusion rate of a species is the same in both the well and the barrier. It has been suggested that this may not always be the case and that the interdiffusion rates for the same species in the well and in the barrier can be significantly different [113]. In such a case Fick's laws of diffusion no longer apply. Fujii et al. [104] have proposed a model for group V interdiffusion where, although the interdiffusion rate is the same for both well and barrier, the interdiffusion rate is actually limited by a much slower interdiffusion across the well-barrier interface, while Mukai et al. [114] have proposed a model where the interdiffusion rate for group V species in the well is not the same as that in the barrier. However, Gillin et al. [115] reporting on thermal interdiffusion of $In_{0.66}Ga_{0.33}As/In_{0.66}Ga_{0.33}As_{0.7}P_{0.3}$ strained QWs where, due to the lack of concentration gradient on the group III sublattice interdiffusion took place on the group V sublattice only, presented experimental results which show that the group V interdiffusion follows Fick's laws.

3.1.1   Effects of Interdiffusion

Consider a III-V semiconductor material $A_xB_{1-x}C_yD_{1-y}$ where A, and B represent two group III atoms, and C, and D represent two group V atoms. After interdiffusion, let $\tilde{x}$, $1 - \tilde{x}$, $\tilde{y}$, $1 - \tilde{y}$ represent the concentration of the corresponding constituent atoms A, B, C, and D respectively. It is assumed here that the group III and group V interdiffusion processes can be modeled by two different interdiffusion rates, but that the interdiffusion rate on each sublattice is the same in both the well and the barrier. The interdiffusion of In and Ga atoms is characterized by a diffusion length L_d, which is defined as $L_d = (Dt)^{1/2}$, where D is the diffusion coefficient and t is the diffusion time; the interdiffusion of As and P atoms is characterized by a different diffusion length L'_d. If the rates of interdiffusion on group III and group V sublattices are identical, then $L_d = L'_d$ and the lattice-matched condition is maintained. If $L'_d \neq L_d$ a strained QW structure results.

Consider an undoped, single $In_xGa_{1-x}As$ well lattice-matched to semi-infinite InP barriers. After disordering, the concentration of the interdiffused atoms across the QW structure is assumed to have an error function distribution. The constituent atom compositional profiles can, therefore, be represented as follows:

(i) in the group III sublattice, the In concentration after interdiffusion, $\tilde{x}(z)$, is described by

$$\tilde{x}(z) = 1 - \frac{(1-x)}{2}\left[\text{erf}\left(\frac{L_z + 2z}{4L_d}\right) + \text{erf}\left(\frac{L_z - 2z}{4L_d}\right)\right] \quad (1)$$

where L_z is the as-grown well width, z is the growth direction, and the QW is centered at $z = 0$.

(ii) in the group V sublattice, the As compositional profile after interdiffusion, $\tilde{y}(z)$, is given by

$$\tilde{y}(z) = \frac{y}{2}\left[\operatorname{erf}\left(\frac{L_z + 2z}{4L'_d}\right) + \operatorname{erf}\left(\frac{L_z - 2z}{4L'_d}\right)\right] \tag{2}$$

where $y = 1$ is the As concentration of the as-grown structure.

The compositional profiles in the disordered QW structure imply that the carrier effective mass, the bulk bandgap, the strain and its effects, if present, vary continuously across the QW. Consequently, the carrier effective mass, $m_r^*(z)$, is now z-dependent and is obtained from $m_r^*(z) = m_r^*(\tilde{x}, \tilde{y})$, where $m_r^*(x, y)$ is the respective carrier A$_x$B$_{1-x}$C$_y$D$_{1-y}$ bulk effective mass, and r denotes either the electron (C), heavy hole (V=hh), or light hole (V=lh). The unstrained (bulk) bandgap in the well, $E_g(\tilde{x}, \tilde{y})$, is also a function of the compositional profile, so that the unstrained potential profile after interdiffusion, $\Delta E_r(\tilde{x}, \tilde{y})$, varies across the well and is given by:

$$\Delta E_r(\tilde{x}, \tilde{y}) = Q_r \Delta E_g(\tilde{x}, \tilde{y}) \tag{3}$$

where Q_r is the band offset and ΔE_g is the unstrained bandgap offset.

When $L'_d \neq L_d$ strain is introduced in the disordered structure. Consequently, the effects of strain on the bandgap, and hence on the carrier confinement profile and subband-edge structure, must be considered.

3.1.2   Effects of Strain

If the QW layer thickness is within the critical thickness regime, the QW structure will be coherently strained after disordering [39], with a biaxial hydrostatic strain parallel to the interfacial plane and a uniaxial shear strain perpendicular to the interfacial plane.

As already indicated, compressive (tensile) hydrostatic strain causes an increase (decrease) in the bandgap energy. The shear strain disrupts the cubic symmetry of the semiconductor and lifts the degeneracy of the heavy hole (hh) and light hole (lh) band edges at the Brillouin zone center Γ. The heavy hole band shifts towards (away from) the conduction band and the light hole band moves away from (towards) the conduction band [34]. In addition, the shear strain causes the lh band to couple with the spin-orbit split-off band [116].

Let $\varepsilon(\tilde{x}, \tilde{y})$ represent the in-plane strain in the well after interdiffusion. Assuming that the growth direction z is along $\langle 001 \rangle$, then for the biaxial hydrostatic stress parallel to the interface the strain components, after interdiffusion, are given by [117]:

$$\begin{aligned}
\varepsilon_{xx} &= \varepsilon_{yy} = \varepsilon(\tilde{x}, \tilde{y}) \\
\varepsilon_{zz} &= -2[c_{12}(\tilde{x}, \tilde{y})/c_{11}(\tilde{x}, \tilde{y})]\varepsilon(\tilde{x}, \tilde{y}) \\
\varepsilon_{xy} &= \varepsilon_{yz} = \varepsilon_{zx} = 0
\end{aligned} \tag{4}$$

where $\varepsilon(\tilde{x}, \tilde{y})$ is defined to be negative for compressive strain, and $c_{ij}(\tilde{x}, \tilde{y})$ are the elastic stiffness constants.

The change in the bulk bandgap due to the biaxial component of strain, $S_\perp(\tilde{x}, \tilde{y})$, is given by:

$$S_\perp(\tilde{x}, \tilde{y}) = -2a(\tilde{x}, \tilde{y})[c_{11}(\tilde{x}, \tilde{y}) - c_{12}(\tilde{x}, \tilde{y})/c_{11}(\tilde{x}, \tilde{y})]\varepsilon(\tilde{x}, \tilde{y}) \quad (5)$$

where $a(\tilde{x}, \tilde{y})$ is the hydrostatic deformation potential calculated from:

$$a(\tilde{x}, \tilde{y}) = \frac{1}{3}[c_{11}(\tilde{x}, \tilde{y}) + 2c_{12}(\tilde{x}, \tilde{y})]\frac{dE_g}{dP}(\tilde{x}, \tilde{y}) \quad (6)$$

where dE_g/dP is the hydrostatic pressure coefficient of the lowest direct energy gap E_g.

The splitting energy, $S_\|(\tilde{x})$, between the hh and lh band edges induced by the uniaxial component of strain is given by:

$$S_\|(\tilde{x}, \tilde{y}) = -b(\tilde{x}, \tilde{y})[c_{11}(\tilde{x}, \tilde{y}) + 2c_{12}(\tilde{x}, \tilde{y})/c_{11}(\tilde{x}, \tilde{y})]\varepsilon(\tilde{x}, \tilde{y}) \quad (7)$$

where $b(\tilde{x})$ is the shear deformation potential. The coupling between the lh and split-off band gives rise to asymmetric heavy-hole to light-hole splitting [118], so that

$$S_{\|HH}(\tilde{x}, \tilde{y}) = S_\|(\tilde{x}, \tilde{y}) \quad (8)$$

$$S_{\|LH}(\tilde{x}, \tilde{y}) = -\frac{1}{2}[S_\|(\tilde{x}\tilde{y}) + \Delta_o(\tilde{x}, \tilde{y})] + \frac{1}{2}\{9[S_\|(\tilde{x}, \tilde{y})]^2 \\ + [\Delta_o(\tilde{x}, \tilde{y})]^2 - 2S_\|(\tilde{x}, \tilde{y})\Delta_o(\tilde{x}, \tilde{y})\}^{1/2} \quad (9)$$

where $\Delta_o(\tilde{x}, \tilde{y})$ is the spin-orbit splitting.

The parameters $a, b, c_{ij}, dE_g/dP$ in Eqs. (4) to (7) above are assumed to obey Vegard's law [119], so that their respective values depend directly on the compositional profiles across the QW.

The QW confinement potential after the disordering process, obtained by modifying the bulk post-processing potential profile with the variable strain effects, is therefore given by:

$$U_r(z) = \Delta E_r(\tilde{x}, \tilde{y}) - S_{\perp r}(\tilde{x}, \tilde{y}) \pm S_{\|r}(\tilde{x}, \tilde{y}) \quad (10)$$

where $S_{\perp r}(\tilde{x}, \tilde{y}) = Q_r S_\perp(\tilde{x}, \tilde{y})$, the +ve sign represents the confined hh profile while the −ve sign represents the confined lh profile, and $S_{\|C}(\tilde{x}, \tilde{y}) = 0$.

Eq. (10) represents the carrier confinement profiles for the general case of a disordered InGaAs/InP QW structure, where interdiffusion on both sublattices takes place, but at different rates. In the two cases of comparable interdiffusion on the two sublattices Eq. (10) reduces to:

$$U_r(z) = \Delta E_r(\tilde{x}, \tilde{y}) \quad (11)$$

In the case of interdiffusion taking place only on the group III sublattice Eq. (10) becomes:

$$U_r(z) = \Delta E_r(\tilde{x}, 1) - S_{\perp r}(\tilde{x}, 1) \pm S_{\parallel r}(\tilde{x}, 1) \tag{12}$$

while for interdiffusion on group V sublattice only the carrier confinement profiles are given by:

$$U_r(z) = \Delta E_r(0.53, \tilde{y}) - S_{\perp r}(0.53, \tilde{y}) \pm S_{\perp r}(0.53, \tilde{y}) \tag{13}$$

3.1.3 Subband-Edge Calculation

The electron and hole subband structure at Γ can be determined by considering the appropriate Schrödinger equation from the BenDaniel-Duke model [120], using the envelope function scheme [121] with an effective mass approximation. The one-electron Schrödinger equation for the disordered QW can be expressed as

$$\frac{-\hbar^2}{2} \frac{d}{dz} \left(\frac{1}{m_r^*(z)} \frac{d\chi_{rl}(z)}{dz} \right) + U_r(z)\chi_{rl}(z) = E_{rl}\chi_{rl}(z) \tag{14}$$

where the growth direction z is the confinement axis, $\chi_{rl}(z)$ is the envelope wavefunction, E_{rl} is the quantized energy level with the subband energy zero at the bottom of the QW, and $l = p$ or q refers to the quantized subband energy levels for the electron and holes, respectively. This equation is solved numerically to obtain the quantized energy levels (E_{Cp}, E_{Vq}), and the envelope wavefunctions (χ_{Cp}, χ_{Vq}). Hence the interband transitions energy can be determined, as well as the overlap integral $\langle \chi_{Vq} | \chi_{Cp} \rangle$ between the p-th conduction subband and the q-th valence subband envelope functions, where

$$\langle \chi_{Vq} | \chi_{Cp} \rangle = \int_{-z_b}^{z_b} \chi_{Vq}^*(z)\chi_{Cp}(z)dz \tag{15}$$

z_b is taken as the boundary where $\chi_{rl}(z_b) \to 0$.

The electron-heavy hole transition energy, E_{chhpq}, and the electron-light hole transition energy, E_{clhpq}, can then be obtained from

$$\begin{aligned}
E_{cphhq} &= E_{ghh}[\tilde{x}(0), \tilde{y}(0)] + E_{cp} + E_{hhq} \\
E_{cplhq} &= E_{glh}[\tilde{x}(0), \tilde{y}(0)] + E_{cp} + E_{lhq}
\end{aligned} \tag{16}$$

respectively, where $E_{ghh}(\tilde{x}(0), \tilde{y}(0))$, $E_{glh}(\tilde{x}(0), \tilde{y}(0))$ are the electron-heavy hole and electron-light hole lowest direct bandgap at Γ after inter-diffusion, respectively.

3.2   Interdiffusion on Group III Sublattice Only

A strained QW structure results when considering interdiffusion on group III sublattice only. The as-grown structure modeled here is a 6 nm thick undoped $In_{0.53}Ga_{0.47}As/InP$ single QW. The compositional distribution, strain profile and carrier confinement potentials after interdiffusion, for $L_d = 1$ nm, are shown in Figure 4 [122].

During the interdiffusion process Ga atoms diffuse into the InP barrier while In atoms diffuse into the QW, forming an InGaP/InGaAs interface. After interdiffusion, the Ga and In concentration profiles are described by the error function distribution, while the As and P concentration profiles do not change, thereby maintaining an abrupt change at the interface, Figure 4(a). Since the InP lattice constant is always larger than that of InGaP, a tensile strain arises in the barrier near the interface, while the InGaAs well becomes compressively strained due to the increased In content. Consequently, the disordering process results in a strained QW structure, with the strain profile across the structure as shown in Figure 4(b). This strain profile affects the shape and separation of the conduction and valence bands, and the hh and lh potential wells no longer coincide. The confinement profile of the disordered structure remains abrupt with a width equal to that of the as-grown QW, as shown in Figure 4(c). This is due to the bandgap change from InGaP to InGaAs at the interface and is in contrast to the graded confinement profiles that result when comparable interdiffusion on both sublattices occurs in $In_{0.53}Ga_{0.47}As/InP$ QW structures, and in the case of the more extensively studied AlGaAs/GaAs [107] and InGaAs/GaAs [112] intermixed QWs.

The confinement profile shown in Figure 4(c) presents interesting features at the top of the well, near the continuum, and at the bottom of the well. A potential build-up occurs in the barrier at the interface because the bulk bandgap of InGaP is greater than that of InP [119]. However, this potential build-up is significantly modified by the effects of the tensile strain in the barrier near the interface, and can even be reversed by strain, as will be shown below for the lh well. Inside the well, the In content at the well center is less than that at the interface so that the bulk (unstrained) bandgap at the center is larger than that at the interface. The compressive strain is much smaller at the well center so that the strain effects on the bandgap are much more pronounced at the interface. The combination of the bulk bandgap and the compressive strain effects on this bandgap results in the potential at the well center being higher than at the interface, and this gives rise to two "miniwells". The strain dependent splitting of the hh and lh subbands in opposite directions causes the hh subband to have the deepest miniwells and the lh subband to ultimately result in an almost flat-bottomed well. The results indicate that in the case of the hh confinement profile, the subband ground state can be supported at energy levels that lie within the miniwells.

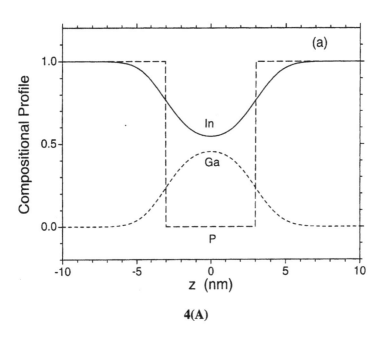

4(A)

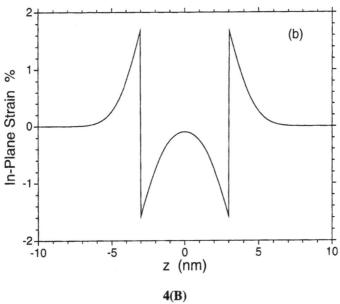

4(B)

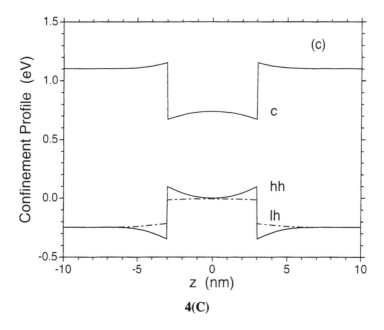

4(C)

Fig. 4. Profiles of (a) composition (b) in-plane strain, (c) electron (c), hh and lh confinement potential for $L_z = 6$ nm, $L_d = 1$ nm, $Q_C = 60\%$, with well center at $z = 0$, considering cation interdiffusion only.

Details of the effects of strain on the carrier confinement profile, for $L_z = 6$ nm and $L_d = 1$ nm, are presented in Figure 5, which shows the effect of the hydrostatic and shear components of strain on the unstrained bandgap, and the final confinement profile. Since only the hydrostatic component of strain modifies the electron confinement profile, the final confinement profile in this case coincides with the profile as modified by the effects of the hydrostatic component of strain. This is shown in Figure 5(a) where it can be seen that the compressive strain in the InGaAs well reduces the depth of the miniwells while the tensile strain in the barrier near the interface reduces the potential build-up at the top of the well. Thus strain reduces the depth of the electron confinement profile. The hydrostatic component of the compressive strain in the InGaAs well again reduces the depth of the miniwells for both the hh and lh cases, as shown in Figure 5(b), and 5(c), respectively, but the shear component of the compressive strain results in opposite shifts of the hh and lh confinement profiles; the hh miniwells now become deeper, while the lh miniwells disappear. The tensile strain in the barrier near the interface also modifies the confinement profile. The hydrostatic component reduces the height of the potential build-up for both the hh and lh confinement profiles.

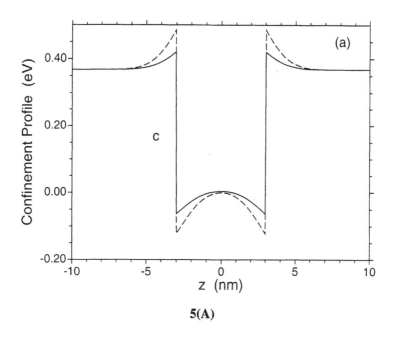

5(A)

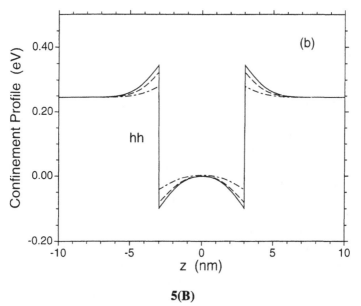

5(B)

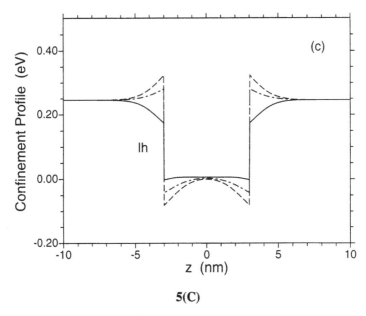

5(C)

Fig. 5.   Effects of strain and disordering on (a) electron, (b) hh and (c) lh confinement profiles in a disordered $In_{0.53}Ga_{0.47}As/InP$ single QW, considering cation interdiffusion only. The dashed lines represent the unstrained bandgap after disordering and the dot-dash lines show how the unstrained bandgap is modified by the effects of hydrostatic strain. The solid lines represent the final carrier confinement profile. The potential is taken to be zero at the well center, $z = 0$, for the unstrained bandgap, in each case. QW details: $L_z = 6$ nm, $L_d = 1$ nm.

The shear component of the tensile strain again produces opposite effects on the hh and lh confinement profiles, so that the potential build-up height increases again in the case of the hh case, while in the lh case this potential build-up disappears and, depending on the extent of the interdiffusion, may actually be reversed, see Figure 5(c). The overall result is that the strain produced in the disordered InGaAs/InP structure increases the depth of the hh confinement profile whilst it decreases the lh confinement profile depth.

The splitting of the hh and lh confinement profiles at Γ caused by the shear component of the compressive strain in the InGaAs well and by the shear component of the tensile strain in the barrier near the interface is shown in Figure 6. As already noted, this splitting strongly modifies the hh and lh confinement profiles and a significant difference in the depth of the hh and lh wells results. For the case considered here, the hh confinement profile depth at the interface is about 440 meV, while the lh confinement profile depth at the interface is about 200 meV.

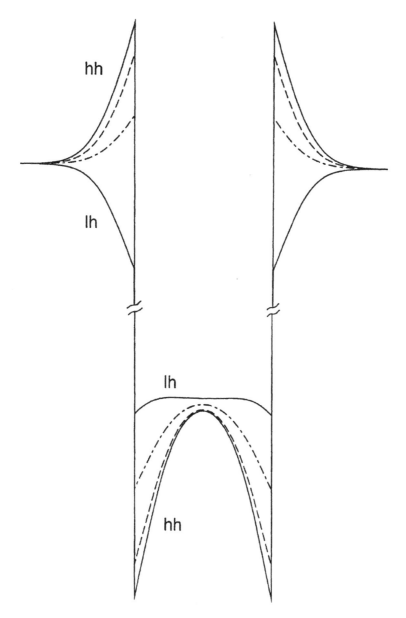

Fig. 6.   The strain induced hh-lh band edge splitting in the well and in the barrier close to the interface, considering group III interdiffusion only, for $L_z = 6$ nm and $L_d = 1$ nm. The dashed lines show the unstrained bandgap after interdiffusion, the dot-dash lines show how this bandgap is modified by the hydrostatic strain and the solid lines represent the final confinement profile.

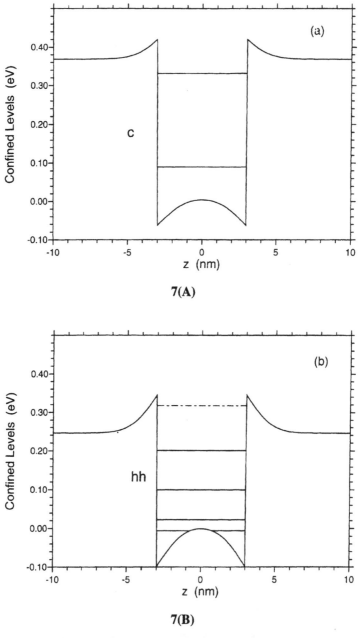

Fig. 7. The confinement profile and the confined subband states for (a) electron, (b) hh, of a disordered $In_{0.53}Ga_{0.47}As/InP$ single QW, considering cation interdiffusion only, for $L_z = 6$ nm, $L_d = 1$ nm and $Q_c = 60\%$. The potential zero is taken at the well center.

The electron (c) and hh subband-edge structure for the numerical case $L_z = 6$ nm, $L_d = 1$ nm, and a conduction band offset $Q_c = 60\%$, are shown in Figure 7. Five subband states result in the hh well with the hh ground state lying near the top of the miniwells. The square of the envelope wavefunction of the first hh subband state, $|\chi_{hh1}|^2$, is shown in Figure 8(a), and indicates that the hh1 carrier couples between the miniwells. The hh5 energy level is at the top of the well where the barrier is sufficiently thin (~ 0.5 nm) for the carrier to tunnel out of the well even in the absence of an applied electric field, and this state can no longer be considered bound. The conduction well, on the other hand, supports two confined subbands. By careful optimization of the disordering process a quasi-bound first excited state at an energy level close to the continuum can be obtained in the conduction well. Because of the presence of the potential build-up caused by the tensile strain in the barrier close to the interface, the envelope wavefunction of the first excited state would still be fairly well confined, as shown in Figure 8(b). With the application of an external electric field, the electron in this first excited state is expected to tunnel out of the conduction well, which may be of interest for device applications.

The variation of strain at a point on either side of, and close to, the interface and at the well center, as the interdiffusion proceeds, is shown in Figure 9(a). In the initial stages of interdiffusion ($L_d \simeq 1$ nm), the Ga atoms diffuse into the InP barrier so that the Ga concentration at a point in the well close to the interface drops sharply while at a point in the barrier close to the interface it rises sharply, and the changes at the well center are insignificant. Subsequent interdiffusion, however, leads to a rapid reduction of the Ga concentration at the well center as the Ga atoms diffuse deeper into the barrier and tend towards a uniform distribution across the structure. This explains the in-plane strain curves of Figure 9(a) where it can be seen that the difference between the strain at the well center and at a point in the well close to the interface is greatest for $L_d \simeq 1$ nm. This strain difference within the well affects the depth of the miniwells. For the well width considered here, the hh miniwells have a maximum depth of about 90 meV, around $L_d = 1$ nm. For $L_d \approx 6$ nm and higher, the strain profile across the well is almost linear, reflecting a more uniform Ga concentration across the well, while for large L_d the miniwells no longer exist. At the interface a lattice mismatch of $\sim 3.1\%$ exists which is almost independent of the extent of the interdiffusion, and is in agreement with experimental results for Zn diffusion reported by Schwarz et al. [39].

The variation of c1-hh1 and c1-lh1 transition energy with interdiffusion are shown in Figure 9(b). The QW bandgap energy is the c1-hh1 transition, reflecting the compressive nature of the strain induced in the QW by the disordering process. The hh1 state occurs in the miniwells for small values of L_d. As the interdiffusion proceeds the bandgap energy decreases, corresponding to a shift to longer wavelengths, which persists for prolonged

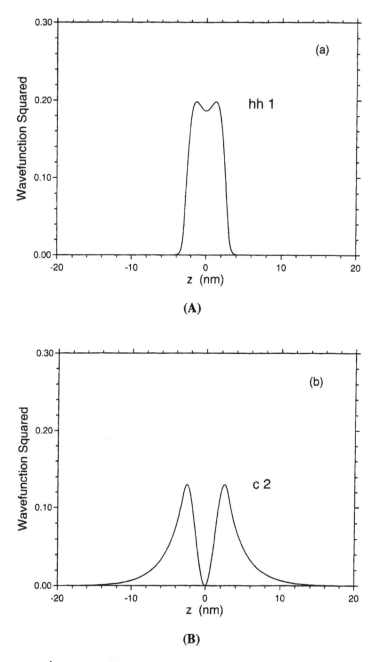

Fig. 8. $|\chi|^2$ for (a) hh1 and (b) c2 subband states for the disordered $In_{0.53}Ga_{0.47}As/InP$ QW, considering group III interdiffusion only.

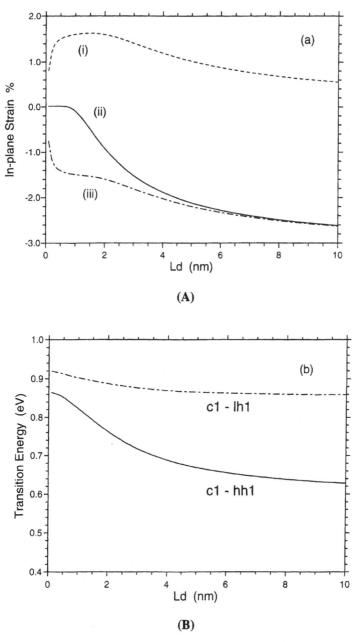

Fig. 9. Cation interdiffusion only. (a) Strain variation with L_d at three points in the QW structure: (i) in the barrier, close to the interface, (ii) at the well center and (iii) in the well, close to the interface; (b) ground state transition energy variation with L_d.

interdiffusion, as shown by experimental results for Zn induced disordering [39,73,101]. The shift of the bandedge wavelength to longer wavelengths with disordering contrasts with the experimental results for disordered AlGaAs/GaAs [107] and InGaAs/GaAs [87], and for InGaAs/InP with a comparable extent of interdiffusion for the two sublattices, where the bandedge wavelength shifts to shorter wavelengths. For the interdiffusion length range in Figure 9(b) the shift in bandgap energy corresponds to a shift in bandedge wavelength from about 1.4 μm to about 1.9 μm. Another contrasting effect is the difference between the c1-hh1 and c1-lh1 transition energy. It can be seen from Figure 9(b) that disordering under these conditions leads to an increased separation between the c1-hh1 and c1-lh1 transition energy, while in disordered InGaAs/GaAs QW structures this separation decreases [123].

3.3   Interdiffusion on Group III and Group V Sublattices

The parameter k is defined as $k = L'_d/L_d$, so that $k = 1$ represents comparable interdiffusion on the two sublattices, while $k \neq 1$ represents different interdiffusion rates. When $k < 1$ the interdiffusion rate on the group V sublattice is less than the interdiffusion rate on the group III sublattice, with $k = 0$ representing interdiffusion on the group III sublattice only, while for $k > 1$, the group V sublattice interdiffusion rate is larger than the group III sublattice one [124].

3.3.1   *Different Interdiffusion Rates*

The compositional profile, in-plane strain, and the carrier confinement potential for $L_d = 1$ nm and $L'_d = 0.25$ nm are shown in Figure 10. In this case $k = 0.25$ and the interdiffusion takes place predominantly on the group III sublattice. In the early stages of the interdiffusion process, represented by $L_d \approx 1$ nm, the Ga atoms near the interface diffuse into the barrier, while In atoms diffuse into the well, but at the well canter the Ga concentration hardly changes. Since $k = 0.25$ the As and P concentration profiles change very slowly and the interface is still quite abrupt, Figure 10(a). It can be seen by comparing Figure 4 and Figure 10 that the compositional, strain and carrier confinement profiles for the two cases $L_d = 1$ nm and $k = 0$, and $L_d = 1$ nm and $k = 0.25$, are quite similar. However, as will be seen shortly, there is a distinct difference between the results of the two interdiffusion processes for prolonged intermixing.

Figure 11 shows the compositional distribution, strain and confinement potential for the case $L_d = 1$ nm, $L'_d = 3$ nm. In this case interdiffusion on the group V sublattice is much faster than that on the group III sublattice. It can be seen from Figure 11(a) that the well is now InGaAsP with the

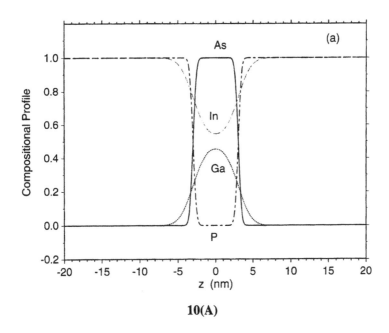

10(A)

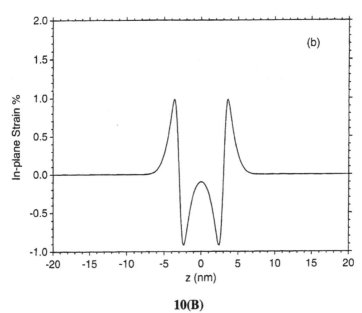

10(B)

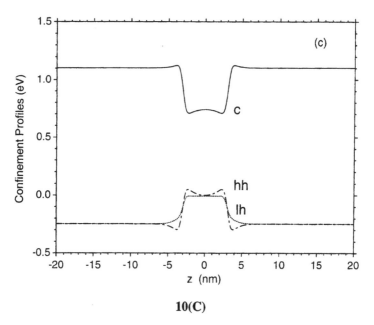

10(C)

Fig. 10. Profiles of (a) composition, (b) in-plane strain and (c) carrier confinement potential for $L_z = 6$ nm, $L_d = 1$ nm, $L_d = 0.25$ nm, $Q_c = 60\%$, with well center at $z = 0$.

As and P concentrations almost at 50% at the well center. The InGaAsP profile across the well is such that a tensile strain is set up in the disordered well, as shown in Figure 11(b). The compositional profile also causes a compressive strain in the barrier close to the interface. The combination of the unstrained bandgap and the effects of the strain profile on this bandgap gives rise to the confinement potentials shown in Figure 11(c). The effects of the tensile strain, which now is at a maximum at the well center, are also clearly evident in the confinement potential. The tensile strain causes the lh potential profile to shift toward the electron potential profile, while the hh potential profile is now shifted away from the electron profile. Both the electron and hh confinement profiles exhibit miniwells, while the lh confinement potential is a monotonically graded profile.

The effects of prolonged interdiffusion have also been investigated. The changes in the strain at the well center with interdiffusion duration, for various values of k, are shown in Figure 12. $k = 0$ represents cation interdiffusion only [122]. The compositional changes at the well center are insignificant in the early stages of interdiffusion ($L_d \approx 1$ nm) so that the strain at the well center is very small. Subsequent interdiffusion, however, leads to a rapid reduction of the Ga concentration at the well center as the Ga atoms diffuse

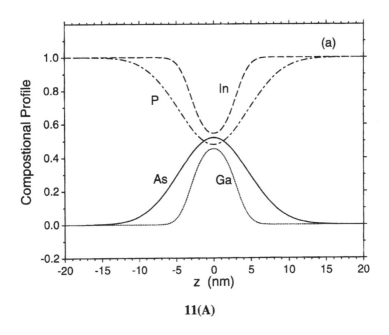

11(A)

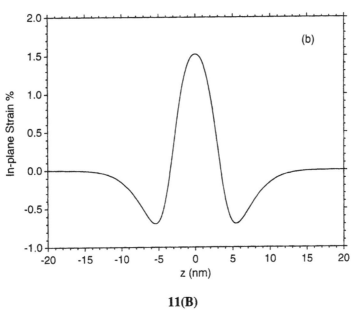

11(B)

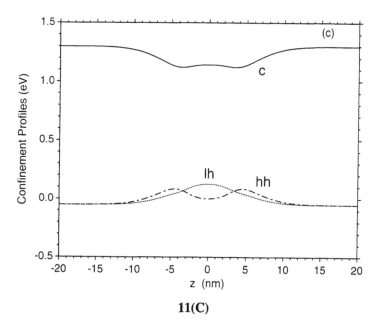

11(C)

Fig. 11. Profiles of (a) composition, (b) in-plane strain and (c) carrier confinement potential for $L_z = 6$ nm, $L_d = 1$ nm, $L'_d = 3$ nm, $Q_c = 60\%$, with well center at $z = 0$.

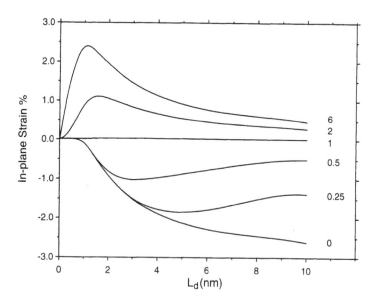

Fig. 12. Strain variation at the well center with interdiffusion, for various values of k.

deeper into the barrier and tend toward a uniform distribution across the structure. The consequent increase in the In content results in a compressive strain at the well center, which increases with further interdiffusion. For $k = 0.25$ and $k = 0.5$, the strain at the well center is still compressive and large, as shown in Figure 12. The rate of interdiffusion on the group III sublattice is significantly faster than that on the group V sublattice, and the InGaAsP composition in the well gives rise to the compressive strain. The lattice mismatch is now less than when $k = 0$ (InGaAs well), and the compressive strain is thus smaller. In contrast with the case of cation interdiffusion only, the compressive strain at the well center for prolonged interdiffusion ($L_d \gtrsim 4$ nm) starts to decrease as a result of the now more pronounced effect of the competing, although slower, interdiffusion on the group V sublattice. When $k = 1$ the InGaAsP well remains lattice-matched to the InP barrier so that there is no strain across the disordered QW structure. For $k > 1$ the rate of interdiffusion on the group V sublattice is higher than that on the group III sublattice. As already indicated, a tensile strain results at the well center. The larger the value of k, the higher is the tensile strain as a result of the increasing interdiffusion rate on the group V sublattice. The tensile strain at the well center peaks at around $L_d \approx 1$ nm for values of $k > 1$. This reflects the rapid change in the compositional profiles that takes place in the early stages of the interdiffusion process. In the later stages of interdiffusion, the compositional profiles of the constituent elements tend to a more uniform distribution, so that the strain decreases progressively and tends towards zero, independently of the value of k. Although only the variation of strain at the well center is shown in Figure 12, it must be remembered that the strain profile across the disordered well is non-uniform, as clearly shown in Figures 10(b) and 11(b), and that the strain profile across the well is modified continuously by interdiffusion. This variation in the in-plane strain with interdiffusion, together with the compositional variation, could provide a technique for shaping confinement profiles for specific purposes.

The variation of the ground state transition energy for electron-heavy hole, c1-hh1, and for electron-light hole, c1-lh1, with interdiffusion, for various values of k, is shown in Figure 13. For $k = 0$, the QW bandgap energy is the c1-hh1 transition, reflecting the compressive nature of the strain induced in the QW by the disordering process. As the interdiffusion proceeds the bandgap energy decreases, corresponding to a large red shift of the effective bandgap of the disordered QW, as shown by experimental results for Zn-induced disordering of InGaAs/InP QWs. When $k = 0.25$, a red shift in the effective bandgap again results. However, for long enough interdiffusion duration, the red shift saturates and even decreases. This corresponds to experimental results reported by Razeghi et al. [96] and by Julien et al. [100] for Zn induced InGaAsP/InP SQW and MQW intermixing. For $k = 0.5$, the red shift again saturates and then decreases, but at an earlier stage of the

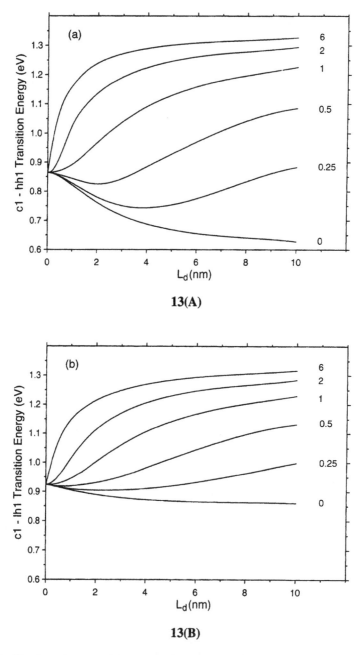

13(A)

13(B)

Fig. 13. Ground-state transition energy variation with interdiffusion, for various values of k, for (a) electron-heavy hole and (b) electron-light hole.

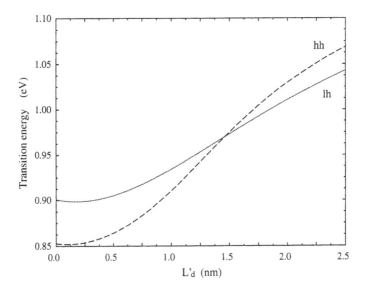

Fig. 14. The variation of heavy hole (hh) and light hole (lh) ground state transition energy with group V sublattice interdiffusion length L'_d, for the condition $L'_d = 2L_d$. At $L'_d = 1.41$ nm the two transition energies coincide.

interdiffusion process, since the competing group V sublattice interdiffusion is now more pronounced.

A blue shift in the ground state transition energy is obtained when $k > 1$. In the early stages of interdiffusion, this shift is significantly larger when compared with the results obtained for $k < 1$. For instance, the change in the c1-hh1 transition energy for $k = 2$ in going from $L_d = 0$ to 2 nm corresponds to a shift to shorter wavelengths of about 300 nm, from about 1.4 to 1.1 μm. The corresponding change in the c1-hh1 transition energy for $k = 0.5$ represents a shift to longer wavelengths of about 100 nm, from about 1.4 to 1.5 μm. The change from a ternary to a quaternary QW is more rapid when the group V sublattice interdiffusion predominates since the group V atoms composition across the as-grown QW structure is 100% As in the well and 100% P in the barrier.

The change with interdiffusion in the hh and lh ground state transition energy for the condition $L'_d = 2L_d$, is shown in Figure 14 [125]. As interdiffusion proceeds the lh transition energy approaches the hh transition energy until the lh ground state is effectively shifted above the hh ground state. This is due to the fact that the QW is now under tensile strain. This is of considerable interest for applications such as semiconductor lasers where the significantly lower effective mass of the light holes approaches better the

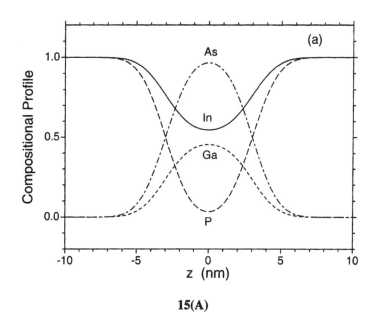

15(A)

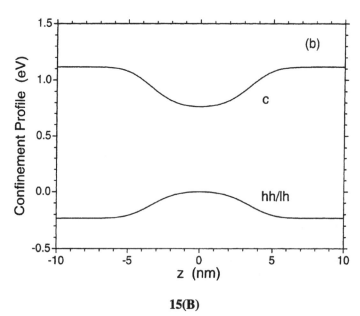

15(B)

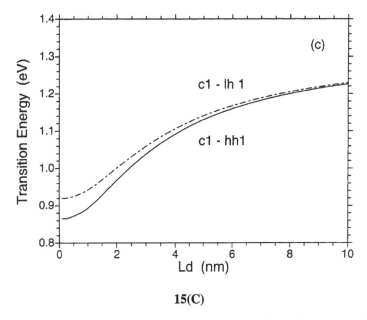

15(C)

Fig. 15. Profiles of (a) composition and (b) electron (c), hh and lh confinement potential for $L_z = 6$ nm, $L_d = 1$ nm and $Q_C = 60\%$, considering identical interdiffusion rates on group III and group V sublattices; (c) ground state transition energy variation with L_d.

ideal requirement of equal conduction and valence band masses. Optimization of the disordering process could thus induce a tensile strain that could be used to obtain desired valence band splitting effects useful in specific photonic applications. At $L'_d = 1.41$ nm the hh and lh ground state transition energies coincide. The energy level splitting of the hh and lh ground states due to the different masses is compensated by the effects of the tensile strain induced in the QW structure. For the specific structure under consideration a 0.6% tensile strain is induced in the centre of the well after interdiffusion. the convergence of the hole ground state transition energies is of significant importance for polarization-insensitive applications.

3.3.2   Comparable Interdiffusion Rates

When the rates of interdiffusion for the two sublattices are comparable, the structure remains lattice-matched. The error function distribution is again used to describe the constituent atom compositional profiles after interdiffusion, which are shown in Figure 15(a). The In$_{0.53}$Ga$_{0.47}$As well now changes to InGaAsP with a larger energy bandgap which is lattice-matched to the InP semi-infinite barrier. The results of Figure 15 show that after interdiffusion, the confinement profiles are no longer abrupt and the hh and lh profiles still coincide since no strain is present. The c1-hh1, c1-lh1 transition energies increase as disordering increases, which results in a bandgap shift to shorter wavelengths, as reported for disordered InGaAs/InP using for example S diffusion [77], as well as for disordered AlGaAs/GaAs [106] where the interdiffusion process is much better understood. The change in the c1-hh1 transition energy in going from $L_d = 0$ to 10 nm corresponds to a shift to shorter wavelengths of about 400 nm, from about 1.4 μm to about 1.0 μm, which is much larger than that for InGaAs/GaAs [123], and is a direct result of the compositional changes brought about by the disordering process. The c1-hh1 and c1-lh1 transition energies converge as L_d increases so that in the disordered InGaAs/InP structure the separation between the two transition energies almost reduces to zero. Since the disordered structure remains lattice-matched, the hh and lh confinement profiles still coincide, and the difference between the two transition energies is simply the result of the different quantization effects for the hh and lh due to their different effective masses. As interdiffusion proceeds, an effectively wider well results [112] and the difference in the quantization effects for the hh and lh therefore decreases. This is in contrast to the results obtained above for the case of cation interdiffusion in InGaAs/InP where the presence of strain results in increased separation between the c1-hh1 and c1-lh1 transition energies.

4 POTENTIAL APPLICATIONS

Strain and interdiffusion both have, independently, important potential applications in several QW photonic devices, particularly in the case of lasers, optical amplifiers, waveguides, and modulators, while interdiffusion also facilitates monolithic integration. The possibility of combining the effects of strain and interdiffusion provides a more powerful tool for the bandgap engineering that is required for device performance optimization and for photonic and optoelectronic integration.

It has been seen that QW intermixing of $In_{0.53}Ga_{0.47}As$/InP can result in simultaneous shifting of the effective bandgap and the introduction of strain in the structure. One important application area is in laser structures, since the single process of post-growth QW intermixing could be used to achieve monolithic integration while at the same time the induced strain would provide the enhanced performance associated with strained QW lasers.

In lattice-matched quantum wells the valence band has a larger density of states than that of the conduction band. This tends to increase the carrier concentration required to obtain the population inversion necessary to have enough optical gain. As already noted, in-plane compressive strain leads to a splitting of the heavy and light hole valence bands and a lowering of the effective hole mass parallel to the interface. This reduction in the density of states in the valence band allows the population inversion needed for lasing to occur at lower threshold currents [27,28], while the lighter hole mass also results in an increased modulation bandwidth compared to lattice-matched QW lasers [126]. Offsey et al. [127] reported significantly improved performance in an InGaAs/GaAs compressively strained QW laser. Tiemeijer et al. [128] reported a much better performance, than predicted by theory, for lasing from the electron-light hole transition in a tensile-strained InGaAs/InP MQW active layer. Experimental results demonstrated a higher differential gain, reflected in a low linewidth enhancement factor, a lower threshold current, and improved modulation bandwidth (compared to a compressively strained MQW laser), which are essential properties for high-speed, narrow linewidth, low-chirp semiconductor lasers for high bit-rate optical-fibre communications.

QW intermixing has also been applied to both GaAs-based [52, 129] and InP-based [54] laser structures for improved performance as well as for wavelength tuning, with high material quality being maintained after the intermixing process.

Thus, since the interdiffusion process in lattice-matched InGaAs/InP QWs can lead to either a compressive or a tensile strain, and the type and amount of strain introduced can be controlled by the process, QW intermixing of In$_{0.53}$Ga$_{0.47}$As/InP QWs is an attractive technique for optimising QW laser structures, both in terms of performance parameters as well as lasing wavelength, and at the same time integrating the laser structure with other photonic devices.

One of the consequences of separating the hh and lh levels due to quantum size effects is that QW structures exhibit strong polarization dependence, since TE polarization interacts mainly with heavy holes while TM polarization interacts with light holes. This polarization sensitivity is, however, an undesirable property in a number of applications. For instance, since optical signals propagating along optical fibres do not usually maintain light polarization, there is a need for polarization independent semiconductor laser amplifiers, as well as modulators, for optical communication systems. New high-speed applications such as in-line optical pulse reshaping and retiming, and optical demultiplexing [130] require polarization insensitivity. By introducing the right type and amount of strain in a QW structure the separation of the hh and lh levels can be counterbalanced, leading back to degeneracy of the hh and lh states, and resulting in polarization insensitivity.

Polarization-insensitive devices have been demonstrated using as-grown tensile-strained structures, including an optical amplifier [131], phase modulator [132], and lasers [128]. Polarization-independent electroabsorption

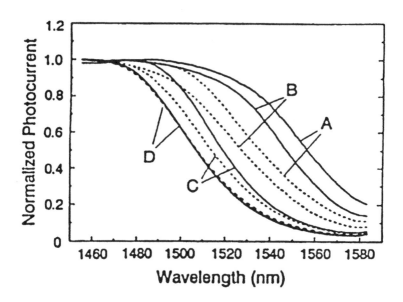

Fig. 16. Photocurrent spectra of four different implantation sections of the waveguides for TE (solid line) and TM (dashed lines) polarization, in the absence of applied field. The curves A, B, C and D correspond to SiO_2 mask layer thicknesses of 2.2, 1.2, 0.65 and 0 μm, respectively.

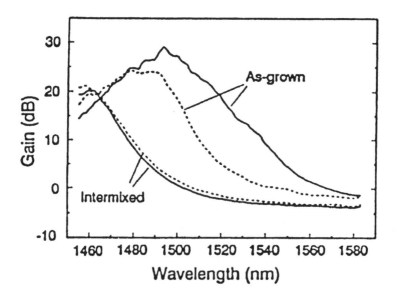

Fig. 17. Signal gain characteristic for TE (solid line) and TM (dashed lines) polarizations for the as-grown material and intermixed section (curve D of Figure 16).

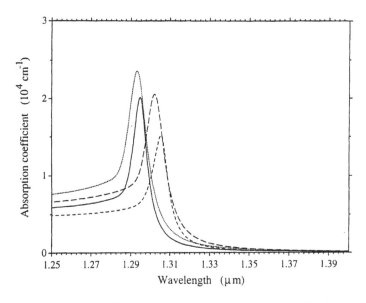

Fig. 18. Absorption coefficient spectra for TE and TM polarization for the interdiffused InGaAs/InP quantum well at $L'_d = 1.41$ nm, calculated for applied electric field $F = 0$ kV/cm (solid line for TE, dotted line for TM) and $F = 100$ kV/cm (short-dash for TE, long-dash for TM).

[133], as well as switching in a directional coupler [134], using as-grown tensile-strained InGaAs/InP QW structures have also been reported. However, a different approach is possible, where, again, the use of strain introduced by QW intermixing is the key to achieving the required property, in this case, polarization insensitivity. The feasibiltiy of this approach has been recently demonstrated by J.-J. He et al. [135]. A polarization-insensitive InGaAs/InGaAsP/InP amplifier using QW intermixing has been implemented. The as-grown structure, consisting of five In$_{0.53}$Ga$_{0.47}$As QWs in an In$_{0.74}$Ga$_{0.26}$As$_{0.57}$P$_{0.43}$ active region clad with InP, was nominally lattice-matched. The structure was implanted with high-energy phosphorus ions at a substrate temperature of 200°C followed by rapid thermal annealing at 700°C for 90 s. Three sections of the QW structure were differently intermixed by varying the energy of the ions reaching the QW structure, through the use of different SiO$_2$ mask layer thicknesses, and their performance compared to the as-grown structure. The degree of intermixing increases from section A (left as-grown) to sections B, C, and D, respectively.

Photocurrent spectra of the intermixed regions of the sample observed under no external bias are shown in Figure 16. It can be seen that the absorption edges in the B, C, and D sections are blue shifted with respect to the as-grown section A, and that the difference in the absorption edges for the TE and TM modes is reduced as the degree of intermixing increases. For section D, the TE and TM absorption edges overlap, indicating that the light and heavy hole states are now degenerate. This is confirmed by the plots of signal gain for the TE and TM modes as a function of wavelength, shown in Figure 17, thus showing the polarization insensitivity of the intermixed QW structure.

Polarization-insensitive electroabsorption using QW intermixing has also been theoretically investigated [125]. Figure 18 shows calculated TE and TM absorption spectra, in the absence of and with the application of, an electric field, obtained using the model presented earlier. The QW intermixing process can be optimized in order to obtain the required insensitivity.

The blue shift in the absorption edge and the reduction in the difference between the TE and TM mode absorption edges, resulting in polarization insensitivity, are consistent with the results of the model discussed above, for a dominant group V interdiffusion process. Since interdiffusion on one sublattice dominates, strain is introduced into the intermixed QW structure, and for group V interdiffusion tensile strain results. It is the presence of this tensile strain which balances the splitting of the hh and lh ground state subbands leading to the condition of degeneracy, and thus polarization insensitivity.

5   SUMMARY AND FUTURE TRENDS

Strain can be introduced into lattice-matched QW structures by means of post-growth QW intermixing. Reported experimental results of $In_{0.53}Ga_{0.47}As/InP$ QW interdiffusion using different intermixing techniques show that both compressive and tensile strain can result from the intermixing process. Detailed modelling of the compositional, strain, and carrier confinement profiles after intermixing is necessary to enable optimization of the intermixing process in order to exploit fully the potential of this technique. This ability of introducing strain during QW intermixing provides a powerful and feasible tool in fabricating high-performance devices and in implementing photonic integration. The potential of the process has already been demonstrated in prototype applications.

Future applications will demand improved performance of photonic devices, a reduced number of process steps in fabricating these devices, and an increased level of photonic and optoelectronic integration. The approach presented in this chapter starts with the relatively simple process of growing a lattice-matched QW structure, and then utilizes the potential of QW intermixing to achieve both localized modification of the bandgap, which is necessary for monolithic integration, as well as to obtain enhanced device performance that results when strain, either compressive or tensile, is introduced into the QW structure.

References

1. G.B. Stringfellow, *Organometallic Vapor-Phase Epitaxy — Theory and Practice*, (Academic, New York, 1990).
2. M.A. Herman and H. Sitter, *Molecular Beam Epitaxy — Fundamentals and Current Status*, in Springer Series Mater. Sci. Vol. 7 (Springer, Berlin, 1989).
3. H. Okamoto, *Jpn. J. Appl. Phys.*, **26**, 315 (1987).
4. S. Takano, T. Sasaki, H. Yamada, M. Kitamura and I. Mito, *Appl. Phys. Lett.*, **53**, 2019 (1988).
5. D.A.B. Miller, *Opt. Quantum Electron.*, **22**, S61 (1990).
6. A.L. Lentine, F.B. McCormick, R.A. Novotny, L.M.F. Chirovsky, L.A. D'Asaro, R.F. Kopf, J.M. Kuo and G.D. Boyd, *IEEE Photon. Technol. Lett.*, **2**, 51 (1990).
7. T.H. Wood, *J. Lightwave Technol.*, **LT-6**, 743 (1988).
8. D.A.B. Miller, D.S. Chemla, T.C. Damen, A.C. Gossard, W. Wiegmann, T.H. Wood and C.A. Burrus, *Phys. Rev. Lett.*, **53**, 2173 (1984).
9. I. Bar-Joseph, G. Sucha, D.A.B. Miller, D.S. Chemla, B.I. Miller and U. Koren, *Appl. Phys. Lett.*, **52**, 51 (1987).
10. M.B. Panish, *J. Cryst. Growth*, **81**, 249 (1987).
11. P.A. Claxton, M. Hopkinson, J. Kovac, G. Hill, M.A. Pate and J.P.R. David, *J. Cryst. Growth*, **111**, 1080 (1991).
12. M. Kitamura, S. Takano, T. Sasaki, H. Yamada and I. Mito, *Appl. Phys. Lett.*, **53**, 1 (1988).
13. H. Temkin, T. Tanbun-Ek and R.A. Logan, *Appl. Phys. Lett.*, **56**, 1210 (1990).
14. H. Temkin, M.B. Panish and R.A. Logan, *Appl. Phys. Lett.*, **47**, 972 (1985).
15. U. Koren, B.I. Miller, T.L. Koch, G.D. Boyd, R.J. Capik and C.E. Soccolich, *Appl. Phys. Lett.*, **49**, 1602 (1986).
16. R.J. Deri, E. Kapon, R. Bhat, M. Seto and K. Kash, *Appl. Phys. Lett.*, **54**, 1737 (1989).
17. U. Koren, T.L. Koch, H. Presting and B.I. Miller, *Appl. Phys. Lett.*, **50**, 368 (1987).
18. I. Bar-Joseph, C. Klingshirn, D.A.B. Miller, D.S. Chemla, U. Koren and B.I. Miller, *Appl. Phys. Lett.*, **50**, 1010 (1987).
19. U. Koren, B.I. Miller, T.L. Koch, G. Eisenstein, R.S. Tucker, I. Bar-Joseph and D.S. Chemla, *Appl. Phys. Lett.*, **51**, 1132 (1987).
20. P.J. Poole, M. Davies, M. Dion, Y. Feng, S. Charbonneau, R.D. Goldberg and I.V. Mitchell, *IEEE Photon. Technol. Lett.*, **8**, 1145 (1996).
21. C.E. Zah, R. Bhat, B. Pathak, C. Caneau, F.J. Favire, N.C. Andreadakis, D.M. Hwang, M.A. Koza, C.Y. Chen and T.P. Lee, *Electron. Lett.*, **27**, 1414 (1991).
22. S. Foroukar, A. Larsson, A. Ksendzov and R.J. Lang, *IEEE Trans. Electron. Devices*,, **39**, 2662 (1992).
23. B.N. Gomatam and N.G. Anderson, *IEEE J. Quantum Electron.*, **28**, 1496 (1992).
24. R. Weinmann, D. Baums, U. Cebulla, H. Haisch, D. Kaiser, E. Kühn, E. Lach, K. Satzke, J. Weber, P. Wiedemann and E. Zielinski, *IEEE Photon. Technol. Lett.*, **7**, 891 (1996).
25. S.R. Forrest, *J. Lightwave Technol.*, **LT-3**, 1248 (1985).
26. H. Künzel, R. Kaiser, W. Passenberg, D. Trommer and G. Unterbörsch, *J. Cryst. Growth*, **111**, 1084 (1991).
27. A.R. Adams, *Electron. Lett.*, **22**, 249 (1986).
28. E. Yablonovich and E.O. Kane, *J. Lightwave Technol.*, **LT-6**, 292 (1988).
29. I.J. Fritz, P.L. Gourley and L.R. Dawson, *Appl. Phys. Lett.*, **51**, 1004 (1987).
30. H. Temkin, D.G. Gershoni, S.N.G. Chu, J.M. Vandenberg, R.A. Hamm and M.B. Panish, *Appl. Phys. Lett.*, **55**, 1668 (1989).
31. G.C. Osbourn, *J. Vac. Sci. Technol.*, **21**, 469 (1982).
32. J.H. van der Merwe, *J. Appl. Phys.*, **34**, 117 (1963).
33. P. Blood, K.L. Bye and J.S. Roberts, *J. Appl. Phys.*, **51**, 1790 (1980).

34. G.C. Osbourn, *J. Appl. Phys.*, **53**, 1586 (1982).
35. D.A. Dahl, *Solid State Commun.*, **61**, 825 (1987).
36. J.W. Matthews and A.R. Blakeslee, *J. Cryst. Growth*, **27**, 118 (1974); *J. Cryst. Growth*, **29**, 275 (1975); *J. Cryst. Growth*, **32**, 265 (1976).
37. R. People and J.C. Bean, *Appl. Phys. Lett.*, **47**, 322 (1985).
38. I.J. Fritz, S.T. Picraux, L.R. Dawson, T.J. Drummond, W.D. Laidig and N.G. Anderson, *Appl. Phys. Lett.*, **46**, 967, (1985).
39. S.A. Schwarz, P. Mei, T. Venkatesen, R. Bhat, D.M. Hwang, C.L. Schwarz, M. Koza, L. Nazar and B.J. Skromme, *Appl. Phys. Lett.*, **53**, 1051 (1988).
40. F.H. Pollack and M. Cardona, *Phys. Rev.*, **172**, 816 (1968).
41. N.G. Anderson, W.D. Laidig, R.M. Kolbas and Y.C. Lo, *J. Appl. Phys.*, **60**, 2361 (1986).
42. E.S. Koteles, *Proc. Mat. Res. Soc.*, **204**, 99 (1992).
43. P.E. Brunemeier, D.G. Deppe and N. Holonyak, Jr., *Appl. Phys. Lett.*, **46**, 755 (1985).
44. S.R. Forrest, P.H. Schmidt, R.B. Wilson and M.L. Kaplan, *Appl. Phys. Lett.*, **45**, 1199 (1984).
45. R.E. Cavicchi, D.V. Lang, D. Gershoni, A.M. Sergent, J.M. Vandenberg, S.N.G. Chu and M.B. Panish, *Appl. Phys. Lett.*, **54**, 739 (1989).
46. M.A. Haase, N. Pan and G.E. Stillman, *Appl. Phys. Lett.*, **54**, 1457 (1989).
47. D. Gershoni, H. Temkin and M.B. Panish, *Phys. Rev. B*, **38**, 7870 (1988).
48. M.D. Camras, N. Holonyak, Jr., R.D. Burnham, W. Streifer, D.R. Scifres, T.L. Paoli and C. Lindstrÿ94m, *J. Appl. Phys.*, **54**, 5637 (1983).
49. W. Xia, S.C. Lin, S.A. Pappert, C.A. Hewett, M. Fernandes, T.T. Vu, P.K.L. Yu and S.S. Lau, *Appl. Phys. Lett.*, **55**, 2020 (1989).
50. Y. Suzuki, H. Iwamura, T. Miyazawa and O. Mikami, *Appl. Phys. Lett.*, **57**, 2745 (1990).
51. W.X. Zou, J.L. Merz, R.J. Fu and C.S. Hong, *Electron. Lett.*, **27**, 1243 (1991).
52. S.Y. Hu, M.G. Peters, D.B. Young, A.C. Gossard and L.A. Coldren, *IEEE Photn. Technol. Lett.*, **7**, 712 (1995).
53. T. Miyazawa, T. Kagawa, H. Iwamura, O. Mikami and M. Nagamura, *Appl. Phys. Lett.*, **55**, 828 (1989).
54. P.J. Poole, S. Charbonneau, M. Dion, G.C. Aers, M. Buchanan, R.D. Goldberg and I.V. Mitchell, *IEEE Photon. Technol. Lett.*, **8**, 16 (1996).
55. O. Wada, A. Furuya and M. Makiuchi, *IEEE Photon. Technol. Lett.*, **1**, 16 (1989).
56. R.C. Alferness, U. Koren, L.L. Buhl, B.I. Miller, R.G. Young, T.L. Koch, G. Raybon and C.A. Burrus, *Appl. Phys. Lett.*, **60**, 3209 (1992).
57. T.L. Koch and U. Koren, *IEEE J. Quantum Electron.*, **QE-27**, 641 (1991).
58. R.L. Thornton, J.E. Epler and T.L. Paoli, *Appl. Phys. Lett.*, **51**, 1983 (1987).
59. H. Ribot, K.W. Lee, R.J. Simes, R.H. Yan and L.A. Coldren, *Appl. Phys. Lett.*, **55**, 672 (1989).
60. T. Miyazawa, H. Iwamura and M. Naganuma, *IEEE Photon. Technol. Lett.*, **3**, 421 (1991).
61. A. Ramdane, P. Krauz, E.V.K. Rao, A. Kamoudi, A. Ougazzaden, D. Robein, A. Gloukhian and M. Carré, *IEEE Photon. Technol. Lett.*, **7**, 1016 (1995).
62. A. Hamoudi, E.V.K. Rao, Ph. Kranz, A. Ramadane, A. Ougazzadin, D. Robein and H. Thibierge, *J. Appl. Phys.*, **78**, 5638 (1995).
63. A. McKee, C.J. McLean, G. Lillo, A.C. Bryce, R.M. De La Rue, J.H. Marsh and C.C. Button, *IEEE J. Quantum Electron.*, **33**, 45 (1997).
64. F. Iikawa, P. Motisuke, J.A. Brum, M.A. Sacilotti, A.P. Roth and R.A. Masut, *J. Cryst. Growth*, **93**, 336 (1988).
65. B. Elman, E.S. Koteles, P. Melman, C. Jagganath, C.A. Armiento and M. Rothman, *J. Appl. Phys.*, **58**, 1351 (1990).
66. H. Temkin, S.N.G. Chu, M.B. Panish and R.A. Logan, *Appl. Phys. Lett.*, **50**, 956 (1987).
67. J.Z. Wan, D.A. Thompson and J.G. Simmons, *Nucl. Instrum. Phys. Res.*, **B 106**, 461 (1995).
68. J.H. Marsh, *Semicond. Sci. Technol.*, **8**, 1136 (1993).
69. D.G. Deppe and N. Holonyak, Jr., *J. Appl. Phys.*, **64**, R93 (1988).
70. D.G. Deppe, L.J. Guido, N. Holonyak, Jr., K.C. Hsieh, R.D. Burnham, R.L. Thomson and T.L. Paoli, *Appl. Phys. Lett.*, **49**, 510 (1986).
71. T. Miyazawa, H. Iwamura, O. Mikami and M. Nagamura, *Jpn. J. Appl. Phys.*, **28**, L1039, (1989).

72. W.D. Laidig, N. Holonyak, Jr., M.D. Camras, K. Hess, J.J. Coleman, P.D. Dapkus and J. Bardeen, *Appl. Phys. Lett.*, **38**, 776 (1981).
73. I.J. Pape, P. Li Kam Wa, J.P.R. David, P.A. Claxton, P.N. Robson and D. Sykes, *Electron. Lett.*, **24**, 910 (1988).
74. D.G. Deppe, D.W. Nam, N. Holonyak, Jr., K.C. Hsieh, J.E. Baker, C.P. Kuo, R.M. Fletcher, T.D. Osentowski and M.G. Craford, *Appl. Phys. Lett.*, **52**, 1413 (1988).
75. H.-H. Park, B.K. Kang, E.S. Nam, Y.T. Lee, J.H. Kim and O'D. Kwon, *Appl. Phys. Lett.*, **55**, 1768 (1989).
76. M.T. Furtado, E.A. Sato and M.A. Sacilotti, *Superlatt. Microstruct.*, **10**, 225 (1991).
77. I.J. Pape, P. Li Kam Wa, J.P.R. David, P.A. Claxton and P.N. Robson, *Electron. Lett.*, **24**, 1217 (1988).
78. D.R. Myers, G.W. Arnold, T.E. Zipperian, L.R. Dawson, R.M. Biefeld, I.J. Fritz and C.E. Barnes, *J. Appl. Phys.*, **60**, 1131 (1986).
79. B. Tell, B.C. Johnson, J.L. Zyskind, J.M. Brown, J.W. Sulhoff, K.F. Brown-Goebeler, B.I. Miller and U. Koren, *Appl. Phys. Lett.*, **52**, 1428 (1988).
80. M.A. Bradley, F.H. Julien, J.P. Gillies, Y. Gao, E.V. Rao, M. Razeghi and F. Omnes, *Electron. Lett.*, **26**, 209 (1990).
81. J.H. Marsh, S.I. Hansen, A.C. Bryce and R.M. De La Rue, *Opt. Quantum Electron.*, **23**, S941 (1991).
82. B.B. Elenkrig, D.A. Thompson, J.G. Simmons, D.M. Bruce, Si Yu, Jie Zhao, J.D. Evans and I.M. Templeton, *Appl. Phys. lett.*, **65**, 1239 (1994).
83. K.K. Anderson, J.P. Donnelly, C.A. Wang, J.D. Wodehouse and H.A. Haus, *Appl. Phys. Lett.*, **53**, 1632 (1988).
84. S.A. Pappert, W. Xia, X.S. Jiang, Z.F. Guan, B. Zhu, Q.Z. Liu, L.S. Yu, A.R. Clawson, P.K.L. Yu and S.S. Lau, *J. Appl. Phys.*, **75**, 4352 (1994).
85. M. Ghisoni, P.J. Stevens, G. Parry and J.S. Roberts, *Opt. Quantum Electron.*, **23**, S915 (1991).
86. B. Elman, E.S. Koteles, P. Melman and C.A. Armiento, *J. Appl. Phys.*, **66**, 2104 (1989).
87. E.S. Koteles, B. Elman, C.A. Armiento and P. Melman, *Superlatt. Microstruct.*, **9**, 533 (1991).
88. A.S. Grove, O. Leistiko and C.T. Sah, *J. Appl. Phys.*, **35**, 2695 (1964).
89. Y. Suzuki, H. Iwamura, T. Miyazawa and O. Mikami, *IEEE J. Quantum Electron.*, **30**, 1794 (1994).
90. J. Beauvais, J.H. Marsh, A.H. Kean, A.C. Bryce and C. Button, *Electron. Lett.*, **28**, 1670 (1992).
91. A. McKee, C.J. McLean, A.C. Bryce, R.M. de La Rue and J.H. Marsh, *Appl. Phys. Lett.*, **65**, 2263 (1994).
92. J. Ralston, A.L. Moretti, R.K. Jain and F.A. Chambers, *Appl. Phys. lett.*, **50**, 1817 (1987).
93. A. Rys, Y. Shieh, A. Compaan, H. Yao and A. Bhat, *Opt. Eng.*, **29**, 329 (1990).
94. C.J. McLean, J.H. Marsh, R.M. De La Rue, A.C. Bryce, B. Garrett and R.W. Glew, *Electron. Lett.*, **28**, 1117 (1992).
95. S.J. Fancey, G.S. Bullen, J.S. Massa, A.C. Walker, C.J. McLean, A. McKee, A.C. Bryce, J.H. Marsh and R.M. De La Rue, *J. Appl. Phys.*, **79**, 9390 (1996).
96. M. Razeghi, O. Archer and F. Launay, *Semicond. Sci. Technol.*, **2**, 793 (1987).
97. S.L. Wong, R.J. Nicholas, R.W. Martin, J. Thompson, A. Wood, A. Moseley and N. Carr, *J. Appl. Phys.*, **79**, 6826 (1996).
98. W.P. Gillin and D.J. Dunstan, *Phys. Rev. B*, **250**, 7495 (1994).
99. T.P. Pearsall, ed., *GaInAsP Alloy Semiconductors*, (Wiley, New York, 1982), p. 295.
100. F.H. Julien, M.A. Bradley, E.V.K. Rao, M. Razeghi and L. Goldstein, *Opt. Quantum Electron.*, **23**, S847 (1991).
101. K. Nakashima, Y. Kawaguchi, Y. Kawamura, Y. Imamura and H. Asahi, *Appl. Phys. Lett.*, **52**, 1383 (1988).
102. G.J. van Gurp, W.M. van de Wijgert, G.M. Fontihn and P.J.A. Thijs, *J. Appl. Phys.*, **67**, 2919 (1990).
103. K. Nakashima, Y. Kawaguchi, Y. Kawamura, H. Asahi and Y. Imamura, *Jpn. J. Appl. Phys.*, **26**, L 1620 (1987).
104. T. Fujii, M. Sugawara, S. Yamazaki and K. Nakajima, *J. Crystal Growth*, **105**, 348 (1990).

105. S.J. Yu, H. Asahi, S. Emura, S. Gonda and K. Nakashima, *J. Appl. Phys.*, **70**, 204 (1991).
106. T.E. Schlesinger and T. Kuech, *Appl. Phys. Lett.*, **49**, 519 (1986).
107. G.W. Arnold, S.T. Picraux, P.S. Peercy, D.R. Myers and L.R. Dawson, *Appl. Phys. Lett.*, **45**, 382 (1984).
108. J.D. Ralston, S. O'Brien, G.W. Wicks and L.F. Eastmen, *Appl. Phys. Lett.*, **52**, 1511 (1988).
109. E.H. Li, B.L. Weiss and K.S. Chan, *Phys. Rev. B*, **46**, 15181 (1992).
110. W.P. Gillin, K.P. Homewood, L.K. Howard and M.T. Emery, *Superlatt. Microstruct.*, **9**, 39 (1991).
111. J. Micallef, E.H. Li and B.L. Weiss, *Superlatt. Microstruct,*, **13**, 125 (1993).
112. E.H. Li, J. Micallef and B.L. Weiss, *Jpn. J. Appl. Phys.*, **31**, L7 (1992).
113. R.M. Cohen, *J. Appl. Phys.*, **73**, 4903 (1993).
114. K. Mukai, M. Sugawara and S. Yamazaki, *J. Cryst. Growth*, **115**, 433 (1991).
115. W.P. Gillin, S.S. Rao, I.V. Bradley, K.P. Homewood, A.D. Smith and A.T.R. Briggs, *Appl. Phys. Lett.*, **63**, 797 (1993).
116. R. People, *Appl. Phys. Lett.*, **50**, 1604 (1987).
117. H. Asai and K. Oe, *J. Appl. Phys.*, **54**, 2052 (1983).
118. R. People, *Phys. Rev.*, **B 32**, 1405 (1985).
119. S. Adachi, *J. Appl. Phys.*, **53**, 8775 (1982).
120. D.J. BenDaniel and C.B. Duke, *Phys. Rev.*, **152**, 683 (1966).
121. G. Bastard, J.B. Brum and R. Ferreira, in *Solid State Physics — Advances in Research and Applications*, Vol. 44, H. Ehrenreich and D. Turnbull, eds., (Academic, New York, 1991), p. 232.
122. J. Micallef, E.H. Li and B.L. Weiss, *J. Appl. Phys.*, **73**, 7524 (1993).
123. J. Micallef, E.H. Li, K.S. Chan and B.L. Weiss, *Proc. SPIE*, **1675**, 211 (1992).
124. W.-C. Shiu, J. Micallef, I. Ng and E.H. Li, *Jap. J. Appl. Phys.*, **34**, 1778 (1995).
125. J. Micallef, J.L. Borg and E.H. Li, *Opt. Quantum Electron.*, **29**, 423 (1997).
126. I. Suemune, L.A. Coldren, M. Yamanishi and Y. Kan, *Appl. Phys. Lett.*, **53**, 1378 (1987).
127. J.P. Offsey, W.J. Schaff, P.J. Tasker and L.F. Eastman, *IEEE Photon. Technol. Lett.*, **2**, 9 (1990).
128. L.F. Tiemeijer, P.J.A. Thijs, A.J. de Waard, J.J.M. Binsma and T.V. Dongen, *Appl. Phys. Lett.*, **58**, 2738, (1991).
129. S. Bürkner, J.D. Ralston, S. Weisser, J. Rosenzweig, E.C. Larkins, R.E. Sah and J. Fleissner, *IEEE Photon. Technol. Lett.*, **7**, 941 (1995).
130. M. Suzuki, H. Tanaka, N. Edagawa and Y. Matsushima, *J. Lightwave Technol.*, **LT-10**, 1912 (1992).
131. M. Joma, H. Horikawa, C.Q. Xu, K. Yamada, Y. Katoh and T. Kamijoh, *Appl. Phys. Lett.*, **62**, 121 (1993).
132. Y. Chen, J.E. Zucker, N.J. Sauer and T.Y. Chang, *IEEE Photon. Technol. Lett.*, **4**, 1120 (1992).
133. K.G. Ravikumar, T. Aizawa, S. Suzuki and R. Yamauchi, *Appl. Phys. Lett.*, **61**, 1904 (1992).
134. T. Aizawa, Y. Nagasawa, K.G. Ravikumar and T. Watanabe, *IEEE Photon. Technol. Lett.*, **7**, 47, (1995).
135. J.-J. He, S. Charbonneau, P.J. Poole, G.C. Aers, Y. Feng, E.S. Koteles, R.D. Goldberg and I.V. Mitchell, *Appl. Phys. Lett.*, **69**, 562 (1996).

CHAPTER 6

Photonic Integration Using Quantum Well Shape Modification Enhanced by Ion Implantation

EMIL S. KOTELES

Institute for Microstructural Sciences, National Research Council of Canada, Ottawa, Ontario, Canada K1A 0R6

1   ABSTRACT

Progress in the development of ion implantation enhanced quantum well shape modification as a technique for monolithically integrating optoelectronic devices of varying functionalities is reviewed. Fundamental issues related to the physics of the technique, both material issues and device issues, are considered and the performance of both discrete and integrated devices is discussed. The main conclusion of this review is that there are no inherent drawbacks to the utilization of this technique and some serendipitous advantages but that more work on reproducibility and reliability is required to ensure commercial viability.

2   INTRODUCTION

The possibility of accurately and precisely modifying the optical properties of epitaxial semiconductor heterostructures in a spatially selective manner has exciting device implications. An important, long range, objective is the monolithic integration of optical, optoelectronic and electronic devices for optical communications and other purposes. This requires precise lateral and vertical modification of the optical parameters and electrical characteristics of such layers. Controlled interdiffusion of barrier and well atoms across quantum well (QW) heterointerfaces could satisfy these requirements by modifying as-grown QW shapes and thereby their quantized energy levels. It has been shown that, by employing such techniques, optical bandgap energies can be altered, the quantum confined Stark effect can be enhanced, and polarization sensitivity of certain InP-based QW structures can be reduced.

However, it should be made clear at the outset that there is an inherent trade-off between fabrication simplicity and device performance. Ideally, a photonic integrated circuit (PIC) should contain optimized optoelectronic devices; that is, each device should have a unique structure (layer thickness, doping, strain, etc.) which permits it to perform at a maximum efficiency. Usually this implies that different devices have different structures. For example, a laser diode may operate most effectively by incorporating several narrow QWs in its waveguide structure, modulators might need a similar waveguide structure but with a higher energy bandgap and, perhaps, a larger number of QWs, whereas a transparent waveguide structure required to transport the optical beam around a planar waveguide circuit would have minimum loss only if it possessed a larger bandgap and no QWs.

The most straightforward method of fabricating such an optimized PIC would be to grow each structure in turn, using etch steps between growths to remove material wherever it is not required and replacing it with a different material structure. Although straightforward in principle, commercial viability is another issue. Growing high quality semiconductor layers in a modern epitaxial growth chamber, whether metal-organic chemical vapour deposition (MOCVD), molecular beam epitaxy (MBE), or any of their variants (CBE, gas-source MBE, etc.) is still something of an art. Adding etching steps between growths on the same substrate usually reduces the yield to low levels, especially if multiple etch-and-regrow processes are needed. Thus, developing a PIC fabrication procedure which requires only a single growth step epi-wafer has an enormous potential advantage — fabrication simplicity leading to higher yields.

This is the lure and the promise of QW shape modification. In principle, multiple etch-and-regrowths are replaced by a single standard epitaxial growth followed by a simple bandgap modification procedure. [Selective area growth for PICs, which also uses a single growth run, requires complex and intricate control of growth parameters, which also impacts yield.] The trade-off is that a single semiconductor waveguide, incorporating QWs, must perform multiple functions in, at least, an adequate fashion. This is not as difficult a task as it may appear. It has long been known, for example, that a reverse biased laser diode structure can perform adequately as a photodetector. In fact, simple PICs consisting of a laser with a rear facet photodetector (essentially the laser structure under reverse bias) have been developed by several research institutes. Further, as will be pointed out later in this review, in some instances QW shape modified waveguide devices, such as quantum confined Stark effect modulators and optical amplifiers, can have enhanced performance compared with standard structures.

2.1 Early Studies

Semiconductor QW shape modification as a technique for altering the optical properties of semiconductor heterostructures in a spatially selective manner has a history which can be traced back many years. In hindsight, controlling QW optical properties with this process is self-evident due to two fundamental realities. First, the optical properties of quantum wells are critically dependent on their exact structure (well width and composition, barrier composition, bandgap offsets, strain, and, of particular importance in this discussion, QW shape). Second, quantum wells are inherently metastable structures. In principle, if we wait long enough, even at room temperature, eventually the atoms of the well layer will completely interdiffuse with those of the barrier layers due to the chemical concentration gradient across the well, producing an amorphous alloy in place of the quantum well. It has been known for some time that this process can be enhanced, given the proper external perturbations. These include the presence of defects and impurities, elevated temperatures, and strain. In 1983, in one of the first studies of this effect, elevated temperatures were employed to blue shift GaAs/AlGaAs QW lasers using layer interdiffusion [1]. The goal was to increase the bandgap of the well material. A more systematic investigation in 1986 determined the interdiffusion rate of Al and Ga in undoped GaAs/AlGaAs QWs annealed at elevated temperatures (650 to 910°C for 1 to 6 hours) [2]. For the first time, fundamental bandgap energy changes, resulting from QW shape modification, were conveniently monitored by measuring low temperature QW photoluminescence (PL) spectra. The PL peak from a QW is generated by fundamental radiative transitions across the forbidden gap (energy bandgap) in energy-momentum space. (An earlier attempt to determine interdiffusion rates by utilizing compositional profiling with sputter Auger techniques was somewhat inconclusive due to experimental complications [3].) Such PL studies have become the standard technique for investigating the bandgap energy changes due to QW intermixing in many different material systems [4].

The GaAs/AlGaAs QW study [2] showed that interdiffusion of Al and Ga atoms across the well/barrier interface is very small at temperatures of 650–750°C but significant for temperatures greater than 800°C. This presented the possibility of spatially selective QW shape modification if some efficient and simple technique for enhancing interdiffusion at lower temperatures (650–750°C) in specified areas of a QW epi-wafer could be found.

At about the same time as these studies, several groups were fabricating buried QW waveguide device structures such as lasers and waveguides by completely disordering QWs on either side of the guiding region [5]. This was accomplished by diffusing impurity atoms, such as Zn, into the structure at elevated temperatures. The presence of the impurity atoms increased the interdiffusion across the well/barrier interface to such an extent that no visage of the QWs could be observed after annealing. The QWs were replaced by an alloy formed from the well and barrier material. Guiding structures could be fabricated using this technique since the bandgap energy, and thus the effective index, of the disordered alloy is different from that of the original QW material. However, the quality of the disordered material was generally lower than that of the starting QWs, as judged by PL intensity, loss, etc. and the advantages of QWs in device structures was lost. Thus interest shifted to developing techniques for modifying the shapes of QWs in a spatially selective manner without destroying their properties altogether. The remainder of this paper will focus on techniques for less drastic alterations of QW shapes which, in principle, should cause minimum degradation of structure quality and permit the device structure to retain or enhance the advantageous properties of quantum wells.

2.2   Ion Implantation Induced QW Shape Modification

One of the fundamental issues in QW shape modification has always been the reliability, reproducibility, and simplicity of the spatially selective enhancement process. Early work, using impurity free QW shape modification in GaAs/AlGaAs QWs, utilized Ga vacancies formed in GaAs at elevated temperatures under a SiO_2 cap layer. The technique was found to produce large bandgap shifts [6]. It is believed that Ga vacancies are the main contributors to the process, although strain (which develops between the dielectric film and the semiconductor) and Si diffusion may also enter into the process (provided that the thermal anneal is short and hot, ion transport is not believed to be an important mechanism for QW shape modification – see later) [7]. The SiO_2 cap absorbs Ga atoms during the anneal process, leaving Ga vacancies behind in the GaAs. These diffuse down into the QW region and enhance the interdiffusion mechanism at the well/barrier interface. Unfortunately, the process, while effective, was not very reproducible. The difficulty appeared to be related to the porosity of the deposited SiO_2 cap layer and/or the semiconductor surface treatment.

A way to bypass this problem is to produce vacancies using ion implantation. The idea is to produce damage, and therefore vacancies (among other entities), near the surface of the structure. These, then diffuse into the depths of the structure during rapid thermal anneal (RTA), as with the SiO_2 cap procedure. Everything in the procedure is identical with the dielectric cap technique, except the technique for generating vacancies near the surface. Ion implantation, as a method for producing vacancies, has the virtues of being simple, of being insensitive to surface morphology or conditions, of being highly reproducible, and of being compatible with any material system. The intermixing effect is due to the thermal diffusion of vacancies generated in the surface region of the sample, not due to deposition of ions deep in the structure in the middle of the QWs as is the case in other ion implantation intermixing techniques [8,9].

An early attempt (1982) to disorder a GaAs/AlAs QW structure by implanting silicon directly into the QWs produced material with a small blue shift and weak PL [8]. A later effort was more successful. Localized Ga^+ implantation led to the enhancement of the Ga/Al interdiffusion by about two orders of magnitude [9] Further, complete recovery of the optical quality, as judged by the cathodoluminescence intensity of the material, was observed after RTA. However, the more general observation is that implantation damage directly introduced into the QW region is not consistent with maintaining optical quality, even if the implantation is followed by an anneal process [10] This issue will be discussed in more detail later.

3   FUNDAMENTAL PHYSICS OF QW SHAPE MODIFICATION

In this section we will consider factors which govern the magnitude of the QW shape modification process and the resultant fundamental bandgap shift. To simplify concepts, many examples will be taken from the lattice matched GaAs/AlGaAs QW material system so as to eliminate the influence of strain on these processes. The specific effects of strain on QW interdiffusion will be discussed later. Also, some examples of SiO_2 cap modified QWs will be used as the influence of vacancy generation and diffusion direction are exceptionally clear in this technique. However, the focus of this paper remains ion implantation enhanced QW shape modification and in all cases, the physics inferred from other techniques, such as SiO_2 capping, will be directly applicable to that technique.

3.1   Parameter Space

The magnitude of the effect of QW shape modification on the optical and electrical properties of a given QW are dependent on a number of parameters, which can be grouped into three areas: those parameters relating to the QW structure itself, those dealing with the details of the generation of the shape modification enhancing entities (e.g., vacancies), and those having to do with the process itself (i.e., the anneal process). Unless otherwise specified, all of the experimental results given below will be based on ion implantation QW shape modification.

3.1.1   QW Structure

As illustrated in Figure 1, the effect of enhanced QW shape modification is, to first order, an increase the bandgap energy of the QW material ($E_g' > E_g$) due to the movement of barrier atoms into the well material and a rounding of the as-grown, abrupt QW interfaces, which are typical of epitaxial growth. This rounding-off process is usually modeled by a symmetric error function [11] or a modified Poschl-Teller potential [12]. Since the net effect of thermal annealing is to move barrier atoms into the well material, this necessitates that a blue shift of the QW PL peak be produced; i.e. the bandgap energy of a single QW is always increased by this mechanism (assuming unstrained material before and after processing). The maximum increase possible occurs when a wide QW (with bandgap energy, E_g^{well}), is completely intermixed with the

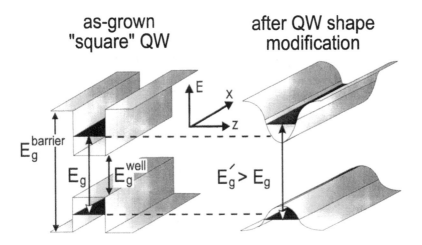

as-grown
"square" QW

after QW shape
modification

E_g^{barrier}

E_g E_g^{well}

$E_g' > E_g$

Fig. 1. Schematic diagram illustrating the effect of well/barrier interdiffusion on the shape and bandgap energy of quantum wells.

barrier material, E_g^{barrier}, yielding a total blue shift of about $E_g^{\text{barrier}} - E_g^{\text{well}}$. Thus the maximum bandgap shift possible in a given QW material system is equal to the total "depth" of the well (i.e., the sum of the conduction and valence band QWs). This is the fundamental limit on the bandgap shift due to QW interdiffusion which can only be modified by changing the initial structure of the QW.

For a given amount of mass transport across the well/barrier interface, the increase in QW bandgap will be a sensitive function of the original amount of material in the well, which, of course, depends on the well width. Thus, there is a strong dependence of the magnitude of bandgap shift on the QW width. However, this dependence is not a monotonic function of width. (see, for example, Figure 2 — in this case, QW shape modification in a sample containing a series of QWs of differing widths was accomplished by the SiO_2 cap technique, so that the source of intermixing-enhancing vacancies was known to be generated at the surface of the structure. Thus, there is no "resonance" effect of bandgap energy shifts due to the location of the generated vacancies with respect to the position of the QWs.). Note that intermediate width QWs ($\sim 4\,\text{nm}$) experience the largest shifts. This behavior is a direct consequence of the dependence of the QW transition energy on QW width. QW energies are weak functions of width for very wide and very narrow QWs, regions where asymptotic bandgap energy values are being approached (E_g^{well} for wide QWs and E_g^{barrier} for narrow QWs). The most rapid change in QW bandgap energy with well width takes place in QWs of

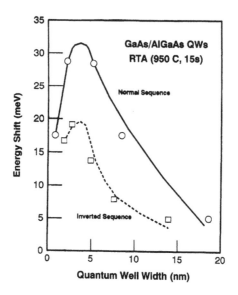

Fig. 2. Energy shifts of a series of GaAs/AlGaAs QWs of varying widths contained in two single samples. Both normal (QWs increasing in width from the surface) and inverted (QW widths decreasing away from the surface) structures are shown.

intermediate width. (For a similar reason, the maximum PL half width of a QW, due to inherent fluctuations in typical QW widths, occurs at intermediate well widths. In the case of GaAs/Al$_{0.38}$Ga$_{0.62}$As QWs, the maximum PL half-width occurs at a well width of about 2 nm [13]. Thus changes in the effective QW width, and depth, induced by intermixing produce the largest changes in QW transition energies in QWs whose bandgap energies lie away from relatively flat asymptotic regions, i.e., those with intermediate widths.

This dependence (but not the absolute magnitude of bandgap shifts) is independent of the depth of QWs from the surface of the structure. This was readily verified by measuring the PL blue shifts of two multiple QW samples (i.e., material which contained a series of QWs of different widths, each of which could be studied independently using PL spectroscopy) which are similar except for the vertical ordering of the QWs with respect to width. The normal sequence had five QWs with widths increasing with increasing distance from the surface while the inverted sequence had decreasing widths. There was no significant difference in the dependence of bandgap energy shifts on the QW width, as is evident in Figure 2 [14]. In both cases, the maximum blue shift occurred for QWs with widths of the order of 4 nm. The difference in the absolute magnitude of the bandgap shifts in these two samples, which were processed under nominally identical dielectric deposition and annealing conditions, is illustrative of the reproducibility problem with the SiO_2 cap technique.

Normally when QW shape modification due to enhanced interdiffusion is modeled, the assumption is made that the quantum well changes shape in a symmetric fashion. Then an error function, or some similar arithmetic function is utilized at each interface to approximate the interdiffused shape and to calculate the resulting change in bandgap energy. However, if the intermixing is due to a unidirectional diffusion of vacancies from near the surface into the depths of the sample structure (the case when low energy ion implantation is used and definitely so for SiO_2 capped interdiffusion), there would be an expectation of some asymmetry present in the shape of the modified QW. Although experimental manifestations of effects due to asymmetry are subtle and difficult to observe in single QWs, in coupled double quantum wells (CDQW) they can be pronounced [15]. CDQWs are heterostructures in which two QWs are separated by a barrier narrow enough to allow overlap between electronic wavefunctions of the quantized levels in the wells and thus to permit interaction and splitting of QW energy levels. In symmetric CDQWs, the wave functions of the split levels are purely symmetric or antisymmetric and selection rules dictate that only transitions between conduction and valence band levels with the same symmetry are allowed [15] However, symmetry-forbidden transitions can dominate the photoluminescence excitation spectrum (PLE — analogous to an absorption spectrum) when moderate symmetry-breaking electric fields are applied [16]. [In addition, the electric field reduces the energy of the lowest energy exciton transition in a CDQW (Stark effect).] Assuming that RTA of SiO_2 capped CDQWs produces the same type of interface rounding as in SQWs, it might be possible that the RTA induced reduction in height of the narrow barrier between the coupled QWs would produce an increase in coupling between the two wells which would, in turn, decrease the energy of the lowest energy exciton transition. In the PLE spectrum, assuming the CDQW remained symmetric, only the four allowed transitions (coupling states with the same symmetry) would be observed but their energies would change as the coupling increased. Two of the transitions (symmetric to symmetric) would decrease in energy and two (antisymmetric to antisymmetric) would increase in energy. Thus the Stark effect (or, equivalently, the presence of asymmetry in the structure) and increased coupling both decrease the energy of the lowest lying exciton transition (which is observed in PL). However, by studying the PLE spectra the presence of asymmetry in the CDQW (giving rise to symmetry forbidden transitions) after processing can be uniquely ascertained.

The effect of heat treatment on the PLE spectra of an undoped CDQW sample is illustrated in Figure 3 [17]. On the left side of the figure, 5K PLE spectra of four pieces of the undoped CDQW wafer either untreated (as-grown) or heat treated (SiO_2 capped RTA at 800, 825 and 875°C) are

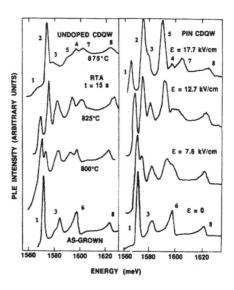

Fig. 3. 5K PL excitation spectra of a p-i-n GaAs/AlGaAs CDQW sample under different electric fields (right) and undoped CDQW samples subjected to enhanced QW interdiffusion via SiO$_2$-capped RTA at different temperatures (left). (reference 17)

presented. For comparison, on the right side, 5K PLE spectra of a p-i-n CDQW sample, subjected to various electric fields, are shown. It is clear that the spectrum of the undoped, as-grown piece (bottom left) is identical to that of the p-i-n CDQW when the internal electric field is zero (bottom right). Only the four symmetry allowed transitions (labeled 1, 3, 6, and 8) are observed. In an electric field (right side), symmetry-forbidden transitions (2, 4, 5, and 7) become possible, and these increase in magnitude to dominate the spectrum when the heterostructure asymmetry is large [for example, when $E = 17.7$ kV/cm (top right)] while symmetry allowed transitions become weak since they become spatially indirect (i.e., they connect electrons in one well with holes in the adjacent well). On the left hand side of Figure 3, as the RTA processing temperature is increased, normally symmetry-forbidden transitions become observable and symmetry-allowed transitions become weak (see top left). Also, the lowest energy transition decreases in energy as the anneal temperature is increased (left) [or as the electric field is increased (right)]. In addition, the lifetime of the lowest energy transition in the CDQW annealed at 875°C is double the lifetime of the as-grown sample and equivalent to the lifetime observed in the p-i-n CDQW subjected to an electric field of 33 kV/cm. These similarities are strong evidence that undoped

CDQWs subjected to vacancy enhanced thermal annealing, due to surface generated vacancies, have their shapes modified in an asymmetric fashion.

It is possible to speculate that this occurs due to the non-reciprocal nature of the vacancy enhancement process at the well/barrier interface. That is to say, vacancies reaching the interface from the barrier side (that is, directly from the large bandgap side of the interface) may have a different magnitude of influence on the interdiffusion process compared with vacancies reaching the interface from within the QW (that is, from the low bandgap side). Further, if significant gettering takes place (that is, if the first interface encountered by the vacancies somehow neutralizes or traps them reducing their number) the interdiffused QW will have skewed or non-symmetric interfaces. Whatever the cause, the experimental evidence from CDQWs seems clear: QWs interdiffused using a uni-directional enhancement technique are not symmetric. This implies that more sophisticated models, taking asymmetry into account, are needed to interpret experimentally measured bandgap shifts due to QW shape modification. (Attempts to fit ion-implantation-enhanced PL shifts from a series of GaAs/AlGaAs QWs with different widths using a symmetric error function with a single diffusivity constant have proven unsuccessful [18]. Note, however, that this was shown to be possible if isotropic enhancement of QW interdiffusion is performed [2]. In this case, only elevated temperatures were utilized to enhance interdiffusion resulting in symmetrically interdiffused barriers.)

3.1.2   Ion Implantation Conditions

The magnitude of the shape modification experienced by a particular QW due to ion implantation is a direct consequence of the ion damage generation process. The number, type, and location of vacancies, interstitials, and other defects have a significant impact on the QW interdiffusion process. In initial studies, ions were implanted directly into the QW region in order to effect as large a change as possible in the QW shapes [8]. However, this had a tendency to inflict such damage on the QWs that they sometimes didn't recover their full optical quality (as judged by the intensity and shape of their PL spectra) even after extensive annealing. Studies of shallow ion implantation demonstrated that significant bandgap energy blue shifting was possible, even when the ion induced damage was well removed from the vicinity of the QWs [6,19]. This was considered a consequence of the large diffusivities of certain of the damage species. The exposition in this section will deal mainly with shallow ion implantation since the complications are fewer than when ions are directly implanted into QWs. Furthermore, since the penetration depth needs to be well controlled, implantations are usually performed off-axis (i.e., away from the crystallographic axis of the epitaxial layer) in order to minimize channeling effects.

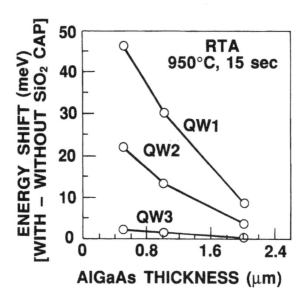

Fig. 4. Blue shifts of the bandgap energies of three GaAs/AlGaAs QWs with differing widths (QW1: 5 nm; QW2: 8 nm; QW3: 20 nm) as a function of the thickness of an AlGaAs overlayer lying between the QWs and the surface. (reference 21)

(Implantation using focused ion beams (FIB) has also been utilized to write spatial patterns directly onto wafers, eliminating the need for a mask [20]. However, this is a slow process unlikely to be commercially viable.)

It is generally understood that, since vacancies and interstitials (among all the types of defects produced by ion implantation) have the largest diffusivities, they are the main players in the interdiffusion enhancement process. There is substantial direct and indirect evidence to support this thinking, as will be pointed out in this section.

First, it is understood that SiO_2 capped QW shape modification relies mainly on vacancies (there is some contradictory evidence of strain playing a role). And since it has been shown to produce large bandgap blue shifts, the presence of vacancies is probably an important component of the mechanism. The implication is that this is true even in the ion implantation technique. The data in Figure 4 (from a SiO_2 capped QW shape modification experiment) demonstrate that vacancies can enhance QW shape modification even in QWs located two microns below the surface of the structure [21]. Clearly this is evidence that vacancies can diffuse long distances when subjected to elevated temperatures. Interstitials (i.e., ions located between lattice planes) are also known to diffuse rapidly and for large distances in semiconductor crystals. Thus it is assumed that they are also

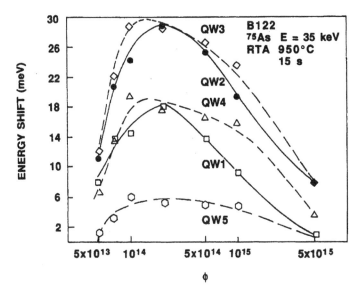

Fig. 5. Fluence dependence of the bandgap energy shift for five single GaAs/AlGaAs QWs after enhanced QW interdiffusion. The nominal widths of QWs 1 to 5 are 2, 4, 7, 11, and 20 nm respectively. (reference 19)

important in the ion implantation induced shape modification mechanism, although there is less direct evidence of their influence.

It is to be expected that the total bandgap energy shift should be a direct function of the number of enhancing species (vacancies, interstitials, etc.) generated by the ion implantation process at a fixed implantation energy. Figure 5 demonstrates that this is the case, but, generally, only for low fluences [19]. In this experiment, 35 keV As$^+$ ions were implanted into the structure to a depth of about 16 nm, which is well removed from the GaAs/AlGaAs QWs located some 300 nms below the surface. It appears to be a general phenomenon that as the total number of ions impacting the samples increases, the bandgap energy blue shift first increases monotonically, reaches a maximum, and then decreases. This behaviour is independent of QW width, as illustrated in Figure 5. It has also been observed in most material systems studied and is thought to be related to amorphization. This is a process by which, at large ion fluences, the high density of generated defects leads to clustering of vacancies (and other enhancing defects) and to the formation of extended defects. These complexes have much lower mobilities than single vacancies and so do not contribute significantly to the interdiffusion process taking place below the implantation region.

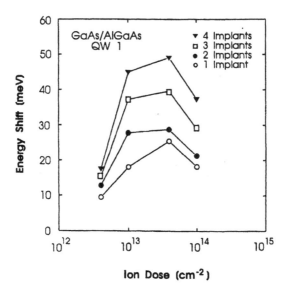

Fig. 6. Bandgap energy shift of a 3.5 nm wide GaAs/AlGaAs QW as a function of ion dose after one, two, three, and four consecutive implants and anneals. (reference 22)

This would tend to reduce the utility of shallow ion implantation induced interdiffusion since the maximum achievable bandgap shift is much smaller than the theoretical maximum discussed above. However, the annealing procedure not only activates the interdiffusion process at the QWs, it also removes much of the damage caused by the ion implantation and returns the material close to its initial state. This suggests that another ion implantation step, followed by an another anneal, would produce a bandgap shift similar to that observed originally. This is indeed the case, as can be seen in Figure 6 [22]. When only an anneal (without an implant step) was performed, no change in PL energies was observed, demonstrating the necessity of the presence of implantation induced defects for the interdiffusion process. In principle, this procedure can be repeated indefinitely until the QW is completed interdiffused with the barrier material. This is a tedious procedure, however. Multiple implants and anneals is more useful as a technique for producing, say, areas with two or three different QW bandgaps on a single chip.

The effects of the identity of the ion species on the implantation process are two fold. First, for a fixed implantation energy, the heavier the ion, the greater the damage and the larger the number of defects produced. However, large numbers of defects may give rise to extended complexes which reduce the interdiffusion process. There is some evidence to suggest that implantation of low mass ions (i.e., H^+, protons) has some advantages [23]. Protons preferentially produce discrete (point) defects which are more efficient in QW shape modification, as discussed above. In fact, up to doses of 5×10^{16} H^+/cm^{-2}, there is no indication of saturation effects. However, being light, protons penetrate large distances, even for low energy implants, and may leave some damage in the active layers after annealing. A significant broadening of the PL peaks may be an indication of this effect. In any case, developing efficient masking material for such long ranging ions may be a challenge.

Secondly, the electrical activity of the implanted ion must be considered. Implanting with an ion species which is normally electrically active in the semiconductor (e.g., Si^+) adds a complication if these species diffuse during the anneal procedure. Their presence, especially if dispersed throughout an active device structure, may degrade its performance. On the other hand, there may be instances in which it is desirable to add active impurities into the structure, to create an ohmic p-contact, for example, and, to increase the bandgap of the QW material. This could be accomplished simultaneously using implantion of electrically active impurity ions. However, little research has been undertaken on this possibility so far. Most work has been performed using electrically neutral ions to generate QW shape modifying defects; generally As^+ for GaAs-based QWs and P^+ for InP-based QWs.

Other parameters which need to be considered include sample temperature during implantation, implantation rate, and implantation energy. There have been studies suggesting that the magnitude and nature of ion-implantation-induced damage is a sensitive function of both the substrate temperature during implantation [24] and the dose rate [25]. Initial studies in the InGaAs/GaAs QW system indicate that sample temperature during implant has little effect on the magnitude of PL shift achieved (only a small increase in PL peak energy was observed) but a large effect on the quality of the QWs after processing (the PL intensity doubled when samples were implanted at temperatures a couple of hundred degrees above room temperature.) [26]. Furthermore, low dose rates also increased PL intensities. However, much more research is needed to completely understand the effects of these parameters on the QW shape modification process. [The procedure described in this review needs to be differentiated from ion mixing, a technique in which ions are implanted into QWs held at elevated temperatures (a significant fraction of the growth temperature). Although this procedure yields significant enhancement of the QW shape modification effect, sample quality suffers significantly [27]].

The situation regarding the effect of ion implantation energy on QW shape modification is even more complicated since an increase in ion energy not only generates more vacancies, interstitials, etc. but also changes their distribution in the structure. Thus, as the ion implantation energy is increased, during implantation into a typical QW structure, defects are initially generated above a QW region, then directly in the QWs, and finally in a region beyond the QWs. The effect of ion energy on the distribution of damage in the structure can be estimated using computer modeling programs such as TRIM. This is illustrated in the lower halves of Figures 7a and 7b [28]. Figure 7a present results for GaAs-based QW structures while Figure 7b relates to InP-based QWs. The two curves at the bottom of each figure present the net production of vacancies in the structure (open circles) and specifically in the QW region (close circles). Comparison of these curves with the measured bandgap energy shift of each QW system in the top of each figure is enlightening. In each structure, an ion implantation energy of about 3–4 MeV is required before the ions generate significant vacancies directly in the QWs. However, in the InP QW case (right side), there is a clear correspondence between the total number of vacancies generated in the whole structure and the bandgap shift. In the GaAs QW case, the correspondence is closer between the number of vacancies generated in the QWs and the bandgap shift. This implies that vacancies in InP-based structures diffuse more freely throughout the material than is the case in GaAs based structures and can enhance QW interdiffusion even if they are generated near the surface, for example, well away from the QWs.

This result appears to contradict the assertion made above that vacancies in GaAs QWs can diffuse long distances from the surface and enable QW shape modification deep within the structure. This anomaly has been known for some time and although its cause is not clear, it appears related to the interface gettering effect alluded to earlier. There is evidence that GaAs/AlGaAs QWs with high Al compositions impede vacancy diffusion through some mechanism. Most of the work sited above (especially that involving SiO_2 capped QW intermixing) utilized GaAs/AlGaAs QWs with relatively low Al compositions (typically 12–20%). When attempts were made to repeat some of the experiments in structures with higher Al compositions (30% and greater), no blue shifting could be observed [29]. There is also evidence that the quality of the substrate material on which QW structures are grown has a significant effect on the magnitude of QW shape modification. In reference 4, for example, it is postulated that the presence of dislocations in the substrate wafer can getter species, such as native point defects or impurities, which are involved in the interdiffusion process. Thus, wafers which have fewer dislocations experience the largest blue shifts during annealing procedures. To reiterate, the cause of this apparent discrepancy between SiO_2 capped and high energy ion implantation QW shape modification in GaAs/AlGaAs QW material is unknown. However, until both types of experiments are performed on samples from the same wafer and under similar conditions, the causes of this dichotomy remain in the realm of speculation.

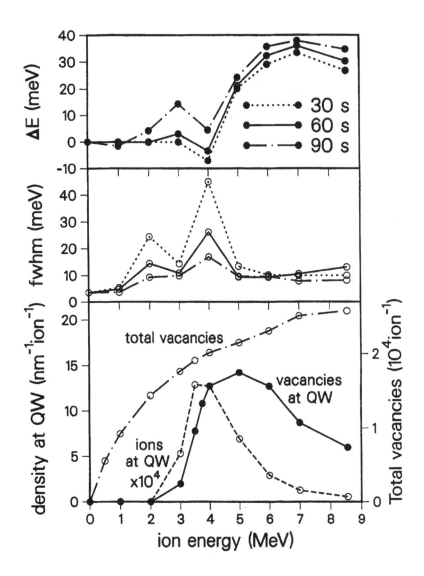

Fig. 7a. Upper curves: Bandgap energy shifts as a function of ion implantation energy for GaAs/AlGaAs QWs after rapid thermal annealing for different anneal times. Middle curves: Full-widths of PL peaks shifts as functions of ion implantation energy for GaAs/AlGaAs QWs after rapid thermal annealing for different anneal times. Lower curves: The calculated relative number of vacancies produced at the depth of the GaAs/AlGaAs QW (closed circles) and the total number of vacancies generated by each ion (open cicles) as functions of ion energy (reference 28)

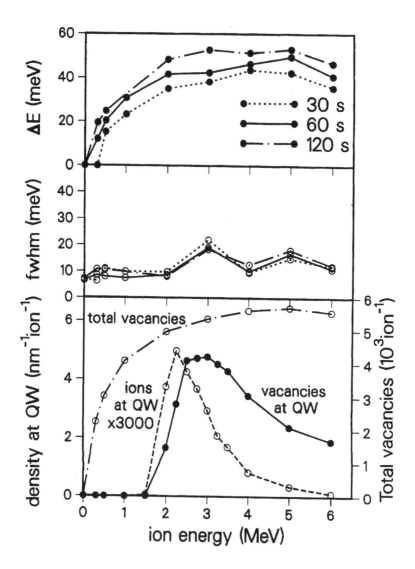

Fig. 7b. Upper curves: Bandgap energy shifts as a function of ion implantation energy for InGaAs/InGaAsP QWs after rapid thermal annealing for different anneal times. Middle curves: Full-widths of PL peaks shifts as functions of ion implantation energy for InGaAs/InGaAsP QWs after rapid thermal annealing for different anneal times. Lower curves: The calculated relative number of vacancies produced at the depth of the InGaAs/InGaAsP QW (closed circles) and the total number of vacancies generated by each ion (open circles) as functions of ion energy (reference 28)

3.1.3 Anneal Parameters

In some ways, the most difficult parameters to control in this QW bandgap shifting technique are associated with the annealing conditions. This is a problem for any QW shape modification technique which requires an annealing step and is especially difficult in the case of rapid thermal annealing. For example, absolute temperatures are notoriously inaccurate and unreliable and the ambient environment can have a significant effect on the success of the anneal. Furthermore, reproducibility and uniformity are always important issues. Unfortunately, no comprehensive study of these issues has been undertaken and so every researcher must determine the unique optimum annealing conditions for his/her own annealing set-up. Not only does this complicate the analysis of experiments but makes problematic comparisons from lab to lab.

The fundamental issue in this thermal process is the trade-off between anneal temperature and anneal time. A long anneal time in conjunction with a relatively low temperature produces similar bandgap shifts due to QW shape modification as short anneal times coupled with high temperatures. A major difference between these two procedures is related to the diffusivity, during RTA, of the enhancing species. It has been demonstrated that, at high temperatures, and relatively short anneal times (20 seconds or less) only vacancies participate in the QW shape modification process [30]. Implanted ions diffuse negligibly during the RTA and therefore play only a minor role. "Impurity" enhanced intermixing requires much longer anneal times and, therefore, lower anneal temperatures, due to the volatility of semiconductor materials.

Thus another major issue, which needs to be considered is the volatility of the various semiconductor compositions in the sample structure. It is especially relevant when ion implantation is utilized to enhance the process since the anneal of these samples usually takes place in the absence of capping layers although, in principle, capping could be added to reduce loss of volatile elements during the anneal process. However, this would increase the number of processing steps and, perhaps, increase the complexity of the anneal by introducing unknown chemical reactions or thermal strain between the cap layer and the sample.

In the case of GaAs and InP based QW structures, most researchers opt for high temperatures and short anneal times due to the relatively high volatility of these III-V materials (As and P are the most volatile elements, respectively). For rapid thermal anneal (RTA) times of a couple of minutes or less, temperatures as high as 1000°C can be tolerated by GaAs QW structures provided that the overpressure of the volatile element can be increased in the anneal chamber. This can be accomplished by placing a sacrificial wafer, which desorbs the same element as the sample, on top of the sample. A better scheme is to place the sample in a small, enclosed, pill-box-like graphite structure, the susceptor, whose function is to average the temperature anomalies of the RTA and to contain gases desorbed by the sample during anneal. Adding a specially treated cover wafer, which increases the volatile element overpressure, can also decrease morphology damage during anneal [31] Further, conditioning the susceptor by heating it up to a very high temperature (much higher than that normally used during RTA) with a sacrificial wafer so as to saturate the graphite with volatile elements also reduces sample surface damage during RTA.

The susceptor is heated by a number of high intensity arc lamps to a high temperature in a very short time. Ramp-up times on the order of seconds are possible with most commercial systems. Temperatures are monitored with optical pyrometers or thermocouples. Each has their advantages and disadvantages but neither is fully satisfactory. It is beyond the range of this article to explore further the intricacies of this processing step [32]

The goal of the annealing step is to raise the temperature of the sample to a level at which significant interdiffusion occurs in the implanted region of the wafer but a negligible amount in the as-grown, untreated region. The operating range is unique to each material system and is a function of the activation energy for self-diffusion. (If, for whatever reason (growth conditions, substrate quality, etc.) there are a large number of defects, dislocations, etc. in the epi-layers, gettering effects can produce a "modified" activation energy, so that blue-shifting is inhibited [4])

For the GaAs/AlGaAs system, activation energies are of the order of 5 to 6 eV [33], implying that significant self-diffusion occurs at temperatures of the order of 850 to 900°C. In the InP based material system, the "elbow" temperature can be 100°C or more lower. The "elbow" temperature is arbitrarily defined as that temperature at which significant QW shape modification occurs in as-grown material as determined by significant blue-shifting of the PL peak wavelength. It is determined experimentally for each material system by subjecting a series of samples (half with a QW enhanced interdiffusion region and half as-grown) to a fixed time anneal at different anneal temperatures. An example of this procedure is given in Figure 8 for the case of InGaAs/GaAs QWs [26]. The behaviour of the bandgap energy of a QW, with respect to that of an unannealed QW, is shown by the triangles for two different anneal times (30 sec, solid line and 60 seconds, dashed line). Up to a temperature of about 850°C, no significant change in the bandgap energy is observed. However, at higher temperatures significant blue-shifting occurs. On the other hand, for the enhanced interdiffusion pieces (ion implanted — circles), a significant blue shift has occurred even at the lowest temperature given in this plot (800°C). The presence of defects introduced by the implantation process has produced a large blue shift at temperatures at which thermally generated defects

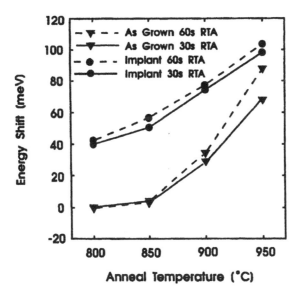

Fig. 8. Bandgap energy shifts of ion implanted (circles) and unimplanted (triangles) InGaAs/GaAs QWs as a function of anneal temperature. The anneal times were 30 (solid lines) and 60 (dashed lines) seconds. (reference 26)

are too few to effect significant interdiffusion. Note, however, that as the temperature is increased, the bandgap energy of QWs in the as-grown material increases more rapidly than that of the implanted QW material. At much larger temperatures self-interdiffusion dominates and both the as-grown and implanted QW material experience similar blue-shifting. Thus the optimum anneal temperature for spatially selective QW interdiffusion in this material system would be between 800 and 850°C. These temperatures produce the maximum blue shift (bandgap difference) between as-grown and ion implanted material.

Note that there is little difference between the measured shifts for 30 and 60 second anneals for the implanted QW material in Figure 8. The dose was kept small enough (2.5×10^{13} ions/cm^2) that vacancy complexes are negligible and a short anneal is sufficient to diffuse them throughout the structure. (However, for the case of the as-grown QW material, longer anneal times at higher temperatures can produce larger blue shifts since the interdiffusion process is a thermally generated process.) On the other hand, for large doses, complexes and extended defects are common. Longer anneal times allow these clusters to anneal out gradually providing a continuous, but not infinite, source of vacancies for further intermixing. This process is illustrated in Figure 9, [19] where the bandgap energy shifts of a series of GaAs/AlGaAs

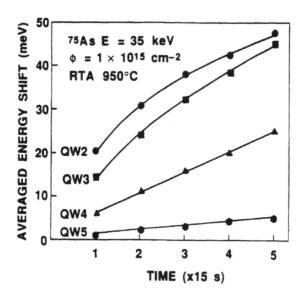

Fig. 9. Average bandgap energy shifts for four single GaAs/AlGaAs QWs (with nominal widths of 4, 7, 11, and 20 nm for QWs 2 through 5) as a function of the number of 15 second, 950°C exposures. (reference 19)

QWs of differing widths as a function of the number of 15 second RTAs is presented. The ion implantation dose was 1×10^{15} As$^+$ ions/cm^2. Note that as the effective length of the RTA is increased from 15 seconds to 75 seconds, the QW bandgap energies continue to increase. There is some indication that the blue shifts of the narrower QWs (QW2 and QW3) are beginning to saturate. This is probably due to the fact that narrow QWs experience a smaller blue shift before reaching the maximum shift possible.

This brief survey of the factors important in quantifying the magnitude of the blue shift in QW materials was not meant to be definitive. Rather, it was meant to present an overview of the issues involved and the complexities of the processes taking place. Clearly, each material system is different and, since the quality of epi-samples is an important factor, it is not unexpected that results and conclusions may vary from laboratory to laboratory. Much more work on reproducibility and reliability will need to undertaken and completed before this procedure is ready to transfer to a commercial setting.

3.2 Material Systems

In this section, I will survey results of QW bandgap energy blue shifting using ion implantation induced shape modification in a number of material systems. In principle, due to the universal nature of the vacancy generation mechanism and its expected effect on QW interdiffusion, this technique is expected to be applicable to any compound semiconductor material system. I will attempt to draw general conclusions where appropriate.

3.2.1 GaAs/AlGaAs QWs (strain free)

GaAs/AlGaAs QWs are unique among the family of III-V heterostructures in that they remain lattice matched, and therefore unstrained, no matter what the aluminum composition of the barrier. (Actually there is a slight mismatch between AlGaAs and GaAs, the substrate material, which is maximum for AlAs, but it is small enough that, for all intents and purposes, its lattice constant can be considered effectively identical that that of GaAs.) Thus, no matter how the QW shape is altered from that of the as-grown, abrupt-sided initial shape ("square" QW), strain will not be a factor when modeling quantum levels. (This is not to be confused with the argument that external strain, generated by dielectric layers such as SiO_2, may be an important factor during intermixing mediated by SiO_2 cap layers.)

Since, in principle, this is the simplest of the QW systems to model, it would be expected that our understanding of the forces and mechanisms in operation during enhanced QW interdiffusion would be the most complete. This is probably true, but there is still much to be learned. Although there is generally good agreement between theory and experiment in the case of as-grown GaAs/AlGaAs QWs (regarding band offsets, PL peak energies as a function of QW parameters, etc.), interdiffused QWs are another matter. As mentioned above, it does not appear possible to determine a unique interdiffusion length independent of QW width (for example). This is probably a consequence of the uni-directional nature of the vacancy diffusion process. Gettering and the non-reciprocal nature of QW interfaces (interfaces produced by growing AlGaAs on GaAs are generally "smoother" than those created by growing GaAs on AlGaAs) presumably play a role. In any case, it is clear that not even in the GaAs/AlGaAs system is QW interdiffusion completely understood at this point.

GaAs/AlGaAs QWs were among the earliest heterostructures studied (from the early 70's) and it was natural that they were used for initial studies of QW interdiffusion [8,1,5]. In fact some of the earliest work centered on producing active devices, buried multiquantum well laser diodes [5]. However, since the focus of most work on QW heterostructural devices is optical telecommunications, interest has shifted to materials with smaller bandgap energies (longer wavelengths) which are compatible with technologically important wavelengths, 1.3 and 1.55 μm.

3.2.2 InGaAs/GaAs QWs (compressive strain)

InGaAs epi-layers grown on GaAs substrates are always compressively strained. However, since QW layers in InGaAs/GaAs QWs are quite thin, even highly strained QWs can be pseudomorphic (that is, strained but still single crystal — not relaxed. It is possible to grow pseudomorphic InGaAs QW layers as thick as 25 nms, even with an indium composition of 20% [34].)

Generally, the behaviour of this QW system is similar to that of the GaAs/AlGaAs system [35]. Most of the relevant parameters (shift dependence on QW width, ion fluence, anneal temperature) are almost identical. The major difference is that any change of the shape of the as-grown QW in this system must alter the QW strain, although there is little direct evidence for this change in strain. (Very wide $In_{0.2}Ga_{0.8}As$/GaAs QWs (22 nms), which were close to the Matthews-Blakeslee boundary between strained and relaxed layers, did appear to be partially relaxed after enhanced QW interdiffusion [34].) There is some evidence, based on the rate of bandgap shift as a function of annealing time, that compressive strain, by enhancing the concentration of vacancies in the structure, influences the magnitude of interdiffusion [36]. However, there is no consensus in the literature on what effect, if any, strain has on QW shape modification (see later).

The main commercial interest in InGaAs/GaAs QW devices lies in their use in laser diode structures operating at 980 nm, which is the wavelength needed to pump erbium-doped fiber amplifiers most efficiently. The main application of QW shape modification in this material system is to enhance the performance of 980 nm laser diodes by, for example, producing transparent windows near the laser facets. This permits laser output powers to be raised without generating catastrophic optical damage (COD) at the facets. An example of this effect will be given later. It may also be possible to "fine-tune" lasing wavelengths so as to more closely match the narrow absorption peak of the erbium doped fiber.

3.2.3 InP-based QWs [lattice matched and pseudomorphic(compressive or tensile)]

There is a great deal of interest in InP-based QW material systems due to their commercial potential. Optoelectronic devices operating in the technologically important 1.3 to 1.55 μm band are possible. However, it is a very complex structural system, involving binary, ternary, and quaternary materials, which can be either lattice matched or strained (compressively or tensilely) with respect to the substrate material, InP. Furthermore, there are tight tolerances on growth and processing procedures, making it very difficult to convincingly make a case that even so-called abrupt, as-grown, QWs are well understood. It is well known, for example, that growth interruptions at QW interfaces (necessitated by the growth procedure) can dramatically alter the QW shape [37]. Relatively long interruptions (generally tens of seconds) are required in order that compositions be switched abruptly from well to barrier material at the interface, rather than gradually, in order that "square" QWs be grown. However, this interruption gives time for residual gases and other impurities, present even in high vacuum systems, to impinge and collect on the growth surface. Thus there is a trade-off between abruptness and cleanliness of the interface. For these reasons, and also because many important parameters are either unknown or in question in these materials (for example, band offsets), it has been very difficult to develop a theoretical structure which can realistically model optoelectronic properties of a real device.

Strain, in particular, is hard to control in these structures, especially if uniformity is required over large substrate wafers. Unlike the GaAs/AlGaAs system in which the lattice constant was "almost" independent of the aluminum composition, any variation in the ternary or quaternary composition during growth or over a wafer will invariably introduce strain. It will also change, in a difficult to predict manner, during QW interdiffusion since the composition gradient across the QW interfaces is modified in this process. Further, since more than one species (group III or V atoms) may move, the relative diffusivities of these atoms becomes important. These considerations make it very difficult to predict, *a priori*, bandgap shifts, for example.

However difficult it may be to predict absolute shifts in the InP-based material system from first principles, it is clear from many reports in the literature that bandgaps in these QWs do change dramatically during enhanced QW interdiffusion [38]. And, for a given QW width, the rule-of-thumb appears to be that the largest blue shifts are observed in compressively strained QWs, followed by lattice-matched QWs, and then tensilely strained QWs [39]. However, in certain material systems, this appears to be more related to the depth of the QW (that is, the difference between the bandgap energies of the barrier and well materials) than to the effect of strain. For example, in Figure 10, the blue shift observed in a series of 3 nm wide $In_{1-x}Ga_xAs/InP$ QWs was observed to be directly related to the QW depth, ΔE [39]. The samples were subjected to a dose of 1×10^{13} Ne^+ cm^{-2} ions implanted at energies of 100 to 200 keV and then annealed at 700°C for 13 seconds (plus an additional anneal at 300°C for 10 minutes). In Figure 10, with increasing ΔE, the QWs are tensilely strained ($+1.5, +1.0, 0.5\%$), lattice matched (0) and then compressively strained ($-0.5, -1.0, -1.5\%$). (Note that a data point from another material system, InAsP/InP, falls neatly on the curve.) In this set of samples, the main effect of QW strain (which also involves a change in composition of the well material) seems to be to

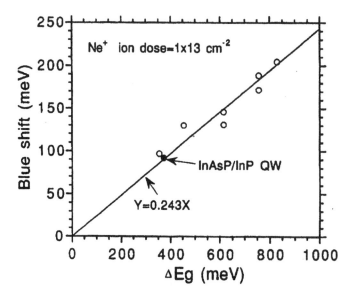

Fig. 10. The bandgap energy increase as a function of bandgap energy difference between the well and barrier materials for InGaAs/InP and InAsP/InP QWs subjected to ion implantation and anneal. The blueshift of the InAsP/InP QW is shown by the solid circle. (reference 39)

change the depth of the wells. The linear relationship between the depth and blue shift is suggestive that QW strain has little or no effect on the magnitude of QW shape modification since a monotonic increase of blue shift with well depth is a trivial consequence for a given change in effective depth, even in the GaAs/AlGaAs QW system where strain plays no role. This result is counter-intuitive since strain is expected to play a more complex role during QW interdiffusion. The slope of this curve is related to the magnitude of the dose, increasing from about 0.18 at low doses (2×10^{12} cm^{-2}) to about 0.25 at moderate doses (2×10^{13} cm^{-2}) and then decreasing again to about 0.2 at larger doses ($> 5 \times 10^{13}$ cm^{-2}). This behaviour mimics that of the dose dependence of the blue shift which, in turn, is related to the magnitude of the QW shape modification (Figure 5).

Some unusual behaviour in the strain dependence of interdiffused In-GaAs/InGaAsP QWs, which leads to polarization independence, will be discussed in some detail later. Suffice it to say, that much more work is required to completely understand the mechanism of QW shape modification in these heterostructures. About all that can be said, in a general way, is that "elbow" temperatures (that is, anneal temperatures at which significant blue shifts are observed) are lower in these QWs (about 750°C) compared with the GaAs/AlGaAs QW system.

3.2.4   Other QW Systems

Only a small amount of work has been done in other QW material systems although, in principle, enhanced QW shape modification is possible in any semiconductor heterostructure. For example, blue shifts have been observed in $InAs_yP_{1-y}/InP$ [39], InGaAs/InGaAlAs [40], and SiGe [41], etc. However, due to space limitations, we will not be discussing these material systems.

4   DEVICE PHYSICS ISSUES

Being able to increase the bandgap of QW materials in a spatially selective manner is a necessary condition for accomplishing photonic integration of optoelectronic devices, but it is not a sufficient condition. Other issues which need to be addressed include the optical quality of the device structures after processing (especially transparency), electrical performance of active devices, fabrication complexities and yield, etc. In this section we will discuss some of the more important of these issues.

4.1   Optical Properties

The intensity of QW photoluminescence peaks has always been assumed to be a simple test of the optical quality of QW material. In contrast to bulk semiconductors in which impurity related luminescence generally dominates, in QW structures the strongest PL signature is due to transitions in the QW between the ground states of the quantized conduction and valence bands (generally the heavy hole). This is especially true at elevated temperatures but even at cryogenic temperatures, luminescence from residual impurities in the QW is generally very weak. Only heavily doped wells exhibit PL structure due to impurity states. Thus the intensity and width of this QW signature peak is a measure, albeit crude, of the "quality" of the QW and thus can be used to indicate the effect of processing on the QW. However, a more sophisticated measure, such as absorption loss in QW waveguides is a more useful sign of the utility of this technology for fabricating PICs (Photonic Integrated Circuits).

Generally, PL intensities can be recovered fully if sufficient care is taken during the processing. This usually means that the QWs not be implanted directly, that the dose not be too high, and that the annealing conditions be optimized (adequate temperatures and sufficient time for vacancy diffusion to produce QW shape modification; but avoiding longer times which can degrade the sample surface due to the volatility of certain atomic species in the structure). Each material system is unique and needs to be calibrated individually. Sometimes, the growth technique (whether MBE, CBE, MOCVD, or gas-source MBE) or the type of substrate used [4], produces material whose properties can vary from reactor to reactor.

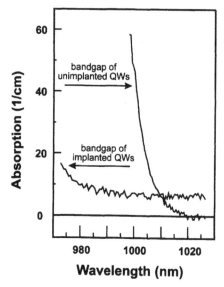

Fig. 11. Absorption of implanted and unimplanted InGaAs/GaAs QW waveguides as a function of wavelength. (reference 42)

Further, each piece of processing equipment (especially the RTA system) in a particular laboratory needs to be calibrated individually due to peculiarities of some experimental apparatus. In many cases, this makes it very difficult to compare experimental results from different laboratories and may explain some of the discrepancies present in the literature.

One of the most important measures of the viability of ion implantation as a technique for increasing QW bandgap energies is waveguide transparency. Intuitively, one would expect that any semiconductor device structure which is implanted will necessarily be of poorer quality (even after annealing) than the as-grown structure. The only question would be: Is the residual quality adequate for fabricating optoelectronic devices? Figure 11 illustrates that if implantation takes place directly into the QWs, the resulting residual damage in the QW waveguide (as measured by the absorption spectra) will increase, but may still be adequate for some uses [42]. In this case, the structure was undoped and the absorption at the bandedge of the unimplanted (as-grown) structure was reduced to 8 cm^{-1}. This would be adequate, for example, for monolithically integrating a laser and electro-absorptive modulator.

However, for truly transparent waveguides, even lower residual losses would be required. This goal is not inherently impossible as can be seen in Figure 12. In this case the ions implanted into an InP based laser structure did not reach the QWs. QW interdiffusion was enhanced solely by vacancies

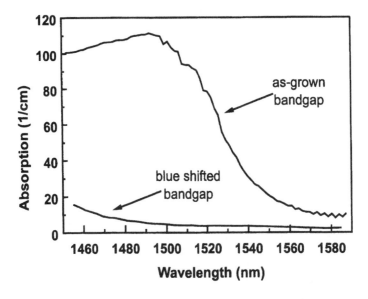

Fig. 12. Absorption of as-grown and blue-shifted InGaAsP/InP QW waveguides as a function of wavelength. (reference 43)

and other mobile defects [43]. At the absorption edge of the unimplanted waveguide (~ 1500 nm), the absorption coefficient was reduced from 110 cm^{-1} to 4 cm^{-1} by QW shape modification. This value is actually lower than that of the unimplanted waveguide, even after the 90 nm shift of the absorption bandedge is taken into consideration. For example, the absorption of the unimplanted waveguide is 8 cm^{-1} at 1580 nm compared with 5 cm^{-1} at 1490 nm in the implanted sample. This decrease is not due solely to effects of annealing on the initial material quality since both the implanted and unimplanted material were annealed. Rather, the decrease could be due to a reduction in Auger effects and intervalence band absorption since the bandgap energy has decreased. Further, the free carrier absorption coefficient is, to first order, proportional to the square of the wavelength and so would be reduced by about 12% over the wavelength range studied. Finally, the guided mode is more strongly confined within the active region as the wavelength is decreased, reducing any scattering losses from imperfections of the ridge guides. Whatever the cause, these results are direct experimental evidence that ion implanted QW waveguides can be made optically transparent. The residual absorption value (4 cm^{-1}) is still too large for use in large planar waveguide geometries. This is due to the large initial absorption in this p-i-n laser structure which was not optimized for this purpose. It is possible to design such structures with absorption values an order of magnitude smaller. Thus, in principle, p-i-n active waveguide devices can be monolithically integrated with transparent passive waveguides using QW shape modification.

With a change in bandgap energy, the effective waveguide index at a particular wavelength is also modified since the wavelength dependence of the index is a strong function of the distance from the QW bandedge. Thus, blue shifts produce a decrease in effective index at a fixed wavelength and this can be quite large (a few percent). This behaviour can be utilized to produce buried waveguides in a simple manner, without multiple growths. This will be discussed in greater detail later.

QW shape modification is expected to have minimal effects on the shape of the optical bandedge, since this is a function only of the properties of the quantized levels in the conduction and valence bands, and not of the energy differences between them. This, indeed, has been observed in the GaAs/AlGaAs QW system [44] and also the InP material system [45]. In Figure 13 [46], the shape of the absorption bandedge for an InGaAs/InGaAsP QW waveguide is essentially unchanged after it has been blue shifted about 50 nm [compare cases A (as-grown) and D (maximum shift)].

A more interesting, and unexpected, effect is the elimination of the polarization anisotropy in the shifted QW material. Anisotropy in QWs is due to the lifting of the heavy-light hole valence band degeneracy in direct bandgap zinc blende III-V compound semiconductors due to the quantum confinement effect. In a QW waveguide, TM polarized light interacts with light-hole-electron transitions only while heavy-hole-electron transitions can interact with both TE and TM polarized light. The light-heavy hole splitting in a QW results in a different bandedge being observed for TE (heavy-hole) and TM (light-hole) polarized light, as can be seen for case A (as-grown QW waveguide) in Figure 13 [46]. This behaviour produces a serious problem for QW waveguide devices in optical telecommunications applications since it is generally impossible to know, *a priori*, the polarization of light impinging on a waveguide optoelectronic device after traveling through an arbitrary length of fiber or other optoelectronic devices. After being blue shifted about 50 nms (case D in Figure 13), the TE and TM bandedges are essentially identical, indicative of polarization insensitivity. The cause and potential applications of this important result will be discussed in greater detail later.

In summary, it is clear that QW shape modification using ion implantation enhanced interdiffusion produces no inherent degradation in the optical quality of QW waveguides provided that the QWs are not directly implanted and that care is taken during processing to anneal out most of the implantation damage. Thus, in principle, reduced optical quality is not a deterrent to the use of this technique for monolithic photonic integration.

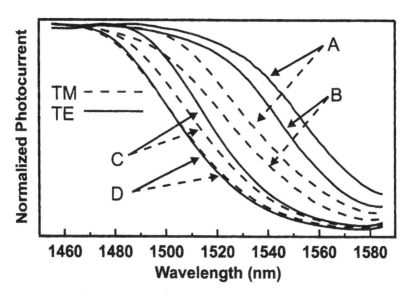

Fig. 13. Photocurrent (absorption) as a function of wavelength for TE (solid lines) and TM (dashed lines) polarized optical modes in four InGaAsP/InP QW waveguides. Sample A is as-grown while samples B, C, and D have experienced different magnitudes of blue shifting. (reference 46)

4.2 Electrical Properties

A major concern in enhanced QW interdiffusion, no matter what the enhancement technique, is the behaviour of dopants. As noted earlier, initial studies of QW intermixing utilized rapidly diffusing impurities such as Zn to enhance the effect. Although optoelectronic devices such as buried waveguide lasers could be fabricated using this technique [5], the presence of large numbers of electrically active impurities in these device structures had serious effects on their performance. As a result, most recent work on QW shape modification focuses on impurity-free techniques for accomplishing the same goals. However, it is not clear that merely by eliminating the incorporation of impurities into the device structure, the situation is rectified. It might be expected that the presence of vacancies, generated by ion implantation for example, might also increase the mobility of dopants in p-i-n structures, thereby altering their distribution throughout the structure, and changing the electrical performance of the device. This has indeed been reported [47]. Somewhat surprisingly, most indications are that dopant diffusion is not a serious problem, at least in some structures. Figure 14 is a plot of current versus voltage for the p-i-n structures illustrated in Figure 13 [46]. Even after ion implantation and annealing, the I-V characteristics of the as-grown and interdiffused diodes are virtually indistinguishable.

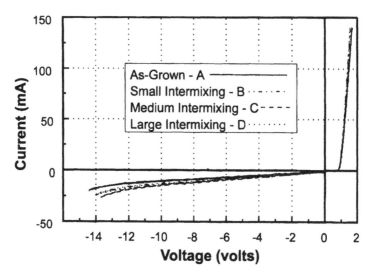

Fig. 14. I-V characteristics of p-i-n diodes fabricated from as-grown and ion implanted and annealed InGaAsP/InP QW waveguides. The curves A, B, C, and D follow the labeling given in Figure 13. (reference 46)

Only in the reverse direction is there a slight difference, a small increase in the "leakage" current in the interdiffused QW diodes. This may, in fact, be related to the change in the shape of the QW after processing. Current-optical output power characteristics of these laser diodes and threshold currents are also unchanged, even when lasing wavelengths were blue-shifted by as much as 64 nm [48].

The preservation of good electrical properties in this structure after QW interdiffusion [48] is probably due to the fact that no damage was directly created in the junction area (QW region) by the implantation and to the fact that a smaller dose and lower annealing temperature were used. The important conclusion to be drawn from these studies is that it is possible to blue shift QW bandgaps using ion implantation and annealing without seriously deteriorating the electrical performance of p-i-n structures.

4.3   Other Issues

Once the important issues of potential optical and electrical quality degradation during bandgap shifting have been resolved, the final "show-stopper" question concerns device lifetimes. In the discussions above, it was demonstrated that it was possible to selectively blue-shift QW bandgaps with no significant deterioration of optical and electrical properties. However, the process does involve the implantation of ions directly into the active device structures (although not necessarily into the QWs themselves) and, it might be expected that, even after heat treatment, not all of the implantation damage is annealed out. Thus, there is a potential for reduced lifetimes of active devices such as photodetectors, modulators, optical amplifiers, and, especially, diode lasers which are subjected to large electric fields or currents.

This possibility was investigated by subjecting blue-shifted InGaAsP/InP 1.3 μm laser diodes to standard accelerated lifetime measurements [49]. The lifetimes of three QW interdiffused lasers with lasing wavelength blue shifts of 29.8, 44.7, and 69.1 nm with respect to the as-grown structure were compared to the lifetime of the standard, unshifted, laser. The reliability of the QW shape modified lasers was equivalent to the standard (nonimplanted) lasers when the wavelength shift was 35 nm or less, and this corresponded to predicted lifetimes in excess of 25 years while operating at 25°C. Since this test was performed using a somewhat arbitrary set of QW blue-shifting process parameters, that is, in the absence of optimum implantation and annealing conditions, it is believed that, with further study, wavelength shifts greater than 35 nm are achievable while retaining the necessary 25 year lifetime required for practical laser diodes.

At least two further issues related to processing need to be addressed before the practicality of this integration technique can be assessed; masking technology and lateral spatial resolution.

For ion implantation enhanced QW shape modification to be a practical technique for monolithic integration, a relatively simple process for assuring spatial selectivity is required. Masking areas of the wafer to be shielded from the implantation is a simple matter if low energy implants are utilized. Standard dielectric films (e.g., SiO_2) up to a micron or so in thickness effectively absorb moderate energy (1–2 MeV P^+, say) ion beams. Higher energy implants require thicker dielectric layers which can strain the structure or develop cracks. Thus, in this case, multi-component masks utilizing SiO_2, other dielectrics, and/or metals are sometimes used, increasing the processing complexity. Fortunately, due to the mobility of vacancies in the technologically important InP — based device structures, low energy implants can be used and, indeed, are desirable. What is important is not the depth of the implant but rather the total number of vacancies and interstitials (and other interdiffusion enhancing defects) produced. Thus only relatively thin layers of SiO_2 are needed for masking. By varying the thickness of this layer, the number of vacancies produced in the semiconductor waveguide structure for a fixed implantation energy can be controlled in a simple manner [48]. In Figure 15, the number of vacancies introduced into an InP-based laser structure by a 1 MeV phosphorus implant through different SiO_2 mask thickness was calculated using a standard ion energy loss model (TRIM-91[50]). For 2.2 μm thick layers, no excess vacancies are produced in the structure (all of the ions are absorbed in the SiO_2 – bottom curve) while

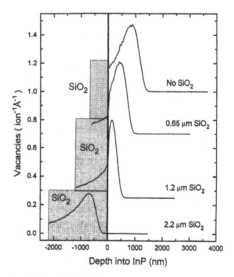

Fig. 15. Vacancy distributions for 1 MeV phosphorus implantation into InGaAs/InGaAsP QW laser structure through different SiO$_2$ mask thicknesses. 0 denotes the surface of the semiconductor material. (reference 48)

for decreasing thicknesses (1.2 and 0.65 μm) increasing concentrations of vacancies are produced in the semiconductor structure. The largest vacancy concentrations are generated in areas with no mask (top curve). In this manner, a monolithic laser diode array with four different lasing wavelengths can be readily fabricated from one structure using a single implant and anneal process. Complex etch-and-regrow technology is replaced by a relatively simple multilevel mask technology [48]. More details will be given later.

The final issue is that of spatial selectivity, which is a function of the lateral resolution achievable. The key question is: how close can interdiffused and as-grown QWs regions be? Modern lithographic techniques are capable of defining sub-micron dimensions and electron and ion beam machines can focus down to dimensions of tens of nms or better. But, energy loss of high energy ions in matter produces straggling effects which degrade the theoretical resolution. This can be minimized by lowering ion energies and reducing the depth of QWs from the surface of the structure.

Some of the earliest work on QW shape modification focused on its use to fabricate submicron periodic structures (i.e., gratings) [10] and quantum wires and dots [51], structures which require very high spatial resolution indeed. For the grating work a focused ion beam (0.1 μm diameter) was used to implant 100 keV Ga$^+$ ions directly into a GaAs/AlGaAs superlattice at the surface of the sample in a line (0.15 μm) and space (0.2 μm) pattern [10]. PL spectra as a function of Ga$^+$ ion dose were analyzed to reveal that ion straggling and defect diffuse limited the resolution to about 0.1 μm in this case.

The QW wire and dot work utilized implantation of 210 keV Ga$^+$ ions through a Ti/Au-Pd mask into GaAs/AlGaAs QWs located 55 nms from the surface [51]. The calculated value of implantation lateral straggling was about 42 nm and that of the defect diffusion length was 20 nm producing an effective lateral extent of interdiffusion of 28 nm. A similar study using 100 keV Ga$^+$ ions implanted into 30 nm deep GaAs/AlGaAs QWs gave an estimate of 20 nm for the lateral extent of QW interdiffusion [52]. In general then, low energy ion implantation into QWs near the surface of the sample produces an effective lateral extent of QW interdiffusion comparable to the depth of the QWs from the surface. A more recent investigation of lateral resolution in a more realistic laser diode structure (in which InGaAs/InGaAsP QWs are located 1.77 μm below the surface) obtained a value of 2.5 μm [53]. This is half the value of the width of a mask under which significant blue shifting of the QW energy is observed due to interdiffusion initiated by lateral diffusion of defects.

As a general rule-of-thumb, the lateral resolution of a structure produced by low energy ion implantation enhanced QW shape modification is comparable to the depth of the QWs from the surface. Since QW waveguide structures have dimensions comparable to the wavelength of light in the material, this implies that devices with lateral dimensions much less than a micron will be difficult to fabricate without more complex processing steps (e.g., interdiffusion of QWs close to the surface followed by an epitaxial overgrowth step). However, for most instances of monolithic integration of active and passive optoelectronic devices (e.g., lasers, photodetectors, modulators, optical amplifiers, etc. with passive waveguides), this is not a serious limitation. Generally, spatial dimensions of the order of microns are adequate for integrating, for example, a 3 μm wide ridge laser diode with a photodetector or transparent waveguide.

5 PERFORMANCE OF QW SHAPE MODIFIED DEVICES

A perusal of the literature on QW shape modification over the last decade or so reveals that many of the reports deal with establishing the reality of QW blue-shifting, investigating various mechanisms for generating blue-shifts, and attempting to understand the physics behind the phenomenon. Although the importance of this effect as a technique for the monolithic integration of optoelectronic devices was realized early on (in fact, some of the earliest work[5,10,51] dealt with the application of QW intermixing for fabricating devices), serious work to establish its commercial viability was slower in coming. In this section, we will briefly review some of this work.

5.1   Discrete Devices

Up to the present time, most reports of the utilization of QW shape modification for device fabrication have dealt with discrete devices. QW interdiffusion was employed either to enhance the performance of a standard device structure (e.g., high power windowed laser diodes, buried waveguide devices, broadband light emitting diodes) or to fabricate unique devices (such as enhanced quantum confined Stark effect modulators and polarization insensitive QW devices). While much of the work has centered on proof-of-concept devices, and, generally, optimized structures have yet to be reported, the results are quite encouraging. In this section, a brief discussion will be given of some of this work. In the following section, PICs (photonic integrated circuits containing two or more devices) will be reviewed.

5.1.1   Buried Waveguides

Some of the first work on devices was concerned with the possibility of using bandgap shifting to produce waveguide structures which possessed both optical and electrical confinement. Although it was well known that buried waveguides possessed superior performance characteristics, their fabrication, requiring a complicated etch-and-regrow procedure, was difficult to implement. It was realized, in the early 80's that, in principle, increasing QW bandgap energies on either side of a narrow "ridge" structure accomplished optical mode confinement (since effective indices were reduced due to the blue shift of the bandgap) at the same time that electrical carriers were more tightly confined (due to the lower bandgap in the "buried" (undiffused) QW waveguide). Much of the earliest work involved impurity diffusion through a mask to completely interdiffuse (and thereby convert into an alloy) QWs on either side of the active region in order to produce, for example, index-guided laser diodes [5]. Although this process did yield laser action, the presence of large numbers of impurities close to the active region and the difficulty of controlling their lateral extent had a tendency to degrade the performance achievable.

The more gentle process of altering the shape of the QW by enhanced interdiffusion and thereby blue-shifting the bandgap energy without completely destroying the QWs, achieves, in principle, the same result in a less drastic manner. It has been suggested that buried waveguides, fabricated using this technique, have relatively little scattering loss from the implanted/nonimplanted interface. In fact, it has been demonstrated that there is no correspondence between the width of a guide and propagation loss, unlike the case of etched waveguides where the loss increase rapidly with decreasing ridge width [54].

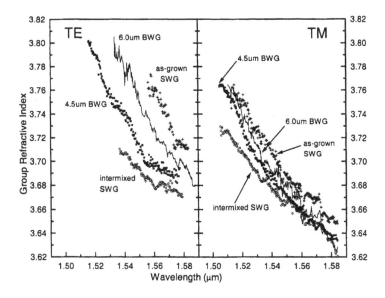

Fig. 16. The group refractive index for InGaAs/InGaAsP QW waveguides as a function of wavelength for TE (left) and TM (right) polarizations. The labels refer to as-grown and intermixed slab waveguides (SWG) and 4.5 and 6.0 mm wide buried waveguides (BWG). (reference 55)

Some recent results, for an InGaAs/InGaAsP lattice matched laser structure, presented in Figure 16, illustrate some of the relevant features of buried waveguides produced with ion implantation enhanced QW bandgap blue-shifting [55]. This figure presents measured group effective indices of slab and buried waveguides (SWG and BWG respectively) as a function of wavelength, measured using the Fabry-Perot technique. Consider, first, the left hand side of the figure which presents the results for TE polarized light. The indices are seen to increase as the wavelength decreases since these data were obtained near the QW bandedge. At a fixed wavelength, the largest values occur in the as-grown (unmodified) SWG, followed by the 6.0 and then the 4.5 mm BWG and, finally, the interdiffused SWG. The decrease in effective index can be as large as several percent (e.g., at 1.56 μm, $\Delta n = 2\%$). The ordering of the group refractive indices is dictated by the progressive blue-shifting of the QW bandgap. The largest shift occurs for the interdiffused slab waveguide. Due to the straggling and lateral diffusion effects (see Section 4.3), the material in the center, as-grown, region also undergoes some blue-shifting. In addition, both the 6.0 and 4.5 μm buried waveguides have modified values since their indices are a mixture of the interdiffused and as-grown slab waveguide values. This is a consequence of the fact that optical modes travelling in the buried waveguide sample both as-grown (in the center) and interdiffused (either side) QW material. Since a mode propagating in the narrower guide (4.5 μm) "sees" more intermixed waveguide material than the wider guide, its effective index is lower than that in the 6.0 mm buried waveguide.

Effective guiding in this type of buried waveguide only occurs over a relatively narrow wavelength range. At shorter wavelengths, absorption due to the as-grown (unmodified) bandgap reduces transmission; at longer wavelengths, the index difference between the as-grown and interdiffused QW material becomes too small to continue confining the optical mode. Furthermore, since the effective index of these buried waveguides is a strong function of wavelength, the confinement and, in fact, the mode properties (e.g., whether single or multiple mode propagation is possible) can change rapidly. Finally, a comparison of the left (TE) and right (TM) hand sides of Figure 16 reveals another potential problem. Due to the quantum confinement effect which lifts the heavy-light-hole degeneracy at the Brillouin zone center, different bandgap energies are observed for TE and TM polarized light. Thus the propagation properties of this buried waveguide will depend on the polarization of the mode. This is not so critical in active devices such as laser diodes since polarizations are well controlled and defined in such structures.

In summary, although it is possible to produce single mode guiding in a slab QW waveguide structure by blue-shifting the QW bandgap on either side of an as-grown QW slab waveguide section, in reality there are several practical disadvantages, in particular, polarization sensitivity. However, it may be possible to utilize this polarization effect in the design of some novel device to perform, for example, polarization rotation or filtering.

5.1.2 Quantum Confined Stark Effect Modulators

Quantum confined Stark effect (QCSE) modulators have superior performance when compared with bulk electro-absorptive modulators. The quantum wells effectively confine the carriers preventing applied voltages from ionizing the QW bound states (excitons) until very large electric fields are generated across the wells. Thus, very large red-shifts of the bandgap are possible in reverse-biased QW waveguide modulators. The modulator structure is very similar in design to that of a laser diode, suggesting that a monolithically integrated laser diode — modulator pair is feasible using a single device structure. For optimum performance, the modulator should have a bandgap energy slightly higher than that of the laser so that, in the absence of applied bias, the modulator is transparent to light from the diode laser. An obvious technique for blue-shifting a laser bandgap in order that it operate as an integrated modulator is ion implantation induced QW shape modification.

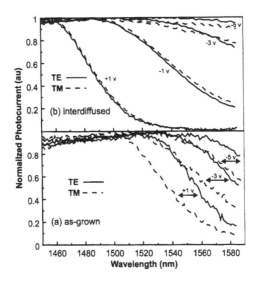

Fig. 17.   Photocurrent (absorption) spectra of as-grown (a) and interdiffused (b)InGaAs/InGaAsP QW waveguides as a function of external bias and for two orthogonal polarizations of the waveguide modes. (reference 57)

This idea has been investigated by a number of groups and shown to be feasible. For example, even after very large blue shifts (138 meV — in this case the well and barrier material were completely interdiffused so that no QWs remained) InGaAs/InGaAlAs material still undergoes a red shift when subjected to a reverse bias and thus, can, in principle, modulate a light beam [56].

A more interesting effect can occur in certain unique QW structures in which the QW is not completely destroyed by interdiffusion [57]. As noted above (Figure 13), lattice matched InGaAs/InGaAsP QWs, when interdiffused using ion implantation enhancement, exhibit interesting polarization properties. They also exhibit an enhanced quantum confined Stark effect, as demonstrated in Figure 17 [57]. In this figure the quantum confined Stark effect in as-grown (abrupt) QWs (Figure 17(a)) is contrasted with that in interdiffused QWs (Figure 17(b) — This corresponds to case D in Figure 13). The normal Stark shift is observed in Figure 17(a) as the bias is increased from about 1 volt (which cancels the built-in field due to the doping in the p-i-n structure and results in an internal electric field of approximately zero) to − 5 volts ($\varepsilon = 160$ kV/cm). The difference between the QW bandgaps for TE and TM polarized light is, again, due to the heavy-light hole splitting resulting from quantum confinement. For the case of the interdiffused QW

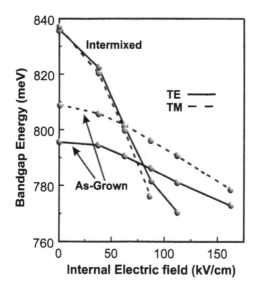

Fig. 18. Bandgap energy for as-grown and interdiffused InGaAs/InGaAsP QW waveguides as a function of internal electric fields for light polarized in two orthogonal directions (TE and TM) with respect to the plane of the waveguide. (reference 57)

waveguide, however, the difference between the TE and TM bandgaps is absent and, furthermore, the red shift observed for a given electric field is dramatically increased. This effect is summarized in Figure 18, a plot of the QW bandgap energy as a function of internal electric field across the QWs [57]. For an internal electric field of about 75 kV/cm, as-grown QW bandgaps are red-shifted by about 10 meV at best due to the quantum confined Stark effect while interdiffused QWs experience a decrease in bandgap energy of about 50 meV. This significant enhancement of the QCSE is probably a consequence of the increased ability of the electric field to polarize the electrons and holes in these "rounded" QWs compared with QWs with "vertical" walls. Another important factor is, undoubtedly, the complex strain arrangement in the interdiffused QWs.

Recall that, in this device structure, the as-grown QWs were nominally lattice matched to the InP substrate (i.e., unstrained). After implantation and annealing, this is probably no longer be the case. In fact, circumstantial evidence suggests that, after processing, the QWs are strained in some complex manner due to the non-uniform movement of group III (Ga and In) and group V (As and P) atoms during the enhanced interdiffusion process. This mechanism has been modeled by assuming different diffusion coefficients, Δ_{III} or Δ_V, for the cations and anions [58]. The parameter of importance is the ratio of diffusion coefficients, $\mathbf{k} = \Delta_V / \Delta_{III}$. Interdiffusion involving only

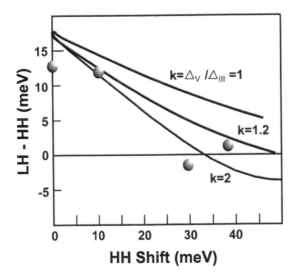

Fig. 19. Comparison of experimental heavy-light hole bandgap splitting as a function of heavy-hole shift in InGaAs/InGaAsP QWs after ion implantation and annealing with theory for different interdiffusion coefficient ratios, **k**. (reference 57)

one sublattice, or where the degree of interdiffusion on the group III and group V sublattices differs considerably ($\Delta_{III} \neq \Delta_V$), will result in a strained material system. The effect of interdiffusion on the energy band profile (including strain effects) and confined states for valence and conduction bands has been modeled using error function distributions for the anions and cations to predict compositional profiles that result after implantation and annealing [58].

For a given QW structure, the LH-HH energy splitting as a function of HH energy can be calculated for various values of **k** and compared with experimental values obtained near flat band conditions as shown in Figure 19 [57]. Within this model, it is clear that our experimental data can only be explained by assuming **k** > 1.5. Scatter in the data and a decreasing sensitivity of LH-HH splitting on increasing **k** values preclude a more accurate determination of the ratio. For such values of **k**, a tensile strain develops at the quantum well center [58]. Larger values of **k** lead to higher tensile strains. Results for a model in which the QW is assumed to remain square and unstrained, as suggested by Temkin et al. [59], are very similar to those shown for **k** = 1 and thus do not explain our data. It has been reported that in tensilely strained QWs the LH-HH splitting can be reduced and even reversed for large values of strain [60]. This is due to the effect of tensile strain which acts on valence bands in a direction opposite to that of quantum confinement. However, to exactly balance the quantum confinement splitting with the negative splitting due to tensile strain requires precise control of strains (i.e., compositions) and QW thicknesses. In contrast, according to the model discussed above and our experimental data (as shown in Figure 19), using QW intermixing to accomplish LH, HH degeneracy is a much easier task since the degeneracy, in this case, is not a very sensitive function of the HH bandgap shift.

The enhancement of the quantum confined Stark effect in these QW waveguides is clearly beneficial for performance although polarization insensitivity is not important if the modulator is integrated with a diode laser (the polarization of a laser is fixed). However, a device with such a large red-shift may find use in an optical switch geometry. In this case, since such devices may be positioned in some arbitrarily location in an optical network, polarization insensitivity is a crucial requirement.

5.1.3   High Power Windowed Laser Diodes

The output power of laser diodes is, in many cases, limited by failure mechanisms of diode laser facets. If the device consists of a single structure, the bandgap energy at the facet is identical to that of the central lasing region. Light, emitted by the gain region, will be absorbed by the material at the facet. Anything, such as defects, dirt on the facet, etc., which increases this absorption, produces a temperature rise at the facet, which, in turn, decreases the bandgap energy of the material, which increases absorption, etc. This positive feedback effect can quickly increase losses at the facet and, in severe cases, can produce catastrophic optical damage (COD) and failure of the laser.

Many techniques have been proposed and demonstrated in order to minimize or eliminate this run-away feedback effect by altering the optical bandgap energy at the facet, including etch-and-regrow, special twin waveguiding structures, etc. Recently, QW shape modification using ion implantation enhanced interdiffusion near the laser facets was shown to significantly reduce laser facet temperatures [61]. The laser structure consisted of two InGaAs/GaAs QWs surrounded by AlGaAs layers for optical confinement. The area of the laser near the facet was implanted with 8.46 MeV As^{4+} ions (passing completely through the QWs) and then annealed. The PL peak position was observed to blue-shift by about 20 nm (Figure 20) [61]. The blue-shifted region was varied from zero to 500 μm. High resolution (1.0 μm) maps of the temperature of the laser facets during lasing were measured using the technique of reflection modulation [62]. It was found that as the length of the blue-shifted region was increased, the heating coefficient (ratio of the temperature rise per unit output power), decreased significantly. For 300 μm, the heating coefficient was essentially zero. The implication is that COD due to temperature induced red-shifting of the bandgap energy of

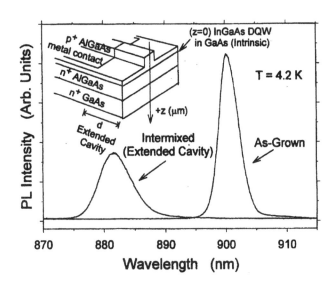

Fig. 20. Low temperature PL spectra for both intermixed (window material) and unimplanted (as-grown — active device material) regions of the laser structure. The insert is a schematic of the device structure. (reference 61)

the QW structure near the facets would be negligible, although the effect of residual damage on the lifetimes of these devices is a cause for concern (see above). In InP-based QW devices, since it is not necessary to implant into the QWs to obtain large blue shifts, this is a less serious problem and long lifetimes are possible [49].

QW shape modification can also be utilized to fabricate discrete laser diodes with specialized performance. For example, integrating an external cavity with a laser is simplified using QW blue-shifting. Such devices can be used to form narrowed Fabry-Perot lasers, mode-locked lasers, or low-loss distributed Bragg reflector lasers. The design of these devices is essentially similar to the insert in Figure 20 with a change in dimensions [63,64]. In the case of an external cavity laser, the passive section can be as long as 2 mm (compared with an active section of 0.5 mm) [63]. Residual modal loss in the passive section is consistent with direct waveguide loss measurements [63]. That is, losses in the blue-shifted passive sections are low enough that the optical performance of the laser diode is not seriously degraded.

5.1.4 Broadband Light Emitting Diodes

Superluminescent diodes (SLDs) which posses broadened spectral widths and thus low optical coherence lengths, are the light source of choice for

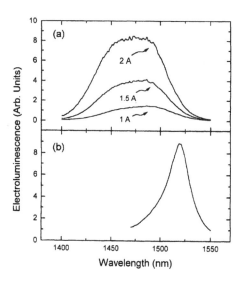

Fig. 21. The output electroluminescence spectra of a laterally modified broad-spectrum light emitting diode for several pump currents (a). This is contrasted with the output electroluminescence spectra of a conventional light emitting diode at a pump current of 900 mA (b). (reference 65)

applications such as fiber-optic gyroscopes and fault location in fiber optic cables. Previous techniques for broadening spectral widths of light emitting diodes, such as incorporating QWs of differing widths into the active structure or varying the width of QWs along the length of the device, suffered from strong input current dependencies. By ion implanting through a SiO_2 mask of varying thicknesses, the emission wavelength of a single device can be adjusted with micron spatial resolution in a single implantation step [65]. The result is a dramatic enhancement in the FWHM (full width — half maximum) of the emission spectrum. In Figure 21 the FWHM is increased from 28 nm to 90 nm and this spectrum is essentially independent of drive current [65].

5.1.5   Polarization Insensitive Optoelectronic Devices

One of the more critical issues which needs to be resolved before QWs can be effectively utilized in optoelectronic devices is that of polarization sensitivity. Since the polarization state of optical signals in an optical telecommunications network is, *a priori*, unknown, any device whose performance is polarization sensitive is of little value.

As noted above, ion implantation enhanced QW shape modification in certain InP-based QW heterostructures can produce polarization independent behaviour due to the generation of a complex strain pattern in the QW (sections 4.1and 5.1.2). This fortuitous situation can be exploited to produce polarization independent electro-absorptive modulators and other useful devices. For example, the QCSE modulator described in section 5.1.2 can also be used as a polarization independent photodetector (see Figure 17(b)). Further, under forward bias and in a non-resonant cavity configuration, it can be utilized to amplify light [66]. Figure 22 compares the measured optical gain in this structure as a function of wavelength for the as-grown QW waveguide (which exhibits the typical polarization dichotomy between TE and TM light) and the intermixed QW waveguide (where the optical gain is independent of polarization) [66]. The decrease of the gain at short wavelengths may be a consequence of the fact that this is the end of the range of the tunable external-cavity laser used in the measurement. Although this structure has not been optimized for optical gain, operating wavelength, or saturation power, these preliminary results are very encouraging.

A polarization insensitive semiconductor QW optical amplifier is unique in itself but this one has the added advantages of easy fabrication and the potential for simple integration with other passive and active optoelectronic devices. The principle difficulty is designing a QW waveguide structure whose operating wavelength is in the technologically important 1.3 and 1.55 μm bands. Since the QW structure must be blue-shifted a considerable distance in order that it be polarization insensitive, the initial bandgap of the unstrained InGaAs/InGaAsP QW must be about 1.6 mm to operate at 1.55 μm. It is unclear from the known material parameters of this system if this requirement can be achieved. However, operation at 1.3 μm using a suitably modified structure should be possible since the constraints are relaxed. In many ways, this wavelength is more commercially interesting in that this is the region of minimum dispersion and furthermore, there are no efficient rival fiber optical amplifiers.

5.2   Integrated Devices

Although, in principle, most of the critical issues related to the use of ion implantation enhanced QW shape modification for monolithic integration of passive and active optoelectronic devices have been successfully addressed, to date there are surprisingly few examples of PICs (photonic integrated circuits) reported in the literature. However, due to the recent increased interest in multi-wavelength optical networks and the inexpensive components and modules required to operate them, this situation is expected to change rapidly in the near future. In this section a brief description will be given of a number of PICs reported on in the literature. We will restrict the

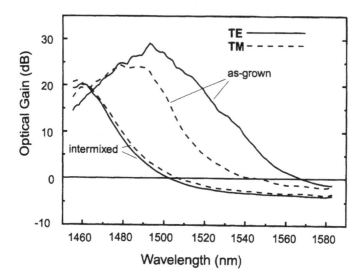

Fig. 22. Optical gain as a function of wavelength for TE (solid lines) and TM (dashed lines) polarized optical modes in as-grown (curves A, Figure 13) and intermixed (curves D, Figure 13) InGaAs/InGaAsP QW waveguides. (reference 66)

discussion to PICs fabricated using only a single growth step followed by QW shape modification. Other, more complicated fabrication procedures (e.g., QW interdiffusion with multiple growth sequences [67]), although undoubtedly important and interesting, will be left for others to review.

5.2.1 Two Wavelength QW Waveguide Demultiplexer

A very simple two wavelength QW waveguide demultiplexer has been demonstrated in both the GaAs/AlGaAs [68] and InGaAs/InGaAsP [69] QW material systems. The principle of the operation of the device is schematically illustrated in Figure 23 [69]. A semiconductor p-i-n QW waveguide is modified by bandgap shifting the first half while leaving the second half unaltered. Ohmic contacts are made to each half (alternatively, metal-semiconductor-metal (MSM) contacts can be used [68]) so that, with reverse bias, the waveguides act as photodetectors. Light is butt coupled into the waveguide at the end containing the largest (blue-shifted) bandgap. The light is absorbed in the first half (photodetector 1) if its energy is larger than the bandgap energy of the blue-shifted QW. The remaining light passes through to photodetector 2, in the as-grown QW waveguide. The performance is shown in Figure 24, a plot of the normalized TE photocurrent detected in regions 1

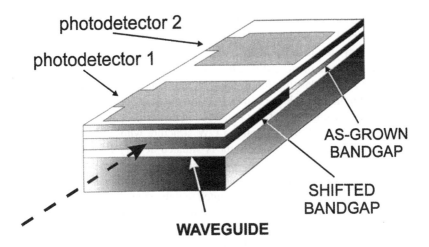

Fig. 23. Schematic diagram illustrating the principle of operation of a two wavelength WDM demultiplexer fabricated using QW shape modification. (reference 69)

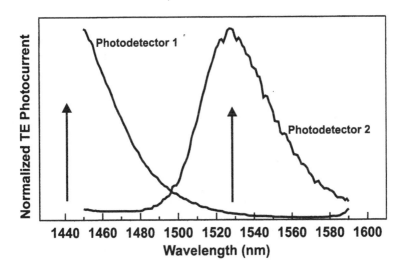

Fig. 24. Normalized room temperature TE photocurrent spectra of as-grown (photodetector 2) and interdiffused (photodetector 1) regions of the two wavelength WDM demultiplexer illustrated in Figure 23. (reference 69)

and 2 as a function of wavelength [69]. High energy (short wavelength) photons are detected in region 1 while low energy (long wavelength) photons are absorbed in region 2. In this manner, an optical beam consisting of two wavelengths (1.44 and 1.53 μm, say) is readily demultiplexed using a device which is very simple to fabricate.

There are some obvious difficulties with the operation of this PIC. First it is polarization sensitive due to the quantum confinement splitting of the heavy and light hole valence bands. In principle, it is possible to ensure that the blue-shifted photodetector is polarization insensitive, as discussed above (Section 5.1.2), and, by designing a QW structure whose absorption edge for both TE and TM polarized light is situated at very long wavelengths (say 1.6 μm) compared with the shifted bandedge, the performance of the demultiplexer will be essentially independent of polarization. A more serious shortcoming of this PIC is that it is limited to only a few wavelengths. This is partially a consequence of the polarization problem but the sharpness of the bandedge will also affect the crosstalk between adjacent wavelength channels. Demultiplexing 1.3 and 1.55 μm, for example, is feasible and it may be possible, by adding intermediate stages, to pick off a couple more wavelengths. However, it is difficult to see this technique being utilized to demultiplex high density multiplexed channels where the channel spacing is a few nms or less.

Although this device has not been optimized (the crosstalk is barely acceptable, even for the two wavelength case), has serious polarization problems, and is not capable of high density demultiplexing, it illustrates the potential of QW shape modification for the fabrication of monolithic PICs. In this case a simple QW waveguide structure was modified to demultiplex two wavelengths which, otherwise, would have required an elaborate micro-optic set-up, involving focusing lenses, dichroic filters, and delicate alignment.

5.2.2   Multiple Wavelength Laser Arrays

One of the first applications of QW shape modification (achieved using simple thermal oven annealing without any enhancing mechanism) was to blue shift the lasing wavelength of a GaAs/AlGaAs QW laser diode [1]. The wavelength was observed to decrease from 820 to 730 nm without appreciable deterioration of the device performance. However, a more useful trick is to be able to modify the wavelength of an individual laser in an array by using a spatially selective QW interdiffusion enhancement technique. Not only are the wavelength shifts increased but, in principle, each laser in an array can have its bandgap, and thus gain curve, tweaked individually. This is very useful if a monolithic DFB or DBR laser array is to be fabricated from a single wafer, as may be required for WDM (wavelength division multiplexed) systems. Lasers at the ends of the range of such an array will have weaker outputs since they are at some distance from the maximum of the gain curve. By adjusting the gain curves so that the wavelength for each laser is situated close to its maximum gain, the output of the laser array can be flattened.

Although there are important advantages to be gained by utilizing such techniques, these must not be achieved by decreasing the performance or the lifetimes of the devices. The serious question which must be answered is, will this process detrimentally affect device performance? For the case of blue-shifting using ion implantation enhanced QW shape modification, the short answer is: no, if proper precautions are taken. For, example, as discussed above, it is generally not a good idea to implant directly into the active (QW) structure since standard rapid thermal anneals may not remove all of the damage. As noted above, for InP-based QW laser structures this is not necessary since the vacancies, interstitials, and other defects generated by the implantation can propagate long distances during thermal anneal.

Bandgap shifting of InGaAs/InGaAsP QW lasers, without significant effect on laser performance, has recently been demonstrated at 1.3 μm [49] and 1.5 μm [48]. Figure 25 presents the room temperature emission spectra of four broad area, Fabry-Perot lasers whose wavelengths have been shifted by ion implantation through a SiO$_2$ mask of varying thicknesses [48]. The spectrum labeled "as-grown" was emitted by a laser from an area on the wafer shielded by a 2.2 μm thick SiO$_2$ mask (see Figure 15) while the other spectra are from lasers fabricated from areas with varying thicknesses of mask

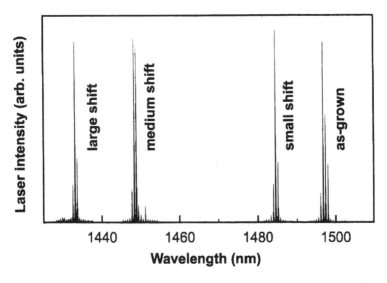

Fig. 25. Room temperature emission spectra of as-grown and bandgap shifted broad area Fabry-Perot laser diodes. (reference 48)

(small: 1.2 μm; medium: 0.65 μm; large: 0 μm). The most important result from this study is that threshold currents do not increase as a consequence of ion implantation and annealing [70]. In addition, as discussed earlier [49], there appears to be little deterioration of laser lifetimes.

In principle, it should be possible to apply this technology to tune gain curves of any QW waveguide structure.

6 SUMMARY

The conclusions of this review can be summarized as follows: spatially selective enhanced QW shape modification using ion implantation has many advantages for the monolithic integration of photonic devices and, at this point, no inherent, "show-stopping" disadvantages. The technique is simple, requiring no elaborate growth or processing techniques and device "friendly", in that it does not significantly alter device performance, as attested to by laser lifetime measurements.

The major issues which still need to be resolved, and which are also present in other QW shape modification techniques, are reproducibility and reliability. These directly impact yield and, therefore, commercial viability. Of all QW interdiffusion techniques, ion implantation enhanced interdiffusion may be the most process tolerate, giving it the best chance to become commercially viable. But this still remains to be demonstrated.

Acknowledgements

The author would like to thank his many collaborators over the years for their support and friendship. These include Paul Melman, Boris Elman, Craig Armiento, Jim Chi, Roger Holmstrom, Xin Wen, Doug Owens, J. Powers, M. Rothman, Masum Choudhury, Barbara Foley, and Andy Silletti of GTE Laboratories, M.A. Koza and R. Bhat of Bellcore and Sylvain Charbonneau and Mike Thewalt of Simon Fraser University for the work performed while at GTE Laboratories in Waltham, MA. His colleagues at NRC include Sylvain Charbonneau, Geof Aers, Phil Poole, Paul Piva, Margaret Buchanan, Garth Champion, Ian Templeton, Robin Williams, Alain Roth, Zbig Wasilewski, Michel Dion, Mahmoud Fallahi, J.J. He, John McCaffrey, Mike Davies, and Yan Feng; Richard Goldberg and Ian Mitchell from the University of Western Ontario, Jacques Beauvais from the University of Sherbrooke; and Jean-Paul Noel from Nortel Technology.

References

1. M.D. Camras, N. Holonyak, Jr., R.D. Burnham, W. Streifer, D.R. Scifres, T.L. Paoli, and C. Lindstrom, *Appl. Phys. Lett.*, **54**, 5637 (1983).

2. T.E. Schlesinger and T. Kuech, *Appl. Phys. Lett.*, **49**, 519 (1986).

3. L.L. Chang and A. Koma, *Appl.Phys.Lett.*, **29**, 138 (1976).

4. For a comprehensive study of thermally induced QW interdiffusion in the InGaAs/InGaAsP system, see R.E. Mallard, E.J. Thrush, R.W. Martin, S.L. Wong, R.J. Nicholas, R.E. Pritchard, B. Hamilton, N.J. Long, S.A. Galloway, A. Chew, D.E. Sykes, J. Thompson, K. Scarrott, J.M. Jowett, K. Satzke, A.G. Norman, and G.R. Brooker, *Semicond. Sci. Technol.*, **8**, 1156 (1993)..

5. See, for example, W.D. Laidig, N. Holonyak, Jr., M.D. Camras, K. Hess, J.J. Coleman, P.D. Dapkus, and J. Bardeen, *Appl. Phys. Lett.*, **38**, 776 (1981); T. Fukuzawa, S. Semura, H. Saito, T. Ohta, Y. Uchida, and H. Nakasima, *Appl. Phys. Lett.*, **45**, 1 (1984).

6. for a review, see Emil S. Koteles, B. Elman, P. Melman, J.Y. Chi, C.A. Armiento, *Opt. and Quant. Electron.*, **23**, S779 (1991).

7. L.J. Guido, N. Holonyak, Jr., K.C. Hsieh, R.W. Kaliski, W.E. Plano, R.D. Burnham, R.L. Thornton, J.E. Epler, and T.L. Paoli, *J. Appl. Phys.*, **61**, 1372 (1987).

8. J.J. Coleman, P.D. Dapkus, C.G. Kirkpatrick, M.D. Camras, and N. Holonyak, Jr., *Appl. Phys. Lett.*, **40**, 904 (1982).

9. J. Cibert, P.M. Petroff, D.J. Werder, S.J. Pearton, A.C. Gossard, and J.H. English, *Appl. Phys. Lett.*, **49**, 223 (1986).

10. Y. Hirayama, Y. Suzuki, and H. Okamoto, *Surface Science*, **174**, 98 (1986).

11. J. Crank, *The Mathematics of Diffusion* (Oxford University, London, 1957).

12. S. Flugge, *Practical Quantum Mechanics* (Springer, Berlin, 1971), pp. 94–100.

13. D.C. Bertolet, J.K. Hsu, K.M. Lau, Emil S. Koteles, and D. Owens, *J. Appl. Phys.* **64**, 6562 (1988).

14. Emil S. Koteles, unpublished.

15. Emil S. Koteles, B. Elman, Johnson Lee, S. Charbonneau, and M. Thewalt, in *Quantum Well and Superlattice Physics III*, G.H. Dohler, Emil S. Koteles, and J.N. Schulman, Editors, **Proc. SPIE 1283**, P.143 (1990).

16. Y.J. Chen, Emil S. Koteles, B.S. Elman, and C.A. Armiento, *Phys. Rev.*, **B38**, 4562 (1987).

17. Emil S. Koteles, B. Elman, C.A. Armiento, P. Melman, J.Y. Chi, R.J. Holmstrom, S. Charbonneau, and M.L.W. Thewalt, *J. Appl. Phys.*, **66**, 5532 (1989).

18. Emil S. Koteles, unpublished.

19. B. Elman, Emil S. Koteles, P. Melman, and C.A. Armiento, *J. Appl. Phys.*, **66**, 2104 (1989).

20. L.B. Allard, G.C. Aers, S. Charbonneau, T.E. Jackman, R.L. Williams, I.M. Templeton, M. Buchanan, D. Stevanovic, and F.J.D. Almeida, *J. Appl. Phys.*, **72**, 422 (1992).

21. Emil S. Koteles, B. Elman, R.P. Holmstrom, P. Melman, J.Y. Chi, X. Wen, J. Powers, D. Owens, S. Charbonneau, and M.L.W. Thewalt, *Superlattices and Microstructures*, **5**, 321 (1989).

22. P.J. Poole, P.G. Piva, M. Buchanan, G.C. Aers, A.P. Roth, M. Dion, Z.R. Wasilewski, Emil S. Koteles, S. Charbonneau, and J. Beauvais, *Semicond. Sci. Technol.*, **9**, 2134 (1994).

23. H.H. Tan, J.S. Williams, C. Jagadish, P.T. Burke, and M. Gal, *Appl. Phys. Lett.* (1995).

24. J.W. Mayer, L. Erickson, and J.A. Davis, *Ion Implantation* (Academic Press, New York, 1970).

25. T.E. Hayes and O.W. Holland, *Nucl. Instrum. Methods*, **B59/60**, 1028 (1991); F.G. Moore, H.B. Dietrich, E.A. Dobisz, and O.W. Holland, *Appl. Phys. Lett.*, **57**, 911 (1990); T.E. Hayes and O.W. Holland, *Appl. Phys. Lett.*, **58**, 62 (1991).

26. S. Charbonneau, P.J. Poole, P.G. Piva, G.C. Aers, Emil S. Koteles, M. Fallahi, J.J. He, J.P. McCaffrey, M. Buchanan, M. Dion, R.D. Goldberg, and I.V. Mitchell, *J. Appl. Phys.*, **78**, 3697 (1995).

27. K.K. Anderson, J.P. Donnelly, C.A. Wang, J.D. Woodhouse, and H.A. Haus, *Appl. Phys. Lett.*, **53**, 1632 (1988); W. Xia, S.A. Pappert, B. Zhu, A.R. Clawson, P.K.L. Yu, S.S. Lau, D.B. Poker, C.W. White, and S.A. Schwarz, *J. Appl. Phys.*, **71**, 2602 (1992); S.A. Pappert, W. Xia, B. Zhu, A.R. Clawson, Z.F. Guan, P.K.L. Yu, and S.S. Lau, *J. Appl. Phys.*, **72**, 1306 (1992).

28. R.D. Goldberg, and I.V. Mitchell, S. Charbonneau, P.J. Poole, Emil S. Koteles, G.C. Aers, and G. Weatherly, MRS (1995).

29. Emil S. Koteles, unpublished.

30. K.B. Kahen, G. Rajeswaran, and S.T. Lee, *Appl. Phys. Lett.*, **53**, 1635 (1988); S. Tong Lee, G. Braunstein, P. Fellinger, K.B. Kahen, and G. Rajeswaran, *Appl. Phys. Lett.*, **53**, 2531 (1988).
31. F.C. Prince, and C.A. Armiento, *IEEE Electron Device Lett.*, **7**, 23 (1986); C.A. Armiento and F.C. Prince, *Appl. Phys. Lett.*, **48**, 1623 (1986); J.D. Woodhouse, M.C. Gaidis, J.P. Donnelly, and C.A. Armiento, *Appl. Phys. Lett.*, **51**, 186 (1987); C.A. Armiento, L.L. Lehman, F.C. Prince, and S. Zemon, *J. Electrochem. Soc.*, **134**, 2010 (1987).
32. There are many books on the theory and practice of rapid thermal annealing. See, for example, "Rapid Thermal Processing of Semiconductors" (Plenum Press, New York, 1997) by V.E. Borisenko and P.V. Heseth and "Rapid Thermal Processing: Science and Technology (Academic Press, Boston, 1993) by R.B. Fair.
33. B. Goldstein, *Phys. Rev.*, **121**, 1305 (1961).
34. Emil S. Koteles, B. Elman, C.A. Armiento, and P. Melman, *Superlattices and Microstructures*, **9**, 533 (1991).
35. Emil S. Koteles, B. Elman, P. Melman, and C.A. Armiento, Effect of Annealing on Strained InGaAs/GaAs Quantum Wells, in *Layered Structures — Heteroepitaxy, Superlattices, Strain, and Metastability, Materials Research Society Symposium Proceedings Series, Volume 160*, 1989 MRS Fall Meeting, Boston, Massachusetts, edited by B.W. Dodson, L.J. Schowalter, J.E. Cunningham, and F.H. Pollak, (Materials Research Society, Pittsburgh, PA, 1990), page 147.
36. S.W. Ryu, I. Kim, and B.D. Choe, *Appl. Phys. Lett.*, **67**, 1417 (1995).
37. B. Elman, Emil S. Koteles, D.G. Kenneson, J.W. Powers, and D.W. Oblas, in *Proceedings of the, Fourth International Symposium on InP and Related Compounds, 1992, Newport, Rhode Island, USA* (IEEE Catalog #92CH3104-7, 1992), page 175.
38. J.H. Marsh, S.L. Habsen, A.C. Bryce, and R.M. De La Rue, *Opt. & Quantum Electron.*, **23**, S941 (1991); J.E. Zucker, B. Tell, K.L. Jones, M.D. Divino, K.F. Brown-Goebeler, C.H. Joyner, B.I. Miller, and M.G. Toung, *Appl. Phys. Lett.*, **60**, 3036 (1992).
39. J.Z. Wan, J.G. Simmons, and D.A. Thompson, *J. Appl. Phys.*, **81**, 765 (1997).
40. J.E. Zucker, M.D. Divino, T.Y. Chang, and N.J. Sauer, *IEEE Photonics Technology Letters*, **6**, 1105 (1994).
41. D. Labrie, H. Lafontaine, N. Rowell, S. Charbonneau, D. Houghton, R.D. Goldberg, and I.V. Mitchell, *Appl. Phys. Lett.*, **69**, 993 (1996).
42. Emil S. Koteles, in *Defining the Global Information Infrastructure: Infrastructure, Systems, and Services*, Stephen F. Lundstrom, Editor, SPIE Vol. CR56, page 28 (1995).
43. J.J. He, Emil S. Koteles, P.J. Poole, M. Davies, R. Goldberg, I. Mitchell, and S. Charbonneau, *Electronics Letters*, **31**, 2094 (1995).
44. J.D. Ralston, W.J. Schaff, D.P. Bour, and L.F. Eastman, *Appl. Phys. Lett.*, **54**, 534 (1989) .
45. Emil S. Koteles, J.J. He, S. Charbonneau, P.J. Poole, G. Aers, Y. Feng, R. Goldberg, and I.V. Mitchell, *Emerging Components and Technologies for All-Optical Networks II*, Emil S. Koteles, Alan E. Willner, Editors, Proc., **SPIE 2918**, 184 (1997).
46. J.J. He, Emil S. Koteles, M. Davies, P.J. Poole, M. Dion, Y. Feng, S. Charbonneau, P. Piva, M. Buchanan, R. Goldberg, and I. Mitchell, *Canadian Journal of Physics*, **74**, S32 (1996).
47. J.E. Zucker, B. Tell, K.L. Jones, M.D. Divino, K.F. Brown - Goebeler, C.H. Joyner, B.I. Miller, and M.G. Young, *Appl. Phys. Lett.*, **60**, 3036 (1992).
48. S. Charbonneau, P.J. Poole, Y. Feng, G.C. Aers, M. Dion, M. Davies, R. Goldberg, and I.V. Mitchell, *Appl. Phys. Lett.*, **67**, 2954 (1995).
49. J.P. Noël, D. Melville, T. Jones, F.R. Shepherd, C.J. Miner, N. Puetz, K. Fox, P.J. Poole, Y. Feng, Emil S. Koteles, S. Charbonneau, R. Goldberg, and I.V. Mitchell, *Applied Physics Letters*, **69**, 3516 (1996).
50. J.F. Ziegler, J.P. Biersack, and U. Littmark, *The Stopping and Ion Range of Ions in Matter* (Pergamon, New York, 1985).
51. J. Cibert, P.M. Petroff, G.J. Dolan, S.J. Pearton, A.C. Gossard, and J.H. English, *Applied Physics Letters*, **49**, 1275 (1986).
52. C. Vieu, M. Schneider, D. Mailly, R. Planel, H. Launois, J.Y. Marzini, and B. Descouts, *J. Appl. Phys.*, **70**, 1444 (1991).
53. J.E. Haysom, P.J. Poole, Y. Feng, Emil S. Koteles, J.J. He, S. Charbonneau, R.D. Goldberg, and I.V. Mitchell, to be published in *J. Vac. Science. Tech. A*.

54. J.E. Zucker, K.L. Jones, B. Tell, K. Brown-Goebeler, C.H. Joyner, B.I. Miller, and M.G. Young, *Electronics Letters*, **28**, 853 (1992).
55. J.E. Haysom, J.J. He, P.J. Poole, Emil S. Koteles, A. Delage, Y. Feng, S. Charbonneau, R.D. Goldberg, and I.V. Mitchell, unpublished.
56. Y. Chen, J.E. Zucker, B. Tell, N.J. Sauer, and T.Y. Chang, *Electronics Letters*, **29**, 87 (1993).
57. Emil S. Koteles, J.J. He, S. Charbonneau, P.J. Poole, G.C. Aers, Y. Feng, R.D. Goldberg, and I.V. Mitchell, in *Emerging Components and Technologies for All-Optical Photonic Systems II*, Emil S. Koteles, Alan Wilner, Editors, Proc. SPIE 2918, 184 (1997).
58. W.C. Shiu, J. Micallef, I. Ng and E.H. Li, *Jpn. J. Appl. Phys.*, **34**, 1778 (1995); J. Micallef, E.H. Li and B.L. Weiss, *J. Appl. Phys.*, **73**, 7524 (1993).
59. H. Temkin, S.N.G. Chu, M.B. Panish, R.A. Logan, *Appl. Phys. Lett.*, **50**, 956 (1987).
60. F. Agahi, Kei May Lau, Emil S. Koteles, A. Baliga, and N.G. Anderson, *IEEE J. Quant. Electron.*, **30**, 459 (1994).
61. P.G. Piva, S. Fafard, M. Dion, M. Buchanan, S. Charbonneau, R.D. Goldberg, and I.V. Mitchell, *Appl. Phys. Lett.*, **70**, 1662 (1997).
62. P.W. Epperlein, *Jpn. J. Appl. Phys.*, **32**, 5514 (1993).
63. J. Werner, E. Kapon, N.G. Stoffel, E. Colas, S.A. Schwartz, C.L. Schwartz, and N. Andreadakis, *Appl. Phys. Lett.*, **55**, 540 (1989).
64. S.R. Andrew, J.H. Marsh, M.C. Holland, and A.H. Kean, *IEEE Photonics Technology Letters*, **4**, 426 (1992).
65. P.J. Poole, M. Davies, M. Dion, Y. Feng, S. Charbonneau, R.D. Goldberg, and I.V. Mitchell, *IEEE Photonics Technology Letters*, **8**, 1145 (1996).
66. J.J. He, S. Charbonneau, P.J. Poole, G.C. Aers, Y. Feng, Emil S. Koteles, R.D. Goldberg, and I.V. Mitchell, *Appl. Phys. Lett.*, **69**, 562 (1996).
67. see, for example, T. Hirata, M. Suehiro, M. Hihara, M. Dobashi, and H. Hosomatsu, *IEEE Photonics Technology Letters*, **5**, 698 (1993).
68. A.N.M. Masum Choudhury, P. Melman, A. Silletti, Emil S. Koteles, B. Foley, and B. Elman, *IEEE Photonics Technology Letters*, **3**, 817 (1991).
69. P.J. Poole, S. Charbonneau, M. Dion, Y. Feng, J.J. He, Emil S. Koteles, I.V. Mitchell, and R.D. Goldberg, in *Emerging Components and Technologies for All-Optical Networks*, Emil S. Koteles, Alan Willner, editors, Proc. SPIE **2613**, 9 (1995).
70. P.J. Poole, S. Charbonneau, M. Dion, G.C. Aers, M. Buchanan, R.D. Goldberg, and I.V. Mitchell, *IEEE Photonics Technology Letters*, **8**, 16 (1996).

CHAPTER 7

Control of Layer Intermixing by Impurities and Defects

DECAI SUN AND PING MEI

Xerox Palo Alto Research Center, Palo Alto, CA

1 INTRODUCTION

III-V semiconductor superlattices have an excellent thermal stability against the inter-diffusion of constituents across the heterojunction under typical processing conditions. For example, the inter-diffusion coefficient of Al in an AlAs/GaAs superlattice at $850°C$ is on the order of 10^{-18} cm^2/sec [1]. This stability, however, can be rapidly reduced by introducing impurities or lattice defects. It was first demonstrated by Laidig et al. [2] that introducing Zn into GaAs/AlAs superlattices could dramatically increase the interdiffusion of Al and Ga by a large factor in 1981. Since then, evidence of superlattice mixing induced by a variety of dopants, both n-type and p-type, has been reported. Examples of these species are Si, Ge, C, Te, Se, S, Sn, and Be [3–9]. These impurities can be introduced either during the growth or post growth from external sources. Besides introducing impurities, superlattice intermixing can be implemented by internal defect migration through an anneal process where a SiO$_2$ surface cap is utilized as a diffusion sink for Ga. The migration of the Ga vacancies from the surface into the structure results in an enhancement of Al and Ga interdiffusion in the superlattice [10]. Alternatively, the mixing of a superlattice can be realized by ion implantation which generates atomic displacement in the superlattice by collision effect.

The layer intermixing technique of semiconductor heterostructures has attracted a great deal of interest in optoelectronic device fabrications since its discovery. The intermixing technique offers a freedom to modify heterostructure after growth. Impurity-induced layer disordering (IILD) has been applied to fabricate high performance buried heterostructure laser diodes and passive optical waveguides [11-18]. Devices fabricated by IILD include single-stripe buried ridge heterostructure lasers [11], coupled multi-stripe buried heterostructure lasers [12] and transparent window lasers [14]. Multi-wavelength lasers have also been demonstrated by layer disordering technique using a variety of surface cap conditions and buried impurity sources [15,16]. In this chapter, we review the mechanism of impurity and ion implantation induced layer intermixing in GaAs/AlAs and InGaAs/InP superlattices. Mixing inhibition in heavily doped and ion implantation damaged regions is discussed. The large difference in diffusion coefficients of group III elements in AlGaAs and AlGaInP QWs is utilized for selective intermixing of the AlGaAs QW in a dual QW heterostructure. We also discuss control of layer intermixing using n-type and p-type modulation doping and surface masks in single and dual modulation doped AlGaAs QWs. We also address fabrication of multi-wavelength laser diodes.

2   LAYER INTERMIXING OF GAAS/ALGAAS SUPERLATTICES

2.1   Impurity Induced Mixing in GaAs/AlAs Superlattices

Several species of n-type dopants, such as Si [3], Ge [4], Se [6], S [7] and Sn [8], have the effect of enhancing Al and Ga interdiffusion in AlGaAs superlattices. In general, the degree of the impurity induced mixing depends on the type of the dopant and the dopant concentration. Among the above species, Si has a high efficiency in terms of enhancing the superlattice mixing at a given concentration level and under a comparable diffusion condition. Since Si diffusion also depends on the concentration level of itself, it tends to have an abrupt diffusion front and a relatively slow diffusion rate. As a result, Si induced mixing may be obtained at a relatively modest doping level and confined to well defined regions of interest, which is important to control the intermixing to a confined region.

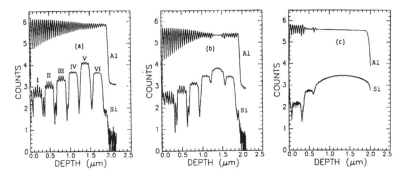

Fig. 1. SIMS depth profiles of Al and Si concentrations in the AlAs-GaAs superlattice samples; (a) as grown, and after 3 hours annealing at (b) 700°C and (c) 900°C.

Although it is convenient and practical to utilize Si implantation for superlattice mixing, it produces complication in distinguishing the impurity effect from the ion collision effect. A quantitative study of Si-induced mixing was conducted on AlAs/GaAs superlattices which were doped with Si during the molecular beam epitaxy (MBE) growth. In the experiment described in reference 19, staircase-like Si doping profiles were introduced in AlAs/GaAs superlattices with various Si concentration plateaus (Figure 1). The magnitude of Al oscillations in an intrinsic sample decreased with the depth due to an intrinsic interface broadening and an additional SIMS-induced broadening. The intrinsic sample was used as a reference to extract the diffusion data in the samples annealed at various temperatures. The anneal was performed in a H_2: Ar atmosphere furnace with proximity GaAs wafers sandwiching the sample. For a low temperature anneal (700°C), Al and Si diffusion was negligible in regions with Si doping levels below 2×10^{18} cm^{-3} (Figure 1 (b)). Si-enhanced Al diffusion is visible in regions IV, V and VI, where the Si doping levels were 2×10^{18}, 3×10^{18} cm^{-3} and 2×10^{18} cm^{-3} respectively. At a higher temperature (900°C), Al and Si diffusion occurred at even lower Si concentrations (Figure 1 (c)). A quantitative study can be carried out by analyzing the Al diffusion as a function of Si concentration.

Al diffusion can be described by Fick's law. At the center of each Si-doped plateau where Si concentration is approximately constant during the diffusion process, the diffusion coefficient for Al is relatively constant. The Al diffusion coefficient can be written as

$$\mathbf{D} = \mathbf{D}_0 e^{-E_a/kT} \tag{1}$$

where D_0, the pre-factor of the diffusion coefficient, is proportional to the atomic hopping frequency at a given temperature; E_a, the diffusion activation

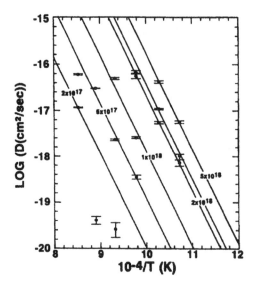

Fig. 2. Arrhenius plot of Al diffusion coefficients at various Si concentrations. Each data point was derived from the peak to valley ratio at the center of the Si doped plateau.

energy, is the sum of the enthalpy of vacancy formation and of atomic migration. The diffusion coefficient of Al at the center of each Si-doped plateau as a function of temperature is shown on an Arrhenius plot in Figure 2. In estimating Al diffusion coefficients, the peak-to-valley ratios of Al oscillations in SIMS depth profiles were reduced to diffusion lengths with error bars corresponding to the rms counting errors in each measurement [20]. These diffusion lengths were corrected with respect to the 500°C data in which there was negligible diffusion and all apparent mixing was SIMS induced.

In Figure 2, the Al diffusion activation energy is about 4.1 eV for regions with Si doping level above 2×10^{17} cm^{-3}. The activation energy is not certain in the region with Si doping level of 2×10^{17} cm^{-3} or below. In comparison, activation energies of 5.6 and 2.6 eV have been measured for atomic Ga [21] and Si [22] diffusion respectively in GaAs. For Al diffusion, activation energies of 4.3 and 3.6 eV have been measured in GaAs and AlAs respectively by a sputter Auger technique [1].

The mechanism of Si-IILD has been debated for a long period of time. It is generally accepted that, in Si induced superlattice mixing, the Al and Si diffusion is assisted by complexes consisting of charged defects. Since the density of the charged defects depends on the Fermi level, the superlattice

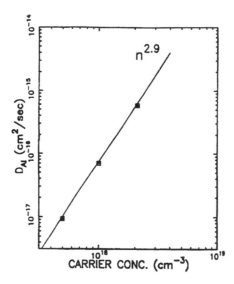

Fig. 3. Plot of the Al diffusion coefficient versus the electron concentration at 800°C. The solid line is the least linear square fitting, indicating that the Al diffusion coefficient varies with the carrier concentration as $n^{2.9}$.

mixing is affected by the Fermi level. The controversial argument, however, has been about the type of the charged defects performing the enhancement for the superlattice mixing. An extensive review of various models can be found in reference 23.

The connection between the electrical properties and the enhanced superlattice mixing for AlGaAs superlattices is described in reference 24. The electrical activation of Si was measured in $Al_{0.01}Ga_{0.99}As/GaAs$ superlattice samples by a profile plotter. In the as-grown samples, the dopants were fully activated up to the Si concentration level of 10^{18} cm^{-3}. At a higher doping level, the carrier concentration was about 40% of the Si-doping concentration, which is commonly noticed in GaAs samples with heavy Si doping. The degree of compensation depends on various factors. In Si doped GaAs samples grown by MBE, for example, the compensation is affected by the As_4/Ga flux ratio and substrate temperature [25,26]. Ishibashi et al. have shown that in MBE grown GaAs, the electron concentration reaches a maximum value for a Si concentration of 6×10^{19} cm^{-3} and decreases at higher Si concentration levels [27]. There are several possibilities which may account for the compensation at a high Si concentration. In general, the compensation ratio is determined in part by the relative numbers of Ga and As vacancies, and the relative free energy. According to a theoretical calculation, the free energy of a Si atom residing on a Ga site is about

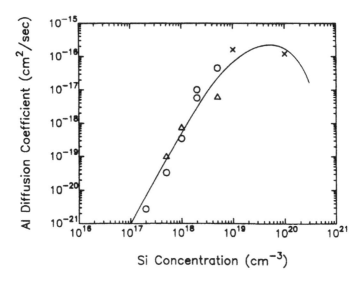

Fig. 4. Plot of the Al diffusion coefficient as a function of the Si concentration. The data points were determined from SIMS measurements of AlAs/GaAs superlattice samples and extrapolated to 700°C. The solid line is proportional to the third power of the electron concentration extracted from carrier concentration measurements of Ishibashi *et al..* [27].

0.3 eV lower than on an As site [28]. Although the concentration of As vacancies in GaAs samples grown by MBE or organometallic vapor phase deposition (OMVPE) may exceed that of Ga vacancies [29], most Si atoms reside on Ga sites as donors at a low Si concentration. At a high doping level, additional Si-doping produces acceptors, which then autocompensate Si donors by forming electrically neutral pairs. Si pairs have been identified by local vibration mode spectroscopy at high doping levels [30]. On the other hand, in heavily Si doped AlGaAs materials, defect complex formation and Si precipitation have been suggested to account for the electrical compensation.

The correlation between Al diffusion rate and the electron carrier concentration is described by Figure 3. The Al diffusion coefficient increases with approximately the third power of the electron carrier concentration. Figure 4 shows a plot of the Al diffusion coefficient versus Si concentration over the range of 2×10^{17} to 10^{20} cm^{-3}. For comparison, also plotted in the figure is the third power of the estimated electron concentration according to the measurements of Ishibashi, et al. for Si-doped MBE materials [27]. It is noticed that even in the heavily compensated region, where the electron concentration decreases with increasing Si concentration, the Al diffusion coefficient is still proportional to the third power of electron concentration. In a heavily Si-doped sample, the decrease of the mixing

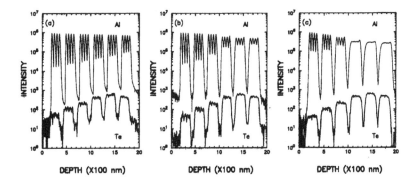

Fig. 5. SIMS depth profiles of Al and Te concentrations in AlGaAs superlattice samples; (a) as grown, and after 30 minutes annealing at (b) 900°C and (c) 1000°C.

correlates to the autocompansation effect described earlier. In addition to Si-induced superlattice mixing, a quantitative study was conducted on Te-doped GaAs/AlAs superlattices [5]. Among the n-type dopant species of Ge, Se, S, Sn and Si, Te has the lowest diffusion coefficient in GaAs [31], and therefore can be used to satisfy the requirements for abruptness and control of doping profiles for device applications. Te-induced AlAs/GaAs superlattice mixing was studied in a similar structure as the Si-induced case. A staircase-like Te-doping profile was introduced with concentration plateaus of 2×10^{17}, 5×10^{17}, 1×10^{18}, 2×10^{18}, 3×10^{18} and 2×10^{18} cm^{-3} (Figure 5). The annealing temperatures ranged from 800 to 1000°C. The minimum temperature for Te-induced mixing is much higher than that for Si. At 1000°C, the mixing enhancement is visible for Te doping level of 2×10^{18} cm^{-3}. As shown in Figure 6, the activation energy for Al diffusion is independent from the Te concentration, similar to that of Si. The activation energy is about 3 eV, which is lower than that of Si induced mixing. The Al diffusion coefficient versus Te concentration at 800°C is plotted in Figure 7. Unlike Si induced mixing, the Al diffusion coefficient shows a near first power dependence on Te concentration. These results indicated that, besides the Fermi level effect, the enhanced superlattice mixing is also affected by the doping species.

2.2   Ion Induced Superlattice Mixing

For practical applications, the ion implantation technique has been commonly employed to introduce impurities into superlattice samples where optical and electrical confinement is desired. During the implantation, energetic ions impact on the target atoms, causing displacement of the target atoms

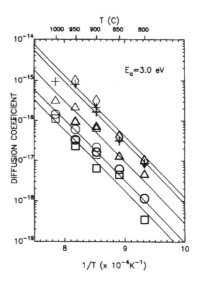

Fig. 6. Arrhenius plots of Al diffusion coefficients at Te concentrations of 2×10^{17} cm^{-3} ($\square$), 5×10^{17} cm^{-3} ($\bigcirc$), 10^{18} cm^{-3} ($\triangle$), 2×10^{18} cm^{-3} ($+$), and 3×10^{18} cm^{-3} ($\diamondsuit$).

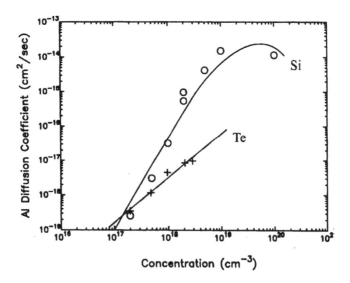

Fig. 7. Plot of the Al diffusion coefficient as a function of Te concentration ($+$). The solid line is a least square fit to the data points with a slope of 1.3 ± 0.1, indicating an approximately linear relationship between Al diffusion coefficient and the Te concentration. As a comparison, the Al diffusion coefficient as a function of the Si concentration at 800 C is also shown.

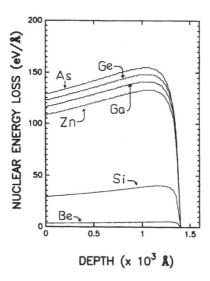

Fig. 8. Plot of the calculated nuclear energy loss of Be (41 keV), Si (150 keV), Zn (332 keV), Ga (334 keV), Ge (355 keV), and As (366 keV) in AlAs/GaAs superlattice samples.

by kinetic energy transfer (the collision effect) through nuclear energy loss. Therefore, the superlattice structure can be mixed by the collision effect alone, and the extent of mixing depends on the nuclear energy loss of the incident ions. Impurity-induced mixing, in comparison, depends on the electronic valence and the chemical environment. In the ion-induced mixing process, both effects are involved. The two distinct effects were demonstrated by comparing the mixing behavior of AlAs/GaAs superlattices for elements of comparable mass but different valence such as Ga, As and Ge, and elements of comparable valence but different mass such as Si and Ge. In this comparison, the implantation energies were chosen to be such that the implantation ranges for the implant species were about the same [9].

   For Ga, As, Ge, and Zn implantation, the atomic displacement of Al can be observed in the as-implanted samples, which is the result of the atomic collisions during implantation. Figure 8 shows the calculated nuclear energy loss of various ion species with incident energy such that the implantation ranges are about 150 nm. As shown in Figure 9, the observed mixing range in the Ga implanted samples is about 150 nm, correlating well with the implantation range calculated using a Monte Carlo technique. Depending on the nuclear energy loss and the dose of the ion implantation, the mean atomic displacement length due to collision mixing can be estimated using a simple diffusion model of Haff and Switkowski [32]. The experimentally observed

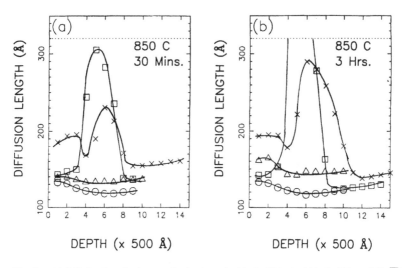

Fig. 9. Al diffusion lengths for samples implanted with Ga (○), As (△), Ge (X), and Si (□) ions followed by a thermal anneal at 850°C for (a) 30 minutes and (b) 3 hours.

magnitude of mixing is in good accord with the prediction of this collision mixing model.

For Ga implantation, a post thermal anneal resulted in a small amount of further incremental mixing. However, the extent of mixing quickly saturates with annealing time (Figure 9). The small amount of additional mixing during the anneal may be attributed to the crystal recovery process. The collision-induced mixing by As implantation behaved in an analogous fashion. In both cases, the mixing regions are confined to the implantation regions, the diffusion lengths are in accord with the collision estimate, and the diffusion lengths saturate with increasing annealing time. Therefore, Ga and As implantation cause primarily collision-induced mixing, where the extent of the mixing agrees well with the theoretical estimates.

In contrast to Ga and As ions induced mixing, Ge and Si implantation with subsequent anneals results in the development of two distinct regions (Figure 10). The first region is a collision-induced mixing region, which is near 150 nm deep, and analogous to that of Ga or As implantation. In this region, the mixing induced by Ge ions is larger than that by Si ions since Ge ions undergo higher nuclear energy loss. The second region is the impurity-induced mixing region, where the mixing of the superlattice has been dramatically enhanced. The range of the mixed region is deeper for longer anneals, and it is correlated with the impurity diffusion front. In addition, comparing with Si-doping induced mixing, the ion damage results

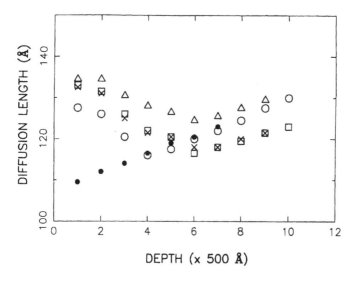

Fig. 10. Al diffusion length as a function of the depth in an as-grown superlattice (•), a sample implanted with 10^{15} ions/cm^2 Ga at 344 keV (○), and samples annealed at 850°C for 30 minutes (X), 3 hours (□), and 6 hours (Δ) subsequent to the Ga implantation.

in a complicated diffusion behavior near the surface. A diffusion barrier was formed in the region where collision-induced mixing occurred prior to annealing. Consequently the impurity-induced mixing was strongly inhibited in the damaged region although there is no saturation of interdiffusion of Al and Ga with increasing anneal time, as is seen in the cases of Ga and As implantation.

The SIMS data for samples implanted by Ge and Si with thermal anneal at 850°C for 3 hours are shown in Figure 11. It can be seen that the impurity-induced mixing region for the Ge implanted sample was extended to a depth of about 550 nm, while that for the Si implanted sample is only about 450 nm. On the other hand, the Al depth profile for the Si implanted sample shows more complete mixing. Therefore, it appears that Ge atoms diffuse faster than Si atoms, resulting in a lower concentration at a given time duration. From the literature [33], it was found that the diffusion coefficient of Ge is larger than that of Si at 850°C, which agrees with this experimental observation. Assuming that Ge and Si-enhanced mixings are dominated by the same mechanisms, the Al total diffusion lengths may depend on the local Ge or Si concentration. Since at a given depth the time-averaged concentration of Ge is lower than that of Si, faster Ge diffusion results in a less Al interdiffusion.

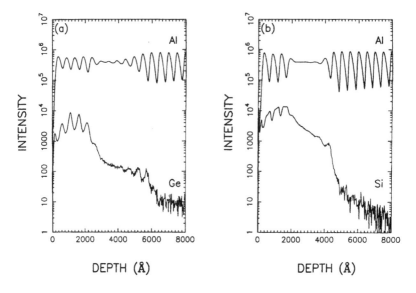

Fig. 11. SIMS depth profiles of (a) Al and Ge in a sample implanted with 355 keV Ge; (b) Al and Si in a sample implanted with 150 keV Si. Both samples were annealed at 850°C for 3 hours after the implantation.

The features of Si ion implantation induced superlattice mixing are examined by TEM and SIMS, and shown in Figure 12 [34]. In this experiment, AlAs/GaAs superlattice samples implanted with 180 keV $^{28}Si^+$ of doses ranging from 3×10^{13} to 3×10^{15} cm^{-2} were examined before and after a 3 hours 850°C anneal. Two distinct features were identified in Figure 12. Near the surface of the sample, the ion damage formed a diffusion inhibition region (region I & II). Beyond this region, Si diffusion resulted in a flat Si concentration plateau and a sharp front. Corresponding to the Si diffusion region, there was an enhanced superlattice mixing (region III).

In contrast, Be ion implantation does not produce significant collision-induced superlattice mixing even at a high dose (Figure 13). The nuclear energy loss for Be implantation is an order of magnitude less than of that for Ga. The effect of the superlattice mixing with post thermal anneals depends on the Be concentration. There is no observable enhanced mixing in the samples implanted at low doses. The mixing occurred at doses above 10^{15} ions/cm^2. The plot of the Al diffusion length resembles the shape of the as-implanted Be profile. The mixing is directly related to the local Be concentration and may be assisted by implantation induced defects. In a similar manner, impurity induced mixing by Zn has a high critical concentration. However, since Zn is much heavier than Be, high dose Zn implantation results in collision induced

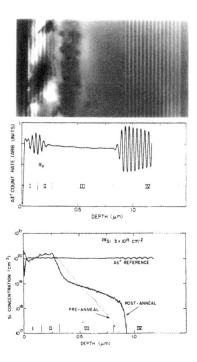

Fig. 12. TEM cross-section image and SIMS depth profiles of Al and Si concentration. The AlAs-GaAs superlattice samples were implanted at 180 keV with 10^{15} Si$^+$/cm^2 (Al depth profile) and 3×10^{15} Si$^+$/cm^2 (Si depth profile) followed by an anneal at 850°C for 3 hours. The Si depth profile before the thermal annealing is also shown.

mixing as well as heavy damage in the implantation region. The high density of the defects inhibit both Al and Zn migration. The mixing inhibition is strong especially for Zn implantation at the dose above 10^{16} cm^{-2}.

2.3   Mixing Inhibition in Heavily Doped and Ion Implanted Regions

The intermixing of Si doped AlGaAs superlattices with the Si concentration well above 10^{20} cm^{-3} has been studied by Guido et al. [35]. In this work, Si atoms were incorporated into the superlattice subsequent to an interruption in the OMCVD growth process, forming an $(Si_2)_x(GaAs)_{1-x}$ alloy. Al and Ga diffusion was found to be inhibited in the region heavily doped with Si. More detailed studies indicated that the mixing inhibition correlates with the formation of dislocation loops and defect clusters in heavily Si-doped AlGaAs superlattices [36].

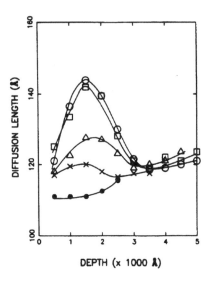

DEPTH (x 1000 Å)

Fig. 13. Al diffusion length as a function of the depth in AlAs/GaAs superlattice samples implanted with Be ions. Be implantation was performed at 41 keV with the dose of 10^{15} cm^{-2} (X), 3×10^{15} cm^{-2} ($\triangle$), 6×10^{15} cm^{-2} ($\square$), and 10^{16} cm^{-2} ($\bigcirc$). The samples were annealed at 850°C for 3 hours after the implantation. The data from an as-implanted sample ($\bullet$) is also shown.

Reference 28 describes an experiment, where AlAs/GaAs superlattices were grown by MBE and doped with silicon over a concentration range of 10^{18} to 10^{20} cm^{-3}. In the as-grown sample, there was no defect clustering observed by TEM in the Si-doped regions. However, Si segregation into the GaAs layers was observed at the Si doped level of 10^{19} cm^{-3} (Figure 14a). A small amount of Al interdiffusion was also observed in the heavily Si-doped region. As described previously (Figure 3), Al diffusion increases with the Si doping level and reaches a maximum value at the Si concentration of mid of 10^{19} cm^{-3}. With further increase of the Si doping level, Al diffusion is retarded. Referring to Figure 14a, in the as-grown sample, it is possible that a brief mixing occurs in a proper doping window. Therefore the degree of the mixing is slightly more in region II compared with that in region III where the doping level is higher. When a post thermal anneal at 750°C was applied to the superlattices, the superlattice layers in region I and region II were completely mixed (Figure 14b). It is also interesting to notice that the Si diffusion in region I is somehow retarded and also segregated into GaAs layers.

A comparison of the Al diffusion at the Si doping levels of 10^{19} and 10^{20} cm^{-3} was performed at an annealing temperature of 700°C. Figure 15

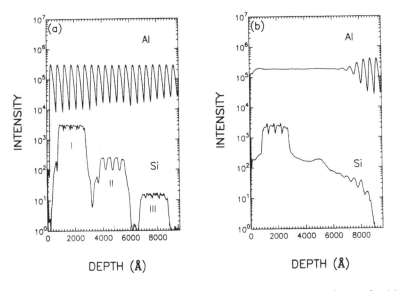

Fig. 14. SIMS depth profiles of Al and Si concentrations in AlGaAs superlattice samples; (a) as grown sample with Si doping level of 10^{20} (I), 10^{19} (II), and 10^{18} cm^{-3} (III); and (b) annealed at 750°C for 3 hours.

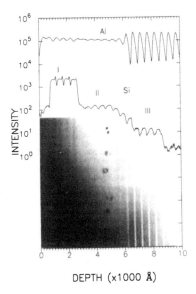

Fig. 15. SIMS depth profiles of Al and Si concentrations in the AlGaAs superlattice sample annealed at 700°C for 3 hours. The corresponding TEM cross section image reveals defect clusters in region I and dislocation loops in region II.

shows a SIMS depth profile and the corresponding TEM cross-section image of a sample annealed at 700°C for 3 hours. At this temperature, more Al diffusion occurs in the region with 10^{19} cm^{-3} of Si doping than at higher or lower Si concentrations. The TEM micrograph reveals that there are prismatic dislocation-loops, 30–40 nm in diameter and lying in (110) type planes in the second Si doped region. In comparison, formation of Si interstitial loops in (111) planes was reported by Guido et al. [37] in Si implanted and annealed samples. A high density of defect clusters is observed in the 10^{20} cm^{-3} of the Si doping region (Figure 15). These defects seem to be preferentially located in the GaAs layers. In the region with Si doping of 10^{18} cm^{-3}, neither mixing nor defect formation was observed.

The inhomogeneity in the distribution of defects in superlattice layers has been reported in Si implantation induced mixing. The Si segregation, preferentially in GaAs observed in SIMS measurements, is probably associated with Si gettering at defect centers in the GaAs layers. The trend in Figure 15 indicates a correlation among the Si concentration, defect formation and Al interdiffusion. In region II (with 10^{19} cm^{-3} Si doping), the growth of dislocation loops, presumably interstitial in nature, results in the creation of equal numbers of Ga and As vacancies which assist Ga interdiffusion. Smaller dislocation loops are also found at a lower annealing temperature (650°C) associated with less Al interdiffusion [38]. In region I (with 10^{20} cm^{-3} Si doping), the sharp Si profile shows a retardation of Si diffusion at a concentration above 10^{19} cm^{-3}, and there is a diffusion shoulder near the surface. This suggests that most Si atoms are trapped which results in the inhibition of Al interdiffusion in this region.

The effect of damage on Si-induced superlattice mixing was examined by implanting 3 MeV Ga ions into AlAs/GaAs superlattices doped with Si [39]. Results from this experiment are illustrated in Figure 16. In the superlattice, region I, II, and III were doped with Si at levels of 5×10^{17}, 10^{18} and 5×10^{18} cm^{-3}. Since the implantation energy was high, most damage in the superlattices in the depth of 3 μm was due to the electronic energy loss and was relatively uniform in this depth (except for a shallow region near the surface). The Si profile in the annealed control sample of Figure 16(b) reveals some key features of the Si induced mixing process. The Si diffusion shoulders and the boundaries of Al diffusion are extremely abrupt, indicating the highly non-linear dependence on the Si concentration. The enhancement of the Al diffusion was strongly inhibited by Ga implantation, as shown in Figure 16(c) and (d). In the implanted superlattices, the Al diffusion activation energy was unchanged despite a two orders of magnitude drop in the Al diffusion coefficient. The results suggest that the "active" Si concentration is proportionately diminished by the Ga implantation. A possible conclusion from these various implantation studies is that implantation damage facilitates the formation of complex defect clusters which

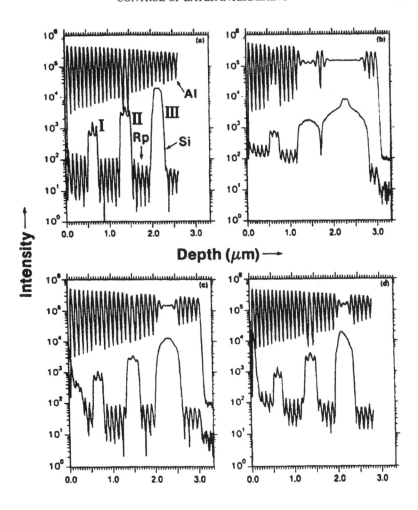

Fig. 16. SIMS depth profiles of Al and Si concentration in AlGaAs superlattice samples with the period of 50 Å; (a) as grown, (b) annealed at 850°C for 3 hours, (c) implanted with 10^{13} cm^{-2} Ga ions at 3 MeV followed by a thermal anneal, and (d) implanted with 10^{14} cm^{-2} Ga at 3 MeV followed by a thermal anneal. The annealing condition for (c) and (d) were the same as that for (b).

immobilize or getter Si atoms and mono-vacancies, thereby inhibiting the mixing process.

3 INGAAS/INP SUPERLATTICE MIXING AND CONVERSION

InGaAsP/InP superlattices are important for optical devices in the long wavelength region. In general, the IILD process developed in the AlGaAs superlattice system can be applied to the InGaAsP superlattice system for device fabrication. The complication in the InGaAsP superlattice system, however, is that the resulting energy band structure from the mixing is determined by both anion (As, P) and cation (In, Ga) interdiffusion. In addition, the sub-lattice interdiffusion causes strain at the interfaces due to a stoichiometric change in the superlattice layers.

Undoped InGaAsP/InP superlattices are stable below 600°C. Evidence of minor P diffusion was observed at 700°C [40]. Zn diffusion induces mixing almost exclusively on the cation (In or Ga) sublattice at 600°C, again with some evidence of anion motion at 700°C [41]. Si implanted InGaAs/InP superlattices revealed partial mixing at 650°C but it was not determined if both sublattices were involved [42]. These results may be contrasted with the AlGaAs system, in which the superlattice mixing produces an unstrained superlattice with a band gap intermediate between those of the original layers. In this case, As diffusion cannot be examined because of its uniform concentration. In the InGaAsP/InP system, cation diffusion results in a superlattice with increased strain and an increased disparity in the layer band gap.

Figures 17 and 18 describe the experimental results of interdiffusion in an InGaAs/InP superlattice annealed in a closed ampoule with Zn_3As_2 as a dopant source [43]. The anneal was performed at 600°C for 1 hour. In this sample, Zn diffusion was fast and resulted in a concentration on the order of 10^{20} cm^{-3}. Figure 17 indicates a substantial interdiffusion of Ga but little movement of As or P. The Ga diffusion profile in the InP substrate indicates a diffusion length of approximately 200 nm. This amount of diffusion is more than sufficient to distribute the Ga uniformly through the superlattice.

Complete Ga redistribution implies a phase stoichiometry of approximately $In_{0.9}Ga_{0.1}As$ from an as-grown $In_{0.53}Ga_{0.47}As$ /InP superlattice with the layer thicknesses of 10nm InGaAs and 40 nm InP. The estimated lattice mismatch for $In_{0.9}Ga_{0.1}As$/ $In_{0.9}Ga_{0.1}P$ is about 3.2%. This mismatch was confirmed by a high resolution TEM lattice image. Figure 18 is a high resolution lattice image of one period of the Zn diffused superlattice illustrating an abrupt 2.1° bend of the (111) lattice fringes through the As layer. The indicated strain may be accurately assessed through measurements of the c axis lattice spacing. Electron diffraction yields a lattice spacing of 0.582 nm in the $In_{0.9}Ga_{0.1}P$ layers; 0.9% smaller than the nearby InP substrate (0.587 nm). The $In_{0.9}Ga_{0.1}As$ layer spacing is 0.620 nm, corresponding to an expansion of 5.6%. No distortion is observed in the a-b growth plane. Poisson's ratio (~ 0.355 in the InAs-InP dilute Ga system [43]) may be used to calculate the bulk lattice parameters, and hence the lattice mismatch, of the two layer materials [44,45]. The lattice mismatch of the $In_{0.9}Ga_{0.1}As$ compound to InP is 2.7%, and that of the $In_{0.9}Ga_{0.1}P$ compound is 0.4% for a net layer mismatch of 3.1%.

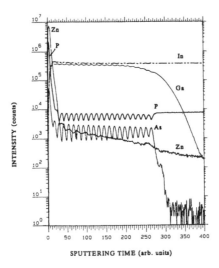

Fig. 17. SIMS concentration depth profile of an $In_{0.53}Ga_{0.47}As/InP$ 20 period superlattice subjected to a 600°C, 1 hour closed-tube Zn diffusion. The period is approximately 40 nm. Mixing is confined to the cation (In and Ga) sublattice. The mixed superlattice has an approximate stoichiometry of $In_{0.9}Ga_{0.1}As/In_{0.9}Ga_{0.1}P$.

Fig. 18. TEM high resolution lattice image of the Zn-diffused superlattice. The (111) planes exhibit an abrupt 2.1° bend at the layer interfaces marked by horizontal lines. The distance between these lines is 7.5 nm. No crystalline defects are observable. A straight line is drawn parallel to the arsenide layer (111) fringes to guide the eye, while the arrows at each interface are parallel to the phosphide layer fringes.

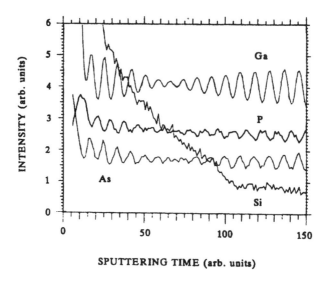

Fig. 19. SIMS concentration depth profile of a Si-diffused superlattice following a 700°C, 3 hour anneal. The lowest Si concentration corresponds to the as-grown background doping level of 10^{18} cm^{-3}. The superlattice period is approximately 50 nm. Comparable cation and anion mixing is observed in the vicinity of the 70th time unit.

The critical layer thickness, beyond which misfit dislocations or microtwins will form to relieve stress, is calculated from the classic formula of Matthews and Blakeslee [45]. For a 2.7% mismatch, the critical layer thickness is approximately 6 nm. The In$_{0.9}$Ga$_{0.1}$As layer thickness in the sample examined by TEM is approximately 7.5 nm. No microtwins or misfit dislocations were observed in any of the superlattice layers corresponding to a viewable interfacial length of tens of microns. Formation of strained layers by mixing therefore has a significant advantage that both layer interfaces are bounded and there is no propagation of defects from the growth surface. Stress is balanced between the layers so that there is no significant accumulation of strain with increasing distance from the InP substrate.

Figure 19 describes the results of Si-induced InGaAs superlattice mixing. In this experiment, Si was diffused from a thin Si surface layer deposited on an as-grown Si-doped sample. The diffusion was performed in a closed-tube at 700°C. The Si diffusion profile exhibits the classical diffusion shoulder, which resembles the profile in the AlGaAs superlattice and suggests that the diffusion mechanism may be similar for these superlattices. The interesting fact is that a comparable mixing occurs on both sub-lattices of Ga and P within a narrow range of Si concentration near the 5×10^{18} cm^{-3} level. This narrow range of the mixing enhancement can be understood, by recalling the mixing in the AlGaAs system, that strong compensation effects and complex defect formation occur at very high Si concentrations near the surface which inhibit mixing enhancement. The mixing enhancement is small when the Si concentration is low in a greater depth. Therefore, a proper level of Si concentration needs to be selected for an optimized mixing enhancement for both cation and anion sub-lattices interdiffusion in InGaAsP superlattices.

In the above descriptions, we have presented the experimental evidence of impurity-induced mixing in AlGaAs and InGaAsP superlattices. For n-type doping, Si is a popular species to induce AlGaAs superlattice mixing. Al diffusion has a fourth power dependence on the Si concentration. Te is another n-type doping candidate examined for impurity induced mixing, which provides enhanced Al diffusion with a near first power dependence on the Te concentration. Be and Zn are common species for p-type dopant for impurity induced layer mixing. For InGaAsP superlattices, Be induces primarily cation interdiffusion, while Si diffusion results in both cation and anion diffusion. In a mixing process induced by ion implantation, both effects of collision and impurity contribute to the superlattice mixing. In addition, ion damage and doping defects may cause mixing inhibition.

In the following section, we will discuss how to utilize defects and localized impurities to control layer intermixing, and apply the technique of layer intermixing control to fabricate multi-wavelength lasers.

4 CONTROL OF LAYER INTERMIXING IN DEVICE FABRICATION

4.1 Differential Al-Ga Interdiffusion in undoped AlGaAs/GaAs and AlGaInP/GaInP Quantum Wells

Intermixing of the column III atoms in III-V semiconductor heterostructures has attracted considerable interest for the fabrication of optoelectronic devices, since it allows modification of the potential profile of a heterostructure after the material growth. Among the techniques for enhancing the interdiffusion of Al and Ga is the use of a SiO_2 cap, which is well-known for altering the electrical and optical properties of AlGaAs/GaAs structures [37,46–51]. Control of the intermixing of an undoped GaAs/AlGaAs QW is possible by patterning the SiO_2 on the surface since SiO_2 mask functions as a sink for Ga outdiffusion, which creates group III vacancies in the structure. In this subsection, we first discuss the Al diffusion coefficient in AlGaAs/GaAs and AlGaInP/InGaP QWs by vacancy enhanced layer intermixing under SiO_2 mask and how to use the large difference in the Al diffusion coefficient to achieve layer intermixing control in a dual stacked AlGaAs and AlGaInP QW structure.

Values of the column III interdiffusion coefficient for samples annealed with an SiO_2 cap have been reported for the $Al_xGa_{1-x}As/GaAs$ and $(Al_xGa_{1-x})_{0.5}In_{0.5}P/Ga_{0.5}In_{0.5}P$ materials systems [37,47,48,51]. The large difference in measured activation energies [48] for AlGaAs/GaAs versus AlGaInP/GaInP suggests that, over a certain temperature range, the interdiffusion of group III atoms in AlGaInP/GaInP samples is much slower than in AlGaAs/GaAs samples. The difference of the interdiffusion of the two types of heterostructures has not only been observed in separate samples, but also in a single sample with a stacked dual-QW structure [52]. A large difference in vacancy-enhanced intermixing rates for the AlGaAs/GaAs and AlGaInP/GaInP materials systems within a single sample was observed, and showed that this can be used to selectively intermix the GaAs/AlGaAs QW only under a SiO_2 cap and leave both QWs intact under a bare surface.

A schematic diagram of such a stacked dual-QW laser structure is shown in Figure 20. The material used in this study was grown by low-pressure OMVPE. The layers form a dual quantum well active region laser diode structure. Between 0.8 μm n- and p-type $Al_{0.5}In_{0.5}P$ cladding layers are an upper 100 Å GaAs QW with $Al_{0.4}Ga_{0.6}As$ confining layers and a lower 80 Å $Ga_{0.4}In_{0.6}P$ QW with $(Al_{0.6}Ga_{0.4})_{0.5}In_{0.5}P$ confining layers. The separation of the QW layers is nominally 1100 Å. All of the confining and QW layers are nominally undoped. Over the upper p-type cladding layer are p-type $Al_xGa_{1-x}As$ layers with $x = 0.8, 0.4, 0.2$, and 0, and thicknesses of 300, 300, 300, and 1000 Å, respectively, with the GaAs layer on the surface. The peak wavelengths of spontaneous emission from the dual-QWs measured by room temperature photoluminescence are 840 and 670 nm respectively. Broad area laser diodes formed from this material have a threshold current density of 200 A/cm^2, and efficiency of 29%/facet at a lasing wavelength of 853 nm. It indicates that the emission is from the GaAs QW region without any thermal annealing.

To fabricate red and infrared side by side lasers or LEDs from the stacked active region laser structure, carrier access to the GaInP QW is needed which means that the bandgap of the GaAs QW needs to be increased to above that of the GaInP QW. This may be achieved by selective intermixing of the GaAs QW. Samples were prepared for annealing by depositing a silicon dioxide layer on the upper surface by chemical vapor deposition and opening 250 μm wide stripes on 500 μm centers. Annealing was carried out in a rapid thermal annealing furnace.

The room temperature photoluminescence spectra are shown in Figure 21 for an SiO_2-capped portion of a sample which was annealed 3 times at 1000°C for a total of 4 minutes. The PL is shown for the as-grown sample and after anneal times of 1, 2, and 4 minutes. Prior to annealing, peaks are present in the PL at 845 and 676 nm, corresponding to the AlGaAs/GaAs and AlGaInP/GaInP quantum wells. The QWs were confirmed as the sources

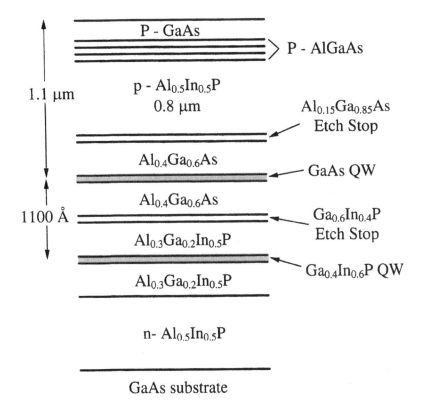

Fig. 20. Schematic diagram of the dual quantum well active region laser structure. The two quantum wells of interest are shaded.

of the emission by measuring PL on an unannealed piece after etching off layers from the surface down to the AlGaAs etch stop, and then to the GaInP etch stop. After annealing, the AlGaAs/GaAs peak shifts to a higher energy, indicating changes of the potential well profile due to the interdiffusion of the Al and Ga, as is commonly seen [46-51] in AlGaAs samples capped with SiO_2. In stark contrast, the position of the peak at 676 nm from the AlGaInP/GaInP QW remains stable, indicating insignificant movement on the column III sublattice at the AlGaInP/GaInP interface. Figure 22 shows the energy shift of the PL peaks for the two QWs with and without the SiO_2 cap for the same sample as in Figure 21. As is usually observed, the shift of the AlGaAs/GaAs QW energy under the SiO_2 cap is much larger than that in regions without SiO_2 cap [46–48,50]. It is interesting to note the opposite behavior seen for the AlGaInP/GaInP QW energy, for which the energy shifts

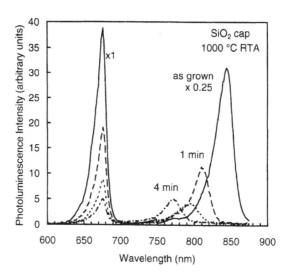

Fig. 21. Room temperature photoluminescence spectra of a SiO$_2$-capped portion of a sample which was annealed at 1000°C. Curves are shown after 0, 1, 2, and 4 min total anneal time.

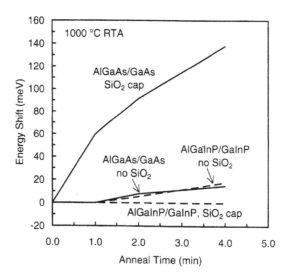

Fig. 22. Energy shifts of the room temperature photoluminescence of an Al$_{0.4}$Ga$_{0.6}$As/GaAs QW and an (Al$_{0.6}$Ga$_{0.4}$)$_{0.5}$In$_{0.5}$P /Ga$_{0.4}$In$_{0.6}$P QW, with and without a SiO$_2$ cap.

only in the regions without the SiO_2 cap. Under either the capped or uncapped surface conditions, the shift is much smaller than the shift of the AlGaAs QW under the SiO_2 cap. This behavior will be discussed further below.

In order to calculate the activation energy of the Al-Ga interdiffusion for the AlGaAs/GaAs QW, samples with an SiO_2 cap were annealed in a furnace at 850, 875, 905, and 925°C, and at 975, 1000, 1025, and 1065°C in the RTA. Anneal times ranged from 5 hr at 850°C to 50 sec at 1065°C.

Values of D were calculated following a procedure similar to that used by others [37,47,48,51]. A square, finite potential well was assumed to represent the QW before annealing. The Al profile of the intermixed QW was calculated as [53,54]

$$ x = x_o \left\{ 1 + \frac{1}{2}\text{erf}\left[\frac{z - L_z/2}{2\sqrt{Dt}}\right] - \frac{1}{2}\text{erf}\left[\frac{z + L_z/2}{2\sqrt{Dt}}\right] \right\}, \tag{2} $$

where $x_o = 0.4$ is the aluminum fraction in the barriers (assumed to be of infinite width), z is the position from the center of the QW of width L_z, D is the Al-Ga interdiffusion coefficient, and t is the anneal time. The resulting aluminum profile was translated into the corresponding conduction and valence band potential profiles using 65:35 splitting of the band gap discontinuity for the conduction and valence bands, respectively. The Schrödinger equation was then numerically integrated in order to calculate the transition energy between the electron and heavy hole ground states. D was adjusted until the calculated transition energy matched the peak of the PL. The values of D determined in this manner are plotted as symbols in Figure 23 along with a line showing the least squares fit. For all data points, the fit is $D = 420\exp(-4.54/kT)$ cm^2/s, where k is the Boltzmann constant in eV/K. Separate fits to the furnace and RTA data give $E_a = 4.5$ eV for the RTA, and 4.8 eV for the furnace data, although both sets of data are described well by $E_a = 4.5$ eV, as shown. This activation energy is somewhat larger than other values reported for SiO_2 capped AlGaAs/GaAs samples, which typically are in the 3.6 to 4.1 eV range [37,47,51]. This may reflect different background doping conditions or the effect of the thick AlInP layers in the present samples. In any case, the value of E_a is considerably lower than the value of 6 eV determined for Al-Ga interdiffusion in intrinsic material without an SiO_2 cap [55].

The value of D at 1000°C calculated for the AlGaAs/GaAs QW in the region not covered by SiO_2 is also shown in Figure 23, and lies a factor of 10 below the value found for the SiO_2-capped portion. Since any wavelength shift of the AlGaInP/GaInP QW under the SiO_2 was not resolved in this experiment, a value for D cannot be calculated for that case. It is only possible to place an upper limit of about 5×10^{-18} cm^2/s at 1000°C, calculated as the value necessary to give a 1 nm wavelength shift for a 4 min anneal. This lies

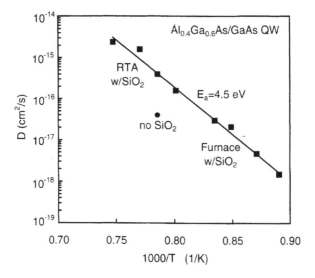

Fig. 23. Calculated values of the Al-Ga interdiffusion coefficient, D, as a function of temperature for the $Al_{0.4}Ga_{0.6}As/GaAs$ quantum well. The calculations are based on the shift in photoluminescence energy. The solid line is a fit to the RTA and furnace anneal data for regions with a SiO_2 cap.

two orders of magnitude below the corresponding value for the AlGaAs/GaAs QW under the SiO_2 cap.

It has been shown that in intrinsic AlGaAs, the Al-Ga interdiffusion most likely involves the triply-charged column III vacancy [55,56], V_{III}^{3-}. The Al-Ga interdiffusion coefficient is then proportional to the concentration and diffusivity [56] of V_{III}^{3-}. The cap is believed to enhance intermixing by introducing additional column III vacancies, V_{III}, at the surface [46,49]. The excess vacancies quickly diffuse into the bulk [49], increasing the Al-Ga interdiffusion. From the shifts seen in Figure 22, it is clear that the SiO_2 cap has greatly enhanced the intermixing of the AlGaAs/GaAs QW in this sample, as is routinely observed in structures composed entirely of AlGaAs. Assuming that the interdiffusion in the AlGaInP follows the same mechanism, either a much lower column III vacancy concentration or diffusivity would be necessary to account for the lack of intermixing of the AlGaInP/GaInP QW under the SiO_2. Experiments conducted on intermixing of AlGaInP/GaInP QWs show that it is difficult to intermix the heterostructure under the conditions employed. Samples containing GaInP/AlGaInP QWs with bare or SiO_2 surfaces have been annealed under phosphorous over pressure at 900°C for 4 hours and showed only a slight bandgap increase of 10 nm [58].

An approach to largely shift the bandgap of a GaInP/AlGaInP QW is by disordering of an ordered or partially ordered alloy structure [50].

To access the GaInP QW in the Red/IR stacked dual-QW structure by carriers, it is necessary to increase the bandgap of the IR QW above the red one. For the structure with $Al_{0.4}Ga_{0.6}As$ confining layers in the IR QW shown in Figure 20, it is found that the intermixing rate slows down dramatically when the emission peak is shifted below 700 nm. In order to intermix the GaAs/AlGaAs QW selectively and move its bandgap above that the InGaP/AlInGaP QW, the interdiffusion coefficient of Ga-Al in the GaAs/AlGaAs QW needs to be increased so that the anneal time needed can be shortened. One approach is to increase the aluminum content in the barrier region next to the QW layer. Figure 24 shows a GaAs/AlGaAs QW structure with $Al_{0.8}Ga_{0.2}As$ sandwiching the QW. We have investigated selective intermixing of two stacked red/IR dual-QW laser structures with GaAs/$Al_{0.8}Ga_{0.2}As$ QW: the IR QW region in the first structure is undoped, and the $Al_{0.8}Ga_{0.2}As$ barrier layer in the second structure is doped with Si at $5 \times 10^{18}/cm^3$. Figure 25 shows the room temperature photoluminescence peak wavelength shifts of such high aluminum content barrier QWs under different anneal conditions. We see the enhancement of the intermixing rate of the GaAs/AlGaAs QW by adding aluminum in the barrier. The bandgap of the undoped infrared QW under SiO_2 is shifted to as far as 640 nm with no sign of saturation. On the other hand, the Si doped GaAs QW annealed under SiN_x at the same anneal temperature shows smaller shifts of peak wavelength. Because SiN_x prevents Ga outdiffusion at the surface, the group III vacancies in the AlGaAs QW region is only affected by the local impurity density. Therefore Si outdiffusion from the QW region is responsible for the decrease of interdiffusion coefficient of Ga-Al as anneal time is increased.

4.2 Selective Intermixing of Single Modulation-doped Quantum Well

One desired degree of control of intermixing is vertically in a heterostructure using impurity doping. As we have shown in the previous sections, the Al diffusion coefficient is dependent on the third power of the electron concentration in the AlGaAs superlattice. Thus the higher the Si concentration, the higher the Al diffusion coefficient and the degree of intermixing. Using the modulation doping technique, it is possible to intermix a heavily doped, n-typed region without intermixing its neighboring undoped or lightly doped regions. Combining the modulation doping with surface mask conditions, one can achieve control of intermixing in both lateral and vertical directions. We will discuss the control of layer intermixing in modulation-doped single and dual QW structures and its application in fabricating advanced optoelectronic devices such as multi-wavelength light emitters.

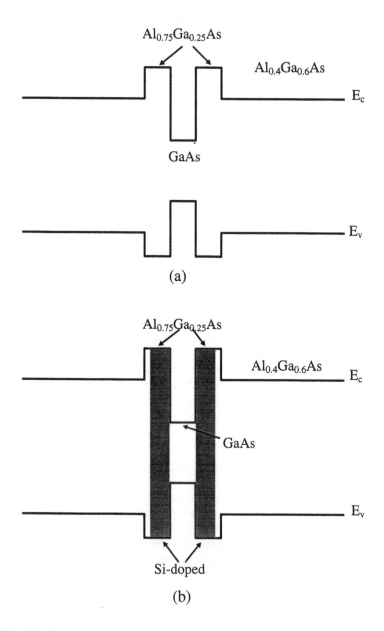

Fig. 24. Schematic band diagram of GaAs/AlGaAs QWs with high Al composition barriers, (a) undoped and (b) Si doped.

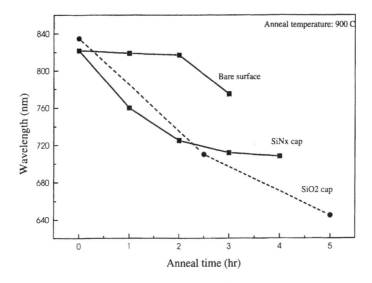

Fig. 25.   Room temperature photoluminescence peak wavelength of GaAs/Al$_x$Ga$_{1-x}$As QWs versus anneal time at 900°C. The solid curves correspond to the Si doped QW structure, and the broken curve corresponds to the undoped QW.

The Al diffusion coefficient of a Si doped AlAs/GaAs superlattice has a fourth power dependence on the Si concentration under thermal annealing when the surface of the sample is capped to prevent As outdiffusion [59]. For a Te-doped AlAs/GaAs superlattice, the Al diffusion coefficient is linearly proportional to the electron concentration. If the surface is bare to allow arsenic atoms to outdiffuse, however, the layer intermixing in the doped region will be suppressed. Thus the Al diffusion coefficient depends not only on the n type impurity doping density but also on the surface cap condition during anneal. In a Mg- or Se-doped GaAs/AlGaAs QW structure, however, it has the opposite effect [56]. The Mg- or Se-doped GaAs/AlGaAs QW is more stable under a SiO$_2$ cap than a bare or a SiN$_x$ surface condition [60]. In Table 1, the states of intermixing of different types of QWs under three types of surface masking conditions are summarized. The table clearly shows that using SiN$_x$/bare surface masks, one can selective intermix a n-type GaAs/AlGaAs QW heterostructure. We will discuss how to use the technique to fabricate multiwavelength side by side laser on a single GaAs substrate.

The band diagram of a modulation doped AlGaAs/GaAs QW separate confinement heterostructure is shown in Figure 26. On each side of the QW, 50 Å sections of the barrier layers are doped with Si. The sample is then annealed with a SiN$_x$ cap at 870°C. The shift of the photoluminescence peak

TABLE 1
Degree of intermixing of semiconductor QWs

	SiO$_2$	Si$_3$N$_4$	Bare surface
undoped QW	intermix	stable	stable
n-type QW	intermix	intermix	stable
p-type QW	stable	intermix	intermix

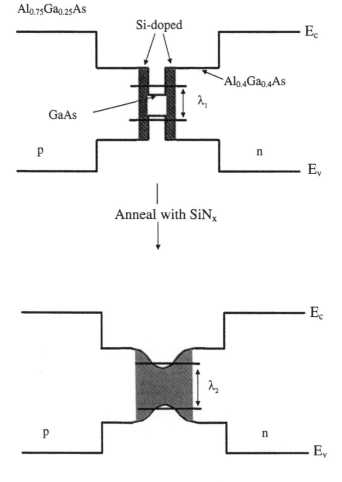

Fig. 26. Schematic band diagram of the active region and part of the cladding layers of an AlGaAs/GaAs separate confinement heterostructure single quantum well laser structure as grown and after annealing. Upon annealing with a SiN$_x$ cap, the Si doping causes localized intermixing of the QW and a bandgap increase.

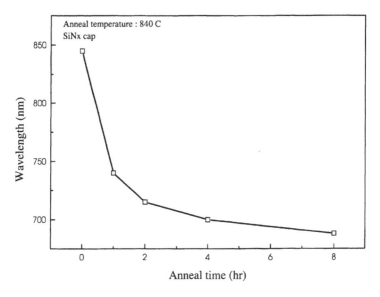

Fig. 27. Energy shifts of the room temperature photoluminescence of the Si-doped $Al_{0.4}Ga_{0.6}As/GaAs$ QW with a SiN_x cap.

is plotted in Figure 27. For eight hours of anneal, the wavelength is shifted as far as 700 nm from the intermixing of Al-Ga. In comparison, a sample with bare surface is also annealed at the same conditions. The wavelength in the bare surface stays almost the same. It demonstrates the possibility of fabricating a multiwavelength laser array with wavelength span of 840 to 700 nm.

A dual-wavelength laser is fabricated from the modulation doped QW laser structure by capping the surface with 300 μm wide SiN_x and 200 μm wide bare surface stripe masks (shown in Figure 28) and annealing the sample at 870°C for one hour. After the anneal, broad area lasers of 100 mm wide and 750 mm long cavity were fabricated in the SiN_x capped and bare surface regions. Devices formed in the SiN_x capped regions have $J_{th} = 325$ A/cm^2 and a differential quantum efficiency of 29%/facet, the emission wavelength is 743 nm, and in the bare surface regions, $J_{th} = 200$ A/cm^2 and a differential quantum efficiency of 33%, the emission wavelength is 806 nm. The L-I curves of the dual wavelength lasers are shown in Figure 29. In this case, the wavelength in the bare surface region is slightly shifted due to the high Si doping in the barrier region of the QW. Lowering the doping can reduce the interdiffusion coefficient of Al-Ga under the bare surface region and minimize the intermixing of the QW. In another modulation doped GaAs/AlGaAs QW laser structure, the Si doping is reduced to 3×10^{18}/cm^3 in the barrier region.

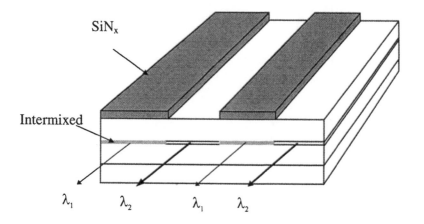

SiN$_x$

Intermixed

λ_1 λ_2 λ_1 λ_2

Fig. 28. Schematic diagram depicting selective intermixing using SiN$_x$ stripes on a Si-doped QW laser structure to fabricate dual-wavelength lasers.

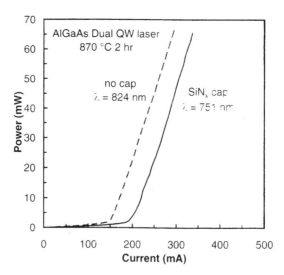

Fig. 29. Light output power as a function of current for representative broad area devices formed in regions annealed with and without the SiN$_x$ cap.

The sample was capped with SiN_x and bare surface stripes and annealed in a rapid thermal anneal furnace at 1000°C for 2 minutes. After the anneal, broad area devices of 100 mm wide and 750 mm long cavity were fabricated from the SiN_x and the bare surface regions. Devices in the SiN_x capped regions showed $J_{th} = 530$ A/cm^2 and a quantum efficiency of 15%/facet, with an emission wavelength of 767 nm, and in the bare surface regions, $J_{th} = 500$ A/cm^2 and a quantum efficiency of 20%/facet, with an emission wavelength of 844 nm. The wavelength in the bare surface region is almost unchanged compared to the as grown structure.

4.3 Selective Intermixing of Modulation Doped Dual AlGaAs Quantum Wells

In fabricating dual-wavelength side by side lasers by selective intermixing of a single modulation doped GaAs/AlGaAs QW active region, the wavelength of the intermixed region is determined by the anneal time and temperature. It is quite difficult to control the final wavelength. One approach to solve the problem is using a stacked dual-QW active region and selectively intermixing the longer-wavelength QW, allowing lasing from the shorter-wavelength and intact QW, and thus making a dual-wavelength side by side laser. The selective intermixing of one QW from another in a closely stacked dual QW structure is achieved by means of a finite Si impurity source next to the longer wavelength QW to be intermixed, which could assist in locally intermixing the thin layer with its neighbors by increasing the Al-Ga interdiffusion coefficient (D_{Al-Ga}) in the region, while leaving other nearby layers intact under SiN_x as shown in Table 1. Under bare surface, both QWs are stable. By controlling surface conditions, the quantum well can be either intermixed or left intact, resulting in a dual-stripe, dual-wavelength laser.

The layers in the structure were grown by low-pressure organometallic chemical vapor deposition on an n-type GaAs substrate and form a separate confinement heterostructure laser with an asymmetric dual quantum well active region. Cladding layers are $Al_{0.75}Ga_{0.25}As$, and the waveguide layers are $Al_{0.4}Ga_{0.6}As$. Figure 30 shows a simplified band diagram (not to scale) of the active region and part of the cladding layers. The upper half of the figure shows the as-grown structure, while the lower part shows the structure after annealing. The 80 Å wide GaAs and $Al_{0.1}Ga_{0.9}As$ QWs are separated by 550 Å, and the total thickness of the waveguide is about 1500 Å. The barrier layers next to QW1 are doped with Si at 5×10^{18}/cm^3 for 50 Å on each side of the QW. The room-temperature photoluminescence wavelengths emitted from the as-grown structure are 831 and 752 nm for QW1 and QW2, respectively. In broad area devices formed from the as-grown material, laser emission is at 832 nm from QW1, with a pulsed threshold current density, J_{th}, of 170 A/cm^2 and external quantum efficiency, η, of 27%/facet

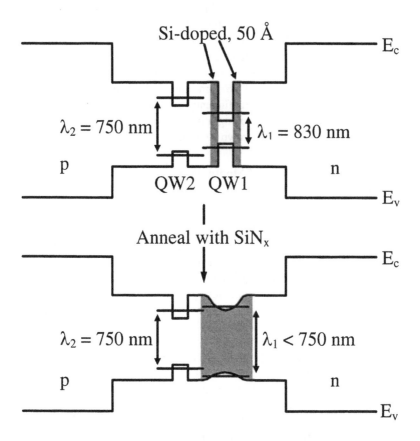

Fig. 30. Schematic band diagram of the active region and part of the cladding layers of the AlGaAs/GaAs separate confinement heterostructure asymmetric dual quantum well laser structure as grown and after annealing. Upon annealing with a SiN_x cap, the Si doping causes localized intermixing of QW1 and a bandgap increase.

for a cavity length of 750 μm. It is notable that the devices are of high quality, despite the presence of the Si doping next to QW1 and QW2 in the active region. Under forward bias, QW1 has a higher population inversion than QW2 in the as-grown structure because of the lower band gap of QW1, and laser emission is at the longer wavelength from QW1. If QW1 were somehow eliminated, then lasing would be from QW2 [61].

In order to effectively remove QW1 from this structure, impurity-induced layer disordering from the grown-in Si is used to intermix QW1 with its neighboring barrier layers until the transition energy of QW1 is higher than that of QW2, as shown in the lower part of Figure 30. During the anneal,

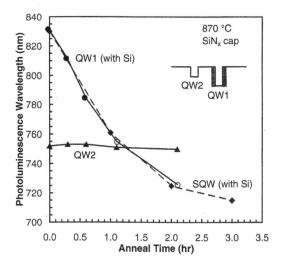

Fig. 31. Room-temperature photoluminescence wavelength vs. anneal time for the asymmetric dual quantum well laser structure and for another structure containing a single QW which is identical to QW1. Circles are for QW1, triangles are for QW2, and the diamonds are for the single QW. Lines are drawn as guides to the eye, and the dashed line corresponds to the single QW.

the Si diffuses outwards and decreases in concentration, resulting in highly localized intermixing [59]. The local nature allows intermixing of QW1 without intermixing QW2, which is only about 550 Å away. Figure 31 shows the measured room-temperature photoluminescence from the two QWs as a function of anneal time at 870°C under a SiN$_x$ cap. Values for QW1 are plotted as closed and open circles, and values for QW2 are plotted as triangles. The diamonds show the wavelengths measured for a structure which contained only a single QW with nominally the same QW and barrier composition, thickness, and doping as QW1. Lines are shown as guides to the eye, with the dashed line corresponding to the single QW structure. The open circles are QW1 points measured as shoulders on the PL peak from QW2, as the two wavelengths are close for those measurements. Thus there is some uncertainty in the assignment of those values. However, the wavelengths measured for the single QW structure, which is nominally the same as QW1, agree quite well with the assignment of the values shown as open circles for QW1. From Figure 31, it is clear that a 2 hr anneal results in a large blue-shift of the PL emission of QW1 to below 730 nm, with negligible shift in the emission from QW2 to 750 nm. This demonstrates both the localized nature of the intermixing and the ability to sufficiently intermix QW1 for the present purpose.

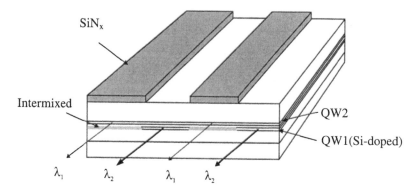

Fig. 32. Schematic diagram of selective intermixing using SiN$_x$ stripes on an asymmetric dual-QW laser structure to fabricate dual-wavelength lasers.

Fabrication of dual-wavelength devices began with deposition of 800 Å of SiN$_x$ on the surface of a sample. Stripes with a width of 305 μm on 510 μm centers were then opened in the SiN$_x$ cap (shown in Figure 32), and the sample was annealed for 2 hr at 870°C. In the uncapped regions, the surface experienced loss of As, and thus an As-poor condition, which was previously shown to inhibit intermixing of an n-type superlattice [56]. Similarly, this bare surface condition was sufficient to largely suppress the intermixing of QW1 in the present case. Broad area lasers were fabricated in the SiN$_x$ capped and bare regions.

The light output power is plotted as a function of current in Figure 33 for representative devices from regions with and without the SiN$_x$ cap. In regions which had the SiN$_x$ cap, lasing is at a wavelength of 751 nm with $J_{th} = 260$ A/cm^2 and $y = 30\%$/facet; and in the regions which had a bare surface, $J_{th} = 195$ A/cm^2 and $y = 32\%$/facet at 824 nm. The emission spectra for the devices are shown in Figure 34 at 1.5 times threshold. Due to the low resolution of the measurement, the longitudinal modes are not resolved. Emission in the regions which had a SiN$_x$ cap is from QW2 at 751 nm, since the emission from QW1 was shifted to below 730 nm. Regions which did not have a cap emit at 824 nm from QW1, shifted only slightly from the as-grown value of 832 nm. The variation in wavelength in the short-wavelength devices is less than 1 nm across a 7.5 mm bar. This is in contrast to the larger variation of over 8 nm on the same size bar from a single-QW structure in which the QW was intermixed to give a wavelength of about 750 nm. There is better wavelength uniformity in the dual-QW structure because the emission is from QW2, which was not shifted during the anneal. Any non-uniformity in the shift of QW1 during the anneal does not effect the

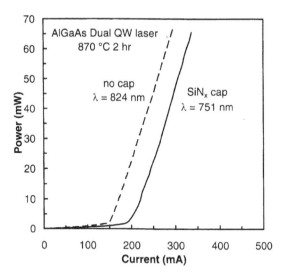

Fig. 33. Light output power as a function of current for representative broad area devices formed in regions annealed with and without the SiN$_x$ cap.

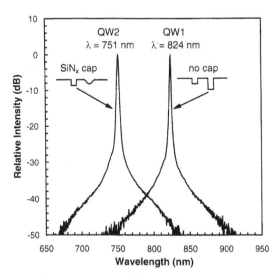

Fig. 34. Emission spectra at 1.5 times threshold for broad area devices formed in regions annealed with and without the SiN$_x$ cap. Emission is at 751 nm from QW2 in the regions annealed with the cap and at 824 nm from QW1 in the regions annealed without the cap.

wavelength, as long as the shift is to a wavelength shorter than that of QW2. Since the short wavelength in the dual-QW structure is from unshifted QW2, there is not only improved uniformity, but also better control over the actual wavelength, since it is determined during the growth, rather than by the details of the anneal. By including additional QWs and properly placed Si doping, along with regions receiving various combinations of capped and uncapped anneals, this method should be extendible to more than two wavelengths.

5 SUMMARY

We have discussed various mechanisms of layer intermixing of III-V semiconductor heterostructures by impurities and defects in this article. Both ion implantation and doped impurities introduce defects and enhance Al diffusion in AlGaAs heterostructures. Si and Te are two n-type impurities examined for IILD. The Al diffusion coefficient is not only dependent on the electron concentration in the structure, but also on the surface cap condition. SiN_x/SiO_2 cap enhances or inhibits Al diffusion depending on the doping property of the impurity, so is the bare surface. Be and Zn are p-type dopants examined in AlGaAs and InGaAsP superlattices, Zn induces primarily cation interdiffusion, and Si induces both cation and anion diffusion. By varying surface cap conditions and modulation doping, control of selective layer intermixing can be realized. This technique can be used to modify the bandgap of a doped QW for wavelength shift or the refractive index of a doped AlGaAs heterostructure for forming optical waveguides. Multi-wavelength side by side lasers are demonstrated by selectively intermixing single or dual modulation Si doped GaAs/AlGaAs and GaInP/AlGaInP QWs.

Acknowledgement

The authors wish to acknowledge Prof. T. Venkatesan, Prof. S. Schwarz, Dr. D.M. Hwang, and Dr. K. Beernink for their contributions to the work.

References

1. L.L. Chang and A. Koma, *Appl. Phys. Lett.*, **29**, 138 (1976).
2. W.D. Laidig, N. Holonyak, Jr., M.D. Camras, K. Hess, J.J. Coleman, P.D. Dapkus, and J. Bardeen, *Appl. Phys. Lett.*, **38**, 776 (1981).
3. J.J. Coleman, P.D. Dapkus, C.G. Kirkpatrick, M.D. Camras, and N. Holonyak, Jr., *Appl. Phys. Lett.*, **40**, 904 (1982).
4. T. Venkatesan, S.A. Schwarz, D.M. Hwang, R. Bhat, H.W. Yoon and Y. Arakawa, *Nuclear Inst. and Meths. in Phys. Res.*, **B19/20**, 777 (1987).
5. P. Mei, S.A. Schwarz, T. Venkatesan, C. Schwartz, E. Colas, *J. Appl. Phys.*, **65**, 2165 (1989).
6. D.G. Deppe, N. Holonyak, Jr., K.C. Hsieh, P. Gavrilovic, W. Stutius and J. Williams, *Appl. Phys. Lett.*, **51**, 581 (1987).

7. E.V.K. Rao, H. Thibierge, F. Brillouet, F. Alexandre and R. Azoulay, *Appl. Phys. Lett.*, **46**, 867 (1985).

8. E.V.K. Rao, H. Thibierge, F. Alexandre, and R. Azoulay, *Appl. Phys. Lett.*, **46**, 867 (1985)

9. P. Mei, T. Venkatesan, S.A. Schwarz, N.G. Stoffel, J.P. Harbison, D.L. Hart and L.A. Florez, *Appl. Phys. Lett.*, **52**, 867 (1988).

10. M.E. Greiner, and J.F. Gibbons, *Appl. Phys. Lett.*, **47**, 1208, (1985).

11. D.G. Deppe, K.C. Hsieh, N. Holonyak, Jr., R.D. Burnham, and R.L. Thornton, *J. Appl. Phys.*, **58**, 4515 (1985).

12. R.L. Thornton, R.D. Burnham, T.L. Paoli, N. Holonyak, Jr., and D.G. Deppe, *Appl. Phys. Lett.*, **47**, 1239 (1986).

13. D.G. Deppe, G.S. Jackson, N. Holonyak, Jr., R.D. Burnham, and R.L. Thornton, *Appl. Phys. Lett.*, **50**, 632 (1987).

14. R.L. Thornton, D.F. Welch, R.D. Burnham, T.L. Paoli, and P.S. Cross, *Appl. Phys. Lett.*, **49**, 1572 (1986).

15. D. Hofstetter, H.P. Zappe, J.E. Epler, and P. Riel, *Appl. Phys. Lett.*, **67**, 1978 (1995).

16. K.J. Beernink, D. Sun, R.L. Thornton, and D.W. Treat, *Appl. Phys. Lett.*, **68**, 284 (1996).

17. J. Cibert, P.M. Petroff, G.J. Dolan, S.J. Pearton, A.C. Gossard, and J.H. English, *Appl. Phys. Lett.*, **49**, 1275 (1986).

18. P.W. Evans, J.J. Wierer, and N. Holonyak, Jr., *Appl. Phys. Lett.*, **70**, 1119 (1997).

19. P. Mei, H.W. Yoon, T. Venkatesan, S.A. Schwarz and J.P. Harbison, *Appl. Phys. Lett.*, **50**, 1823 (1987).

20. S.A. Schwarz, T. Venkatesan, R. Bhat, M. Koza, H.W. Yoon, Y. Arakawa, and P. Mei, *Proc. Mater. Res. Soc.*, **56**, 321 (1986).

21. J. Cibert, P. Petroff, D.J. Werder, S.J. Pearton, A.C. Gossard, and J.H. English, *Appl. Phys. Lett.*, **49**, 223 (1986).

22. K.L. Kavanagh, J.W. Mayer, C.W. Magee, J. Sheets, J. Tong, and J.M. Woodall, *Appl. Phys. Lett.*, **47**, 1208 (1985).

23. T.Y. Tan, U. Gosele, and S. Yu, *Critical Rev. in Sol. Sta. Mat. Sci.*, **17**, 47 (1991).

24. P. Mei, T. Venkatesan, S.A. Schwarz, N.G. Stoffel, J.P. Harbison, and L.A. Florez, Epitaxy of Semiconductor Layered Structures, *Proc. of MRS Sym*, Vol. **102**, 161 (1987) (published by Material Research Society, Pittsburgh).

25. K. Akimoto, M. Dohsen, M. Arai and N. Watanabe, *Appl. Phys. Lett.*, **43**, 1062 (1983).

26. Y.G. Chai, R. Chow, and C.E.C. Wood, *Appl. Phys. Lett.*, **39**, 800 (1981).

27. T. Ishibashi, S. Tarucha and H. Okamoto, *Jpn. J. Appl. Phys.*, **21**, L476 (1982).

28. I. Teramoto, *J. Phys. Chem. Solids*, **33**, 2089 (1972).

29. H.C. Casey and G.L. Pearson, *Point Defects in Solids*, J.H. Crawford, Jr. and L.M. Slikin, eds., Plenum Press, New York 1975.

30. W.G. Spitzer, A. Kahan, and L. Bouthillette, *J. Appl. Phys.*, **40**, 3398 (1969).

31. Y. Houng and T.S. Low, *J. Crystal Growth*, **77**, 272 (1986).

32. P.K. Haff and Z.E. Switkowski, *J. Appl. Phys.*, **48**, 3383 (1977).

33. K.L. Kavanagh, Ph.D. Thesis, Cornell University (1987).

34. T. Venkatesan, S.A. Schwarz, D.M. Hwang, R. Bhat, M. Koza, H.W. Yoon, P. Mei, Y. Arakawa and A. Yariv, *Appl. Phys. Lett.*, **49**, 701 (1986).

35. L.G. Guido, N. Holonyak, Jr., K.C. Hsieh, R.W. Kaliski, J.E. Baker, D.G. Deppe, R.D. Burnham, R.L. Thornton, and T.L. Paoli, *J. Electron. Mater.*, **16**, 87 (1986).

36. P. Mei, S.A. Schwarz, T. Venkatesan, C.L. Schwartz, J.P. Harbison, L. Florez, N.D. Theodore, and C.B. Carter, *Appl. Phys. Lett.*, **53**, 2650 (1988).

37. L.J. Guido, K.C. Hsieh, N. Holonyak, Jr., R.W. Kaliski, V. Eu, M. Fengand R.D. Burnham, *J. Appl. Phys.*, **61**, 1329 (1987).

38. N.D. Theodore, C.B. Carter, S.A. Schwarz, J.P. Harbison, and T. Venkatesan, *Advances In Materials, Processing And Devices In III-V Compound Semiconductors Symposium* (USA: Mater. Res. Soc, 1989. pp. 139–44).

39. T. Venkatesan, S.A. Schwarz, P. Mei, H.W. Yoon, Materials Modification and Growth Using Ion Beams Symposium, (USA: Mater. Res. Soc., 1987. pp. 171–185).

40. K. Nakashima, Y. Kawaguchi, Y. Kawamura, H. Asahi and Y. Imamura, *Jap. J. Appl. Phys.*, **26**, L1620 (1987).

41. M. Razeghi, O. Acher, and F. Launay, *Semicond. Sci. and Tech.*, **2**, 793 (1987).

42. B. Tell, B. C. Johnson, J.L. Zyskind, J.M. Brown, J.W. Sulhoff, K.F. Brown-Goebeler, B.I. Miller, and U. Koren, *Appl. Phys. Lett.*, **52**, 1428 (1988).
43. S.A. Schwarz, P. Mei, T. Venkatesan, R. Bhat, D.M. Hwang, C.L. Schwartz, M. Koza, L. Nazar, and B.J. Skromme, *Appl. Phys. Lett.*, **53**, 1051 (1988).
44. K. Kamagaki, H. Sakashita, H. Kato, M. Nakayama, N. Sano, and H. Terauchi, *J. Appl. Phys.*, **62**, 1124 (1987).
45. J.W. Matthews and A.E. Blakeslee, *J. Cryst. Growth*, **27**, 118 (1974).
46. D.G. Deppe, L.J. Guido, N. Holonyak, Jr., K.C. Hsieh, R.D. Burnham, R.L. Thornton, and T.L. Paoli, *Appl. Phys. Lett.*, **49**, 510 (1986).
47. J.D. Ralston, S. O'Brien, G.W. Wicks, and L.F. Eastman, *Appl. Phys. Lett.*, **52**, 1511 (1988).
48. S. O'Brien, D.P. Bour, and J.R. Shealy, *Appl. Phys. Lett.*, **53**, 1859 (1988).
49. K.B. Kahen, D.L. Peterson, G. Rajeswaran, and D.J. Lawrence, *Appl. Phys. Lett.*, **55**, 651 (1989).
50. E.S. Koteles, B. Elman, R.P. Holmstrom, P. Melman, J.Y. Chi, X. Wen, J. Powers, D. Owens, S. Charbonneau, and M.L.W. Thewalt, *Superlattices and Microstructures*, **5**, 321 (1989).
51. I. Gontijo, T. Krauss, J.H. Marsh, and R.M. de La Rue, *IEEE J. Quantum Electron.*, **30**, 1189 (1994).
52. K.J. Beernink, D. Sun, D.W. Treat, and D.P. Bour, *Appl. Phys. Lett.*, **66**, 3597 (1995).
53. J. Crank, *The Mathematics of Diffusion* (Clarendon Press, Oxford, England, 1957).
54. M.D. Camras, N. Holonyak, Jr., R.D. Burnham, W. Streier, D.R. Scifres, T.L. Paoli, and C. Lindström, *J. Appl. Phys.*, **54**, 5637 (1983).
55. T.Y. Tan and U. Gösele, *Appl. Phys. Lett.*, **52**, 1240 (1988).
56. D.G. Deppe and N. Holonyak, Jr., *J. Appl. Phys.*, **64**, R93 (1988).
57. L.J. Guido, N. Holonyak, Jr., K.C. Hsieh, and J.E. Baker, *Appl. Phys. Lett.*, **54**, 262 (1989).
58. Decai Sun, unpublished.
59. K.J. Beernink, R.L. Thornton, G. B. Anderson, and M.A. Emanuel, *Appl. Phys. Lett.*, **66**, 284 (1995).
60. R.W. Kaliski, D.W. Nam, D.G. Deppe, N. Holonyak, Jr., K.C. Hsieh, and R. D. Burnham, *J. Appl. Phys.*, **62**, 998 (1987).
61. K.J. Beernink, R.L. Thornton, and H.F. Chung, *Appl. Phys. Lett.*, **64**, 1082 (1994).

CHAPTER 8

Quantum Well Intermixing by Ion Implantation and Anodic Oxidation

H.H. TAN[a], S. YUAN[a], M. GAL[b] AND C. JAGADISH[a]

[a]Department of Electronic Materials Engineering, Research School of Physical Sciences and Engineering, Institute of Advanced Studies, The Australian National University, Canberra, ACT 0200, Australia
[b]School of Physics, University of New South Wales, Sydney, NSW 2052, Australia

1 INTRODUCTION

The focus of this chapter is to understand the role of defects and damage on intermixing. Ion implantation is a precise and reproducible way of introducing controllable amount of impurities and/or defects in the near surface region. Thus, defect- and/or impurity-assisted interdiffusion process can be easily controlled. The two major mechanisms for intermixing (disordering) are direct collision and diffusion processes [1]. However, the effect of direct collisional disordering is only appreciable at high implantation doses. Although the doses of protons used in this study could be quite high, it is expected that direct collisional disordering is negligible due to the small mass of this ion. Arsenic and oxygen implantations in this work fall into the diffusion regime because of the small doses used. The diffusion process, on the other hand, relies on impurities and/or native defects and their diffusion rates. A high temperature treatment step is usually required to initiate the interdiffusion process and at the same time remove or minimise the undesirable defects. Although an abrupt GaAs-AlGaAs interface can be grown easily, due to the small diffusion coefficient of Al-Ga [2,3], the presence of impurities and other defects may enhance this diffusion coefficient by orders of magnitude [1–6].

Implantation creates point defects (vacancies and interstitials), complexes and other extended defects. The accumulation of these defects increases the system's free energy and during annealing, the free energy is lowered by annihilation of these excess point defects through the process of diffusion. During the diffusion process, these defects migrate across the GaAs-AlGaAs heterointerface [1–10], thereby enhancing the interdiffusion between the Al-Ga atoms. Thus, the degree of intermixing is dependent on the diffusion rates of these defects. Although a variety of mechanisms have been proposed to help understand the defect-enhanced interdiffusion process [6–10], these models share a common feature where point defects (interstitials, vacancies or antisites) are assumed to be responsible for the interdiffusion. The simplest model assumes gallium vacancies, V_{Ga}, controlling the intermixing process.

Anodic oxidation is one of the traditional ways of making native oxides [11]. The pulsed anodization technique has recently attracted attention as a new way of creating current blocking layers for ridge-waveguide quantum well laser fabrication, since it was simple, reliable, and cost-effective [12,13]. On the other hand, pulsed anodic oxide can be used as an impurity-free defect source to enhance intermixing in GaAs/AlGaAs quantum well structures [14]. We have recently applied this technique to enhance the photoluminescence (PL) and cathodoluminescence (CL) signals from V-grooved quantum wires and obtained both spectrally and spatially well-resolved light emission from quantum wire [15]. Like other intermixing techniques [16], this interdiffusion technique can be conveniently applied to the fabrication of multiple wavelength lasers and non-absorbing mirrors for high power laser applications.

2   ION IMPLANTATION

Earlier studies of ion implantation-induced intermixing have also concentrated on the use of dopants or other impurities, such as Si, Zn, S, Se and O [1,5–7,17–26]. More recent studies have used chemically inert ions (inert gases) [23,27] or ions that are identical to the constituent atoms of the material system (Al, Ga, As) [23,28,29]. This so-called impurity-free intermixing method is of more interest, particularly in device fabrication because complications and reliability problems that may be associated with dopants or impurities can be avoided. Nevertheless, the introduced defects must be removed in device applications. Again, these previous studies have been focussed on the use of medium to heavy mass ions and these ions are known to cause more damage in the form of complexes and extended defects, which are more difficult to remove during annealing. Furthermore, with increasing ion dose, the accumulation of large damage clusters and extended defects may cause a saturation in the concentration of point defects, which evolve during annealing. The stronger thermal stability of these defects also inhibit the interdiffusion process, thus leading to a reduction (saturation) in intermixing, as indeed reported by several groups [20–23,27,29–32].

In this section, ion implantation-induced intermixing processes in GaAs-AlGaAs quantum wells will be systematically studied in a set of undoped quantum well structures grown by Metal Organic Chemical Vapor Deposition (MOCVD). The effect of ion species (H, As and O) are compared. Other parameters, such as implantation dose, implantation temperature, annealing time/temperature and the role of Al content in the barriers on intermixing are also investigated. The intermixing technique is then applied to blue shift the emission wavelengths of quantum well lasers.

2.1   Proton Implantation

The PL results for a sample implanted with protons to a dose of 5×10^{15} cm^{-2} and annealed at 900°C, 30 s are shown in Figure 1. Also shown is the unimplanted (but annealed) spectrum for reference. The inset is a schematic of the quantum well structure with $Al_{0.54}Ga_{0.46}As$ barriers. Four peaks corresponding to the four QW of different thicknesses (1.4, 2.3, 4.0 and 8.5 nm) are clearly observed. The sharp peak at 820 nm is due to the excitonic transition from the GaAs buffer/substrate region while the smaller peak at higher wavelength is due to donor-acceptor related transition. Very large blue shifts in the PL peaks are observed after implantation and annealing, in particular QW2 and QW3 (intermediate well thickness) having shifts >100 meV. Although the intensities have degraded somewhat, their recovery are quite significant, only about a factor of two lower than those of the reference sample even at such a high dose. This indicates that most of the implantation-induced defects have been annealed. It is worth mentioning that no PL signal is detected prior to annealing due to the ion implantation-induced defects that act as non-radiative recombination centres. The PL linewidths have broadened significantly, indicating a highly non-abrupt interface due to the interdiffusion process. Cross-section transmission electron microscopy, XTEM, results show that the GaAs-AlGaAs interfaces are very much diffused and somewhat 'broaden' as a result of intermixing, particularly in the two narrowest wells (QW1 and

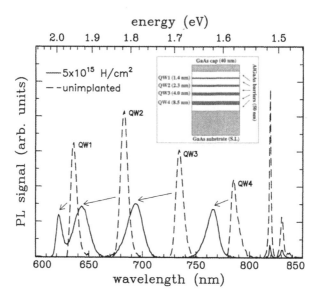

Fig. 1. Low temperature PL spectrum for a sample implanted with 5×10^{15} H/cm^2 and annealed at 900°C for 30 s. Also shown is the spectrum for an unimplanted (but annealed under same conditions) sample for reference. The inset shows schematic of the sample.

QW2). Double crystal X-ray diffraction, DCXRD (Figure 2) results also indicate the highly non-abrupt interfaces after implantation and annealing. The main peak in Figure 2 (unimplanted) is due to the GaAs substrate while the secondary peak at lower angle is due to AlGaAs with an 'average' Al content. The smaller satellite peaks are due to the diffraction from the interference of the abrupt GaAs-AlGaAs interfaces. After implantation oscillation are observed between the main and the secondary peaks as a result of increased strain due to the ions. Upon annealing most of the additional strain is relieved but the original satellite peaks have either disappeared or somewhat decreased in intensity due to the diffused interfaces.

A plot of energy shift as a function of implantation dose is shown Figure 3. Very large energy shifts of up to about 200 meV are observed for the two intermediate wells. However, for the narrowest and widest QWs, the energy shifts are smaller. This effect is similar to that observed by Elman et al., who attributed it to the saturation of the exciton energy at the barriers [29]. As the well becomes progressively narrower, the ground state energy level of electrons (holes) moves closer to the conduction (valence) band of the barriers. Hence, for very narrow wells, there is a smaller range in which this level may move for a perturbation in the well. On the other hand, in wide wells there is only a small range in which the ground state can move for a change

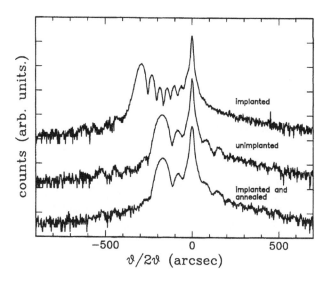

Fig. 2. Double crystal X-ray diffraction rocking curves for implanted (5×10^{15} H/cm^2) but unannealed, unimplanted and implanted (5×10^{15} H/cm^2) and annealed (900°C for 30 s) samples.

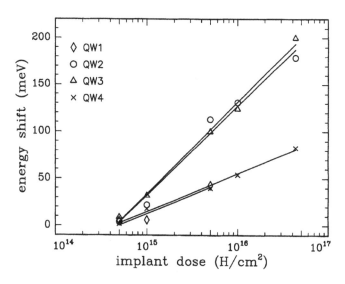

Fig. 3. Magnitude of energy shifts as a function of dose. All samples have been annealed at 900°C for 30 s.

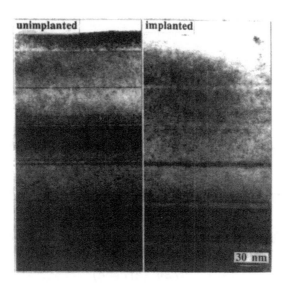

Fig. 4. Cross-sectional TEM of (a) unimplanted and (b) implanted sample (1×10^{16} H/cm^2, annealed 900°C, 30 s).

in the well shape due to the continuum limit for wide wells. Thus, in wells of intermediate width, the exciton energy is most sensitive to any perturbation in the well and hence, the large energy shifts. In all cases, magnitude of energy shifts increases linearly with implantation dose with no saturation observed. This effect is in stark contrast to several reports, which indicated a reduction in energy shift with implantation dose [20–23,27,29–32]. However, the results in those reports were for experiments using heavier ions such as Al, Ga or As. Thus, the use of protons is expected to create only point defects and dilute defect cluster [33,34] and hence, very efficient in creating intermixing. Even at high proton doses ($> 1 \times 10^{15}$ cm^{-2}) a large concentration of point defects may still exist as indicated by the large energy shifts with no apparent saturation. These observations are similar to a very recent study where it was reported that large energy shifts were obtained by protons [35]. In addition, proton bombardment is not expected to result in appreciable formation of more stable defects such as defect clusters and extended defects during annealing. This is further supported by the XTEM results in Figure 4 where no extended defects are observed after annealing even for a high implantation dose of 1×10^{16} cm^{-2}. This suggests that almost all of the point defects created by proton implantation are involved in the intermixing process. The lack of defect agglomeration during annealing thus improves intermixing and optical recovery.

It is also interesting to note that at higher implantation doses ($> 5 \times 10^{15}$ cm^{-2}), the PL signal from the narrowest well (QW1) disappears completely. Two immediate explanations for this effect could be that 'complete' intermixing has occurred at this stage (ie. complete destruction of QW1) or the merging of the PL signal with that of the adjacent well (QW2) to form a broader peak. The latter scenario is unlikely because the linewidth of the adjacent QW at 1×10^{16} cm^{-2} is comparable to that implanted at lower doses. Complete intermixing of the narrowest QW has also been ruled out as indicated by the cross-sectional electron microscopy results shown in Figure 4 for a sample implanted to a dose of 1×10^{16} cm^{-2} and annealed at 900°C, 30 s. Although the QWs in the implanted (and annealed) sample appear to be 'diffused', due to intermixing, all four QWs are still observable. Hence, the more likely explanation of the disappearance of the PL signal from the narrowest well is that, as the well becomes progressively more intermixed, the ground state energy level of the Γ-band in the QW is raised to higher energies. When this level exceeds that of the X-band in the barrier the probability of direct transition decreases, making the QW indirect, similar to the type I to type II transitions reported in some very narrow and indirect quantum well [36–38]. This effect is further enhanced by the additional scattering of the photoexcited carriers due to some residual defects left in the sample after annealing. The disappearance of the PL signal at high implantation doses has also been reported recently by Kupka and Chen [31]. Although, they attributed this effect to $\Gamma - X$ scattering, the damage created in their case was quite extensive (3×10^{15} Ga/cm^2). It is most likely that amorphization of the GaAs QW has occurred at this dose and the disappearance of the PL signal could be due to 'complete' intermixing in their case.

The results of energy shifts as a function of annealing temperature and time are shown in Figure 5(a) and (b), respectively. It can be seen from these graphs that annealing above 900°C or 30 s does not further enhance intermixing in QW2 and QW3. Furthermore, no significant improvement in the PL intensities is observed above these conditions. The annealing results of proton bombardment in GaAs-Al$_{0.54}$Ga$_{0.46}$As quantum wells indicate that the annealing conditions can be adjusted to provide optimum shifts with acceptable PL signal recovery in the temperatures of 900–950°C for 30–45 s anneal. At lower temperatures (~ 850°C), not all of the dilute point defect clusters can break up to liberate point defects which contribute to the interdiffusion process. Thus, the restoration of crystalline perfection is not efficient at these temperatures. Indeed, the PL intensities of the samples annealed at 850°C are about a factor of 2–3 lower than those annealed at 900 or 950°C. At higher temperatures on the other hand, very little improvement in the magnitude is observed, most likely due to the exhaustion of the available point defects. However, for the deepest well (QW4), a slight enhancement in intermixing may be achieved above 900°C or 30 s anneal. This effect can be

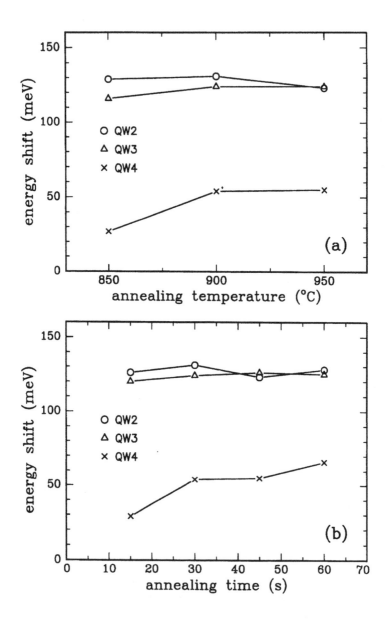

Fig. 5. Dependence of energy shifts on (a) annealing temperature for 30s and (b) annealing time at 900°C.

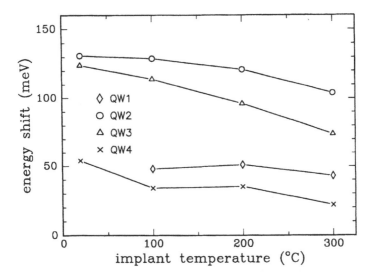

Fig. 6.   Effect of implantation temperature on energy shifts (annealed at 900°C, 30s).

associated with the deepest well being closest to the projected range of the ions, where thermally stable defect loops and clusters that are more likely to be formed towards the end-of-range of the ions. Hence, higher/longer annealing temperatures/times are required to remove these defects.

It has been reported in several studies that increasing the implantation temperature enhances the degree of intermixing due to the increase in mobility of defects [17,26,39]. However, the results with proton implantation in the temperature range of 20 to 300°C do not support this phenomenon. Instead, there is a consistent reduction in the degree of intermixing with increasing implantation temperatures as shown in Figure 6. Furthermore, implantation at elevated temperatures does not significantly improve the PL intensities or linewidths. Almost a factor of two reduction in energy shift is observed at 300°C. Although, these results are in contradiction to reports by other groups, it should again be emphasised that their experiments were based on the use of heavier ions. It is well known that the nature and types of defects are very different for light and heavy ions [33,34,40,41]. It is also known that if implantation is carried out at elevated temperatures, dynamic annealing is enhanced, whereby the defects created by the ions may annihilate *in situ* to restore crystallinity and hence the residual defect concentration may be lowered [33,34,41–43]. However, the type of residual defects may also be quite different at elevated implantation temperatures [33,34,41–43]. Although the samples used in this chapter consist of multilayered structures, they are predominantly AlGaAs layers. The result of the strong dynamic annealing in AlGaAs is that ion beam-induced damage builds up in the form of point defects and point defect clusters [44,45]. Being the lightest ion, the defects created by protons in AlGaAs are predominantly point defects. Although at elevated implant temperature additional *in situ* repair of crystalline damage may be achieved, these mobile point defects may also acquire more energy to coalesce to form more stable loops/clusters, thereby lowering the system's free energy. The formation of loops and clusters consumes point defects and as a result, less intermixing is achievable during annealing. Furthermore, the higher thermal stability of these loops and clusters also act to retard intermixing during annealing. The results of reduction in the magnitude of energy shifts with increasing implantation temperature suggests that the predominant point defects created in AlGaAs by proton implantation agglomerate into more stable defects and/or annihilate at elevated implantation temperatures.

2.2   Arsenic and Oxygen Implantations

Large energy shifts may also be obtained with heavier ions, such as arsenic, but at significantly lower implantation doses. From TRIM [46] calculations, the collisional displacement density caused by arsenic ions is expected to be a factor of ~ 1000 higher than that of protons. Thus, the ion doses must be reduced by about 3 orders of magnitude to maximise optical recovery upon annealing as shown by the PL spectrum in Figure 7 for an implantation dose of 1×10^{12} cm^{-2} (after 900°C, 30 s anneal). It can be seen from this figure that the energy shifts for all the quantum wells are significant and also recovery of PL intensity is reasonable for this dose. A plot of the dose dependence of the magnitude of energy shift is shown in Figure 8. Energy shifts in excess of 200 meV are achieved at a dose of 1×10^{14} cm^{-2}. However, at this dose the PL intensities are quite weak and also the disappearance of the two narrower wells (QW1 and QW2) is observed, similar to the proton implantation case. A major difference between this result and that of proton implantation is that a saturation of the energy shift is observed in the case of arsenic ions. Although the implantation doses have been lowered significantly, it is known that the nature of defects caused by heavier ions is quite different to those from light ions [33,34,40,41]. The damage cascade caused by an individual ion track is expected to be denser and larger in size for heavier ions and, hence, increases the chance of agglomeration of the surrounding point defects into clusters and other extended defects by the overlap of these damage cascades [33,34,40,41].

An implantation-temperature study with arsenic ions shows that additional improvement in the energy shift may be obtained by arsenic implantation at elevated temperatures, although this improvement is only marginal (Figure 9

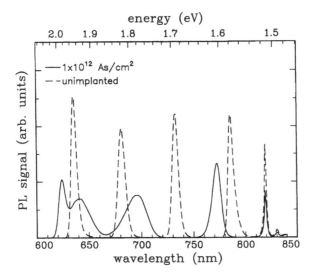

Fig. 7. Low temperature PL spectrum for a sample implanted with 1×10^{12} As/cm^2 and annealed at 900°C for 30 s. Also shown is the spectrum for an unimplanted (but annealed under same conditions) sample for reference.

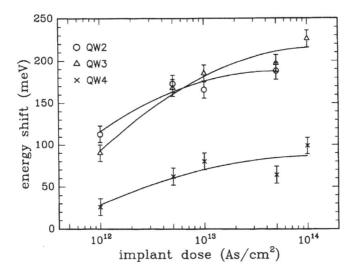

Fig. 8. Dose dependence of energy shift of arsenic implanted samples. All samples have been annealed at 900°C for 30 s.

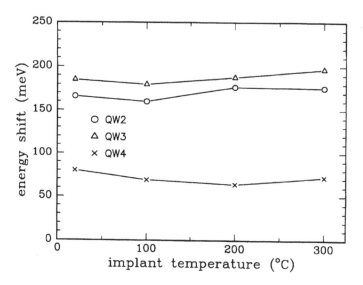

Fig. 9. Energy shift for arsenic implanted samples as a function of implantation temperature. Annealing was carried out at 900°C for 30 s.

for an implantation dose of 1×10^{13} cm^{-2}). However, the results do not show any significant improvement in the recovery of the PL intensities at elevated implantation temperatures. This behaviour has the opposite trend to that observed with proton implantation where increasing the implantation temperature retards the intermixing. Thus, as mentioned previously, the nature of defects created by heavy and light ions are fundamentally quite different and intermixing is governed by the type and stability of the defects created during and after implantation.

Figure 10 shows the dose dependence of energy shift in the oxygen implanted sample. The implantation doses have been scaled according to TRIM calculations, which indicate that the displacement density due to oxygen ions is about a factor of 100 greater than that created by protons. Similar to the results of arsenic ions, a saturation effect is observed at higher doses but the magnitude of energy shift is somewhat reduced from either that of arsenic or proton implantations. Nevertheless, moderate energy shifts of up to 130 meV may still be obtained with oxygen ions. Since the displacement density for oxygen implantation has been scaled appropriately, the observed smaller energy shifts are due to the effect of the ions. Oxygen is known to act as an impurity and pinning defects and bonds with Al to form complexes [47,48], thereby reducing the concentration of mobile point defects available for intermixing. These complexes are also more thermally stable than simple

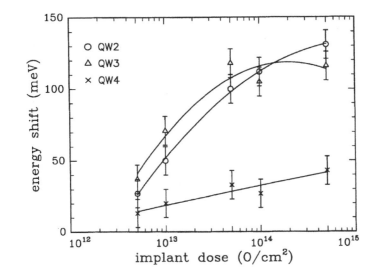

Fig. 10.   Dose dependence of energy shift of oxygen implanted samples. All samples have been annealed at 900°C for 30 s.

point defects and hence act to retard intermixing. Indeed, by annealing at higher temperatures, a further improvement in the energy shift is observed (Figure 11), particularly for the deeper wells, due to the breaking up of these complexes into point defects to enhance intermixing during annealing. It is also worth mentioning that elevated implantation temperature results for oxygen ions are similar to those of arsenic ions, again suggesting that defects created by medium to heavy mass ions are fundamentally different from those formed by protons.

2.3   Comparison of Proton, Oxygen and Arsenic Implantations

Figure 12 summarizes the comparative results of protons, oxygen and arsenic implantations. The energy shift is plotted against the collisional displacement density (as calculated by TRIM) across QW3 so that direct comparison may be made on the influence of defect density (since the ion doses are different). The displacement density should only be taken as a guide and not the actual defect concentration since dynamic annealing is known to occur at room temperature in GaAs and especially in AlGaAs [44,45]. Two main features can be noted from this figure. Firstly, for heavier ion (As, O) implants, a saturation effect in the energy shift (intermixing) is observed at higher displacement density but not for the case of protons.

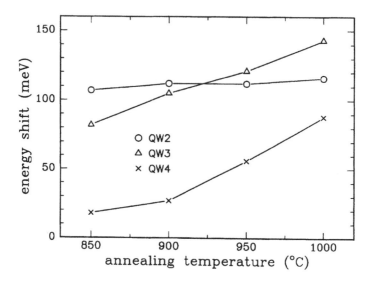

Fig. 11. Plot of energy shift as a function of annealing temperature for 30 s anneals. Samples have been implanted with 1×10^{14} O/cm^2.

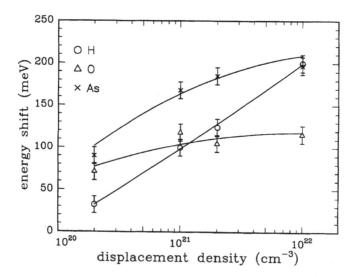

Fig. 12. Comparison of the energy shifts of QW3 for H, O and As implantations as a function of displacement density. The implantation doses were chosen such that they resulted in the same displacement density (as calculated by TRIM) in the QW for all the three ions.

The saturation effect is much more pronounced with oxygen implantations due to the chemical effect of oxygen, which is discussed below. Secondly, at lower displacement densities, heavier ions are more efficient at intermixing as indicated by the larger energy shift. However, with increasing implantation dose (displacement density), protons become more efficient in intermixing. The latter is not surprising since dilute cascades of light ions allow more efficient dynamic annealing, leaving fewer point defects for intermixing. However, as the implantation dose (displacement density) is increased, the agglomeration of point defects into more stable defects increases significantly for heavier ions, due to the larger and denser damage cascades caused by the heavy ions. Hence, a saturation in the energy shift is observed in heavier ions with increasing implantation dose.

Comparative study with heavier ions shows that arsenic may also create large energy shifts with reasonable optical recovery but at significantly lower doses (2–3 orders of magnitude lower than protons), in proportion to the higher displacement density of the heavier ions. However, with increasing implantation dose, a saturation effect in the degree of intermixing is observed for the heavier ions. It is also observed that at lower displacement densities, heavier ions are more efficient for intermixing. Unlike proton implantation, the results for heavier ions show that intermixing can be improved slightly with elevated implantation temperature, similar to some previous reports [17,26,39]. However, the issue of implantation temperature dependence in intermixing is not very well understood and indeed, there have been several reports that the ion beam mixing rate in II-V compound semiconductors is a complicated function of implantation temperature, whereby it increases until a critical temperature is reached after which it drops markedly [49,50]. Although the results in this chapter do not resolve this issue completely, they provide further insight into this issue.

It is proposed that the difference in the elevated implantation temperature results between heavy and light ions is related to the fine balance between the defects created by the incoming ions and the strong dynamic annealing in AlGaAs. The size of the damage cascades created by heavier ions is reduced at elevated temperature and hence, the overlap of these cascades created by the individual ion track to form more stable defects is proportionately reduced. The strong dynamic annealing in AlGaAs further enhances this effect. Thus, a small improvement in intermixing is observed with increasing implantation temperature. With protons, the strong dynamic annealing in AlGaAs and the small size of the damage clusters means that increasing the implantation temperature has very little effect on the overlap of these cascades, since the defects created are predominantly point defects. However, at higher temperatures, these point defects become extremely mobile to return to their lattice sites and/or agglomerate into more stable defects to lower the system's free energy. Hence, intermixing is retarded in this case.

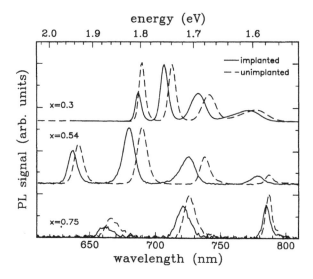

Fig. 13. Low temperature PL spectra for $Al_{0.3}Ga_{0.7}As$, $Al_{0.54}Ga_{0.46}As$ and $Al_{0.75}Ga_{0.25}As$ samples implanted with 1×10^{15} H/cm^2 and annealed at 900°C for 30 s. Also shown are the spectra for an unimplanted (but annealed under same conditions) samples for reference.

The results for oxygen ions are somewhat similar to that of arsenic ions, with a saturation in the energy shifts observed with increasing dose and the implant temperature dependence on intermixing are similar. However, the lower degree of intermixing (for a fixed displacement density) may be attributed to the chemical effect of oxygen, which can easily form complexes with defects and/or Al bonds in II-V materials [47,48]. The formation of these complexes consumes point defects and hence less is available for intermixing during annealing. The thermal stability of the complexes implies that higher annealing temperatures and/or longer annealing times are required to break these bonds to free the Al atoms.

2.4   Effects of Al Content in the Barriers

The results shown thus far are for GaAs quantum wells with $Al_{0.54}Ga_{0.46}As$ barriers. What is the role of Al concentration in the barriers on the effect of intermixing? This section will highlight these effects of proton irradiations for structures with $Al_{0.3}Ga_{0.7}As$ and $Al_{0.75}Ga_{0.25}As$ barriers. A comparison of the PL spectra for different Al composition barriers is presented in Figure 13. In the sample with $Al_{0.75}Ga_{0.25}As$ barriers, there is no PL emission from the narrowest well (QW1) even in the unimplanted sample. Calculations show that this effect is due to the ground state energy level of the QW in the Γ-band being higher than that of the barrier in the X-band by about $110\,\mathrm{meV}$ at $12\,\mathrm{K}$, thereby making the QW indirect. Despite better confinement, the PL intensities in this sample are also quite weak and this is attributed to unintentional incorporation of O during the growth of high Al content AlGaAs. In all cases, the widths of the QWs are identical and a blue shift is observed after implantation and annealing. Very good recovery of the PL intensities is also observed in the three samples after annealing.

A plot of the magnitude of energy shift versus implantation dose for the two Al content barriers are shown in Figure 14. The trend for the $Al_{0.75}Ga_{0.25}As$ barriers (lower figure) is very similar to that of $Al_{0.54}Ga_{0.46}As$ barriers, in as much as very large energy shifts of up to $\sim 200\,\mathrm{meV}$ are obtainable with no apparent saturation. It is also interesting to note that above the dose of $1 \times 10^{15}\,\mathrm{cm}^{-2}$, the PL signal from QW2 has disappeared (c/f. QW1 in the case of $Al_{0.54}Ga_{0.46}As$ barriers), again due to the transition from direct to indirect QW. The trend of energy shift with annealing temperature/time is similar to that of the previous case, with no further improvement observed above $900°C$ or for longer than $30\,\mathrm{s}$ annealing time. Although much larger energy shifts are expected from this system due to the higher concentration gradient of Al across the (barrier-well) interfaces, the magnitude of energy shift is approximately similar to that of the samples with $Al_{0.54}Ga_{0.16}As$ barriers. This effect can be understood in terms of the increased incorporation of impurities, such as oxygen, in AlGaAs of high Al content in MOCVD grown material. It is well known that MOCVD-grown AlGaAs with standard trimethylaluminum (TMA) precursors will result in the incorporation of substantial amounts of oxygen and this effect is increased with increasing Al content [51,52]. Furthermore, oxygen is known to be a non-radiative deep donor in AlGaAs, which degrade the luminescence efficiency [51,52]. This is further supported by the weaker PL intensities in samples with $Al_{0.75}Ga_{0.25}As$ barriers in comparison to that of $Al_{0.54}Ga_{0.46}As$ barriers. It is also worth noting that no PL signal was detected in the QW samples with AlAs barriers. The presence of oxygen may also affect the intermixing mechanism by forming complexes with the point defects during implantation and/or annealing [47,48]. Since point defects are responsible for intermixing, the formation of oxygen-related complexes will deplete the point defect concentration and hence, retard the diffusion of the point defects. Thus, intermixing is somewhat suppressed for the samples with higher Al content. Moreover, the higher thermal stability of these complexes implies that less point defects may be liberated from these complexes to cause intermixing during annealing. This effect is not unlike the effect described earlier for samples implanted with oxygen ions where intermixing was retarded.

For samples with $Al_{0.3}Ga_{0.7}As$ barriers, the dose dependence results are quite different than in samples with $Al_{0.54}Ga_{0.46}As$ barriers. The lower degree

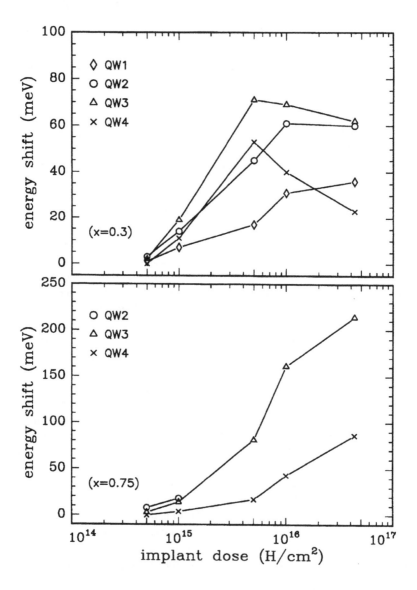

Fig. 14. Dose dependence of energy shift for $Al_{0.3}Ga_{0.7}As$ and $Al_{0.75}Ga_{0.75}As$ samples. Annealing was at 900°C, 30 s.

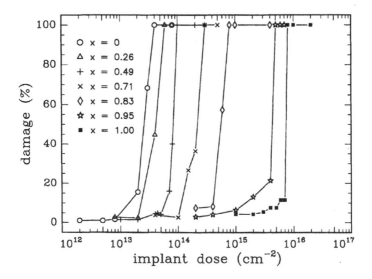

Fig. 15. Damage build-up in thick $Al_xGa_{1-x}As$ layers for various Al contents (x) versus implantation dose. Implantations were performed at liquid nitrogen temperatures with Si ions. The 100% level in damage corresponds to the onset of amorphization.

of intermixing in this case is expected due to the smaller Al gradient across the GaAs-AlGaAs interface. More significantly, saturation in energy shift with increasing dose is observed in the samples with lower Al content. The saturation effect is more significant for the two deepest wells (QW3 and QW4). In fact for these two wells, the energy shifts increase to about 5×10^{15} cm^{-2} and then begin to drop above this dose. Indeed, by annealing at higher temperatures, where more point defects can be liberated from these more complex defects, further improvement in the energy shifts is observed. This effect can be understood in terms of smaller point defect concentrations available during annealing due to the increased likelihood of the formation of more stable defects in AlGaAs of lower Al content. Indeed, it has previously been shown that the threshold dose for amorphization of AlGaAs increases with Al content [44,45]. More than two orders of magnitude difference in dose are required to amorphize AlAs in comparison to GaAs (Figure 15). Also, the predominant type of implantation-induced defects in AlGaAs may be quite different for different Al composition. For low Al content AlGaAs, the damage consists of complex defects such as large clusters and small pockets of amorphous zones created by the individual ion tracks. On the other hand, for high Al content AlGaAs, it is the accumulation of secondary defects such as point defects and other extended defects [44,45].

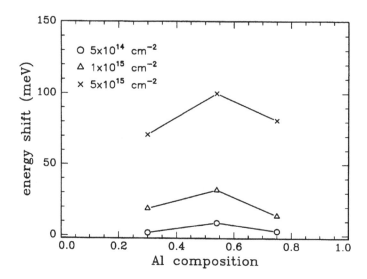

Fig. 16. Energy shift as a function of Al composition in the $Al_xGa_{1-x}As$ barriers for QW3 for 3 different implantation doses and annealed at 900°C, 30 s.

Furthermore, dynamic annealing is very strong in AlGaAs even at liquid nitrogen temperatures and this process is more efficient with increasing Al content [44,45]. Thus, this effect becomes even more pronounced at room temperature where the defects created in AlGaAs have higher mobility. However, the dynamic annealing process is not perfect and accumulation of defects in the form of point defect clusters may still occur. Thus, in AlGaAs of lower Al content (as in $Al_{0.3}Ga_{0.7}As$), the defect accumulation process becomes more efficient due to weaker dynamic annealing.

As a summary of the comparison of the effect of different barrier composition Figure 16 shows a plot of the energy shift of QW3 versus the Al composition of the barriers for three different implantation doses. A peak in the magnitude of energy shift is observed for Al mole fraction at around 0.54 but at higher and lower composition, a retardation of the energy shift is observed. Thus, there is an optimal composition in this material system where maximum intermixing may be achieved with protons. The effect of implantation temperature in these two exhibit similar behaviour as that of the $Al_{0.54}Ga_{0.46}As$ barriers, where increasing the implantation temperature reduces the degree of intermixing.

Based on these results, the following phenomenological model is proposed to explain the observed behaviour. The use of light ions enables a large concentration of point defects to be accumulated in AlGaAs and this effect is enhanced with increasing Al content. During annealing these defects may then be injected (diffused) across the QW (heterointerface) to induce a larger degree of intermixing. However, the accumulation of these point defects in AlGaAs may reach a critical concentration level at a certain implantation dose (higher for higher Al content), and ultimately agglomerate into more stable defects such as point defect clusters. Thus, for $Al_{0.3}Ga_{0.7}As$, the point defect saturation (defect agglomeration) occurs at a lower dose than, say, $Al_{0.54}Ga_{0.46}As$. Hence, during annealing, less point defects are available for the intermixing process, and consequently less energy shift occurs. If the annealing temperature/time is increased, these point defect clusters may break up to liberate more point defects to cause further intermixing, as observed in samples with $Al_{0.3}Ga_{0.46}As$ barriers. In this study, no saturation in the point defect concentration is observed in samples with $Al_{0.54}Ga_{0.46}As$ and $Al_{0.75}Ga_{0.25}As$ up to a dose of at least 4×10^{16} cm^{-2}, but a saturation is observed in $Al_{0.3}Ga_{0.7}As$ at around 1×10^{16} cm^{-2}, consistent with the proposed model. For the $Al_{0.75}Ga_{0.25}As$ samples, the energy shift is less than that of $Al_{0.54}Ga_{0.46}As$ due to impurity (oxygen) incorporation during MOCVD growth. Thus, it is foreseeable that in 'purer' $Al_{0.75}Ga_{0.25}As$ samples (for example those grown by using low O content TMA source or by Molecular Beam Epitaxy), greater than 200 meV shifts may be achieved.

2.5   Intermixed Lasers

In this section, the application of intermixing is implemented on quantum well lasers. Proton implantation is used due to the superior properties mentioned above. Furthermore, in actual laser structures the active regions are often quite deep (about 1 μm from the surface) and protons have the additional advantage of larger penetration depth. Wavelength shifting of the devices is of particular interest, especially in wavelength-division-multiplexing (WDM) applications where multiple lasers of different wavelengths may be integrated onto a single chip as the WDM source. Ion implantation is a precise and controllable technique, may offer a practical post-growth alternative to selective- area growth or other layer intermixing technique.

The material used for this study was a GRaded-INdex Separate-Confine-ment-Heterostructure (GRINSCH) laser structure grown by MOCVD. $Al_{0.6}Ga_{0.4}As$ layers of 1 μm thickness were used as the cladding layers. The active region consisted of three 6 nm GaAs QWs separated by 20 nm $Al_{0.3}Ga_{0.7}As$ barriers. Only the top contact layer (0.2 μm) was p-doped with Zn to a level of about 1×10^{19} cm^{-3}. These conditions were chosen to minimize the effect of Zn diffusion on intermixing during annealing. The ion energy was chosen to produce maximum displacements (damage) around the active region. Annealing was carried out in proximity cap conditions at 950°C for 30 s.

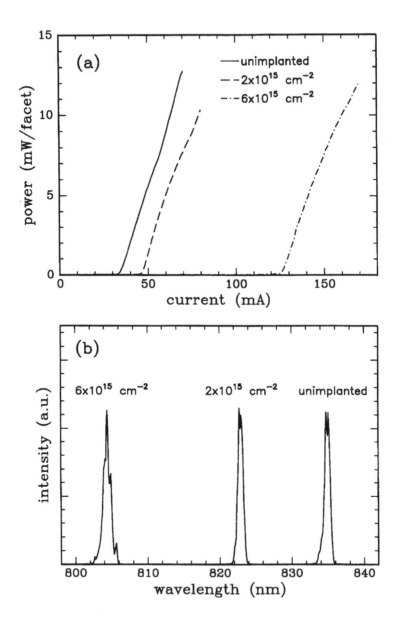

Fig. 17. Light output versus injected current characteristics of 3 different lasers on the same bar (a) and the corresponding spectra for the 3 lasers at $1.5I_{th}$. (b).

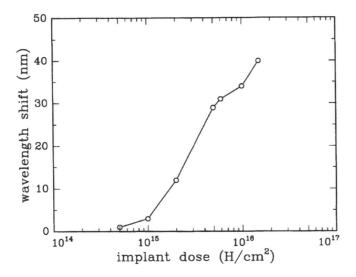

Fig. 18.   Wavelength shift for the lasers at various implantation doses.

The light-current (L-I) and spectral characteristics of the lasers are tested without any mirror coating and additional heat sinking under pulsed conditions with 2 μs pulses of 1% duty cycle. Figure 17 shows the (a) L-I curves and (b) lasing spectra of 4 μm ridge waveguides (400 μm long) devices implanted with doses of 2×10^{15} and 6×10^{15} cm^{-2} at $1.5 I_{th}$. Also shown is the L-I curve of the unimplanted (but annealed) reference laser. The reference device has a lasing threshold and wavelength of 33 mA and 835 nm, respectively. A slight increase in the lasing threshold is evident in devices implanted at a dose of 2×10^{15} cm^{-2}. However, the emission spectrum has been blue-shifted to 823 nm (22 meV shift). The wavelength shift is significant considering there is very little change in the slope efficiency. At a higher implantation dose of 6×10^{15} cm^{-2}, a very large shift of 31 nm is observed although the threshold current has increased to 125 mA. The differential efficiency decreases slightly from 47% (unimplanted) to about 35%. The increase in lasing threshold after implantation can be understood in terms of poorer carrier confinement due to the 'diffused' quantum wells. Figure 18 shows the wavelength shift of these devices as a function of implantation doses. Up to 40 nm shift is observed corresponding to an implantation dose of 1.5×10^{16} cm^{-2}. The values of wavelength shift are about the order of shift expected for 6 nm GaAs QWs with Al$_{0.3}$Ga$_{0.7}$As barriers from photoluminescence results in earlier sections. These results show that ion implantation can achieve very large blue shifts in GaAs quantum well GRINSCH lasers and could be very promising for integrating lasers of different wavelengths for optical communication applications, particularly as WDM sources.

3 ANODIC OXIDATION

The sample structures used for this work were grown by low pressure MOCVD on p^+-GaAs substrates. The substrates were Zn-doped at 1×10^{19} cm^{-3} concentration and $2°$ off $\langle 100 \rangle$ towards $\langle 110 \rangle$. All epilayers were nominally undoped. Pulsed anodization was carried out at room temperature in an electrolyte consisting of ethylene glycol : de-ionized water : phosphoric acid (40:20:1 by volume). Half of the samples were masked during anodization for reference. The pulse width of 1 ms with a period of 12 ms was used and total anodization time was 4 mins. Anodization current density was in the range of 40 to 200 mA/cm^2 as determined by the leading edge of the pulse.

3.1   GaAs/AlGaAs Quantum Wells

Figure 19 shows PL spectra from a 8.6 nm thick GaAs/Al$_{0.54}$Ga$_{0.46}$As quantum well structure for an as-grown sample, an anodized but not annealed sample, a sample annealed at 900°C for 120 s without anodic oxide, and an anodized sample which was then annealed at 900°C for 120 s. The QW PL peak for the anodized but unannealed sample does not show energy shift with respect to the as-grown sample. The unanodized but annealed sample shows blue shift with respect to the as-grown sample as a result of thermal intermixing [53], while the anodized and then annealed sample shows larger blueshift due to anodic-oxide-induced intermixing. The unanodized quantum well is relatively thermally stable up to 925°C, then it starts to show a strong blueshift at 950°C. PL from anodized samples show continuous blue shift and significant intensity increase for annealing temperatures above 925°C. This trend was observed for all samples anodized at different current densities. No linewidth broadening was observed from the PL spectra of the anodic-oxide-induced interdiffused samples. The anodic-oxide-induced intermixing in GaAs/AlGaAs quantum well structures is not sensitive to the anodic current density in the range from 40 mA/cm^2 to 200 mA/cm^2 in the temperature range studied [14]. This is due to the fact that the anodic oxide thickness is not sensitive to the anodization current density. The anodic oxide thicknesses for samples anodized at 80, 160 and 200 mA/cm^2 for 4 mins are determined to be 125, 110 and 125 nm, respectively.

In Figure 20, the relative PL energy shift for anodized GaAs/Al$_{0.54}$Ga$_{0.46}$As four quantum well structure (similar to ion implanted structure) annealed

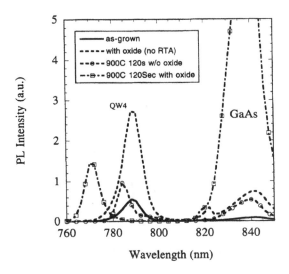

Fig. 19. PL spectra of GaAs/AlGaAs multiple quantum well samples. For clarity, only PL peaks from QW4 are shown.

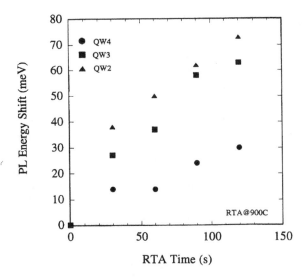

Fig. 20. PL energy shift of anodized GaAs/AlGaAs quantum well structures (with respect to unanodized reference sample) annealed at 900°C as a function of RTA time.

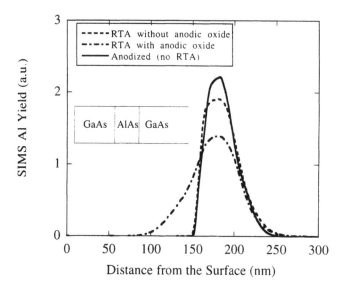

Fig. 21.   Al profiles of a GaAs/AlAs/GaAs structure measured by SIMS. The solid line is for
an anodized but unannealed sample. The profiles for another sample annealed at 950°C, 60 s
with the anodized half (dashed-dotted line) and the unanodized half (dashed line) are also shown.

at 900°C is plotted as a function of rapid thermal annealing (RTA) time.
The relative PL shift is the PL energy difference between anodized area
and unanodized area on the same piece of sample annealed together. PL
peaks from the QWs of unanodized areas (the reference samples) show
marginal blueshift after RTA, in agreement with thermal stability studies of
GaAs/AlGaAs single quantum wells using photoreflectance [53]. Anodized
samples, however, show significant blueshift of PL peak energies with
increasing RTA time and temperature. In this figure, PL energy shifts for
QW1 are not shown as this QW becomes indirect even for shortest annealing
times used in this study (similar to the ion implanted case).

Figure 21 shows the SIMS profiles of Al atoms from three samples with
an AlAs marker layer, one annealed without anodic oxide, one annealed
with anodic oxide, and another anodized but not annealed. The annealing
temperature is 950°C and annealing time is 105 s. For these profiles, the
annealed without oxide sample shows same broadening with respect to
the anodized but unannealed sample due to thermal intermixing, while the
sample annealed with oxide shows clear migration of Al atoms towards the
surface. This indicates that at elevated temperature, the anodic oxide cap
layer at the surface enhances the Al atom diffusion towards the surface in this
sample.

From the PL and SIMS studies, it is believed that at elevated temperatures the anodic oxide cap layer on a GaAs/AlGaAs quantum well structure enhances the intermixing through the enhanced diffusion of Al atom towards the surface. Without the anodic oxide layer, PL blueshift is much smaller and similar to the shifts reported due to thermal interdiffusion suggesting that the dopant (Zn) in the substrates does not play any direct role in the anodic oxide induced intermixing.

It is known that at elevated temperatures, the oxides of Ga and As are thermodynamically stable relative to GaAs oxide [54]. The interdiffusion results imply that the concentration of a native point defect, i.e., a group III vacancy or interstitial, is increased at elevated temperatures when annealing under an anodic oxide. For the impurity-free enhanced interdiffusion induced in GaAs by a deposited SiO_2 layer, it is often assumed that an increased Ga vacancy concentration enhances the interdiffusion. However, the chemical interactions between GaAs and a hydrated mixture of Ga and As oxides are presumably quite different than those between GaAs and SiO_2. Based upon previously reported measurement [56], it is likely that the residual water in an anodically oxidized layer induces further oxidation of the GaAs at elevated temperatures [14]. This effect is further supported by another study where two samples were anodized and annealed (900°C, 75 s) under the same conditions. The only difference is that one sample had been annealed (prebaked) at 600°C for 5 mins before the 900°C rapid thermal annealing to drive out residual water in the anodic oxide. Results show that residual water in the anodic oxide induces further interdiffusion, but was not the main cause of the anodic oxide-induced interdiffusion. However, further studies are required to further understand the mechanism of interdiffusion caused by anodic oxide.

A technique was developed to determine the Al profile from the fringes of the TEM images [14], and the results are shown in Figure 22 for an as-grown sample and a sample annealed with oxide at 950°C for 105 s. The diffusion coefficient of the QW4 annealed at 950°C was calculated to be 1.76×10^{-16} cm^2/s, and the Al profile of the quantum well after 105 s annealing at 950°C was evaluated using this value and compared to that obtained from TEM measurements as shown in Figure 22.

3.2 GaAs/AlGaAs Quantum Wire Structures

Anodic oxide-induced intermixing is applied to the luminescence studies of GaAs/AlGaAs quantum wire (QWR) samples [15]. The GaAs/AlGaAs quantum wire structures grown on patterned GaAs substrates are shown on the micrographs in Figure 23 for (a) the reference sample and (b) after intermixing. At the bottom of the V grooves, quantum wires are formed due to self-ordering growth process [57,58]. Figure 24 shows the PL spectra of these

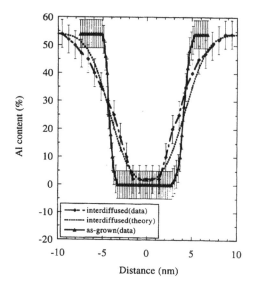

Fig. 22. Aluminium profiles of two anodized (at 40 mA/cm² for 4 mins) quantum well samples; one as-grown and the other annealed at 950°C, 105 s. The calculated profile is also shown for comparison.

QWR structures. As-grown samples showed a broad peak corresponding to emission from the side wall quantum wells (QWL). This is mainly due to the large volume of side wall quantum wells with respect to that of the quantum wires in these structures. However, upon annealing of these quantum wire structures with anodic oxides on the surface, the side wall quantum wells and the quantum wires are blue shifted. Intermixing of side wall quantum wells enhances the lateral potential confinement for the quantum wires and hence the PL from the quantum wire regions is also enhanced.

4   CONCLUSIONS

The results from this chapter show that very large energy shift (intermixing), up to ~ 200 meV, may be achieved with proton implantation. Although higher implantation doses are required, good recovery in the optical properties may be achieved by standard annealing conditions. The use of heavier ions, such as arsenic, can also achieve similar results but at significantly lower doses. However, with increasing dose, heavier ions may inhibit the interdiffusion process due to the formation of more stable clusters and extended defects which consume point defects that are required for

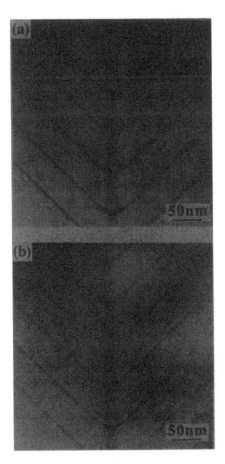

Fig. 23. Cross-sectional TEM micrographs for multiple GaAs-AlGaAs quantum wire structures grown on non-planar substrate. The reference (as-grown) sample is shown in (a) and the intermixed sample in (b).

intermixing. Although oxygen is a good candidate for electrical isolation, it should be avoided for intermixing applications, where oxygen-related complexes may be easily formed during bombardment and/or annealing which then retards intermixing.

Proton irradiation with subsequent rapid thermal annealing is shown to blue shift the lasing wavelengths of GaAs quantum well GRINSCH lasers. Up to 40 nm (75 meV) shift may be obtained for devices implanted to a dose of 1.5×10^{16} cm^{-2}, but below a threshold dose of 5×10^{14} cm^{-2}, no significant wavelength shift is observed.

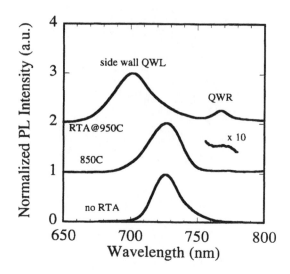

Fig. 24. PL spectra from the V-grooved quantum wire structure : anodized but before RTA (lower curve), anodized and annealed at 850°C for 30 s (middle curve), and anodized and annealed at 950°C for 30 s (upper curve).

Anodic oxide is shown to blue shift photoluminescence emission peaks in GaAs/AlGaAs quantum wells during rapid thermal annealing due to enhanced intermixing. Anodic oxide induced intermixing has been used to enhance photoluminescence from the quantum wire regions of structures grown on patterned substrates.

These techniques, ion implantation and anodic oxidation may be very promising methods of integrating lasers of different wavelengths for wavelength-division-multiplexing applications and could be readily extended to the fabrication of photonic integrated circuits.

Acknowledgements

This research is partly supported by the Australian Research Council, Research Grants Council of Hong Kong, Australian Agency for International Development through IDP Education Australia under the Australia — China Institutional Links Program. The authors would like to thank Professors Rick Cohen, Peter Zory, E. Herbert Li, David Cockayne, Drs. Yong Kim, Mladen Petravic, Zou Jin, L.V. Dao and M.B. Johnston, P.T. Burke and M.Y.C. Chan for various discussions and measurements.

References

1. C. Vieu, *Defect and Diffusion Forum*, **119–120**, 127 (1995).
2. P.M. Petroff, *J. Vac. Sci. Technol.*, **14**, 973 (1977).
3. R.M. Fleming, D.B. McWhan, A.C. Gossard, W. Wiegmann and R.A. Logan, *J. Appl. Phys.*, **51**, 357 (1980).
4. W.D. Laidig, N. Holonyak, Jr., M.D. Camras, K. Hess, J.J. Coleman, P.D. Dapkus and J. Bardeen, *Appl. Phys. Lett.*, **38**, 776 (1981).
5. Y. Hirayama, Y. Horikoshi and H. Okamoto, *Jpn. J. Appl Phys.*, **23**, 1568 (1984).
6. D.G. Deppe and N. Holonyak, Jr., *J. Appl. Phys.*, **64**, R93 (1988).
7. J. Cibert, P.M. Petroff, D.J. Werder, S.J. Pearton, A.C. Gossard and J.H. English, *Appl. Phys. Lett.*, **49**, 223 (1986).
8. K.B. Kahen, D.L. Peterson, G. Rajeswaran and D.J. Lawrence, *Appl. Phys. Lett.*, **55**, 651 (1989).
9. J.A. Van Vechten, *J. Appl. Phys.*, **53**, 7082 (1982).
10. S. Mitra and J.P. Stark, *J. Mater. Sci.*, **26**, 6650 (1991).
11. S.K. Ghandhi, *VLSI Fabrication Principles*, 487, John Wiley & Sons, Inc., New York, (1994).
12. M.J. Grove, D.A. Hudson, P.S. Zory, R.J. Dalby, C.M. Harding and A. Rosenberg, *J. Appl. Phys.*, **76**, 587 (1994).
13. S. Yuan, G. Li, H.H. Tan, F. Karouta and C. Jagadish, *Proceedings of the IEEE/LEOS '96 Meeting*, vol. 2, 132 (1996).
14. S. Yuan, Y. Kim, C. Jagadish, P.T. Burke, M. Gal, J. Zou, D.Q. Cai, D.J.H. Cockayne and R. M. Cohen, *Appl. Phys. Lett.*, **70**, 1269 (1997).
15. Y.Kim, S. Yuan, R. Leon, C. Jagadish, M. Gal, M. Johnston, M. Phillips, M. Stevens Kalceff, J. Zou, and D. Cockayne, *J. Appl. Phvs.*, **80**, 5014 (1996).
16. R.M. Lammert, G.M. Smith, D.V. Forbes, M.L. Osowski, and J.J. Coleman, *Electron. Lett.*, **31**, 1070 (1995).
17. E.A. Dobsiz, B. Tell, H.G. Craighead, M.C. Tamargo, *J. Appl. Phys.*, **60**, 4150 (1986).
18. S.-T. Lee, G. Braunstein, P. Fellinger, K.B. Kahen and G. Rajeswaran, *Appl. Phys. Lett.*, **53**, 2531 (1988).
19. E.P. Zucker, A. Hashimoto, T. Fukunaga and N. Watanabe, *Appl. Phys. Lett.*, **54**, 564 (1989).
20. K.B. Kahen and G. Rajeswaran, *J. Appl. Phys.*, **66**, 545 (1989).
21. S. Chen, S.-T. Lee, G. Braunstein and T.Y. Tan, *Appl. Phys. Lett.*, **55**, 1194 (1989).
22. L.J. Guido, K.C. Hseih, N. Holonyak, Jr., R.W. Kaliski, V. Eu, M. Feng and R.D. Burnham, *J. Appl. Phys.*, **61**, 1329 (1987).
23. H. Leier, A. Forchel, G. Hörcher, J. Hommel, S. Bayer, H. Rothfritz, G. Weimann and W. Schlapp, *J. Appl. Phys.*, **67**, 1805 (1990).
24. E.G. Bithell, W.M. Stobbs, C. Phillips, R. Eccleston and R. Gwilliam, *J. Appl. Phys.*, **67**, 1279 (I 990).
25. B.L. Weiss, I.V. Bradley, N.J. Whitehead and J.S. Roberts, *J. Appl. Phys.*, **71**, 5715 (1992).
26. R. Kalish, L.C. Feldman, D.C. Jacobson, B.E. Weir, J.L. Merz, L.-Y. Kramer, K. Doughty, S. Stone and K.-K. Lau, *Nucl. Instrum. Methods*, **B80/81**, 729 (1993).
27. K.B. Kahen, D.L. Peterson and G. Rajeswaran, *J. Appl. Phys.*, **68**, 2087 (1990).
28. K. Kash, B. Tell, P. Grabble, E.A. Dobsiz, H.G. Craighead and M.C. Tamargo, *J. Appl. Phys.*, **63**, 190(1988).
29. B. Elman, E.S. Koteles, P. Melman and C.A. Armiento, *J. Appl. Phys.*, **66**, 2104 (1989).
30. P.J. Poole, P.G. Piva, M. Buchanan, G.C. Aers, A.P. Roth, M. Dion, Z.R. Wasilewski, E.S. Koteles, S. Charbonneau and J. Beauvais, *Semicond. Sci. Technol.*, **9**, 2134 (1994).
31. R.K. Kupka and Y. Chen, *J. Appl. Phys.*, **78**, 2355 (1995).
32. E.S. Koteles, B. Elman, P. Melman, J.Y. Chi and C.A. Armiento, *Opt. Quantum Electron.*, **23**, S779 (1991).
33. J.S. Williams and J.M Poate, eds., Ion Implantation and Beam Processing, Academic Press, Sydney(1984).
34. J.S. Williams, *Trans. Mater. Res. Soc. Jpn.*, **17**, 417 (1994).
35. G.F. Redinbo, H.G. Craighead and J.M. Hong, *J. Appl. Phys.*, **74**, 3099 (1993).
36. Y. Matsumoto and T. Tsuchiya, *J. Physical Soc. Jpn.*, **57**, 4403 (1988).

37. D.B. Holt, C.E. Norman, G. Saiviati, S. Franchi and A. Bosacchi, *Mater. Sci. Eng.*, **B9**, 285 (1991).
38. C.N. Yeh, L.E. McNeil, L.J. Blue and T. Daniels-Race, *J. Appl. Phys.*, **77**, 4541 (1995).
39. S. Charbonneau, P.J. Poole, P.G. Piva, G.C. Aers, E.S. Koteles, M. Fallahi, J.-J. He, J.P. McCaffrey, M. Buchanan, M. Dion, R.D. Goldberg and I.V. Mitchell, *J. Appl. Phys.*, **78**, 3697 (1995).
40. F.L. Vook, Defects in Semiconductors, *Inst. Phys. Conf. Ser.*, **16**, 60 (1972).
41. D.K. Sadana, *Nucl. Instrum. Methods*, **B7/8**, 375 (1985).
42. J.S. Williams, R.G. Elliman, S.T. Johnson, D.K. Sengupta and J.M. Zemanski, *Mater. Res. Soc. Symp. Proc.*, **144**, 355 (1989).
43. W. Wesch, *Nucl. Instrum. Methods*, **B68**, 342 (1992).
44. H.H. Tan, C. Jagadish, J.S. Williams, J. Zou, D.J.H. Cockayne and A. Sikorski, *J. Appl. Phys.*, **77**, 87(1995).
45. H.H. Tan, C. Jagadish, J.S. Williams, J. Zou and D.J.H. Cockayne, *J. Appl. Phys.*, **80**, 2691 (1996).
46. J.F. Ziegler, J.P. Biersack and U. Littmark, The Stopping and Range of Ions in Solids, vol., **1**, Pergaman Press, New York (1989).
47. M. Skorowski, Deep Centers in Semiconductors, S.T. Pantelides, ed., Gordon and Breach Publishers, Switzerland (1992).
48. S.J. Pearton, M.P. Iannuzzi, C.L. Roberts, C.L. Reynolds, Jr., and L. Peticolas, *Appl. Phys. Len.*, **52**, 395 (1988).
49. J.L. Klatt, R.S. Averback, D.V. Forbes and J.J. Coleman, *Appl. Phys. Lett.*, **63**, 976 (1993).
50. D.V. Forbes, J.J. Coleman, J.L. Klatt and R.S. Averback, *J. Appl. Phys.*, **77**, 3543 (1995).
51. T.F. Kuech, *Mater. Sci. Rept.*, **2**, 1 (1987).
52. G.B. Stringfellow, Organometallic Vapor-Phase Epitaxy: Theory and Practice, Academic Press, San Diego (1989).
53. P.J. Hughes, E.H. Li and B.L. Weiss, *J. Vac. Sci. Technol.*, **B13**, 2276 (1995).
54. C.D. Thurmond, G.P. Schwartz, G.W. Kammlott and B. Schwartz, *J. Electrochem. Soc.*, **127**, 1366 (1980).
55. J. H. Marsh, *Semicond. Sci. Technol.*, **8**, 1136 (1993).
56. P. Murarka, *Appl. Phys. Lett.*, **26**, 180 (1975).
57. Eli Kapon, *Optoelectronics: Devices and Technologies*, **8**, 429 (1993).
58. Eli Kapon, *Microelectronics Journal*, **26**, 881 (1995).

CHAPTER 9

Impurity-Free Vacancy Disordering of GaAs/AlGaAs Quantum Well Structures: Processing and Devices

J.H. MARSH AND A.C. BRYCE

1 THE IFVD TECHNIQUE

Quantum well intermixing (QWI) using the impurity free vacancy disordering (IFVD) technique is one of the simplest and most versatile ways of altering the bandgap of a quantum well (QW) structure after growth, particularly in GaAs/AlGaAs structures [1]. The technique involves depositing a SiO_2 cap onto the surface of the GaAs and annealing at temperatures of $900°C$ or higher. At these temperatures Ga has a very high diffusion coefficient in SiO_2, an effect which was has been reported as far back as 1957 [2], although the detailed mechanism for this diffusion is still unknown. Secondary ion mass spectroscopy (SIMS) profiles [1] of quantum well samples, which have an uppermost layer of GaAs, annealed with SiO_2 caps indicate, first of all, that there is minimal Si diffusion from the SiO_2 into the GaAs layer. Furthermore, there is a significant level of Ga detected in the SiO_2 cap after annealing with large Ga accumulation at the SiO_2 free surface. The outdiffusion of the Ga atoms into the SiO_2 during annealing generates group III vacancies, V_{III} in the GaAs, which subsequently diffuse through the structure. Due to the concentration gradient between Ga in the QW and Al in the barriers, the Ga vacancies promote the diffusion of Al into a buried GaAs QW and Ga into AlGaAs barriers, hence changing the QW bandgap to a higher energy by partially intermixing the QW. One of the attractions of this process is that no impurities are introduced in the process, only vacancies, unlike impurity induced disordering.

1.1   Silica as a Mask for Inducing Quantum Well Intermixing

1.1.1   Bandgap changes

Large increases in the exciton energy (over 200 meV) have been observed in single quantum well structures which were capped with SiO_2 and annealed at 950°C for 30 s [3]. Later work by Gontijo et al. [4] shows the variation of the bandgap shift with annealing temperature and time. The structure used in the study was a double quantum well laser; it was found that at an annealing temperature of 900°C the PL peak shifted 17 nm for a 15 s anneal. On increasing the annealing temperature to 950°C this shift increased to over 40 nm, the increase being a linear function of temperature. When annealing at 950°C the bandgap shift in these structures was about 25 nm for a 15 s anneal increasing, more or less linearly, to over 60 nm when the annealing time was increased to 60 s. Similar results from a similar structure were reported by Hofstetter et al. [5].

Studies of the IFVD effect in intrinsic, p- and n-AlGaAs samples have been carried out [6]. The samples were annealed, using an RTP, at a temperature of 925°C for times between 15 and 90 s. The time dependence of the bandgap shift (relative to the as-grown material) reported is shown in Figure 1. It can be seen that both the intrinsic and n-AlGaAs samples shifted by about the same amount, but p-AlGaAs material exhibited lower degrees of bandgap shifts under the same RTP conditions.

The IFVD process involves the diffusion of V_{III} generated by the SiO_2 cap. V_{III} is suppressed and I_{III}, interstitial Ga or Al, is favoured in p-type material, therefore, smaller bandgap shifts were obtained. The intrinsic and n-AlGaAs samples both support V_{III} generated by the SiO_2 layer, hence larger intermixing degrees were observed. This effect may be attributed to the crystal Fermi level effect [7], through which the equilibrium Ga vacancy concentration is larger in n-type material than in p-type material due to a

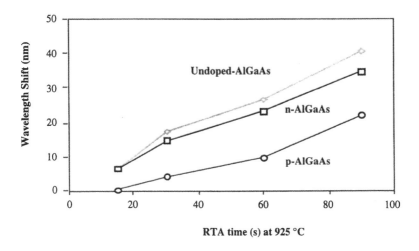

Fig. 1. The wavelength shifts as a function of annealing time obtained from SiO$_2$ capped intrinsic, n- and p-AlGaAs samples annealed at 925°C.

reduction in the formation energy of group III vacancies. In the n-i-p sample, the n doped contact and top cladding layers support Ga vacancies generated by the SiO$_2$ layer, hence larger degrees of intermixing were observed. From the integrated photonic devices point of view, the above results suggest that the growth of n-i-p structures would give a higher degree of intermixing than conventional p-i-n structures.

Complications may occur due to chemical reactions between the dielectric cap and the semiconductor. This is a particularly serious problems with Al-containing semiconductor alloys, such as AlGaAs, because Al is a very effective reducing agent. So when the SiO$_2$ is in direct contact with AlGaAs, the O and the Al react to form Al$_2$O$_3$ freeing the Si so that it can diffuse through the semiconductor and cause impurity induced disordering. The presence of Si in the semiconductor also dopes it n-type The O-Al reaction can be suppressed by the formation of a native oxide on the AlGaAs surface; and such a process has been used in the fabrication of lasers [8]. However, the usefulness of this technique is limited by the doping introduced by the process.

1.1.2 Refractive index changes

Systematic studies of the effect of disordering on the refractive index have not been reported for IFVD, but have been reported for implantation induced disordering by Hansen et al. [9]. The MQW structure, shown in Figure 2,

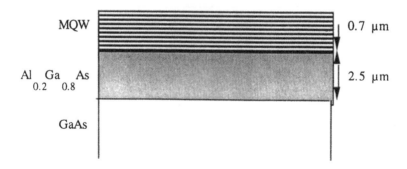

Fig. 2. The MQW waveguide structure used for refractive index measurements.

was designed to be suitable for waveguide studies. An MQW high-index region consisting of 58 periods of 60 Å GaAs wells and 60 Å $Al_{0.26}Ga_{0.74}As$ barriers was grown as the top part of the structure. Optical confinement was provided by air above the core region and by a thick $Al_{0.20}Ga_{0.80}As$ lower cladding layer. The structure formed a single vertical mode waveguide in the wavelength range of interest. The detailed parameters of the structure were chosen to ensure that only the lowest order depth mode would propagate in the waveguide experiments. The absence of an upper cladding layer ensured a high optical field at the semiconductor-air interface for output coupling purposes.

Samples were uniformly implanted throughout the depth of the MQW layer with boron or fluorine ions and were capped with a 1200 Å thick layer of plasma-deposited SiO_2 prior to annealing. The thickness of the capping layer was designed to give a reasonable output coupling efficiency when used in the fabrication of a grating output coupler, as well as to give added protection against arsenic desorption from the material. Annealing conditions used a temperature of 890°C with times of 90 minutes for the fluorine implanted sample and 120 minutes for the boron implanted sample. Photoluminescence measurements showed an energy shift of 28 meV and 40 meV for boron and fluorine, respectively, in the quantum well excitonic peak after annealing. This shift is consistent with previous results and is a good measure of the degree of disordering. Complete disordering of these particular samples would result in an increase in the bandgap energy of 90 meV.

Output grating couplers were then fabricated in the SiO_2 capping layer present on top of the slab waveguides. The grating pattern was produced in photoresist by laser holography and transferred by reactive ion etching to the SiO_2 film. The complete process is described in detail elsewhere [10]. The grating pitch of 285 nm was designed to give output coupling angles in the

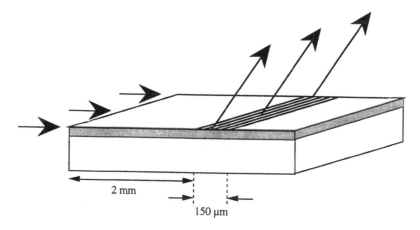

Fig. 3. The structure of the device used for the measurement of the refractive indices of waveguides.

region of 10° to 45° with respect to the normal to the sample. Conventional optical lithography was then used to remove the grating from the samples except in stripes 150 μm wide parallel to the grating. The samples were then cleaved parallel to the stripes to give coupler devices of 3 mm in length, with a 150 μm long grating region and a 2 mm long waveguide region before the grating to strip off leaky modes. The device is shown in Figure 3.

Waveguide index measurements were performed in the range 820 to 920 nm using a titanium:sapphire solid state laser. End-fire coupling was used to excite the waveguide mode(s) in the slab waveguide coupler device and the output coupling angle from the grating was measured. The polarisation dependence of the refractive index was investigated by varying the polarisation of the input laser beam. The modal refractive index of the slab waveguide was then found from the simple relation [11]:

$$n_g = \sin\theta + m\frac{\lambda}{\Lambda} \tag{4}$$

where n_g is the effective index of the waveguide, θ is the output coupling angle, λ is the free space wavelength of the laser light, Λ is the pitch of the grating. These results were used to find the material index of the MQW core using an iterative slab index procedure, while a Lorentzian oscillator model for 2-D excitons [12] provided a theoretical fit. The basic form of this model has previously been used for the modelling of MQW waveguide refractive indices:

$$\varepsilon_{MQW}(\omega) = \varepsilon_g(\omega) + 4\pi\beta_{l,h}\frac{\omega_{l,h}^2}{(\omega_{l,h}^2 - \omega^2 - i\omega\Lambda_{l,h})} \tag{5}$$

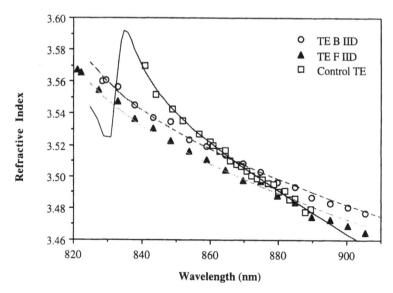

Fig. 4. The refractive index of the MQW layer before and after disordering for TE polarised light.

In the above equation, ε_g represents the background dielectric constant, containing contributions from all interactions except the exciton(s) in question. $\beta_{l,h}$ is the oscillator strength of the exciton transition, $\omega_{l,h}$ is the exciton centre frequency, $\Gamma_{l,h}$ is the linewidth of the exciton, and $\varepsilon_{MQW}(\omega)$ is the dielectric constant for the MQW material. To model ε_g, a semi-empirical Sellmeier equation was employed to calculate the dielectric constant of $Al_xGa_{1-x}As$ [13].

The refractive index of the MQW layer before and after disordering is shown in Figures 4 and 5. The range of results was limited by the tuning range of the laser in the case of the disordered material, whilst in the case of the control sample it was also limited by the absorption edge of the material at shorter wavelengths. At 840 nm, the limiting wavelength for TE measurements on the control sample, the TE refractive index change is 0.027 and 0.036, and for the TM polarisation at the corresponding wavelength of 835 nm, 0.014 and 0.021 for boron and fluorine respectively. At longer wavelengths, the disordered samples are observed to have a higher refractive index than that of the starting material, this being particularly evident in the case of boron. The TM crossover occurs at 850.1 nm for boron and at 861.6 nm for fluorine, with TE crossovers at the longer wavelengths of 864.5 nm and 885.9 nm for boron and fluorine respectively. An increase in refractive index at longer wavelengths is expected, since the background dielectric

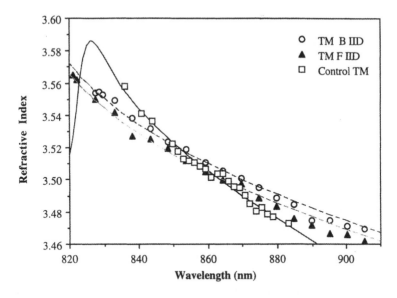

Fig. 5.   The refractive index of the MQW layer before and after disordering for TM polarised light.

contribution increases with disordering whilst the excitonic contribution decreases, making the disordered MQW waveguide appear to behave, in the limit of total disordering, as a compound semiconductor of $Al_xGa_{1-x}As$ composition with the Al-fraction equal to the material average of the MQW region. Despite being more disordered, the refractive index of the fluorine disordered material remains below that of the boron disordered material at longer wavelengths.

1.2   Masks for Inhibiting QW Intermixing

For the technique to be useful in the fabrication of monolithic photonic integrated circuits, it is necessary that it can be used to control the bandgap of the QWs across a wafer to allow the fabrication of integrated lasers, modulators, and low-loss waveguides. Also, in order to prevent As desorption at the high annealing temperatures necessary for IFVD, it is necessary to cap the areas which are not to be intermixed. The desorption of As not only causes degradation of the material quality but, also, leads to intermixing due to the generation of vacancies on the group V sublattice which cross over onto the group III sublattice. Due to these two considerations it is necessary to find a cap which will protect the surface and prevent intermixing.

1.2.1 Silicon nitride

Early studies made use of Si_3N_4 [14] as a protective cap, however such films were found to have poor reproducibility in the suppression of intermixing. Furthermore, high purity Si_3N_4 films are difficult to obtain because of the systematic incorporation of O_2 in the film, resulting in SiO_xN_y, which can be an effective cap for inducing Ga out-diffusion [14].

1.2.2 Strontium fluoride

Strontium fluoride was first reported to suppress intermixing by Beauvais et al. [3]. Samples of GaAs/AlGaAs MQW that had been capped with SrF_2, SiO_2 and Si_3N_4, and some that were uncapped were annealed at 950°C for 15 s. The thicknesses of the caps were 185, 90 and 135 nm respectively. The PL signals from the different samples after annealing were compared to that from an as-grown sample. As expected, large shifts were observed in the SiO_2 (87 meV) capped samples and also in the Si_3N_4 (77 meV) capped samples. The SrF_2 capped samples exhibited the smallest shift of 7 meV, however there was some surface damage. Using a dual cap of 185 nm of SrF_2 and 150 nm of SiO_2 overcame the problem of surface damage and the bandgap shifts were still small (4 meV). There was no broadening or reduction in intensity of the PL signal from the SrF_2 and dual capped samples when compared to the signal from the as-grown sample.

However, it has been found that there can be surface damage to samples with patterned areas of SrF_2 after the thermal processing [15], particularly when the devices being fabricated require large areas of SrF_2. It is thought that the damage is due to strain induced in the SrF_2, SiO_2 film and the GaAs substrate, because of the large differences in the coefficients of expansion between these three compounds which lead to stresses during the annealing step. Despite this, SrF_2 has been used to produce simple photonic integrated circuits, which are discussed later in this chapter.

The use of other fluoride compounds as annealing caps has been investigated, namely BaF_2, CaF_2 and MgF_2 [4]. It was found that these caps also suppressed intermixing, however it proved difficult to remove them without damaging the surface of the sample which put them at a disadvantage compared to SrF_2.

1.2.3 Doped silica layers

Initially, the use of a SiO_2 layer doped with phosphorus (SiO_2:P) was reported by Rao et al. [16] to be an 'universal' intermixing source for III-V compounds. SiO_2:P with 1% by weight of P [17] was used to induce intermixing in nominally undoped GaAs/AlGaAs shallow multiple QW structures by furnace annealing at a relatively low temperature, (850°C), and therefore a masking dielectric cap to prevent intermixing and for surface protection was not required during annealing.

More recently, it has been reported that SiO_2 doped with 5% P by weight could be used as a cap to suppress intermixing in laser structures [18]. As can be seen from Figure 6, the SiO_2:P suppresses the intermixing at temperatures up to 950°C whilst the sample with the SiO_2 cap shows bandgap shifts up to 40 nm. Also shown are the shifts of an uncapped sample that are caused by the desorption of As, as discussed above.

There are at least two possible explanations for the masking properties of SiO_2:P films. Firstly, it is well known that SiO_2:P films are more dense and void-free than SiO_2 [19]. Films of SiO_2:P with a weight ratio P_2O_5/SiO_2 of 4% have been used [20] as a capping material for open-tube thermal activation of Si implants in GaAs, and it has been found that the diffusion coefficient of implanted Si in semi-insulating GaAs is about one order of magnitude smaller for SiO_2:P than for SiO_2. This was attributed to the presence of a reduced number of group III vacancies due to less Ga out-diffusion. Secondly, due to the difference in thermal expansion coefficients at the annealing temperature, a strain effect exists at the interface between GaAs and SiO_2 during the annealing stage. The thermal expansion coefficient of GaAs is about ten times larger than SiO_2 and, as a consequence, the SiO_2 film is under tensile strain and the GaAs surface layer is under compressive strain. Under this condition, because of the high diffusion coefficient of Ga in SiO_2, the out-diffusion of Ga atoms into the SiO_2 film is an energetically favourable process because it minimises the strain in the system. The addition of P into the SiO_2 film leads to an increase in the thermal expansion coefficient and a decrease in the glass softening temperature [19]. Therefore, less compressive strain will be induced during the annealing step in the GaAs surface layer and, as a result, the number of Ga vacancies will be reduced due to less Ga out-diffusion.

The results of Cusumano [18] do not necessarily conflict with those reported by Rao et al. [16] because the addition of small amounts of P (1% wt) does not drastically change [21] the properties of SiO_2:P as compared with SiO_2. Hence both SiO_2 and SiO_2:P 1% could have a promoting effect on intermixing while SiO_2:P 5% suppresses intermixing. Also to be considered, however, is that the structure used by Rao et al. had only a very thin layer of GaAs (8 nm) between the dielectric cap and the AlGaAs. It has been shown that such a thin layer can allow Al to diffuse into the SiO_2 in high enough concentrations as to cause a reaction between the O in the cap and the Al. This reaction leads to the creation of free Si and O which then diffuse through the epitaxial layers, resulting in intermixing through impurity induced disordering [22].

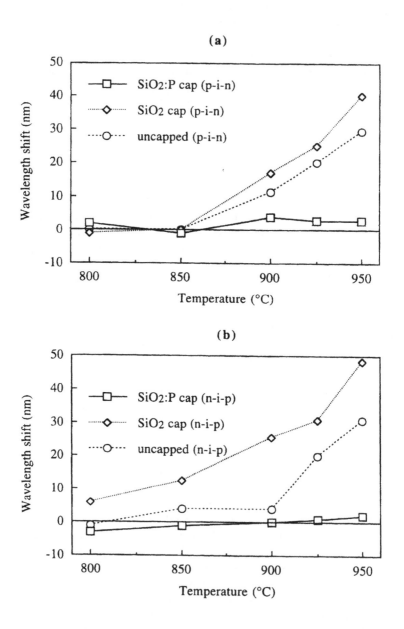

Fig. 6. Wavelength shifts of the 77 K PL peak for SiO_2 capped, SiO_2:P capped and uncapped (a) *p-i-n* and (b) *n-i-p* samples as a function of annealing temperature with a duration of 60 s.

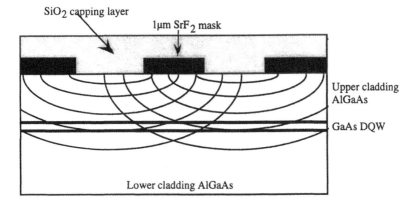

Fig. 7. Schematic representation of vacancy diffusion profile in SrF_2 masked QW material (not to scale). The high mobility of the vacancies leads to a uniform distribution at the DQW for patterns of a smaller dimension that the diffusion length (3 μm).

1.2.4 The 'Selective Intermixing in Selected Areas (SISA)' Technique

The selective intermixing in selected areas technique [23] allows different bandgap shifts to be achieved over a sample in a controllable manner using a single annealing step. The effect is achieved by patterning the semiconductor, using electron beam lithography, with submicron to 1 μm sized features of SrF_2 to act as a bandgap control mask, followed by deposition of SiO_2 over the sample to act as an intermixing source. The SrF_2 mask pattern has to have dimensions smaller than, or comparable with, the diffusion length of the point defects to allow uniform intermixing at the quantum well depth by overlapping of the vacancy diffusion fronts (Figure 7). From TEM work on MQW structures with a similar composition, this diffusion length was found to be 3 μm for 1 μm deep quantum wells, indicating a large diffusion constant and high mobility for the vacancies. As a result, spatial control of the bandgap shift can be achieved using a single rapid thermal annealing (RTA) step. The degree of intermixing is dependent upon the area of sample in direct contact with the SiO_2 layer. This technique is a simple, one-step process and is reproducible.

The SISA technique was applied to a DQW laser structure [24]. Six fields with size of 2×2 mm^2 each containing 1 μm $\times$ 2 mm stripes of SrF_2 were written onto e-beam resist on top of the sample. The SrF_2 was thermally evaporated onto the sample followed by lift-off. A layer of 200 nm SiO_2 was deposited by PECVD after the SrF_2 lift-off, and annealing was performed using an RTA at 925°C for 30 s.

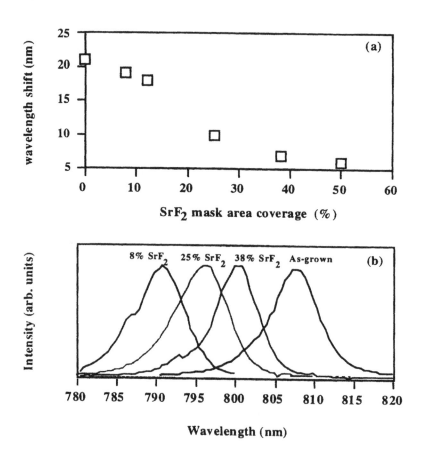

Fig. 8. (a) Wavelength shift and (b) PL spectra observed from the DQW material disordered using SISA technique. The PL measurements were performed at 77 K.

Figure 8 shows the wavelength shift in the exciton peak from a quantum well laser structure processed using the SISA technique as measured from the 77 K photoluminescence. The SrF_2 area coverage studied varied between 8 and 50% (8%, 12%, 15%, 38%, 50%), which was achieved by varying the spacing between 1 μm stripes of SrF_2. A 100% SrF_2 region was also prepared and the 0% reference was taken from the area outside the written fields. Intermixing was inhibited in the 100% SrF_2 region (not shown in Figure 8(a)). It can be seen that, although the bandgap shift is not linear, the degree of bandgap shift is dependent on the area coverage of the SrF_2. It was also reported that the PL signals did not exhibit any significant broadening of their full wave at half maximum (FWHM) when compared to as-grown

PL and remained a single peak, indicating that the intermixing is even and homogeneous.

2   INTEGRATED DEVICES REALISED USING IFVD

2.1   Mutliwavelength Lasers

The SISA technique is important for the fabrication of devices such as those needed for WDM, in which a number of independent wavelengths can be transmitted simultaneously via a single optical fibre. Such components are key for high-capacity information networks.

For WDM applications the ability to produce lasers emitting at several wavelengths on the one chip is desirable. It has been demonstrated that this is possible using IFVD [5,24].

In [5], ridge waveguide Fabry-Perot lasers, with their emission wavelength selectively shifted by 20 nm were integrated with unshifted lasers on the same chip. This was achieved by patterning deposited SrF_2 on the sample from which the lasers were to be fabricated, then depositing SiO_2 over the sample. The sample was then annealed at 960°C for 24 s. Lasers were fabricated in both the shifted and unshifted regions and in as-grown material. The as-grown, unshifted (SrF_2 capped) and shifted (SiO_2 capped) devices exhibited peak spontaneous emission wavelengths of 826, 823 and 805 nm, respectively, i.e. a wavelength shift of 3 nm in the area capped with SrF_2 and 21 nm in the area capped with SiO_2 during annealing.

The light-current and current-voltage characteristics of the devices were measured. It was found that the as-grown lasers had threshold currents of 12 mA and a slope efficiency of 0.52W/A. In comparison the SiO_2 capped lasers had lower threshold currents of 10 mA, whilst the SrF_2 capped devices had somewhat higher threshold currents of 14 mA. The slope efficiencies of the SiO_2 capped lasers (0.42 W/A) and the SrF_2 capped lasers (0.36 W/A) were lower than that of the as-grown lasers. The I-V characteristics of the as-grown and SiO_2 capped lasers were comparable, with voltages at the lasing threshold of 2.1 V and 2.15 V and series resistances of 30 and 24 Ω respectively. The higher threshold voltage of 3.3 V and series resistance of 37 Ω of the SrF_2 capped lasers were attributed to the exposure of the contact region to three plasma processes during fabrication, the other two sets of lasers having experienced only one plasma exposure. The higher threshold voltage and series resistance of the SrF_2 capped lasers may explain the higher threshold current and reduced slope efficiency.

The fabrication of lasers with 5 different wavelengths on a single chip was achieved using the SISA technique [24]. A sample was patterned with different coverage fractions of SrF_2, 0, 15, 25, 50 and 100%, then covered

TABLE 1
Summary of the laser characteristics from samples with different degrees of intermixing.

SrF$_2$ coverage (%)	Threshold current density (for infinite length) (A cm^{-2})	Transparency current (A cm^{-2})	Internal quantum efficiency (%)	Loss (cm^{-1})
(as-grown)	218	62	71	14.7
0	227	62	68	14.5
15	235	64	68	14.5
25	251	68	66	14.0
50	217	60	64	12.1
100	237	63	65	12.0

with SiO$_2$ and annealed. After annealing the SrF$_2$/SiO$_2$ was removed and a new layer of 200 nm SiO$_2$ was deposited. 75 μm wide oxide stripe lasers were then fabricated using standard photolithographic techniques. The lasers were then cleaved into various lengths and pulse tested.

Lasers fabricated on a single chip using these SISA patterns were observed to operate at five well separated wavelengths (861 nm, 855 nm, 848 nm, 844 nm and 840 nm), whilst the lasers fabricated from the as-grown material showed a lasing wavelength of 863 nm. The light-current characteristics were measured, and the threshold current and the slope efficiency were analysed. A summary of the laser characteristics from samples intermixed to different degrees is tabulated in Table 1. Only small changes in their infinite length threshold current density were observed, and no correlation could be found between the degree of intermixing and the current density. The internal quantum efficiency of the intermixed material is a few percent lower than the as-grown material. However, these changes are very small, which indicates that the material quality has not been deteriorated by the SISA technique. The smaller changes in the laser characteristics induced by the SISA process as compared to the changes observed by Hofstetter et al. [5] along with the fact that the SISA lasers had been subjected to fewer plasma processes suggests that the plasma processes used for the fabrication of the laser can cause degradation of laser performance. Since the lasers produced by SISA showed only small differences in their lasing parameters, this technique is potentially very useful in the production of WDM components.

2.2   Four-channel Waveguide Photodetectors

Two wavelength demultiplexers have been demonstrated using IFVD and other QWI related techniques [24–26]. In this section we describe the use of the SISA technique in the fabrication of four-channel waveguide

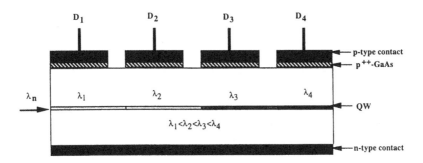

Fig. 9. Schematic cross-section of the four-channel demultiplexing waveguide photodetector. The length of an individual diode is 300 μm and the spacing between diodes is 100 μm.

photodetectors. The device consists of four diodes, with their bandgaps shifted by different degrees. The intermixed regions serve both as absorbing layers for shorter wavelength light and as passive waveguides for the long wavelength light to pass into the diodes disordered to lesser degrees. The schematic diagram of this device is given in Figure 9.

Four different SrF$_2$ coverage fractions, 0, 25, 50, and 100% were used. The spacing between the diodes was 100 μm and the diode length was 300 μm, which has to be made long enough to absorb the optical signal. The p-type contacts were defined using lift-off and the top p$^+$-GaAs layer between the electrodes was removed to improve electrical isolation.

Photocurrent measurements were carried out using an argon-ion pumped tuneable Ti:sapphire laser that was end-fire coupled into the cleaved edge of the photodetector which had been intermixed to the largest degree. The photocurrent spectra measured from the individual diodes suggest that the optimum operating wavelengths for this device to be 832 nm, 843 nm, 855 nm and 868 nm for diodes which were processed with 0, 25, 50 and 100% SrF$_2$ coverage, respectively.

The spontaneous emission spectra under pulsed current conditions were also measured from these each of these diodes. Lasing, or stimulated emission, was observed from the diode in the 100% SrF$_2$ coverage region at injected current of around 500 mA, whilst only spontaneous emission was observed from the other three diodes. This is due to the fact that, for the longest wavelength diode, the device acts as a broad area extended cavity laser. In addition, this diode absorbs shorter wavelength emission and suppresses the lasing in other shorter wavelength diodes. The emission spectra were well spread, peaking at 828 nm, 837 nm, 847 nm, and 861 nm. The peak positions were located at shorter wavelengths than the photocurrent spectra because the diodes were reverse biased during the photocurrent measurements.

2.3   Integrated Modulators/Waveguides

Electroabsorption modulators integrated with passive waveguides have been fabricated using the IFVD technique [27]. The structure used was a double quantum well laser. The areas to form the modulators were capped with SrF_2 and the rest of the sample was capped with SiO_2. On annealing at 900°C for 60 s, the quantum wells in the area under the SiO_2 had blue shifted by 12 nm, while the PL peak wavelength of the quantum wells under the SrF_2 had not changed. The losses in the passive waveguide, fabricated in the intermixed areas, were measured to be 28 dB/cm at a wavelength of 858 nm, close to the band edge of the modulator sections. On measuring the losses in the modulators at this wavelength it was found that, on intermixing, a reduction of the propagation losses by a factor of 8 had been achieved. Measurements showed that, for a modulator length of 400 μm at the wavelength of 861.6 nm, an ON/OFF ratio of more than 35 dB was achieved with a reverse voltage of 3 V, but also at 854.3 nm, where the losses in the ON state (no voltage applied) increased by $\sim$ 10 dB, an ON/OFF ratio of 39 dB was obtained with only 1.5 V reverse bias. No significant degradation of the QW structure after the IFVD process in the areas coated with SrF_2 was observed. The authors point out that the losses in the passive sections are mainly due to the proximity to the band edge and can be substantially reduced by increasing the amount of disordering in the passive waveguides.

2.4   Extended Cavity CW Lasers

The first extended cavity lasers fabricated using IFVD reported were broad area devices which used SrF_2 as the cap to suppress intermixing in the active areas of the devices [15,28]. In the first of these reports the active section was gain guided and the passive section was a slab waveguide. The losses measured in the passive section of these devices were 17 dB/cm, these losses being measured using subthreshold electrical pumping of the devices and comparing the output power from both facets [29,30]. The optical power output from the extended cavity facet was lower than the output from the facet at the active side due to waveguide losses in the passive section. The later devices were fabricated as very wide (45 μm) ridge waveguides and the losses in the passive section were reduced to 10 dB/cm. This apparent reduction was attributed to the better confinement of the light provided by the multimode rib waveguide structure used for these lasers, making the loss measurement more accurate.

More recently, single mode ridge waveguide extended cavity lasers have been fabricated using P:SiO_2 to suppress the intermixing in the active region [31]. A differential shift of 30 nm between the passive and active sections was obtained after RTA at 940°C for 60 s. Both discrete and extended cavity lasers were cleaved and characterised in CW operation.

Light-current characteristics and related spectra were measured for a discrete 400 μm long laser and for an integrated device with 400 μm/2.73 mm long active/passive sections, reflecting the typical behaviour of the fabricated devices. The extended cavity lasers showed equal power from both active and passive ends with emission wavelengths between 853 and 855 nm. The average threshold current of a batch of discrete lasers was 6.7±0.5 mA and for a batch of integrated devices was 9.0 ± 1.0 mA. Due to the losses introduced by the integrated waveguide, the mean threshold current increases by 33%, from 6.8 mA for the discrete laser to 9 mA for the extended cavity laser, and the slope efficiency decreases by 40%, from 0.32 W/A per facet (measured into an NA of 0.65) to 0.19 W/A per facet. The losses in similar bandgap widened passive waveguides have been measured directly as a function of the wavelength and for TE polarisation, using the Fabry-Perot resonance method, giving a propagation loss of about 1.9 cm^{-1} throughout the wavelength range from 850 nm to 900 nm [32]. Below 850 nm, the loss increases due to resonant absorption in the partially intermixed QW.

Assuming a logarithmic dependence [33] of the gain versus current density,

$$g = g_0 \ln \left(\frac{\eta_i J}{n J_T} \right),$$
(1)

and from the balance between gain and losses at threshold, the following formulae are obtained:

$$\frac{I_{ex}}{I_{nor}} = \exp \left(\frac{\alpha_p L_p}{n \Gamma g_0 L_a} \right)$$
(2)

$$\frac{\eta_{ex}}{\eta_{nor}} = \frac{g_{nor}^{th}}{g_{ex}^{th}} = \frac{\alpha_a + \frac{1}{L_a} \ln \frac{1}{R}}{\alpha_a + \frac{1}{L_a} \ln \frac{1}{R} + \alpha_p \frac{L_p}{L_a}}$$
(3)

where α_a, L_a and α_p, L_p are the losses and length, respectively, of the active and passive sections, Γ is the optical overlap factor per well, g_0 the QW gain constant, J_T the transparency current density, n the number of wells and η_i is the internal quantum efficiency.

In deriving (2) and (3), coupling losses between the active and passive waveguide have been assumed to be zero because no mode mismatch due to refractive index change is introduced by the IFVD process. The change in refractive index produced by intermixing the QWs will be at most 3%, and the optical overlap of the guided mode with the QWs [9] is 2.75% per well. The effective refractive index step will therefore be $\sim 10^{-3}$, giving an interface reflection coefficient of only $\sim 10^{-6}$.

In order to determine the material parameters, broad area lasers of width 75 μm and with cavity lengths in the range 400 to 1000 μm were fabricated from the same wafer as the extended cavity devices. From the usual plots of reciprocal of the external slope efficiency against cavity length and log of threshold current against reciprocal cavity length, the following material parameters were deduced: $\alpha_a = 5.2\,\mathrm{cm^{-1}}$, $g_0 = 865\,\mathrm{cm^{-1}}$, $J_T = 68\,\mathrm{A\,cm^{-2}}$ and $\eta_i = 70\%$, the values of g_0 and J_T being in close agreement with theoretical values for 100 Å QWs. The threshold current density for infinite cavity length was 216 A cm^{-2}. The efficiency of the ridge devices is reduced because of current spreading.

From (2) we obtain a value for α_p of 2 cm^{-1}, in close agreement with the linear loss measurement of 1.9 cm^{-1} discussed above [32]. Assuming a modal reflectivity R of 0.32, the mirror losses are 28.5 cm^{-1} and the losses introduced by the integrated waveguide amount to 13.6 cm^{-1} due to the quite high ratio L_p/L_a of 6.87. From (3), we then predict a drop in the slope efficiency due to the integrated waveguide of 30%, which is reasonably close to the observed drop of 40% given the approximate nature of equation (1).

Transparency could be also achieved by current injection in a non-intermixed QW but this, for long integrated devices, would require a second contact, electrically insulated from the active region, and the subsequent increase in the total injected current not always is compatible with CW operation. The behaviour of a long all-active laser (AAL) is compared with an ECL in Section 2.5, and it is seen that the threshold current is a factor of around 5 larger for the AAL.

Losses in the integrated waveguides are limited mainly by p-type free carrier absorption from the grown-in dopants with a possible contribution from acceptor-like Ga vacancies [34] that have diffused into active region. A simple way to reduce the losses could be by tailoring the doping profiles together with the structural parameters of the laser structure [35].

The ability to produce integrated low loss passive sections within the laser cavity has allowed the fabrication of lasers with a non-absorbing, passive DBR grating [36,37]. The lasers fabricated were single mode ridge waveguide devices, both emitting at a wavelength around 850 nm. Comparison of the laser performance with that of Fabry Perot lasers fabricated from the same epilayer by Hofstetter et al. [37] showed an increase in the threshold current, from 14 mA to 25 mA, and a decrease in the slope efficiency, from 0.36 W/A to 0.2 W/A. Lasers with a grating section around 100 μm long showed improvement of the side mode suppression ratio (SMSR), from 20 dB to 25 dB. Lasers with a longer grating section, 1000 μm had a higher SMSR of 31 dB. Hofstetter et al. measured a tuning coefficient of 0.05 nm/°C and observed no mode hops as the temperature was varied from 7 to 25°C. Both sets of DBR devices exhibited high threshold current densities which both sets of authors attributed to mismatch between the Bragg peak of the grating and the laser gain peak.

2.5   Extended Cavity Mode-locked Lasers

The IFVD process has been used to fabricate ECLs for use as mode-locked (ML) sources; these devices have been characterised and their performance compared to that of mode-locked (ML) all active lasers (AAL) [38]. The QW intermixing was carried out using SiO_2 and $P:SiO_2$ with the annealing conditions being 925°C for 60 s. 77 K PL measurements showed that the intermixed regions, had been blue shifted by 33 nm with respect to the as-grown material, and the active areas, under the $P:SiO_2$, had shifted by 6 nm. The differential shift between active and passive regions was, therefore, 27 nm. Single mode ridge waveguide lasers were fabricated and the p-type contact to the active section of the device was split, using lift off, dividing the active region into different sections, namely a gain section and a saturable absorber section. The contacts were separated by a 10 μm gap, which gave an isolation resistance of 5–6 kΩ when the highly p-doped GaAs contact layer was removed from between them. The lasers were then cleaved. A 5 mm long AAL and a 4 mm long ECL were studied, the former working at around 8 GHz and the latter at around 10 GHz. The AAL devices had a 4910 μm long gain section and an 80 μm long saturable absorber and the ECLs had a 440 μm long gain section, a 50 μm long saturable absorber and a 3500 μm long extended cavity for the ECL, as shown in Figure 10.

Measurement of the light-current characteristics with the saturable absorber floating showed that the threshold current for the ECL was 32 mA, less than a third of that of the AAL which was 105 mA. A greater improvement in the threshold current was measured when the saturable absorber was short circuited with the gain section. The threshold current for the ECL, with both sections forward biased, was reduced to 18 mA, while the AAL threshold current was only reduced to 100 mA, more than a factor of five larger than that of the ECL.

Apart from the improvement in threshold current, the ECL also demonstrated a major advantage compared to the AAL, in the absence of self-pulsating regimes accompanying mode locking. Depending on the negative bias applied to the saturable absorber and the current supplied to the gain section, the AAL operated in several distinct dynamic regimes, which included pure ML, pure self-pulsation, and ML combined with a deep self-pulsing envelope [39]. For all device conditions investigated, the ECL showed no self-pulsing modulation of the ML pulse train.

Figure 11 shows temporal measurements taken with a high speed streak camera together with autocorrelation traces. In Figure 11(a) and (c), 500 ps long streak camera profiles are depicted. Figure 11(a) and (b) measured from the 5 mm long AAL, show a ML pulse train at approximately 8 GHz with 10.3 ps pulses, while Figure 11(c) and (d), measured from the 4 mm long ECL, shows a ML pulse train at around 10 GHz with 3.5 ps pulses.

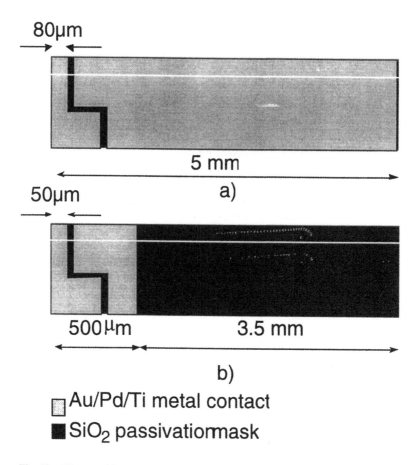

Fig. 10. Diagram of the a) All-active mode-locked semiconductor laser and b) Extended cavity mode-locked semiconductor laser.

The pulse lengths were measured using a two-photon absorption semiconductor waveguide autocorrelator [40], which is sufficiently sensitive to allow us to acquire autocorrelation measurements even under pulsed excitation. These measurements show that the ECL pulse width varied from 3.5 to 6.5 ps (FWHM assuming hyperbolic secant square shape), depending on the biasing conditions, while the AAL pulse width varied from 10.2 to 13.8 ps, again depending on the biasing conditions. The increase in the pulse width in the AAL is most probably due to gain dispersion [41]. Gain dispersion effects tend to broaden the pulses more effectively when the optical pulse propagates through long sections of active waveguide.

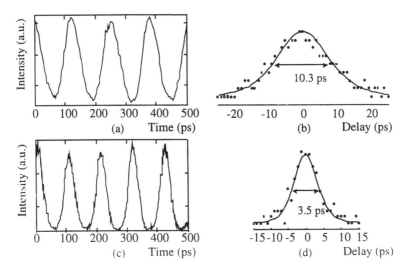

Fig. 11. Temporal measurements from the AAL (upper graph) and ECL (lower graph).

Fluctuations in mode-locked lasers include variations in both pulse intensity and pulse timing [42]. Carrier density fluctuations modulate the round trip time for the optical pulses inside the laser cavity, and cause jitter in ML devices. The root-mean-square (rms) timing jitter, under CW operation, was estimated by the frequency domain technique [43]. The AAL jitter (10 kHz–10 MHz) was found to be around 15 ps, while the ECL jitter (10 kHz–10 MHz) was just 6 ps. The ECL shows the expected reduction in jitter levels predicted by Derickson et al. [41]. The jitter levels are larger for all-active waveguide configurations than for extended cavity configurations due to the fact that, for similar carrier density levels in each active waveguide, the phase noise level will be larger in the AAL than in the ECL roughly by the ratio of the active waveguide lengths, the former being 5 mm long and the latter being 500 μm long.

Optical spectra were measured using an optical spectrum analyser with a resolution of 0.1 nm. Figure 12 shows spectra taken from the AAL and ECL when a) the saturable absorber was not biased and b) the devices were mode-locked. In both cases the optical spectra suffer a shift of more than 2 nm to longer wavelengths when the laser is mode-locked, they are very asymmetric and the spectral width increases. For the AAL, the spectral width increases from 0.4 nm at zero absorber bias to almost 2 nm when the saturable absorber is reverse biased and for the ECL from 0.2 nm to 2 nm. The optical spectrum of light pulses travelling in the laser cavity is distorted considerably during the amplification process if the refractive index

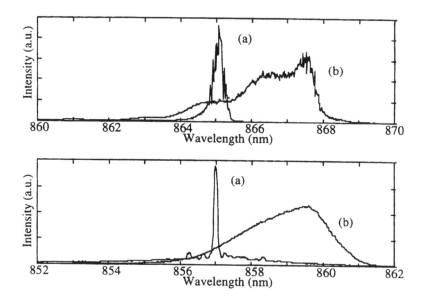

Fig. 12. Spectral measurements taken with a) the saturable absorber floating and b) the saturable absorber reversed biased from the AAL (upper graph) and ECL (lower graph).

becomes nonlinear, even when the pulse shape remains unchanged. Gain and absorption saturation produce a shift and distortion in the optical spectrum. The physical mechanism responsible for this shift and distortion is self-phase modulation (SPM) [44]. The time dependence of the saturated gain leads to a temporal modulation of the phase, i.e. the pulse modulates its own phase as a result of gain saturation. The multi-shouldered structure shown in the AAL optical spectrum has been observed previously in semiconductor laser amplifiers [44] and it was shown that SPM was responsible. The spectrum from the ECL is more symmetric than that from the AAL suggesting that the pulse chirp is more linear from the ECL, a very important factor since linearly chirped pulses lend themselves well to pulse compression techniques [41].

The time-bandwidth product for the AAL was calculated to be 7, while for the ECL it was just 2.5. Previously reported extended cavity mode-locked lasers, made using regrowth techniques [45], show similarly large time-bandwidth products to that of the AAL, around 20 times larger than the theoretical value, most probably due to reflections between the active and passive section. In our case, because of the negligible reflection at the interface active/passive waveguide of the ECL, the time-bandwidth product is reduced to 2.5, and the overall performance is, to best of our knowledge,

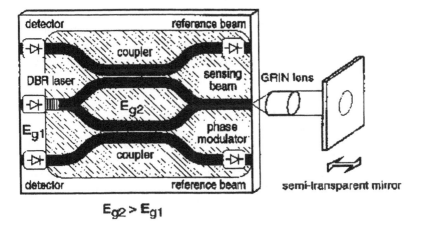

$$E_{g2} > E_{g1}$$

Fig. 13. Schematic of the monolithically integrated double Michelson displacement sensor with direction determination.

the best achieved so far with a monolithic Fabry-Perot diode laser at repetition frequencies around 10 GHz [38].

The performance of these extended cavity mode-locked lasers demonstrates that, not only can the IFVD technique be used to fabricate integrated devices, but also that these devices can show improved performance over similar devices fabricated using other integration techniques.

2.6 Fully Functional Photonic Integrated Circuits

A monolithically integrated optical displacement sensor fabricated in the GaAs/AlGaAs material system has been reported [46,47]. The device was fabricated on a single quantum well laser structure grown by MOCVD and is shown in Figure 13. The single chip consisted of a distributed Bragg reflector laser, two photodiodes, two phase modulators, two Y-couplers and two directional couplers. All of the chip, except the active section of the laser and the photodiodes, was bandgap widened by IFVD, SiO_2 being used to promote the intermixing and SrF_2 to suppress the intermixing. The bandgap difference between the intermixed and nonintermixed sections was 45 meV. The absorption loss measured in the intermixed section at the lasing wavelength was 35–45 dB/cm in 3 μm ridge waveguides. A 200 μm long grating for the DBR laser was fabricated in the intermixed section of the sample. It had a reflectivity of 95%; the other facet of the laser was cleaved. The emission wavelength was 822 nm and the spectrum showed a SMSR of approximately 25–30 dB. The modulator, also fabricated in the intermixed section of the device, was 340 μm long and a one-pass phase shift of π was measured when a reverse bias of 12 V was applied. The device was configured as a double Michelson interferometer and allowed the determination of both the magnitude, with resolution better than 100 nm, and the direction of displacement of a movable mirror using only one external element for beam collimation.

3 PLASMA PROCESS

It will be appreciated from the foregoing that, although IFVD is an elegant technique with many advantages, no one set of caps has proven to be universally suitable for all III-V material systems. While, for example, considerable success has been achieved for the GaAs-AlGaAs material system, the poor thermal stability of the InGaAs-InGaAsP system has made intermixing problematic, due largely to difficulties encountered in inhibiting the intermixing process. Also, as yet, there has been no demonstration of impurity-free QWI in the GaInP-AlGaInP system, which is widely used to fabricate visible laser diodes. Intermixing may prove highly valuable for this material system as it may enable the fabrication of non-absorbing mirrors (NAMs), providing a means for improving the output of high power visible lasers by increasing the threshold for catastrophic breakdown of the laser facets. Recent research [48] has led to the development of a new intermixing technique which has so far been used to intermix a number of different material systems, including GaAs-AlGaAs, GaInP-AlGaInP, InGaAs-InGaAsP and InGaAs-InGaAlAs multiple quantum well (MQW) structures and has in each case yielded very promising results.

The GaAs-AlGaAs and GaInP-AlGaInP (for brevity hereafter referred to as AlGaAs and AlGaInP respectively) structures studied were standard p-i-n MQW laser structures grown on GaAs substrates by metal-organic vapour phase epitaxy (MOVPE). Both structures included an n-type GaAs buffer layer doped with Si ($1 - 3 \times 10^{18}$ cm^{-3}), approximately 1 μm thick $Al_{0.42}Ga_{0.58}As$ or $(Al_{0.7}Ga_{0.3})_{0.5}In_{0.5}P$ n- and p-type cladding layers doped with Si ($6 - 8 \times 10^{17}$ cm^{-3}) and Zn ($2 - 4 \times 10^{17}$ cm^{-3}) respectively, an undoped active region, and a 25–100 nm p-type (1×10^{19} cm^{-3}) GaAs cap. The active region of the AlGaAs structure contained 2×100 Å GaAs QWs, centrally placed within a 230 nm undoped $Al_{0.2}Ga_{0.8}As$ waveguide region, separated by a 100 Å $Al_{0.2}Ga_{0.8}As$ barrier, while that of the AlGaInP structure comprised 2 strained 70 Å $Ga_{0.41}In_{0.59}P$ QWs within an undoped $(Al_{0.3}Ga_{0.7})_{0.5}In_{0.5}P$ waveguide core, separated by a 250 Å $(Al_{0.3}Ga_{0.7})_{0.5}In_{0.5}P$ barrier. The InGaAs-InGaAsP and InGaAs-InAlGaAs (hereafter referred to as InGaAsP and InAlGaAs) structures were both standard unstrained MQW laser structures grown on n-type InP substrates. The InGaAsP structure was grown by MOVPE and contained 5×65 Å QWs

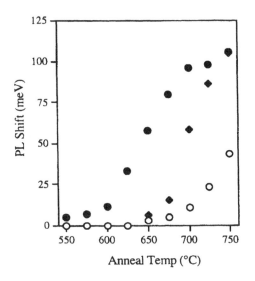

Fig. 14. PL peak energy shifts as a function of anneal temperature for InGaAsP with 200 nm sputtered (•) and PECVD (○) SiO₂ caps and InAlGaAs capped with 200 nm sputtered SiO₂ (◇).

separated by 120 Å InGaAsP ($\lambda = 1.26$ μm) barriers within compositionally graded InGaAsP inner cladding layers. The p and n-type InP outer cladding layers were doped with Si and Zn respectively ($0.5 - 2 \times 10^{18}$ cm^{-3}). The InAlGaAs structure was grown by solid source molecular beam epitaxy and comprised an active region containing 5×70 Å InGaAs QWs separated by 80 Å $In_{0.53}Al_{0.2}Ga_{0.27}As$ barriers centrally located within a 350 nm undoped waveguide layer of the same composition as the barrier material, surrounded by a lower InAlAs n-type cladding layer doped with Si (5×10^{17} cm^{-3}) and an upper InAlAs p-type cladding layer doped with Be (5×10^{17} cm^{-3}). Both structures were capped with a 200 nm p-type ($0.5 - 1 \times 10^{19}$ cm^{-3}) InGaAs layer.

A 200 nm sputtered SiO₂ cap was deposited upon all sample structures. The deposition was carried out at an RF power of 100 W and a DC bias of 1 kV, using a 9:1 Ar:O₂ mixture to generate the sputtering plasma. The wafers were then cleaved into 2×2 mm^2 samples and annealed in a rapid thermal annealer (RTA) at temperatures between 550 and 750°C for 60 s, along with samples capped by 200 nm SiO₂ deposited by plasma enhanced chemical vapour deposition (PECVD). During the anneal samples were placed between two Si wafers to prevent desorption of the Group V elements. The bandgap shifts induced by the above procedure were measured using 77 K photoluminescence (PL) measurements.

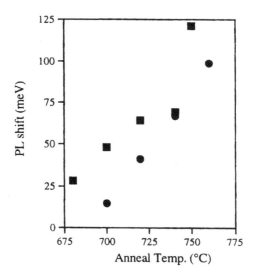

Fig. 15.   PL peak energy shifts as a function of anneal temperature for AlGaAs (●) and AlGaInP (■) capped with 200 nm sputtered SiO$_2$.

Figure 14 shows the PL energy shifts obtained for the InGaAsP and InAlGaAs structures as a function of RTA temperature. The InGaAsP samples capped with sputtered SiO$_2$ exhibit bandgap shifts at temperatures as low as 550°C. The bandgap shifts rapidly increase with increasing anneal temperature, up to over 100 meV at temperatures above 700°C. Meanwhile, samples capped with PECVD SiO$_2$ do not exhibit any PL shift up to an anneal temperature of 650°C. The use of sputtered rather than PECVD SiO$_2$ clearly leads to a substantial lowering of the threshold temperature for intermixing, well below the temperature for thermally induced intermixing of the material (∼ 700°C). For InAlGaAs samples with sputtered SiO$_2$ caps, bandgap shifts are first observed at a temperature of 650°C and increase rapidly with increasing anneal temperature, again up to over 100 meV for an anneal temperature of 750°C. Samples capped with PECVD SiO$_2$ show no bandgap shift over the temperature range shown and for clarity are not included in the figure.

Figure 15 presents PL shifts obtained for both AlGaInP and AlGaAs samples capped with 200 nm sputtered SiO$_2$. Bandgap shifts are initially found to occur at temperatures of 680°C and 700°C for the AlGaInP and AlGaAs materials respectively. Above these threshold temperatures the bandgap shifts increase rapidly up to over 100 meV at 760°C. This temperature is considerably lower than that required for IFVD (∼ 850°C) in AlGaAs [49,50] and, over the temperature range shown, bandgap shifts obtained under PECVD SiO_2 are negligible for both material systems. Following the intermixing process, all samples retain a high optical quality and good surface morphology, factors of considerable importance for subsequent device fabrication.

The processing stages of the technique we report here are very similar to those of IFVD, in which a SiO_2 cap is also used to enhance intermixing. As described earlier, the IFVD process in AlGaAs has been ascribed to the out-diffusion of Ga into the SiO_2 cap during the annealing stage. This results in an increase in Ga vacancies close to the semiconductor surface which thermally diffuse into the active region during the anneal stage and promote the interdiffusion of the well and barrier atoms. However, the technique we report here yields large bandgap shifts at significantly lower temperatures than those required for IFVD. This implies that, despite the processing stages being similar to those of IFVD, the fundamental intermixing mechanism is very different to that of IFVD. This is confirmed by removal of the sputtered SiO_2 cap prior to the annealing stage and the deposition of a replacement PECVD SiO_2 film of the same thickness. In this case we find that the bandgap shift following the anneal is the same for the sample coated with PECVD SiO_2 as that for a sample coated with sputtered SiO_2. This implies that the enhanced intermixing we observe is dependent upon the process used to deposit the SiO_2 cap rather than any particular property of the dielectric film. We postulate that the intermixing process can be largely ascribed to the creation of point defects at the sample surface, induced by atomic bombardment during the sputtered SiO_2 deposition, and their subsequent diffusion during the anneal stage. The presence of an increased density of point defects may also promote atomic out-diffusion, however, and thereby provide an additional enhancement of the intermixing rate. The contributory role of such out-diffusion is currently under further investigation.

The possible damage generated during the sputtering process has been investigated using a MQW probe sample, designed specifically for the study of dry etch damage in the InGaAsP system. The sample comprised a number of InGaAs QWs with well widths, in order of increasing depth, of 20, 40, 60 and 80 Å, separated by 200 Å InGaAsP barriers, the uppermost (20 Å) QW being located 300 Å below the sample surface. In addition a 120 Å well was situated at a depth of approximately 4000 Å, which should be immune to the damage generated during the sputtering and therefore provides a PL reference signal. A schematic diagram of the sample structure is given in Figure 16.

Figure 17(a) shows the 5K PL spectrum obtained for the as-grown sample. This clearly exhibits five peaks, one for each of the QWs, as labelled. Figure 17(b) shows the spectrum obtained for the sample after the deposition of 200 nm of sputtered SiO_2. The PL intensity from the two shallowest QWs

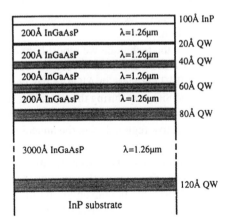

Fig. 16.   Structure of the MQW probe sample used to investigate damage generated during the sputtering process.

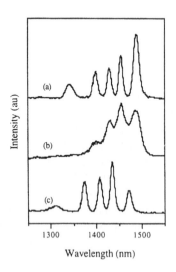

Fig. 17.   PL spectra obtained for a MQW damage probe sample before (a) and after deposition of sputtered SiO_2 (b) and after annealing at 500°C (c).

(20 and 40 Å) is clearly reduced following the deposition, suggesting that the sputtering process does generate damage, in the form of mobile point defects, leading to an increase in non-radiative recombination and a resultant decrease in PL intensity. Analysis of Figures 16 and 17 suggests that most of the damage extends to a depth beyond that of the second well (approximately 550–750 Å).

Figure 17(c) shows the PL spectrum from a sample with a 200 nm sputtered SiO_2 cap following an anneal at $500°C$ for 60 s. Compared to Figure 17(b), the PL signal from the shallowest wells is clearly recovered, due to the diffusion of the high density of point defects towards the surface and deeper into the sample. In addition, the PL from all the QWs is found to undergo a significant blue shift (18 nm), due to the initiation of the intermixing process. These spectra strongly indicate that point defects are generated during the sputtering process and that the thermal diffusion of such defects can lead to an enhancement in the rate of intermixing.

The use of QWI for monolithic integration requires the ability to control the bandgap shift selectively over a single wafer. For the technique we report here, such spatial control can be easily achieved using a 2 μm thick layer of photoresist to protect the sample surface from the sputtering induced damage. The resist is photolithographically patterned prior to the sputtering stage, to mask the regions in which suppression of the intermixing process is required. Following the deposition of sputtered SiO_2, the resist and the overlying layer of SiO_2 are removed by lift-off in acetone. A 200 nm thick layer of PECVD SiO_2 is then deposited on the sample surface to protect the exposed regions during the subsequent high temperature anneal. We have found that the resist film provides an effective mask for suppressing the intermixing process, with, in most cases, complete intermixing suppression in masked regions. This enables large differential bandgap shifts to be obtained on a single wafer, of up to 75 meV for the InGaAsP system and over 100 meV in the other material systems investigated.

In conclusion, a universal technique for QWI has been developed. The technique is closely related to IFVD, but also appears to rely on creating point defects during the deposition of a sputtered SiO_2 film. The technique allows large bandgap shifts to be obtained for a range of different material systems, covering the wavelength range 600–1600 nm. By using a photoresist film as an inhibiting mask, the intermixing process can, in most cases, be completely suppressed. The technique appears to be universal, simple and reliable and is likely to be an important tool for photonic integration.

References

1. D.G. Deppe, L.J. Guido, N. Holonyak Jr., K.C. Hsieh, R.D. Burnham, R.L. Thornton and T.L. Paoli, "Stripe-geometry quantum well heterostructure $AI_xGa_{1-x}As$-GaAs lasers defined by defect diffusion." *Appl. Phys. Lett.*, **49**, 510–512 (1986).

2. C.J. Frosch and L. Derick, "Surface protection and selective masking during diffusion in silicon", *J. Electrochem. Soc.*, **104**, 547–552 (1957).

3. J. Beauvais, J.H. Marsh, A.H. Kean, A.C. Bryce and C. Button, "Suppression of bandgap shifts in GaAs/AlGaAs quantum wells using strontium fluoride caps", *Electron. Lett.*, **28**, 1670–1672 (1992).

4. I. Gontijo, T. Krauss, J.H. Marsh and R.M. De La Rue, "Postgrowth control of GaAs/AlGaAs quantum well shapes by impurity free vacancy disordering", *IEEE J. Quantum Electron.*, **30**, 1189–95 (1994).

5. D. Hofstetter, H.P. Zappe, J.E. Epler and P. Reil "Multiple wavelength Fabry-Perot lasers fabricated by vacancy enhanced quantum well disordering", *Appl. Phys. Lett.*, **67**, 1978–1990, (1995).

6. B.S. Ooi, A.C. Bryce, J.H. Marsh and J.S. Roberts: "Effect of p and n-doping on neutral impurity and SiO_2 dielectric cap induced quantum well intermixing in GaAs/AlGaAs structures", *Semiconductor Science and Technology*, **12**, 121–127 (1997).

7. D.G. Deppe and N. Holonyak Jr., "Atom diffusion and impurity-induced layer disordering in quantumwell III-V semiconductor heterostructures", *J. Appl. Phys.*, **64**, R93–R113 (1988).

8. R.S. Burton, T.E. Schlesinger, D.J. Holmgren, S.C. Smith and R.D. Burnham, "High-performance diffusion disordered $Al_x Ga_{1-x} As$ lasers via a self-aligned process and conventional open-tube annealing", *J. Appl. Phys.*, **73**, 2015–2018 (1993).

9. S.I. Hansen, J.H. Marsh, J.S. Roberts, R. Gwilliam, "Refractive index changes in a GaAs multiple quantum well structure produced by impurity induced disordering using boron and fluorine", *Appl. Phys. Lett.*, **58**, 1398–1400 (1991).

10. S.I. Hansen, J.H. Marsh, J.S. Roberts, "Polarization dependence of the refractive index of MQW waveguides" — *IEE Proc. Pt. J. (Optoelectronics)*, **138**, 309–312 (1991).

11. M.L. Dakss, L. Kuhn, P.F. Heidrich and B.A. Scott, *Applied Physics Letters*, **16**, 525 (1970).

12. M. Dagenais and W.F. Sharfin, "Linear-optical and nonlinear-optical properties of free and bound excitons in CdS and applications in bistable devices", *Journal of the Optical Society of America*, **B2**, 1179–1187 (1985).

13. M.A. Afromowitz, *Solid State Communications*, **15**, 59 (1974).

14. M. Kuzuhara, T. Nozaki, T. Kamejima, "Characterization of Ga out-diffusion from GaAs into $SiO_x N_y$ films during thermal annealing", *J. Appl. Phys.*, **66**, 5833–5836 (1989).

15. I. Gontijo, T. Krauss, R.M. De La Rue, J.S. Roberts and J.H. Marsh, "Very low loss extended cavity GaAs/AlGaAs lasers made by impurity free vacancy diffusion", *Electron. Lett.*, **30**, 145–146 (1994).

16. E.V.K. Rao, A. Hamoudi, Ph. Krauz, M. Juhel, and H. Thibierge, "New encapsulant source for III-V quantum well disordering", *Appl. Phys. Lett.*, **66**, 472–474 (1995).

17. E.V.K. Rao, Private Communication.

18. P. Cusumano, B.S. Ooi, A. Saher Helmy, S.G. Ayling, B. Vögelle, A.C. Bryce, J.H. Marsh and M.J. Rose, 'Suppression of quantum well intermixing in doped and undoped GaAs/AlGaAs structures using phosphorus-doped SiO_2 encapsulant layer', *J. Appl. Phys.*, **81**, 2445–7 (1997).

19. S.K. Ghandhi, *VLSI fabrication principles*, (John Wiley & Sons, New York, 1994), pp. 530–532 and references therein.

20. S. Singh, F. Baiocchi, A.D. Butherus, W.H. Grodkiewicz, B. Schwartz, L.G. Van Uitert, L. Yesis, and G.J. Zydzik, "Electron-beam evaporated phosphosilicate glass encapsulant for post-implant annealing of GaAs", *J. Appl. Phys.*, **64**, 4194–4198 (1988).

21. S.K. Ghandhi, *VLSI fabrication principles*, (John Wiley & Sons, New York, 1994), pp. 530–532 and references therein.

22. L.J. Guido, J.S. Major, Jr., J.E. Baker, W.E. Plano, N. Holonyak, K.C. Hsieh and R.D. Burnham, "Column III vacancy- and impurity-induced layer disordering of $Al_x Ga_{1-x} As$-GaAs heterostructures with SiO_2 or $Si_3 N_4$ diffusion sources", *J. Appl. Phys.*, **67**, 6813–6818 (1990).

23. S.G. Ayling, J. Beauvais and J.H. Marsh "Spatial control of quantum-well intermixing in GaAs/AlGaAs using a one-step process", *Electronics Lett.*, **28**, 2240–2241 (1992).

24. B.S. Ooi, M.W. Street, S.G. Ayling, A.C. Bryce, J.H. Marsh and J.S. Roberts, "The application of the selective intermixing in selected area (SISA) technique to the fabrication of photonic devices inGaAs/AlGaAs structures", *International Journal of Optoelectronics*, **10**, 257–263 (1996).

25. B.C. Johnson, J.C. Campbell, R.D. Dupuis, and B. Tell, *Electron. Lett.*, "2-wavelength disordered quantum-well photodetector", **24**, 181–182 (1988).

26. T. Miyazawa, T. Kagawa, I. Iwamura, O. Mikami and M. Naganuma, *Appl. Phys. Lett.*, "2-wavelength demultiplexing p-i-n GaAs/AlAs photodetector using partially disordered multiple quantum well structures" **55**, 828–829 (1989).

27. P. Cusumano, T. Krauss and J.H. Marsh, "High extinction ratio GaAs/AlGaaAs electroabsorption modulators integrated with passive waveguides using impurity free vacancy diffusion", *Electron. Lett.*, **31**, 315–317 (1995).

28. J. Beauvais, S.G. Ayling and J.H. Marsh, "Low loss extended cavity lasers by dielectric cap disordering with a novel masking technique", *Photon. Technol. Lett.*, **4**, 372–373 (1993).

29. S.A. Gurevich, E.L. Portnoi and M.E. Raikh, "Absorption of light in GaAs/Al$_x$Ga$_{1-x}$As film waveguides and its influence on threshold characteristics of heterojunction lasers with Bragg mirrors", *Sov. Phys. Semicond.*, **12**, 688–693 (1978).

30. J. Werner, E. Kapon, N.G. Stoffel, E. Colas, S.A. Schwarz, C.L. Schwartz and C.N. Andreakis, "Integrated external cavity GaAs/AlGaAs lasers using quantum well disordering", *Appl. Phys. Lett.*, **57**, 540–542 (1989).

31. P. Cusumano, J.H. Marsh, M.J. Rose and J.S. Roberts, "High-quality extended cavity ridge lasers fabricated by impurity-free vacancy diffusion with a novel masking technique", *Photon. Technol. Lett.*, **9**, 282–284 (1997).

32. B.S. Ooi, K. McIlvaney, M.W. Street, A. Saher Helmy, S.G. Ayling, A.C. Bryce and J.H. Marsh, "Selective quantum well intermixing in GaAs/AlGaAs structures using impurity free vacancy diffusion", *IEEE J. Quantum Electron.*, **33**, 1784–1793 (1997).

33. P.W.A. McIlroy, A. Kurobe, and Y. Uematsu, "Analysis and application of theoretical gain curves to the design of multiquantum well lasers", *IEEE J. Quantum Electron.*, **QE-21**, 1958–1963 (1985).

34. S. O'Brien, J.R. Shealy, F.A. Chambers, and G. Devane, "Tunable (Al)GaAs lasers using impurity-free partial interdiffusion", *J. Appl. Phys.*, **71**(2), 1067–1069 (1992).

35. R.G. Waters, D.S. Hill and S.L. Yellen, "Efficiency enhancement in quantum well lasers via tailored doping profiles", *Appl. Phys. Lett.*, **52**(24), 2017–2018 (1988).

36. S.G. Ayling, J. Beauvais and J.H. Marsh "A DBR laser using dielectric cap disordering and strontium fluoride masking." *Proc. ECIO 1993*, pp. 7.10–7.11.

37. D. Hofstetter, H.P. Zappe and J.E. Epler, "Ridge waveguide DBR laser with nonabsorbing grating and transparent integrated waveguide", *Electron. Lett.*, **31**, 980–982 (1995).

38. F. Camacho, E.A. Avrutin, P. Cusumano, A Saher-Helmy, A.C. Bryce and J.H. Marsh, "Improvements in mode-locked semiconductor lasers using monolithically integrated passive waveguides made by quantum well intermixing", *IEEE Photonics Technology Letters*, **9**, 1208–1210, 1997.

39. K. Lau and J. Paslaski, "Condition for short pulse generation in ultrahigh frequency mode-locking of semiconductor lasers", *IEEE Photonics Technol. Lett.*, **3**, 974–976 (1991).

40. F.R. Laughton, J.H. Marsh, D.A. Barrow and E.L. Portnoi, "The two-photon absorption semiconductor waveguide autocorrelator.", *IEEE J. Quantum Electron.*, **30**, 838–845 (1994).

41. D.J. Derickson, R.J. Helkey, A. Mar, J.R. Karin, J.G. Wasserbauer and J.E. Bowers, "Short pulse generation using multisegment mode-locked semiconductor lasers.", *IEEE J. Quantum Electron.*, **28**, 2186–2202 (1992).

42. D.J. Derickson, P.A. Morton, J.E. Bowers and R.L. Thornton, "Comparison of timing jitter in external and monolithic cavity mode-locked semiconductor lasers.", *Appl. Phys. Lett.*, **59**, 3372–3374 (1991).

43. D. von der Linde, "Characterisation of the noise in continuously operating mode-locked lasers", *Appl. Phys. B*, **39**, pp. 201–217 (1986).

44. G.P. Agrawal and N.A. Olsson, "Self-phase modulation and spectral broadening of optical-pulses in semiconductor laser amplifiers.", *IEEE J. Quantum Electron.*, **25**, 2297–2306 (1989).

45. P.B. Hansen, G. Raybon, U. Koren, P.P. Iannone, B.I. Miller, G.M. Young, M.A. Newkirk and C.A. Burrus, "InGaAsP monolithic extended-cavity lasers with integrated saturable absorber for active, passive, and hybrid mode locking at 8.6 GHz.", *Appl. Phys. Lett.*, **62**, 1445–1447 (1993).
46. D. Hofstetter, H.P. Zappe and R. Dändliker, "Monolithically integrated optical displacement sensor in GaA/AlGaAs", *Electron. Lett.*, **31**, 2121–2122 (1995).
47. D. Hofstetter, H.P. Zappe and R. Dändliker, "A monolithically integrated double Michelson interferometer for optical displacement measurement with direction determination", *Photon. Technol. Lett.*, **8**, 1370–1372 (1996).
48. O.P. Kowalski, C.J. Hamilton, S.D. McDougall, J.H. Marsh, A.C. Bryce, R.M. De La Rue, B. Vögele and C.R. Stanley, C.C. Button and J.S. Roberts, "A Universal Damage Induced Technique for Quantum Well Intermixing" — C.C. Button and J.S. Roberts, *Appl. Phys. Lett.*, **72**, 581–583 (1998).
49. J.H. Marsh, P. Cusamano, A.C. Bryce, B.S. Ooi and S.G. Ayling, *Proc. SPIE*, **2401**, p. 74, 1995.
50. W.D. Laidig, N. Holonyak, Jr., M.D. Camras, K. Hess, J.J. Coleman, P.D. Dapkus, J. Bardeen, "Disorder of an AlAs-GaAs super-lattice by impurity diffusion". *Appl. Phys. Lett.*, **38**, 776–778 (1981).

CHAPTER 10

Selective Interdiffusion of GaAs/AlGaAs Quantum Wells Through SiO$_2$ Encapsulation — Comparison with the Ion Implantation Approach

A. PÉPIN AND C. VIEU

Laboratoire de Microstructures et de Microélectronique (L2M)/CNRS, 196 Avenue Henri-Ravéra, BP 107, 92225, Bagneux, France

1 INTRODUCTION

For the past decade, extensive work has been dedicated to semiconductor quantum well (QW) intermixing as an avenue to monolithic optoelectronic integration [1–3]. Interdiffusion of well and barrier atoms creates a electronic compositional disorder which results in a modification of the electronic and optical properties of QWs. In GaAs/AlGaAs heterostructures for instance, Ga-Al interdiffusion taking place at the QW heterointerfaces induces a penetration of Al inside the QW, and hence gives rise to an increase in the effective bandgap energy which is accompanied by changes in the refractive index. This thermally activated phenomenon can be further enhanced by introducing impurities or additional defects into the heterostructure. When intermixing is stimulated only in selected areas of a wafer, permanent built-in lateral barriers, e.g. localized regions of larger effective bandgap, can be created inside the QW. Selective intermixing therefore represents an attractive post-growth process for the fabrication of a variety of integrated devices involving the confinement of photons and/or carriers. In addition, when defects are injected in QW heterostructures to stimulate layer disordering, investigation of interdiffusion becomes a unique mean to study the diffusion properties of the injected defects during annealing. QWs can thus be regarded as sensitive probes for defect diffusion.

Among the number of existing intermixing stimulation techniques, impurity-free techniques have gained a high level of interest in the last few years because of their ability to tailor the properties of multilayered structures without introducing any dopants [4–39]. Impurity-free layer disordering is based upon the introduction of point defects which are able to enhance interdiffusion at heterointerfaces in localized regions of a sample during heat treatment. Different sources of point defects have been investigated for this purpose. In III-V semiconductors, the control of the vapour pressure of the anionic element at the surface of the sample allows the nature and concentration of elemental point defects to be controlled. The patterning of an appropriate capping layer deposited on the surface of the sample enables the spatial localization of the defect injection [40] but long annealing durations are then required for the defects to reach the heterointerfaces of interest at a given distance from the surface. These techniques of injection under equilibrium conditions seem to be very reproducible and the nature of the injected elemental point defects can be changed in order to activate interdiffusion on a given sublattice. However, for very long annealing times intrinsic defects present in the sample after growth can interact with the injected defects and large variations from sample to sample can be observed, especially for layers with a high dislocation density [41]. Besides, very little is known about the lateral resolution which can be achieved with such a mechanism only based on diffusion. A local increase of the native point defect concentration can also be achieved using focused laser annealing [42–44]. However, the lateral resolution of this method is limited by the finite laser spot size and is strongly dependent on the mechanical stability of the experimental set-up. Ion implantation is on the other hand a very precise and well-controlled defect injection tool. Implantation of inert species such as Ga^+ in AlGaAs/GaAs heterostructures has proven to be an effective impurity-free intermixing enhancement technique [8,11,15,19–22,45–55]. Lateral selectivity can be achieved through the use of masks patterned on the surface of the sample or by means of a focused ion beam. Very high lateral resolution can be achieved with this technique which was successfully exploited for the realization of quantum wires in AlGaAs/GaAs heterostructures [11,45–55]. However, the main drawback of this technique is the presence of unavoidable residual extended defects after anneal which can dramatically affect the electronic properties of the intermixed material.

Dielectric encapsulant layers, such as silicon dioxide and silicon nitride, have long been used for passivation in III-V semiconductor device fabrication technology. More recently, their utility as intermixing sources in specific III-V systems was demonstrated. In GaAs/AlGaAs quantum well (QW) heterostructures, SiO_2 encapsulation of samples was found to significantly enhance Ga-Al interdiffusion during high temperature annealing [26–39]. On the other hand, capping undoped GaAs/AlGaAs samples with Si_3N_4 proved to effectively inhibit interdiffusion [26–28]. For its relative simplicity of implementation and moreover for its believed impurity-free vacancy-driven disordering behavior, silicon dioxide capping and annealing has raised significant interest over the past few years. This distinctive property has been attributed to the special affinity of SiO_2 for gallium. During high temperature annealing, preferential absorption of Ga atoms by the SiO_2 capping layer occurs, which causes the generation of excess Ga vacancies (V_{Ga}) under the oxide/semiconductor interface. These additional point defects then diffuse to the QW heterointerfaces where they promote interdiffusion of group III species.

Although SiO_2 encapsulation-induced intermixing was investigated by several groups and quickly led to the fabrication of novel built-in opto-electronic devices such as low-loss optical waveguides, modulators, and quantum well laser diodes [26,31], the basic understanding of the specific diffusion mechanism involved has suffered from a lack of reproducibility and discrepancies due to the wide variety of experimental conditions used (differences in oxide deposition techniques, annealing conditions, epitaxial layer composition, etc.). Moreover, when optical confinement is no longer of concern but instead quantum confinement of carriers is wanted, the problem of the lateral selectivity of the intermixing technique is then of paramount importance because structures of dimensions close to electron de Broglie wavelengths (<50 nm) have to be fabricated.

The work presented in this paper was initiated to further explore the interdiffusion mechanisms governing this promising in this promising SiO_2 capping technique, and evaluate its applicability to the realization of low dimensional quantum nanostructures and as a possible alternative to Ga^+ implantation-induced selective intermixing.

A great care was given to the choice of efficient characterization techniques. Our results were confronted through the use of the following complementary techniques: low temperature optical spectroscopy, including photoluminescence (PL), photoluminescence excitation spectroscopy (PLE) and linear polarization analysis, secondary ion mass spectroscopy (SIMS) and finally, cross-sectional transmission electron microscopy (XTEM).

Low temperature PL spectroscopy is a powerful tool of investigation for semiconductor QW interdiffusion. Indeed, upon intermixing the shape of quantum wells (QWs) is modified leading to an energy shift of their PL emission. For GaAs/(Ga,Al)As structures the situation is rather simple since only group III elements are involved and the interdiffusion does not modify the stress in the layers due to the very small lattice mismatch between GaAs and GaAlAs compounds. In this system, a blue energy shift of the PL peak energy will be the signature of Ga-Al interdiffusion, and through a very simple modelling, the interdiffusion length at heterointerfaces can be deduced from the amplitude of this blueshift [11,56]. Qualitative informations on the integrity of the electronic properties can be obtained from the PL signal intensity. Finally, the broadening of the PL peak of interdiffused QWs can be regarded as an indication of the spatial homogeneity of the intermixing mechanism. When regions of a QW are mixed with different degrees of intensity, on a length scale relevant with the exciton diameter, a distribution in energy of the electronic levels is generated leading to a broadening of the PL peak. PL characterizations were carried out at 2 K using an Ar laser (=488 nm) with a power of 1 mW for a 40 μm spot size, at different stages throughout the process. PL excitation spectroscopy (PLE) was also carried out, using either a dye laser or a tunable Ti-sapphire laser. Confinement effects were more specifically investigated through PLE measurements in coupling with linear polarization anisotropy analysis. In these experiments, the $(I_\perp - I_{//})/(I_\perp + I_{//})$ ratio was measured for each energy of the PLE spectrum, where $I_\perp (I_{//})$ is the collected intensity when the exciting light is polarized in the direction perpendicular (parallel) to the wire axis [57].

However, PL experiments are not a direct measurement of the Al profile through the GaAs/(GaAlAs) heterostructure but the final result of a complicated sequence of events such as carrier relaxation, exciton formation and radiative recombination. Other experimental techniques must be absolutely implemented in complement to PL investigations to avoid any problem of interpretation. By using secondary ion mass spectrometry (SIMS), the depth profile of the Al concentration can be obtained unambiguously but the depth resolution of this technique of a few nanometers limits the accuracy for very thin layers or for low intermixing rates [58]. SIMS characterization can also be used before the annealing step in order to separate the effect of the encapsulation from the diffusion of the various species during annealing, i.e. Ga inside the oxide layer, Si or O inside the heterostructure, etc. SIMS analysis was carried out with a 5.5 keV Cs^+ primary beam, using standard conditions for depth profiling.

However, this technique does not give any indication on the nature and spatial distribution of residual defects after annealing. Cross-sectional electron microscopy can provide the missing structural information. PL and XTEM techniques are particularly complementary for the analysis of layer disordering in semiconductor heterostructures. If defects are present in the vicinity of QWs, a decrease in the intensity of the emitted luminescence will be observed. However, this can be the source of misleading interpretations of PL data, in particular when a high concentration of defects are present in the material. For example, when no PL signal coming from the interdiffused QW can be detected, this technique does not allow us to distinguish between a complete mixing situation and the formation of a high quantity of non radiative centers quenching the PL signal. In the case of selective interdiffusion where patterned regions experience different mixing rates, PL can detect several emission peaks corresponding to the spatial distribution of the interdiffused areas. It is thus possible to be sure that the mixing is not homogeneous but selective (parts of the QW are interdiffused, others are not), and the lateral interdiffusion length measuring the lateral selectivity of the process can be deduced from the energy position of the different PL peaks [49,51]. However, PL characterization is not able to localize the origin of the emitted lines when nanometric patterns are probed. The identification of the different peaks is usually deduced from common sense considerations. For example, when nanometric trenches are patterned in a SiO_2 capping layer in order to achieve selective QW interdiffusion and two PL peaks are observed after annealing, the low energy peak can usually be attributed to weakly interdiffused regions located under the trenches while the high energy peak can be attributed to the strongly interdiffused regions covered with SiO_2. We will show in this work that such simple interpretations of the spatial localization of interdiffused areas from PL data can be completely wrong in the case of SiO_2/Si_3N_4 combined depositions. It then becomes absolutely necessary to systematically couple PL characterizations with XTEM observations. If accurate quantitative measurement of the interdiffusion length, deduced from the contrast in the electron image due to the difference in the atomic scattering factors between GaAs and GaAlAs layers, remains difficult, qualitative information can be readily obtained in XTEM [59]. Moreover, the presence of defects which can affect the PL properties of QWs can be detected by XTEM and the localization of the interdiffused areas as a function of the patterning can be unambiguously determined on a simple dark-field (DF) image. XTEM analysis was performed on a Philips CM20 FEG microscope at an operating voltage of 200 kV.

The comparative work presented in this paper provides a comprehensive explanation for the various behaviors reported in the literature. We highlight the critical role played by the thermal stress distribution, generated in the heterostructure by the capping layers, on vacancy diffusion. In the first section of this paper, the interdiffusion mechanisms under uniform SiO_2 encapsulation of GaAs/(Ga,Al)As QW heterostructures are investigated. In a second part, we study the lateral selectivity of SiO_2 capping-induced intermixing and demonstrate the fabrication of quantum well wires (QWRs). In a final section, the possibility of using Si_3N_4 in conjunction with SiO_2 in patterned structures in order to achieve selective intermixing is explored.

2 INTERDIFFUSION UNDER UNIFORM SIO_2 ENCAPSULATION

2.1 Principle

The active role played by silicon dioxide encapsulation has been attributed through various studies to the special affinity of SiO_2 for gallium. During high temperature annealing, preferential absorption of Ga atoms by the SiO_2 capping layer occurs, which causes the generation of excess Ga vacancies (V_{Ga}) under the oxide/semiconductor interface. These additional point defects then diffuse to the QW heterointerfaces where they promote interdiffusion of group III species.

The work presented in this section was initiated to further explore the interdiffusion mechanisms involved in this promising technique by performing a thorough study of specific key parameters: dielectric deposition technique, SiO_2 layer thickness, annealing temperature, distance of the QW from the surface.

Six undoped GaAs/AlGaAs QW heterostructures of three different types were grown by molecular beam epitaxy (MBE) on (001) semi-insulating GaAs substrates for this study. The first type of structure contains a single GaAs quantum well (SQW), of thickness 3 nm (structure A) or 6 nm (structure B), inserted in between 10 nm-thick $Al_{0.33}Ga_{0.67}As$ barriers further embedded in a symmetrical set of short period 14 nm-thick GaAs/AlAs superlattices (SLs). The second type of SQW heterostructure (structure C) was designed to enhance the introduction of Al inside the QW, and thus the effects of intermixing, and comprises of a 6 nm-thick GaAs SQW bordered by two 1.5 nm-thick AlAs layers, further embedded in a set of $Al_{0.33}Ga_{0.67}As$ barriers (100 nm on the bottom, 20 nm on the top). The third type of heterostructure (structures D, E and F) contains several individual QWs and were used to investigate the dependence of the intermixing rate upon depth. Structure D is a multi-quantum well (MQW) structure comprising ten identical uncoupled GaAs (5 nm) / AlAs (5 nm) QWs, embedded in 20 nm-thick upper and lower $Al_{0.33}Ga_{0.67}As$ barriers. Structure E contains three SQWs of nominal widths 6.2, 3.1 and 1.7 nm, separated by $Al_{0.33}Ga_{0.67}As$ barriers of approximate thickness 25 nm, with the thinnest QW located farthest from the surface. Structure F is a 0.45 μm-thick GaAs (5 nm)/ $Al_{0.33}Ga_{0.67}As$ (10 nm) MQW. In order to prevent possible reduction of the SiO_2 by Al and resulting in-diffusion of silicon and/or oxygen inside the epitaxial layers as reported by various authors [27,28], all of our structures are capped with a final 5 nm-thick GaAs layer.

Four different dielectric deposition techniques were used to coat samples with a thin layer of silicon dioxide:

(1) Radio frequency (RF) sputtering carried out at 100°C, under Ar pressure (10 Torr). Power density was about 1.3 W/cm. Deposition rate is close to 20 nm/min.

(2) Plasma-enhanced chemical vapor deposition (PECVD) performed at 300°C using SiH_4 and N_2O as reactive gases (10 sccm and 80 sccm respectively), plasma power was 10 W and pressure 200 mTorr. Deposition rate is about 65 nm/min.

(3) Rapid thermal chemical vapor deposition (RTCVD), an original technique developed in the Centre National d'Etudes des Télécommunications in Bagneux, France, involving the accelerated pyrolitic decomposition of SiH_4 and O_2 gases (10 and 100 sccm respectively) in an inert flowing N_2 atmosphere assisted by the rapid heating of the sample holder up to 700°C by a set of halogen lamps [35]. Total pressure was 5 Torr. Very high deposition rates (500 nm/min) are achieved with this technique.

(4) Ultra violet induced chemical vapor deposition (UVCVD), which is a cold process (250°C) involving photolysis of SiH_4 and O_2 gases (15 and 150 sccm respectively) under UV illumination by low pressure Hg lamps, in a inert flowing N_2 atmosphere. Chamber pressure was 10 Torr. Deposition rate is about 10 nm/min. A shot redensification anneal at 500°C immediately followed the deposition step in our experiments.

Conventional thermal anneals were carried out in a flowing $Ar:H_2$ (10% H_2) ambient at either 850°C or 900°C. During annealing the sample is sandwiched in between two GaAs clean samples and inserted in a graphite box filled with fine GaAs powder. This original set-up ensures both constant temperature and an As overpressure around the sample being annealed. Samples were characterized before and after SiO_2 deposition, as well as after heat treatment (with or without SiO_2 capping). Our experimental observations were systematically validated through the use of the three complementary characterization techniques described earlier: low temperature photoluminescence, secondary ion mass spectroscopy and cross-sectional transmission electron microscopy.

The results presented in this section reveal that an optimized, strong, homogeneous, and reproducible V_{Ga}-driven intermixing can be achieved under a SiO_2 capping layer of high density and purity. A simple interdiffusion model is proposed to explain our results, which includes the role played by the stress distribution, created during annealing in the heterostructure by the SiO_2 capping, on the vacancy diffusion [38,39,61].

2.2   Influence of the SiO_2 Deposition Technique

A variety of technological processes can be used to deposit thin silicon dioxide films on AlGaAs-based heterostructures. Several techniques had been reported in the previously published works, and it quickly appeared to us that a possible cause for the disparities encountered between these various studies could be the type of dielectric deposition technique used. Indeed, whatever the exact mechanism leading to enhanced interdiffusion, it obviously is intimately related to the activity of the SiO_2 film on the semiconductor surface. Deposition conditions have a strong influence on the quality of a silicon dioxide film and therefore different techniques are expected to produce films differing in composition, density, impurity incorporation, built-in stress, and quality of the SiO_2/GaAs interface. We chose to compare the efficiency of the four techniques described above, with respect to intermixing rate, homogeneity and reproducibility. Samples from structure A (3 nm-thick SQW) were thus separately coated with SiO_2 using either RF-sputtering, PECVD, RTCVD or UVCVD. An approximate thickness of 100 nm was used for all films except for PECVD where a 500 nm-thick film was deposited.

Infrared transmission spectroscopy experiments performed on our samples after deposition (data reported elsewhere [39]) revealed a higher compactness and purity in RTCVD SiO_2 films, as compared to sputtered films where a lesser density due to under-stoichiometry (SiO_x with $x < 2$) was observed, and to UVCVD and PECVD films in which traces of hydrogen were recorded.

PL spectra obtained after a 3 hour anneal at 850°C are shown in Figure 1. Emission from the as-grown SQW of structure A occurred at an energy of 1.69 eV. As can be seen in Figure 1(a), the uncapped sample shows a great stability against heat treatment. In contrast, all coated samples (Figures 1(b)

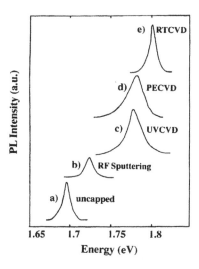

Fig. 1. Typical PL spectra obtained at 2 K on different samples from heterostructure A annealed for 3 hours at 850°C. a) uncapped sample; and samples capped with a layer of SiO_2 obtained by b) RF sputtering, c) UVCVD, d) PECVD, e) RTCVD.

to 1(e)) exhibit a blueshift of their QW emission, showing evidence of enhanced intermixing. However, intermixing rates as well as intensity and linewidth of the PL peaks vary from one deposition technique to the other.

The largest blueshift (110 meV) is achieved with RTCVD. In spite of this strong intermixing, electronic properties are retained in the RTCVD sample as evidenced by the high signal intensity and narrow linewidth. Significant blueshifts are also induced with PECVD and UVCVD (80–90 meV). However in the case of PECVD, the higher oxide thickness — the influence of which will be discussed in the next section — should be taken into consideration. Emissions are broadened for both techniques which indicates a spatial inhomogeneity of the induced interdiffusion. Furthermore, significant differences in the energetic position of the PL peak could be observed when probing different locations on the PECVD sample, and to a lesser extent, on the UVCVD sample. Such spatial inhomogeneities can be attributed to porosity of the oxide structure or/and to the incorporation of impurities. The presence of pores, or pinholes, in CVD oxide layers was indeed previously reported by some authors as being a cause for spatial inhomogeneity [27,28]. A higher intermixing rate was observed under the area surrounding a pinhole. This effect was attributed to the fact that under As overpressure annealing conditions the As-rich surface condition at the pinhole location will slightly facilitate the creation of column III

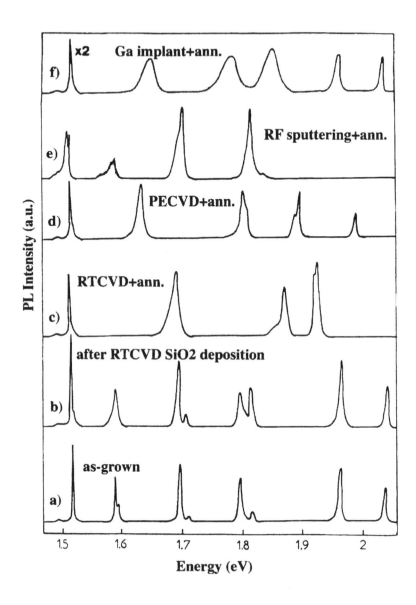

Fig. 2. 2 K PL spectra of different samples from heterostructure E. a) as-grown b) after capping with RTCVD SiO$_2$, c) after subsequent 10 minute anneal at 900°C, d) after a 10 minute anneal at 900°C under SiO$_2$ capping obtained by PECVD, e) after a 10 minute anneal at 900°C under SiO$_2$ capping obtained by RF-sputtering, and f) uncapped sample after Ga$^+$ implantation-induced intermixing (dose = 10^{13} ions/cm^2, energy = 50 keV, 10 minute anneal at 900°C).

vacancies [27,62,63] and thus the diffusion of Al towards the surface. The subsequent reduction of SiO_2 by Al around the pinhole will provide free Si and O for in-diffusion and impurity-induced disordering of the underlying heterostructure. The rapid redensification anneal immediately following UVCVD is expected to reduce film porosity, which could explain the lesser spatial inhomogeneities in comparison to PECVD SiO_2. Lastly, the RF-sputtered SiO_2 film induces very little intermixing (35 meV blueshift) but strongly affects signal intensity. Deposition techniques involving plasmas and moreover involving the presence of energetic particles, such as RF-sputtering, usually induce sample surface damage which can affect the electronic properties of QWs located close to the surface, as seems to be the case here. Furthermore, the higher content of Si inside the RF-sputtered silica does not seem to have favored any kind of impurity-induced intermixing process.

Even more instructive are the comparative PL results obtained on various samples from structure E (containing 3 SQWs of different thickness) presented in Figure 2. For this experiment, 400 nm-thick SiO_2 films were deposited, either by sputtering, PECVD or RTCVD. The as-grown PL spectrum, displayed in Figure 2(a), exhibits six main peaks. Starting from the lower-energy side, we observe the emissions originating from the GaAs substrate, QW_1, QW_2, QW_3, the $Al_{0.31}Ga_{0.69}As$ alloy and the short-period SL. Furthermore, we note that all three QWs exhibit clearly resolved double-line peaks. The line splittings which correspond closely to a difference of one-monolayer (ML) in the thicknesses of each QW, are found to vary only slightly from location to location over the sample (4 meV splitting $\pm$ 0.5 meV variation over the sample for QW_1, 19 meV $\pm$ 1.5 meV for QW_2, 27 meV $\pm$ 1.0 meV for QW_3). We therefore attribute these splittings to the formation during growth of large ML-flat islands (larger than the exciton diameter). In the unstrained GaAs/AlGaAs system, MBE growth interruptions indeed tend to smooth the interfaces particularly the top AlGaAs on GaAs interface and create wide terraces [64,65] differing in thickness by one ML. The small variations measured when changing the probing location, and the slight departure of our recorded PL energetic positions from the exact positions expected for QWs of thicknesses corresponding exactly to an integer number of MLs, can most likely be explained by the sensitivity of our data analysis to composition inhomogeneities in the AlGaAs barriers, inaccuracies in the value of the bound exciton energy, and possibly by the existence of some roughness on the atomic scale at the lower heterointerface (GaAs on AlGaAs) [66]. Within the uniformity of our sample, dielectric deposition using RTCVD (Figure 2(b)) causes minute changes to the PL spectrum, relative intensity changes being attributable to a change of location in the probed region.

In contrast, after a 10 minute anneal at 900°C under SiO_2 capping (Figures 2(c) through 2 (e)), the PL spectra suffer significant changes. After annealing under sputtered SiO$_2$, practically no blueshift is induced but the emission originating from QW$_1$, located closest to the surface, is greatly attenuated (Figure 2(e)). These results indicate both a very weak intermixing and a degradation of the material in the near surface region. After annealing under PECVD SiO$_2$, more significant blueshifts are observed in the QW emissions, which indicates a stronger intermixing rate, and signal intensity is also retained (Figure 2(d)). However, the spectra obtained vary significantly over the sample, once again revealing spatial inhomogeneities over distances larger than 1 μm. After annealing under RTCVD SiO$_2$ capping, QWs experience enormous blueshifts in their emission energies: approximately 110 meV for QW$_1$, 180 meV for QW$_2$, and 130 meV for QW$_3$. Despite these dramatic energy increases, PL peaks roughly maintain both their narrowness and intensity. Most strikingly, the peaks corresponding to the two narrowest wells (QW$_2$ and QW$_3$) still exhibit line splittings related to their initial one ML variation in well thickness. This unique result has important consequences and will be further analyzed in Section 2.4). For comparison, the PL spectrum obtained for a Ga-implanted sample taken from the same MBE-grown layer and annealed under the same conditions as our SiO$_2$-capped samples is also presented, in Figure 2(f). The comparison between Figures 2(c) and 2(f) will be further discussed in Section 2.4. Let us simply point out for now that while large blueshifts are also obtained, QW PL peaks on Figure 2(f) have suffered significant linewidth broadening and intensity loss and no longer exhibit one-ML difference-related splittings, which indicates spatial inhomogeneity at a length scale shorter than a micron.

PL results can be correlated with the XTEM micrographs presented in Figure 3. We recall that in the dark field (DF) imaging obtained by selecting the (200) diffracted beam, GaAs layers appear as dark regions, while layers rich in Al appear as lighter areas. Figure 3(a) is a (200) DF image obtained after a 3 hour anneal at 850°C on a sample from structure A capped with sputtered SiO$_2$. Numerous defects can be seen under the SiO$_2$/semiconductor interface, extending throughout the upper SL. Meanwhile, the QW heterointerfaces appear only slightly blurred and the lower short-period SL is clearly visible. These observations validate the fact that samples capped with sputtered silica undergo very limited intermixing. The loss of radiative efficiency can easily be related to the observed extended defects. We believe most of these defects to be generated during the sputtering process, due to the bombardment of the sample surface with energetic particles. The presence of irregularities in sputtered SiO$_2$, as can be seen on the bright field micrograph of Figure 3(b), also attests to the lower quality of the deposited oxide.

In contrast, the high resolution image displayed in Figure 3(e), obtained after a 3 hour anneal at 850°C on a sample from structure B capped with RTCVD SiO$_2$, shows the high quality of the RTCVD SiO$_2$/GaAs interface.

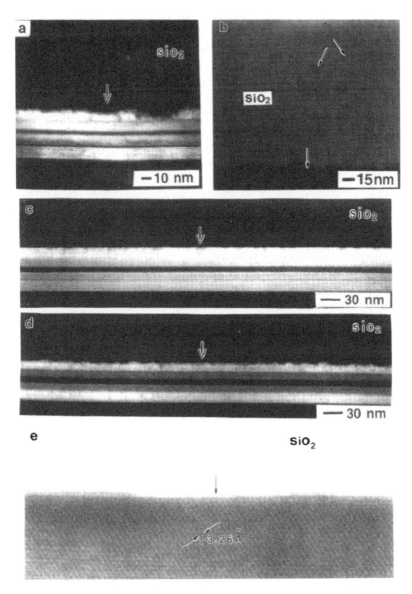

Fig. 3. a) DF (200) XTEM image of a sample from structure A annealed for 3 hours at 850°C under RF-sputtered SiO₂; b) BF XTEM image of the same sample as on a), the tilted arrows point at defects in the RF-sputtered SiO₂; c) DF (200) XTEM image of a sample from structure B coated with a layer of RTCVD SiO₂ before anneal; d) DF (200) XTEM image of the same sample after a 3 hour anneal at 850°C; e) High resolution XTEM image of the RTCVD SiO₂/GaAs interface on the same sample as on d). On each micrograph, the vertical arrow points at the SiO₂/GaAs interface.

The dense, regular, yet amorphous structure of the RTCVD SiO_2 is clearly visible and the interface appears to be atomically flat. Micrographs presented in Figures 3(c) and 3(d) show (200) DF images of RTCVD-capped samples also from structure B, after deposition and after anneal respectively. While contrasts due to differences in Al concentration are very pronounced in the unannealed sample, they are much weaker in the annealed sample, and heterointerfaces appear blurry. However, no extended residual defects are observed under the SiO_2/GaAs interface or elsewhere in the heterostructure. The material after interdiffusion shows very high structural quality.

These XTEM observations therefore agree with both the higher intermixing rate and the good conservation of electronic properties observed in PL under RTCVD SiO_2 capping. Surprisingly, we also note that the lower SL has been layer-averaged to form a $Al_{0.5}Ga_{0.5}As$ alloy while the Al modulation in the upper SL is still visible after annealing. This result is indeed unexpected for a technique where the point defects responsible for the enhancement of the interdiffusion are created at the SiO_2/GaAs interface. We will discuss this interesting observation in Section 2.4).

In summary, parallel characterizations in PL and XTEM revealed that the dielectric deposition technique had a crucial influence on the intermixing process. SiO_2 layers obtained by RTCVD clearly appear to be the best suited to achieve a strong, spatially homogeneous intermixing of GaAs/AlGaAs QW heterostructures in a reproducible fashion. We believe there exists a direct relation between the structural quality of RTCVD silicas and their high efficiency as interdiffusion sources. Non-porosity, purity, and stoichiometry of the deposited films as well as quality of the oxide/semiconductor interface appear to be the most critical parameters in order to achieve quality intermixing. Most of the results presented hereafter will thus focus on RTCVD SiO_2 caps.

2.3   Interdiffusion Enhancement Mechanism

In order to determine the origin of the defects responsible for the enhancement in the interdiffusion rate, our samples were analyzed using SIMS. Changes in the stoichiometry of the SiO_2 film upon annealing were examined. We also looked for possible penetration of impurities, in particular dopant impurities such as Si, inside the semiconductor layers. Figure 4 displays the Ga, Si, and O secondary ion counts, plotted as a function of depth, which were obtained on three samples from structure B capped with a thin layer of RTCVD SiO_2. Since all measurements were made under the same conditions, relative signal levels can be compared in terms of concentration. The SIMS profile presented in Figure 4(a) was obtained after oxide deposition. Profiles shown in Figures 4(b) and 4(c) were recorded on samples annealed for 3 hours at 850°C and 10 minutes at 900°C, respectively.

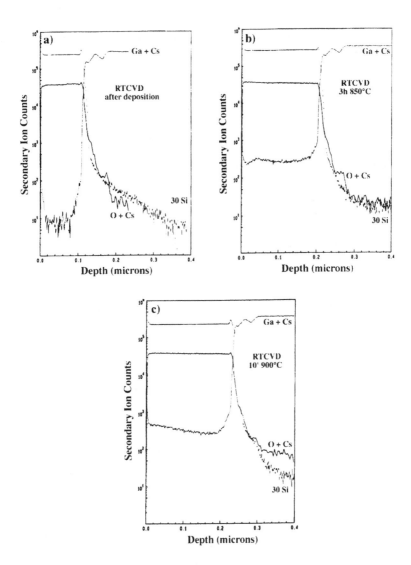

Fig. 4. Ga, Si and O SIMS profiles as a function of depth for three samples from structure A capped with a film of SiO$_2$ obtained by RTCVD a) after deposition of a 100 nm-thick layer, b) after a 3 hour anneal at 850°C under a 200 nm-thick layer, c) after a 10 minute anneal at 900°C under a 200 nm-thick layer.

The rapid decrease of Si and O signals around a depth of 0.1 μm in Figure 4(a), indicates the position of the abrupt SiO$_2$/semiconductor interface, between the 100 nm-thick oxide and the GaAs cap layer. Modulations of the Ga signal inside the semiconductor give us information on the layer composition of heterostructure B.

After annealing, whether at 850°C or 900°C, a drastic change occurs in the Ga profile. In both samples of comparable SiO$_2$ thicknesses around 200 nm, a massive increase of the Ga concentration inside the SiO$_2$ layer is detected. Ga appears to have been literally pumped inside the oxide layer, with accumulation towards the surface. Assuming that erosion speeds and ionization rates are comparable in the dielectric and in the semiconductor, a rough estimate of the Ga concentration inside the SiO$_2$ layer after anneal can be deduced from these profiles. Under such assumptions, Figure 4(c) indicates an average 5×10^{19} Ga atoms/cm^3. The main effect of this Ga pumping by SiO$_2$ will be the generation of excess point defects under the SiO$_2$/GaAs interface. We believe, as it has been strongly suggested by other authors, that most of these defects will consist of Ga vacancies (V_{Ga}). If the SiO$_2$/GaAs interface is of high enough quality that trapping of V_{Ga} does not occur, these point defects will then be free to diffuse inside the heterostructure, thus leading to an enhancement of the Ga-Al interdiffusion at the heterointerfaces. We can thus assess the concentration of V_{Ga} in the near surface region to be nearly 500 times larger than the average thermodynamic equilibrium V_{Ga} concentration in GaAs at 900°C (10^{17} atoms/cm^3) [67]. We also note that the Ga accumulation observed near the surface of the oxide layer is higher in the case of the shorter 900°C anneal, as compared to the sample annealed for 3 h at 850°C which even exhibits a small dip in the Ga concentration right under the oxide surface. We believe this effect could be due to the formation of a volatile gallium oxide at the surface of the SiO$_2$ layer. Such an oxide will easily evaporate at high temperature, as seems to be the case for the longer 3 hour anneal, while the 10 minute anneal is too short to allow significant oxide evaporation to occur. Signatures of interdiffusion can be observed in the Ga profile inside the heterostructure, where the peak corresponding to the QW is slightly damped after anneal as compared to the unannealed situation. This effect is again more pronounced in Figure 4(c). Meanwhile, neither the Si nor the O profiles appear to be affected by the high temperature anneal. No diffusion of Si and/or O species inside the semiconductor is induced. In other words, the intermixing process is impurity-free for this type of silica. Furthermore, we expect to find a correlation between the amount of Ga pumped by the oxide layer and the magnitude of intermixing induced, due to the larger number of vacancies generated. Thus, according to our SIMS analysis, the sample annealed at 900°C should show a larger interdiffusion rate than the one annealed at 850°C.

As we mentioned in Section II, the intermixing rate can be extracted from PL characterization. PL experiments were carried out on all three samples

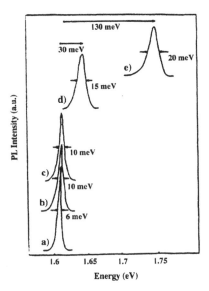

Fig. 5. 2 K PL spectra of different samples from structure B. a) as-grown b) after capping with RTCVD SiO$_2$, c) after a 3 hour anneal at 850°C without capping, d) after a 3 hour anneal at 850°C under a 200 nm-thick RTCVD SiO$_2$ capping, e) after a 10 minute anneal at 900°C under a 200 nm-thick RTCVD SiO$_2$ capping.

prior to SIMS analysis. Results are presented in Figure 5. Once again, we see from spectra (a), (b) — corresponding to the SIMS data of Figure 4(a) — and (c), that the deposition step does not affect the QW emission, nor does the high temperature anneal when no dielectric capping is used. In contrast, the blueshifts exhibited by both SiO$_2$-capped samples after anneal clearly indicate an enhanced intermixing. However, the sample annealed at 900°C experiences a blueshift four times larger than the sample annealed at 850°C (133 meV as compared to 33 meV), which is indeed in agreement with the conclusions drawn from SIMS analysis. Such correlation between the higher Ga content inside the SiO$_2$ after annealing and the larger interdiffusion rate was observed on several RTCVD SiO$_2$-capped samples. In this case, the conservation of signal intensity and the small linewidth broadening, reported earlier for SiO$_2$-capped samples after anneal, seem to agree well with an impurity-free mechanism. XTEM performed on both of these samples further reinforces this diagnostic. The results obtained on the sample annealed for 3 hours at 850°C, in an area which was not probed in SIMS, have already been presented in Figure 3(d). XTEM performed on the sample annealed 10 minutes at 900°C (data not shown here [39]) showed very similar results to that displayed in Figure 3(d). As a consequence of the higher intermixing rate, interfaces appeared more blurry, but yet no extended residual defects were visible. As before, a stronger intermixing rate in the lower SL was observed.

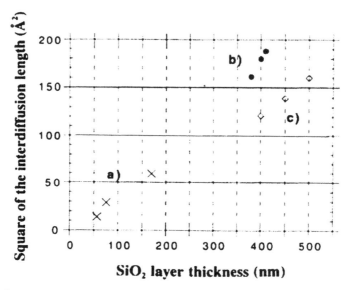

Fig. 6. Plot of the square of the interdiffusion length as a function of the thickness of the RTCVD SiO₂ encapsulant layer for a) samples from structure B annealed for 3 hours at 850°C, b) samples from structure C annealed for 10 minutes at 900°C, c) samples from structure C annealed for 1 hour at 850°C.

However, the difference in intermixing rates obtained between both annealing conditions is a surprising result. Indeed, with respect to intrinsic thermal interdiffusion, e.g. self-diffusion of Ga inside GaAs, both annealing conditions are equivalent, i.e. lead to the same interdiffusion length. The departure we observe therefore indicates that the activation energy of the interdiffusion process under SiO₂ capping is significantly lower than the 6 eV intrinsic interdiffusion activation energy reported in the literature [62]. Moreover, since both samples had the same SiO₂ thickness (~200 nm), the fact that more Ga was pumped during the higher temperature — yet shorter — anneal, seems to indicate that the activity of the silicon dioxide layer is sensitive to temperature. We attribute this effect to a consequence of the increase of Ga solubility in SiO₂ with temperature [68]. The oxide layer somewhat acts as a Ga sponge, absorbing Ga until it reaches saturation, which is higher for higher temperatures.

Another way to increase the amount of Ga pumped would be to increase the oxide thickness. Such experiments were indeed carried out on several samples coated with a layer of RTCVD SiO₂, PL results have been summarized in Figure 6. We have plotted the square of the interdiffusion length, Δ^2, as a function of the thickness of the deposited SiO₂ layer. As mentioned earlier, Δ is the relevant parameter to evaluate the interdiffusion rate. Δ^2 is equal to the product of the interdiffusion coefficient D by the anneal duration t.

And D itself is directly proportional to the V_{Ga} concentration, according to a vacancy-driven mechanism. The three series of experimental data displayed in Figure 6 correspond to three different annealing conditions and/or different heterostructures. Not only does it show very clearly that for a given annealing condition intermixing rate increases with increasing oxide thickness, but Δ^2 seems to increase linearly with SiO_2 thickness. In particular, we do not observe any saturation effect. A similar result had been reported by other authors previously [32] on a shorter range of thicknesses. It therefore appears that for a fixed Ga solubility inside SiO_2 e.g. for a given anneal temperature the thicker the SiO_2 layer, the more Ga atoms it can absorb, and thus the more excess V_{Ga} can be generated, thus leading to a stronger stimulation of the interdiffusion process. However, the existence of a critical thickness over which the quality of deposited films will be deteriorated due to high intrinsic stress is expected to set an upper limit on the thickness that can be used.

SIMS measurements were also carried out on samples coated with UVCVD and sputtered SiO_2. Presence of Ga inside the oxide film after anneal was also observed, but at a somewhat lower scale. In the case of UVCVD, a slight penetration of Si inside the semiconductor could be observed [39], which seems to agree with the aforementioned influence of the presence of pinholes in CVD layers and their influence on the origin of interdiffusion stimulation. Therefore, Ga pumping by SiO_2 and subsequent creation of excess Ga vacancies appears to be a relatively universal phenomenon which depends only slightly on the dielectric deposition technique used. Film density, purity, as well as film/substrate stress state, although they may have some influence on the pumping process, do not seem to play a major role. On the other hand, the diffusion of the created vacancies inside the semiconductor appears to be significantly affected by the quality of the SiO_2/GaAs interface. It is highly likely that extended surface defects induced by the sputtering process act as traps for the V_{Ga} and therefore prevent these point defects from reaching the QW heterointerfaces. Likewise, the presence of pinholes in the SiO_2 film will favor the penetration of Si and/or O and this local impurity-induced intermixing will give rise to spatial inhomogeneities.

Yet another implication of the high intermixing rate achieved even after a short 10 minute anneal at 900°C, is the fact that Ga pumping and subsequent V_{Ga} diffusion seem to occur rather quickly in the early stages of the anneal, instead of continuously throughout heat treatment. To appreciate better the kinetics of SiO_2 capping-induced intermixing, we studied more specifically the influence of anneal duration on a MQW heterostructure (structure F) coated with a 480 nm-thick layer of RTCVD SiO_2. Several samples were annealed at a fixed temperature of 850°C but for different durations varying from a few seconds (using either rapid thermal annealing or a push-pull procedure in a conventional furnace) up to 4 hours. SIMS was used to monitor the evolution of the Ga distribution inside the oxide layer as well as the changes in the Al depth profile within the heterostructure. Ga and Al profiles are shown in Figures 7 and 8, respectively. We see in Figure 7 that even for the shortest anneals a significant quantity of Ga has been pumped inside the oxide cap as opposed to the as-deposited situation. We also note an accumulation of Ga towards the SiO_2 surface on most of the samples except for the longer 4 hour anneal. As suggested to account for the similar SIMS profiles of Figure 4, this behavior appears to be consistent with the formation of a Ga oxide at the surface and its subsequent evaporation during prolonged anneals. Besides this small loss of Ga near the surface in the longer anneal durations, it appears that the total amount of Ga inside the SiO_2 capping layer is roughly the same after a push-pull, a 10 minute or a 4 hour anneal. The prolonged anneals only seem to render the Ga distribution more uniform within the SiO_2 cap with no further increase in Ga concentration. Concurrently, the Al depth distributions of Figure 8 reveal a striking fact: for all anneal durations, a nearly uniform damping of the amplitude of Al oscillations throughout the thickness of the heterostructure (~ 450 nm) is observed. In other words, a nearly uniform and depth-independent disorder throughout the MQW is generated very early in the annealing stage. Longer anneal durations only uniformly increase the interdiffusion rate. However, we note that a smaller interdiffusion rate is observed for the very first QWs located right under the $SiO_2/GaAs$ interface (down to ~ 30 nm).

All our results converge towards the following conclusions regarding the interdiffusion enhancement mechanism: during the early stages of the anneal, a fixed quantity of Ga, limited both by the solubility limit at a given temperature and by the oxide thickness, is quasi-instantaneously — at the resolution of our annealing time measurement — pumped into the SiO_2 encapsulant. The resulting pulse generation of V_{Ga} is quasi-simultaneously followed by the abnormally rapid diffusion of these excess point defects deep inside the semiconductor and their uniform distribution throughout the heterostructure. In a second stage, the vacancies can diffuse normally around their crystalline sites and stimulate Ga-Al exchanges across the heterointerfaces. Intermixing then proceeds normally, increasing with anneal duration. The activation energy of the process only requires E_m, the V_{Ga} migration energy and is thus much lower than the 6 eV energy corresponding to $E_f + E_m$, where E_f is the V_{Ga} formation energy. The intermixing process will then progressively decrease due to the presence of V_{Ga} traps in the structure. These V_{Ga} sinks are located in the bulk of the sample, most likely deep inside the heterostructure at the epilayer/substrate interface, and also in the near $SiO_2/GaAs$ interface region.

2.4   Characteristics of the Interdiffused Material

One of our main motivations for this work was to find an intermixing process which would be more homogeneous than commonly-used ion implantation. Indeed, let us go back to Figure 2 and consider the PL spectrum of the

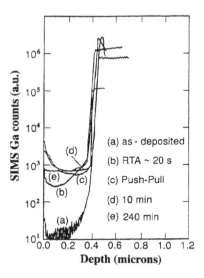

Fig. 7. SIMS Ga concentration profiles obtained in the RTCVD SiO$_2$ capping layer of several samples from structure F: a) after deposition and before annealing; b) after a 20 s rapid thermal anneal at 850°C; c) after a push-pull anneal at 850°C in a conventional furnace; d) after a 10 min. conventional anneal at 850°C; e) after a 4 hour conventional anneal at 850°C.

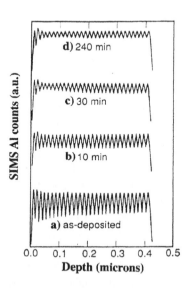

Fig. 8. SIMS Al concentration profiles obtained inside the same samples as above (samples from structure F capped with RTCVD SiO$_2$) : a) after removal of the oxide layer before annealing; b) after a 10 min. anneal at 850°C and subsequent removal of the oxide layer ; c) after a 30 min. anneal at 850°C and subsequent removal of the oxide layer ; d) after a 4 hour anneal at 850°C and subsequent removal of the oxide layer.

Ga-implanted sample (dose = 10^{13} ions/cm^2, energy = 50 keV) displayed in Figure 2(e). While large blueshifts are obtained with this technique, we note that the QW PL peaks have suffered significant linewidth broadening and intensity loss and that the initial one-ML difference-related splittings observed after anneal have been completely washed out. These effects indicate the existence of spatial inhomogeneities i.e. variations in the degree of disordering across the QW which are attributable to the specific defect injection mode involved with this technique. Collision cascades are typically generated during implantation [69] and give rise to inhomogeneities in the plane of the layers, thus creating a microroughness on the scale of the exciton diameter [70]. Ion implantation is also intrinsically inhomogeneous in the depth direction since its range of action from the surface is determined by the energy of the incident ions [71]. In addition, ion implantation [8,11,15,18–22,45–55] as well as laser-induced disordering [42-44] also generate non radiative recombination centers that can significantly affect electronic properties of QWs and homogeneity.

Further insight into the behavior of RTCVD SiO$_2$ encapsulants with regard to homogeneity can be gained from the PL spectrum of Figure 2(c). We commented earlier on the good electronic properties conserved after anneal by the RTCVD SiO$_2$-capped sample. Perhaps the most striking feature is the fact that the peaks corresponding to the two narrowest wells still exhibit line splittings related to their initial one ML-thickness difference, even after strong intermixing. But these splittings are somewhat reduced as compared to the as-grown situation and now give 7.8 meV for QW$_2$, and 6.8 meV for QW$_3$. Although the measured blueshifts seem to indicate a stronger intermixing rate for QW$_2$, we extract the following interdiffusion lengths: $\Delta_1 = 1.9$ nm, $\Delta_2 = 3.2$ nm, $\Delta_3 = 4.8$ nm, for QW$_1$, QW$_2$, and QW$_3$, respectively. We stress the fact that for narrow wells undergoing significant interdiffusion, flattening of the barriers and effective widening of the QW subsequent to Al penetration inside the well will both tend to reduce carrier confinement, and therefore affect the measured blueshift. Since the interdiffusion length does not depend upon the QW's width or the barrier height but solely upon the Ga-Al interdiffusion process itself, the extracted interdiffusion lengths reveal that the intermixing rate is in fact higher for QW$_3$ than for QW$_2$, and therefore appears to be increasing with depth.

The simulation of the blueshift as a function of the well width for two different interdiffusion lengths (3.2 and 4.8 nm) is presented in Figure 9. The blueshift increases with decreasing well widths until it reaches a certain value below which it rapidly decreases. The significantly flattened error function-shaped potential obtained for QW$_3$ according to our measurements is also displayed on the inset of Figure 9. However, for very narrow QWs and very high intermixing rates, our data analysis can become strongly sensitive to uncertainties in the well width prior to annealing, in the observed blueshift, or in the exciton binding energy.

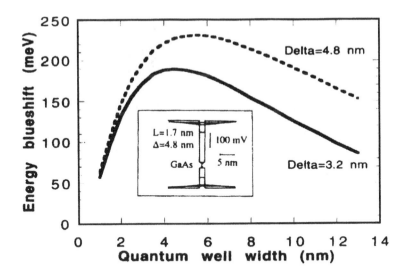

Fig. 9. Simulation of the energy blueshift as a function of the initial QW width for a given interdiffusion length. $\Delta = 3.2$ nm (solid line), $\Delta = 4.8$ nm (dashed line). Inset: conduction and valence bands, and first electron and hole subbands calculated for QW_3, in the as-grown case (square well) and after intermixing (erf-shaped well) assuming $\Delta = 4.8$ nm.

Following PL characterization, the sample was cleaved into two pieces in order to allow SIMS analysis and XTEM observations to be carried out in parallel. Shown in Figure 10(b) is a BF XTEM image of the annealed sample. The vertical arrow points to the clean SiO_2/GaAs interface. The three horizontal arrows indicate the positions of QW_1, QW_2, and QW_3, respectively from top to bottom. Figure 10(c) offers a DF XTEM image of the same region. Contrasts are fairly weak. The deeper the QW, the more blurry are its heterointerfaces. QW_3 has almost totally merged with the adjacent alloy barriers, but a wide strip extending over 1.7 nm in thickness still indicates the presence of a highly interdiffused QW. XTEM observations therefore qualitatively corroborate our PL analysis.

Additional evidence of the increasing interdiffusion rate with depth can be found in the SIMS profiles presented in Figure 11. We have superimposed the Al profiles obtained on an as-grown sample and on the annealed sample once the oxide layer had been removed from the surface in dilute HF. The three minima observed in the Al concentration, corresponding to the three QWs, are less pronounced after annealing, which is a signature of intermixing. The interdiffusion length can be extracted from such SIMS profiles. A specific interdiffusion length must be associated to the SIMS resolution, due to the fact that sputtering of the sample induces some mixing of the signals originating from the different layers probed [52]. We can then extract the

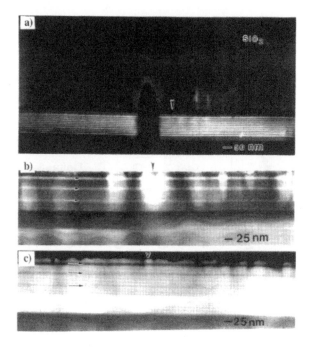

Fig. 10. a) DF (200) XTEM image of a sample from structure D annealed for 30 minutes at 850°C under RTCVD SiO$_2$ capping; b) BF XTEM image of a sample from structure E annealed for 10 minutes at 900°C under RTCVD SiO$_2$ capping, the horizontal arrows point at the three QWs; c) DF (200) XTEM image of the same region as on b) On each micrograph, the vertical arrow shows the SiO$_2$/GaAs interface.

following interdiffusion lengths from Figure 11 : $\Delta_1 = 2.1$ nm, $\Delta_2 = 3$ nm, $\Delta_3 = 3.5$ nm, for QW$_1$, QW$_2$, and QW$_3$, respectively. These values are once again in good agreement with the values deduced from PL data.

We noticed that the energetic splittings observed in the QW emissions had been significantly reduced after intermixing. In order to study the expected evolution of these PL splittings with the interdiffusion length, we performed a theoretical calculation, plotted in Figure 12 of the splitting measured after interdiffusion between the emission of a QW of initial width L and the emission of a QW of initial width L + (one ML), as a function of Δ. Both the L = 5 MLs and L = 9 MLs cases, corresponding to QW$_3$ and QW$_2$ respectively, have been considered. The same exciton binding energy for QWs of width L and L + (one ML) was assumed for our calculation. We note that the splittings measured prior to annealing agree well with the values calculated for $\Delta = 0$. For both cases, the splittings decrease with increasing interdiffusion length and eventually tend towards the same value, corresponding to band-to-band recombination in the barrier material.

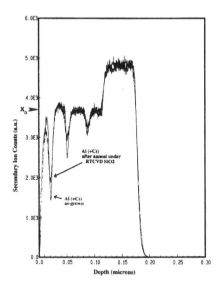

Fig. 11. Superimposed Al concentration profiles obtained in SIMS on two samples from structure E a) as-grown b) after a 10 minute anneal at 900°C under RTCVD SiO$_2$ capping and subsequent removal of the oxide layer. x_o indicates the average Al concentration contained in the AlGaAs barriers.

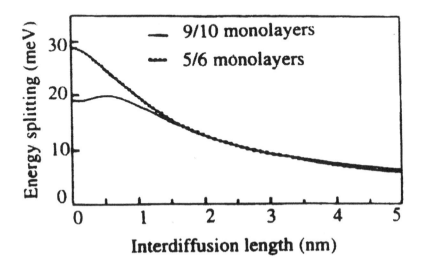

Fig. 12. Theoretical calculation of the energy splitting obtained between the emissions of a QW of initial width L and a QW of initial width L + (one ML) as a function of the interdiffusion length, for both L = 5 MLs (broken curve) and for L = 9 MLs (solid curve).

Deduced interdiffusion lengths compare successfully with our PL data: for our measured splittings, the same value of 3.2 nm is extracted for the L = 9 MLs case, and 4 nm is found instead of 4.8 nm for the L = 5 MLs case. These results thus appear to validate the large one-ML thickness difference islands hypothesis. Furthermore, they offer a convenient and reliable way of evaluating Δ in such samples, since measuring the energetic splitting frees us from having to take into account possible differences between the exciton binding energy in the square as-grown QW and in the intermixed QW.

In summary, our results indicate that, unlike ion implantation or laser-induced disorderings, very little broadening is observed with RTCVD SiO_2 encapsulation even though PL peaks have been permanently blueshifted by up to 180 meV. Indeed, for QW_3 a quasi-complete intermixing was achieved, as depicted on the inset of Figure 9 and on the DF micrograph of Figure 10(c). Nevertheless, a strong carrier transfer from the barrier inside the altered QW is still effective, as demonstrated by the very narrow and intense PL signals observed. Such results attest to the higher lateral homogeneity of this disordering process. Due to the higher quality of RTCVD deposits and moreover to the fact that only point defect diffusion is involved in this mechanism, highly interdiffused QWs show in their PL spectra a memory of their initial one-ML thickness variations. Furthermore, the conservation of ML-related splittings in PL spectra highlights the fact that narrow linewidths and the appearance of the latter splittings cannot be interpreted as a figure of merit for the abruptness of the QW heterointerfaces (along the direction of epitaxial growth), since the QW compositional profile is on the contrary highly graded after intermixing [38].

With respect to homogeneity in the depth direction, a parallel can be made between these results and both the XTEM observation in Figure 3(d) of a weaker intermixing rate in the SL located closest to the surface as opposed to the deeper located SL, and the SIMS observation of Figure 8 where we observed a smaller intermixing rate in the near surface region of a MQW even though an extremely uniform intermixing was achieved at depths of about half a micron. Additional PL and XTEM measurements were also performed on a sample from structure D (GaAs (5 nm)/AlAs (5 nm) MQW). Only one fairly broadened peak, blueshifted by over 100 meV, was observed in PL. Figure 10(a) shows a DF image of our structure after a 30 minute anneal at 850°C under RTCVD SiO_2 encapsulation. No significant depth dependence can be evidenced. All QWs seem indeed to have undergone the same interdiffusion rate. Moreover, no extended residual defects are observed. The achieved intermixing is not only homogeneous in depth, it also offers great homogeneity laterally, in the plane of the layers. In view of this new data, we trust that the peculiar surface effect observed on several other samples capped with a layer of RTCVD SiO_2 is indeed related to a local depletion in the number of V_{Ga} available in the near surface region (extending typically 30 nm below the SiO_2/GaAs interface) due to the presence of traps in the proximity of the SiO_2/GaAs interface.

2.5   Interdiffusion Model

We have demonstrated, through our study of the stimulation of intermixing by optimized RTCVD SiO_2 capping, that the strong enhancement effect observed was due to a significant decrease in the activation energy of the Ga/Al interdiffusion process under SiO_2 encapsulation. More specifically, we have attributed this decrease to a decrease in the vacancy formation energy as a consequence of the spontaneous generation of excess Ga vacancies under the SiO_2/GaAs interface following the fast out-diffusion of Ga inside the SiO_2 layer. A clear correlation could be established between the amount of Ga pumped inside the oxide encapsulant and the interdiffusion rate. We also showed that the special affinity of SiO_2 for Ga, which seems to depend solely on its chemical composition, could be interpreted as a sponge-like behavior. Indeed, presumably owing to the very fast diffusivity of Ga inside SiO_2, the oxide layer rapidly sucks Ga atoms up to its solubility limit. As expected, increasing the SiO_2 thickness, which has a direct consequence on the total amount of Ga the SiO_2 layer can absorb, was found to linearly increase the interdiffusion rate. Moreover, an increase in the limit of solubility was observed with increasing anneal temperature. Finally, we have shown how the interdiffusion enhancement mechanism could be separated into two distinctive diffusion phases, a very short first phase corresponding to the Ga pumping and simultaneous V_{Ga} penetration, and a second more prolonged phase corresponding to the Ga/Al interdiffusion process itself. Such a two-phase process had been tentatively suggested before by Guido and coworkers [27] but no explanation was proposed for the abnormally fast vacancy diffusion occurring in the transient phase, which results in a uniform distribution of vacancies throughout the depth of the heterostructure and establishes a quasi-stationary equilibrium. We propose to explain this phenomenon by the effect of the thermal stress imposed on the semiconductor by the encapsulating layer during high temperature anneal.

The thermal stress σ_{th} generated in a thin film deposited on a thicker substrate for a temperature variation ΔT is given by the following expression [72]:

$$\sigma_{th} = \left[\frac{E_f}{(1-\nu_f)}\right]\Delta\alpha\Delta T \tag{1}$$

where E_f and ν_f are respectively Youngs modulus and Poisson coefficient of the thin film, and $\Delta\alpha$ measures the difference in thermal expansion coefficient between substrate and film. Table 1 gives the elastic coefficients and thermal expansion coefficients found in the literature for SiO_2, Si_3N_4 and GaAs [73–75].

TABLE 1

Thermal expansion coefficient α and elastic coefficients of SiO_2, GaAs and Si_3N_4. E_f and ν_f are the Young modulus and Poisson coefficient, respectively.

Material	α(in C^{-1})	$E_f/(1 - \nu_f)$(in GPa)
SiO_2	0.52×10^{-6}	0.854×10^2
GaAs	6.86×10^{-6}	1.24×10^2(100)
Si_3N_4	2.8×10^{-6}	3.7×10^2

Thin SiO_2 films obtained by RTCVD are generally under compressive stress at room temperature [60]. A value of -0.2 GPa (where the $-$ sign indicates compression) was measured for our samples. The stress state during anneal will be the sum of this residual stress and the thermal stress as given from equation (1). For a 900°C anneal, we obtain a value of $+0.27$ GPa. We have assumed for this calculation that the increase in temperature does not modify the chemical nature of the oxide film or its intrinsic stress. In particular, we have disregarded the effect that the Ga pumping can have on the stress. Indeed, the presence of a large amount of interstitials in a material is known to induce a slight compressive stress. We thus describe here the stress state involved precisely at the moment of the V_{Ga} generation, not after. Moreover, we have assumed that the adhesion of the SiO_2 film to the GaAs substrate is not affected by the anneal, as seems to be the case through our XTEM observations. Due to its low thermal expansion coefficient as compared to GaAs, the stress state in the SiO_2 film changes sign during a 900°C anneal with respect to its value at room temperature. We emphasize the fact that our strain measurement set-up is not suited for high temperatures characterization and the exact value of this tensile stress could therefore not be directly measured. Thus, at the annealing temperature, the SiO_2 capping layer is under tensile stress and the surface of the semiconductor is under compressive stress. For equilibrium purposes [76,77], a stress gradient exists inside the semiconductor. Its compression decreases with depth until it reaches a neutral line near mid-thickness. The substrate is under increasing tension below that line.

Because of the small distortions they induce in the crystal lattice, point defects such as vacancies will elastically interact with a macroscopic external stress field. This effect is at the origin of the so-called Cottrell atmospheres [78] which take place in the highly stressed regions located around dislocations. It was observed that vacancies segregated towards the highly compressive regions formed around the heart of dislocations. It indeed seems quite intuitive that vacancies, which induce a slight tensile stress around their sites, would diffuse preferentially towards compressive regions where they can partially relax the imposed stress. By analogy, in our case, an elastic interaction will exist between the generated V_{Ga} and the external stress gradient. In the early stages of the anneal, V_{Ga} are injected into a compressive region which is thus in favor of their penetration. A driving force originating from the elastic interaction between the macroscopic thermal stress gradient and the vacancies will thus quickly push the V_{Ga} into the depth of the semiconductor. In this transient phase, the interdiffusion rate is very small. Vacancy-assisted atomic exchanges across heterointerfaces indeed require that vacancies visit the interfaces several times, or in other words that the vacancy diffusion be isotropic, which is clearly not the case in the highly directional diffusion induced in this transient phase. Once the end of first phase is reached, a dynamic equilibrium develops between the transport due to the stress gradient, the diffusion due to the concentration gradient, and the presence of traps under the SiO_2/GaAs interface. A steady V_{Ga} concentration, quasi-constant in depth is established, except for a small depletion under the SiO_2/GaAs interface. The vacancies are now free to diffuse around their sites and Ga/Al exchange is promoted. Intermixing develops with time until excess vacancies are trapped by undetermined sites of either recombination or agglomeration, located most likely at the epilayer/substrate interface.

We underline here that although this simple, qualitative, model is in agreement with all of our experimental results and provides an explanation for the guided diffusion of the excess V_{Ga} toward the depth, it also appears to be consistent with an influence of the thermal stress on the Ga pumping process itself. Indeed, a layer under tension — SiO_2 here — will tend to accept interstitials in order to relax its stress state. The Ga pumping effect was observed for all types of deposited SiO_2 layers, and it is clear that, regardless of the deposition technique, all SiO_2 films are under tension during anneal. The PECVD process we tested actually gives SiO_2 films which are already under slight tension at room temperature. The high temperature anneal will only accentuate further this state. We would therefore expect a larger amount of Ga to be pumped by these strongly tensile layers. The effect of thermal stress on V_{Ga} creation was also supported more recently by the work of Cusumano et al, [79] who found out that phosphorus-doped silica containing 5 wt% P effectively inhibited Ga-Al interdiffusion in GaAs/AlGaAs heterostructures, and attributed this property to a strain relaxation effect of the cap layer during annealing. Furthermore, the kinetics of the pumping should also be affected, the larger stress difference between the film and the surface of the substrate inducing a faster out-diffusion of Ga towards SiO_2. Such a behavior was indeed observed by Katayama et al. [80] in rapid thermal anneals performed for the same duration but at different heating rates, and they attributed this effect to the spreading and possibly the cut-off of the Si-O bonds due to the larger stress which caused an easier pass for Ga out-diffusion. However, in our experiments no significant difference in the Ga concentration between PECVD and RTCVD or RF-sputtered films could be evidenced in SIMS, while a smaller interdiffusion rate as compared to RTCVD layers was almost systematically observed.

As the SiO_2/Si_3N_4 patterning experiments presented in Sections 3 and 4 will later come to confirm, we believe the results reported here are a signature of the stress-dependence of vacancy diffusion, rather than a proof of the effect of strain on Ga out-diffusion and subsequent V_{Ga} creation. Although a tensile stress state in the SiO_2 layer is not essential for V_{Ga} creation, a compressive stress state in the substrate clearly appears to be indispensible for intermixing to occur. The dielectric deposition technique plays a major role, above all because it can affect the quality of the key $SiO_2/GaAs$ interface. Some V_{Ga} can be trapped below the $SiO_2/GaAs$ interface due to interface defects and roughness and thus become ineffective for the interdiffusion process, and pores in the SiO_2 layer can change the nature of the intermixing mechanism involved by inducing Si diffusion inside the semiconductor, presumably through reduction processes.

2.6   Advantages Compared to Ion Implantation

Ion implantation is a standard process in the semiconductor industry and has been widely investigated for intermixing purposes. Its main advantages are a good control and an excellent reproducibility. The depth distribution and the amount of implanted species are completely determined by the choice of the ion beam energy, the implanted dose and the nature of the target material. The penetration and progressive slowing down of the implanted ions inside the material give rise to collisions with the target atoms. If the striken atoms receive enough energy to be kicked out of their cristalline site, secondary collisions take place and collision cascades develop, thus giving rise to the formation of a large number of defects including both vacancies and interstitials. The injected defect distribution is also predetermined by the implantation parameters but will be modified by a subsequent thermal annealing. This post-implant anneal plays a crucial role in the intermixing process by allowing implanted species and point defects to diffuse. Intermixing is stimulated and point defects can recombine therefore repairing part of the damage induced by the ion bombardment. At the same time, some defects agglomerate and form extended residual defects which are highly prejudicial to the electronic properties of the interdiffused material.

Here we will mostly center our comparison on Ga^+ implantation-induced intermixing. Ga^+ is one the ions most investigated for the interdiffusion of GaAs/(Al,Ga)As heterostructures [8,11,15,19–22,45–55] and was the subject of in-depth studies by C. Vieu and coworkers [22,51–53,69]. Ga being a matrix element in Ga(Al)As. Ga^+ implantation-induced intermixing is thus an impurity-free technique and the comparison between this technique and

SiO_2 capping-induced intermixing is thus particularly relevant. In addition to its non-dopant character, Ga^+ is also attractive because of its good nuclear efficiency with respect to the collisional mechanisms of disordering in AlGaAs-based heterostructures, and also because a stable Ga liquid metal ion source (LMIS) can be used in a focused ion beam (FIB) system.

Like SiO_2 encapsulation and annealing, Ga^+ implantation is an effective V_{Ga} source. However, point defect injection is fundamentally different in both cases. Although Ga^+ implantation allows a precise amount of excess Ga vacancies to be created at a desired depth, many other defects less useful intermixing-wise are also created in the process. The most undesirable ones being extended defects such as dislocation loops, mostly of interstitial nature, and resulting from the agglomeration of point defects during the thermal anneal or even during irradiation. Moreover, in contrast with SiO_2 encapsulation which induces a very homogeneous distribution of the generated vacancies inside the heterostructures, the collision cascades created by ion implantation lead to a heterogeneous distribution of the defects.

Typically, Ga^+ implantation of AlGaAs-based heterostructures is carried out at energies between 20 keV and 400 keV and doses ranging from 10^{13} ions/cm^2 to 10^{15} ions/cm^2. As expected, the higher the implanted ion dose, the higher the amount of injected point defects and therefore the higher the intermixing rate. However, two distinct behaviors can be observed depending on the implanted dose. At low doses ($< 1 \times 10^{14}$ ions/cm^2), the collisional intermixing generated during implantation is weak and interdiffusion is enhanced by the diffusion of excess V_{Ga} during thermal annealing. The concentration of residual defects after annealing is small. Due to the statistical distribution of implantation cascades, at low doses the post-implantation distribution of defects is not homogeneous and the induced intermixing is thus non uniform. Intermixing rate remains low but recovery of the electronic properties is fair. In contrast, at high implanted doses ($> 1 \times 10^{14}$ ions/cm^2), collisional intermixing becomes preponderant. The diffusion of active defects during annealing is limited due to a rapid coalescence into extended defects which are unavailable to help in the interdiffusion, and are responsible for PL quenching in interdiffused QWs. However, overlapping of individual collision cascades leads to a more uniform intermixing. Intermixing rate increases rapidly with dose and complete intermixing of the layers can be achieved.

We have given evidence for the outstanding homogeneity of SiO_2 capping-induced intermixing, both laterally and in the depth direction. Laterally homogeneous Ga^+ implantation-induced interdiffusion is feasible but at the cost of extended defects. Homogeneity in the depth direction is on the other hand difficult to obtain via ion implantation. However very deep implants can be performed and buried intermixed regions achieved by using high energy ions in the meV range [81], which can a priori not be obtained via SiO_2

encapsulation. It was recently suggested [82] in the case of low implantation doses of Ga, As and Kr in InGaAs/GaAs heterostructures that injected defects were actually diffusing very quickly in the very early stages of the anneal and building a uniform concentration throughout the heterostructures, thus, leading to an even intermixing. However, interdiffusion behaviors are slightly different in AlGaAs/GaAs and InGaAs/GaAs systems, as reported by Allard et al. who observed a much lower threshold dose for ion-induced intermixing in InGaAs/GaAs than in GaAs/AlGaAs [83].

Another consequence of the presence of residual extended defects is the loss of electronic properties of the implantation-intermixed material. Indeed, if residual defects are present in the vicinity of QWs, non radiative recombination channels are available causing a decrease of the intensity of the emitted luminescence. High concentrations of non radiative centers completely quench the PL signal. In addition, broadening of the PL peak of interdiffused QWs is an indication of the spatial inhomogeneity of the mixing mechanism. Moreover, due to the induced disorder, there is no strong evidence that after interdiffusion the mechanism of luminescence still presents an excitonic character. While this change can be of little importance for application to the fabrication of passive devices, it becomes redhibitory for the realization of active structures. Implantation of lighter ions could be used to attenuate this major drawback. For instance, Tan and coworkers have shown that proton irradiation and subsequent RTA could be effectively used to generate a high concentration of point defects with minimal formation of extended defects and therefore attain both a high intermixing and good recovery in PL intensities [84]. However, as the conservation of one ML-related PL signatures after annealing demonstrates, SiO_2 capping-induced intermixing guarantees the conservation of luminescence mechanisms. As we will show in the next section, the presence of clear excitonic features in strongly intermixed QWs further reinforces this point.

While SiO_2 capping-induced intermixing is highly material-dependent and closely linked to the quality of the deposited oxide layer, ion implantation-assisted intermixing is basically material-independent. However in the case of Ga^+ implantation, the distribution of implantation damage as well as the microstructure of residual defects after annealing were found to be very sensitive to the thickness and composition of the different layers forming the heterostructure [20,51]. For instance, implantation damage accumulates rapidly in GaAs layers, while a large amount of self-annealing is promoted in AlAs layers. This phenomenon combined with intermixing effects enables a delay in the damage build-up in a GaAs layer when adjacent AlAs layers are deposited at each heterointerface. Many devices use QWs as active layers and the presence of agglomerated defects in the QWs will of course strongly alter their electronic properties. Nevertheless, the ability of AlAs layers to prevent damage accumulation in the GaAs QWs during implantation and to drain the defects far away from the QW during implantation can be exploited in the design of optimized heterostructures.

In conclusion, high quality interdiffusion induced by SiO_2 capping-induced intermixing, in particular when RTCVD SiO_2, makes it an attractive alternative to widely-used ion implantion for various applications. Furthermore, when a strong selectivity between intermixed and non-intermixed regions, for instance for quantum wire fabrication where a high barrier potential is absolutely indispensable to achieve a strong carrier confinement, the high implantation dose regime will have to be chosen. As will be detailed later in Section 3.7, the density of extended residual defects can then become very harmful for the observation of quantum confinement signatures. The latter point was one of our strongest motivations for investigating the lateral selectivity of SiO_2 capping-induced intermixing.

3   SELECTIVE INTERDIFFUSION UNDER PATTERNED SIO_2 CAPPING LAYER AT A NANOMETRIC SCALE

3.1   Purpose and Principle

Fabrication of low dimensional semiconductor quantum structures has attracted great attention in recent years since new physical phenomena leading to significant improvements for lasers and other optoelectronic devices are expected [85–88]. Among the various techniques investigated to realize these quantum wires or dots, spatially selective intermixing of quantum well (QW) heterostructures is a post-growth technique presenting attractive potentialities for future integration. For GaAs/AlGaAs QWs, the effective band gap is enlarged in intermixed regions, which can thus provide built-in compositional lateral barriers for carriers.

As studied in Section 2, under controlled conditions of annealing, GaAs/AlGaAs QW heterostructures exhibit an enhancement of the Ga-Al interdiffusion coefficient when capped with a layer of silicon dioxide. We have shown that high quality SiO_2 films obtained by rapid thermal chemical vapor deposition (RTCVD) could promote a strong and homogeneous impurity-free layer disordering of GaAs/AlGaAs QWs located more than 100 nm below the surface, after a 10 min. anneal at 900°C. While several experimental studies have been devoted to disordering under uniform SiO_2 layers or to spatially-selective intermixing involving patterned areas of dimensions of relevance to the confinement of photons [26,29–31], no work has yet been reported on the lateral resolution of this process when trenches of nanometric dimension are etched into the SiO_2 film, at a scale compatible with the lateral quantum confinement of electrons and holes.

In this section we give optical evidence of the realization of quantum wires (QWRs) by SiO_2 capping-enhanced intermixing, and show how the high lateral selectivity of our process can be related to the effect on V_{Ga} diffusion of the stress field distribution generated during anneal under the SiO_2 film edges [38–39,89–91].

3.2   Nanofabrication Processes

Two different undoped QW heterostructures were grown by molecular beam epitaxy for this study. The first structure (structure G) comprises a single 3.7 nm-thick GaAs QW bordered by two 1 nm-thick layers of AlAs, further embedded in two 50 nm-thick $Ga_{0.67}Al_{0.33}As$ barriers. The structure being capped with a final 10 nm-thick GaAs layer, the distance from the QW to the surface is 60 nm. The second structure, already described in Section 2 (structure D), is a multi-QW (MQW) consisting of a 20 nm-thick $Ga_{0.67}Al_{0.33}As$ barrier, 10 GaAs (5 nm)/AlAs (5 nm) QWs, a 20 nm-thick $Ga_{0.67}Al_{0.33}As$ barrier, and a final 5 nm-thick GaAs cap layer. Samples were coated with a 400 nm-thick layer of higher quality RTCVD SiO_2.

High resolution electron beam lithography (HREBL) of a positive organic resist (PMMA) in association with a lift-off process of a 70 nm-thick Al film was used next to generate arrays of metallic lines on previously mesa-etched regions (80 μm × 80 μm) of the samples. The SiO_2 film was then patterned via reactive ion etching (RIE) in a SF_6/CH_3 gas mixture, using the Al lines as an etch mask. Typically, trenches of width 100 nm or less were etched into the SiO_2 film, at periods varying from 200 nm to 20 μm. After Al mask removal in dilute NaOH, thermal annealing was performed in a conventional furnace, for 15 minutes at 850°C, using the enhanced As overpressure GaAs proximity cap technique described earlier.

Samples were characterized by low temperature (2K) photoluminescence (PL) using an argon laser ($\lambda = 488$ nm, $P \sim 1$ mW for a 50 μm spot size), before and after oxide deposition as well as after heat treatment with and without the SiO_2 encapsulation layer. Additional signatures of the lateral confinement of carriers into QWRs were investigated through PLE coupled with linear polarization anisotropy analysis.

3.3   Conditions for Laterally Selective Intermixing

Figure 13 illustrates the effect of the distance between trenches etched in the SiO_2 film upon the lateral selectivity of the intermixing process. Plotted are the energetic position of the PL emission peaks, recorded for arrays of approximately 100 nm-wide trenches, as a function of the SiO_2 surface coverage. Figure 13(a) corresponds to arrays located on different mesas of the single QW (SQW) sample and Figure 13(b) to arrays on the MQW sample. All PL measurements reported here were obtained after annealing with the oxide film still on the surface. The spectra collected prior to annealing with the patterned SiO_2 film on top did not reveal any modification of the

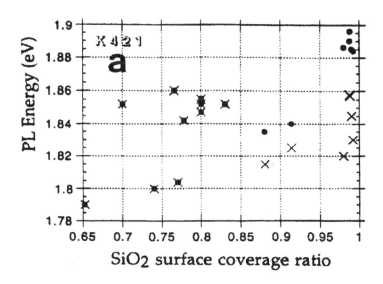

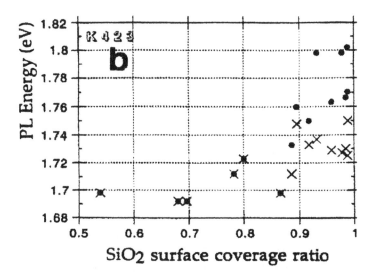

Fig. 13. Energetic position of the PL peaks obtained after anneal as a function of the SiO_2 surface coverage ration for arrays of approximately 100 nm-wide trenches. (a) was obtained on the SQW structure (structure G) and (b) was obtained on the MQW structure (structure D). Crosses and dots represent the low and high energy peak, respectively. Overlapping symbols indicate that a single PL peak was observed.

PL emission as compared to as-grown samples. Likewise, PL properties were not affected by the annealing step alone as could be verified through measurements performed on uncapped regions of the samples. In contrast, for both samples, a strong disordering occurs under uniformly capped regions: for the SQW structure, the PL peak is blueshifted from 1.767 eV to 1.896 eV with respect to the as-grown situation. Similarly, a blueshift from 1.690 eV to 1.800 eV is observed for the MQW, and we do not observe a strong broadening which would have indicated that various QWs of the structure could have experienced different intermixing rates. We thus conclude that the disordering is quite homogeneous in the depth direction down to at least 130 nm below the surface. As can be seen in Figure 13, different situations are encountered after annealing when the SiO_2 film is patterned. When the SiO_2 coverage ratio is lower than 0.87, namely when the period of the array is shorter than 750 nm, only one PL peak is observed, while at larger separation between wires, two emission peaks are clearly indentified. This behavior turns out to be very reproducible and is observed for both the SQW and the MQW structures.

In order to determine the spatial origin of each of the luminescence peaks observed on selectively interdiffused arrays, we performed spatially resolved photoluminescence (SRPL) experiments on several arrays from the MQW sample. For SRPL measurements, an argon laser of spot size significantly smaller (~ 2 μm in diameter) than the argon laser used in our standard PL experiments is focused on a chosen region of the sample. A piezoelectric system is used to move the sample and high precision control of the positioning is obtained via servo control and a CCD camera/Nomarski filter ensemble. A precise cartography of the sample can thus be obtained using this technique. Typically, a transversal scan is performed across the array of interest at a fixed detection energy, and corresponding changes in the PL signal intensity are recorded. This technique is similar to cathodoluminescence in that it allows regions emitting at a given energy to be localized independently. The resolution of SRPL is limited by the laser spot size and arrays of period larger than 4 μm have thus been investigated. Figure 14 displays the intensity modulations observed on an array of period 8 μm and trench width 120 nm.

Two distinct PL emissions had been observed beforehand at 1.716 eV and 1.765 eV. Two 15 μm-long scans were carried out, fixing the detection energy at 1.716 eV (Figure 14(a)) first and 1.765 eV (Figure 14(b)) next. As can be clearly seen, both curves are complementary: low energy and high energy emissions originate from distinct regions of the array. Furthermore, the two low energy emissions are indeed 8 μm apart. Moreover, the FWHM of the corresponding peaks is roughly 2 μm, which is the resolution of the experimental set-up. The actual width of the regions emitting at low energy is thus very likely much smaller than 2 μm.

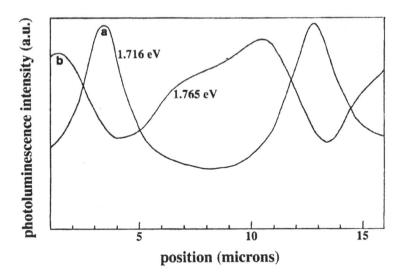

Fig. 14. Spatially resolved PL spectra obtained by scanning an array of period 8 μm and trench width 120 nm from structure D. a) detection energy was fixed at 1.716 eV b) detection energy was fixed at 1.765 eV.

These results therefore unambiguously indicate that in arrays of well separated wires the low energy PL peak originated from the narrow regions located under the trenches, while the high energy PL was emitted by the large regions covered by the SiO_2 strips. Thus, for relatively distant wires, selective interdiffusion can be achieved. As expected, the regions of the QWs located under the openings are preserved from the interdiffusion, while the adjacent regions capped by the SiO_2 strips are strongly interdiffused barriers. When the distance between wires is reduced below 650 nm, the interdiffusion is no longer laterally selective. Intermixing rates under the openings and under the SiO_2 strips are comparable, leading to a single PL peak slightly blue-shifted with respect to the reference.

3.4   Influence of SiO_2 Removal After Annealing

The true post-anneal situation can only be observed once the SiO_2 has been removed from the surface. Effects of SiO_2 removal on PL spectra are illustrated in Figure 15, obtained for an array of period 5 μm and trench width 70 nm from structure D. While the spectrum before oxide removal exhibits only 2 peaks, we distinctly see a third peak appearing at higher energy on the spectrum obtained after SiO_2 removal which reveals that there are actually three regions of different intermixing rates in our arrays. Moreover, the initial

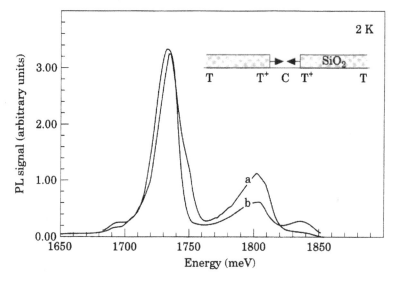

Fig. 15. PL spectra obtained after anneal on an array of 70 nm-wide trenches an period 5 μm realized on structure D, (a) with the SiO_2 film still on top, (b) after SiO_2 removal. Inset: schematic illustration of the stress field distribution generated at 2 K in the heterostructure by the patterned oxide layer.

high energy peak is slightly blueshifted while the initial low energy emission is slightly redshifted.

These spectral changes can be well understood if we take into account stress relaxation subsequent to the removal of the patterned surface oxide layer. Due to thermal expansion mismatch between GaAs and SiO_2, the stress field distribution schematically depicted on the inset of Figure 15 is generated below the patterned oxide film at 2K. Under the openings, the semiconductor is under compression while it is under tension below the oxide strips, with an enhanced region of tensile stress located under the film edges. It is well-known that stress affects the electronic properties of GaAs/AlGaAs QWs [92] and that a QW under tension (or respectively, compression) will experience an effective bandgap shrinkage (or respectively, increase). Thus, removing the surface stressor will allow stress relaxation and, as observed, regions initially under compression — wires — will have their PL shifted towards lower energies while regions initially under compression — barriers — will show a blueshift. Hence, it seems likely that two interdiffused barrier regions of different intermixing rates, indicated by the presence of two high energy PL peaks after SiO_2 removal, would have merged luminescence prior to SiO_2 removal due to the difference in their initial tensile states: regions of strongest intermixing rate located under the edges will experience a stronger PL redshift

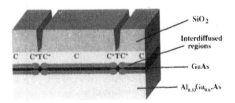

Fig. 16. Schematic cross-sectional view of the quantum wire structure and stress field distribution generated during anneal. C and T indicate regions under compression and tension, respectively.

than regions of both lower intermixing rate and tensile state located under the SiO_2 strips farther away from the edges. A major implication of such a result is that this technique spontaneously generates so-called double-barrier QWRs, where an increased barrier height exists at the wires' borders, hence providing improved lateral confinement.

3.5   Role of Stress on the Interdiffusion Mechanism

If we consider the thermal stress distribution generated during annealing in the heterostructure, we find it is roughly inverted with respect to the 2 K situation: a tensile region is generated under the trenches while regions below SiO_2 are under compression, with a higher compressive state at the edges. We now show that the presence of regions of higher intermixing rates under the films edges and the existence of adjacent non-interdiffused wires of width down to 20 nm can be explained by taking into account a driving force acting on the diffusion of point defects, originating from the thermal stress field distribution generated at the annealing temperature in the heterostructure. It is indeed well known that the presence of a vacancy in a lattice creates a slight local tensile state. Hence an elastic interaction exists between the vacancy and the macroscopic stress field. We believe that excess V_{Ga} created under the SiO_2/GaAs interface will therefore be drained towards the higher compressive regions located under the film edges while the adjacent strongly tensile regions located under the trenches will act as barriers to this vacancy diffusion. Such a stress-driven anisotropic diffusion can thus give rise to high lateral resolution of the intermixing process i.e. steep lateral potential modulation which would be impossible to obtain through a simple isotropic diffusion process. This situation is illustrated in Figure 16 which gives a schematic representation of the stress state generated during annealing temperature in our arrays and shows the effect of the induced intermixing on the QWs shape.

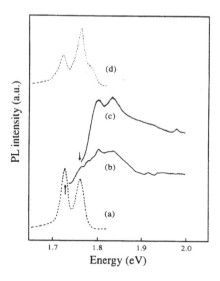

Fig. 17. PL spectra (dashed lines) and PLE spectra (solid lines) obtained after anneal on an array of 85 nm-wide trenches and period 3 μm from the MQW structure. (a), (b) and (c) were measured with the SiO$_2$ still on the surface while (d) was recorded after SiO$_2$ removal.

For our larger period arrays, intermixing is laterally selective, thus diffusion of the excess V$_{Ga}$ generated under the SiO$_2$/GaAs interface is highly anisotropic, while in shorter period arrays, the stress fields of adjacent wires begin to overlap and V$_{Ga}$ diffusion becomes more isotropic, and thus lateral selectivity is lost. In accordance with our experimental observations, calculations of film edge-induced stress [93] show that pattern geometry is a crucial parameter since more pronounced stress states, and thus expected anisotropy, are obtained best for very narrow openings and large separation between adjacent trenches.

3.6   Fabrication of Quantum Wires and Their Optical Properties

Typical PL and PLE spectra obtained on selectively interdiffused arrays are shown in Figure 17. Figures 17(a)–(c) were obtained on a 3 μm period array of 85 nm-wide trenches from the MQW sample with the patterned oxide layer still on the surface. The PL spectrum obtained after oxide removal on that same array is shown in Figure 17(d) for comparison. Two distinct emission peaks are clearly identified in the PL spectrum of Figure 17(a). As otherwise confirmed by SRPL, the low energy peak — which is slightly blueshifted with respect to the PL peak corresponding to large uncapped and thus non-intermixed reference regions — is the wire emission originating from the narrow regions located under the trenches, and the high energy peak corresponds to the strongly interdiffused barriers created under the SiO_2 strips. We notice first that the PL spectrum exhibits two separate peaks of comparable intensities. Since the wire to barrier surface coverage ratio of these arrays is typically lower than 5% (less than 1% for the 20 μm period array where intense luminescence originating from four isolated wires is measured), this high wire emission indicates a very efficient collection of carriers by the wires. The spectral changes observed after removing the patterned SiO_2 layer, which we discussed in the previous sections, are again clearly visible in Figure 17(d).

Results of photoluminescence excitation measurements are particularly striking. As can be seen in Figure 17(c) when the detection energy is adjusted at the position of the high energy PL peak, we observe two clear resonances in the PLE spectrum corresponding to the heavy hole (hh) and light hole (lh) excitonic transitions. Such conservation of excitonic properties in strongly interdiffused QWs attests for the outstanding homogeneity of the disordering induced under SiO_2 and for the higher quality of the material forming the lateral barriers. In other selective interdiffusion techniques such as ion implantation [8,11,15,18–22,45–55] or laser-induced intermixing [42–44], non radiative recombination centers — often extended residual defects — alter both the quality and the homogeneity of intermixed regions and excitonic peaks are lost in PLE. When detection corresponds to the low energy PL peak (Figure 17(b)), we continue to observe excitonic absorption due to the interdiffused regions which demonstrates an effective carrier transfer from the lateral barriers towards the wire. Furthermore, striking features appear at lower energy on the absorption edge, which can be attributed to the formation of one-dimensional (1D) subbands.

Signatures of the lateral confinement of carriers into quantum wires were futher investigated through PLE coupled with linear polarization anisotropy analysis. Experiments were carried out after removal of the SiO_2, on the MQW structure. Figure 18(a) shows the spectrum obtained on the same array as that of Figure 17, with the detection energy adjusted at the position of the lower energy PL peak, corresponding to the wire emission. The distinctive features appearing on the absorption edge and attributed to transitions due to 1D subbands, are once again observed in the PLE spectrum after SiO_2 removal and appear to be very reproducible. At the same time, we observe a strong anisotropy of the linear polarization along the wire axis for the first 1D transition, as predicted by theory. In contrast, Figure 18(b) presents the spectra obtained on an uncapped reference region. The PLE spectrum exhibits the two traditional heavy hole and light hole exciton peaks while the polarization signal is flat, thus indicating no significant polarization anisotropy as expected for a 2D QW.

A simple calculation of the lateral confining potential obtained in our structures can be performed from PL data. The generally accepted model treats confinements along the vertical and the lateral directions independently

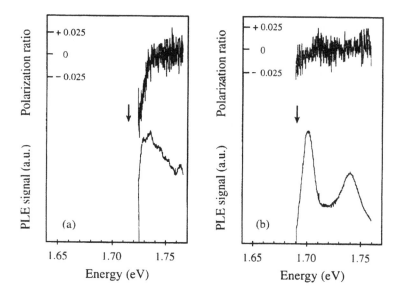

Fig. 18. PLE and linear polarization spectra obtained on the MQW sample after anneal and SiO$_2$ removal on (a) an array of 85 nm-wide wires and period 3 μm, (b) a 2D reference region. Arrows recall the position of the corresponding PL peak.

and assumes error-function-like shapes for both potentials [46]. The transfer matrix procedure [94] is used to calculate the electronic levels in such a potential, with a band offset of 0.67 [95]. It should be noted than an exact calculation of the electronic levels in our wires would however require a 3D solution of the Schrödinger equation for the real spatial modulation of the potential induced by the disordering of the QW with different degrees of intermixing across the wire. The lateral potential is characterized by a lateral interdiffusion length Δ_l which is an indicator of the anisotropy of the diffusion i.e. of the steepness of the lateral bandgap modulation, and by a lateral barrier height V_B given by the experimental PL blueshift observed for the border barrier emission. We can then distinguish between two components inducing the blueshift observed in the QWR emission: a ground-state quantization energy E_{1D} and an energy E_{Al} which represents the effects due to the penetration of Al at the center of the wire. Typical values obtained for these parameters are reported in Table 2.

High confinement energies in the 10–20 meV range are obtained for the narrowest wires while lateral interdiffusion lengths as short as 6 nm show the achievable steepness of the potential modulation, as compared to the best value of 8 nm reported in the literature for a selective interdiffusion technique [54,55]. We outline the fact that in contrast with other intermixing techniques where Δ_l is typically a constant, here anisotropy varies with pattern geometry:

TABLE 2
Typical parameters obtained for the lateral confining potential.

Sample	Array (nm) opening/period	V_B (meV)	Δ_t (nm)	E_D (meV)	E_{Al} (meV)
SQW	20/5000	137	6	15.5	32.5
SQW	40/4000	130	9.5	10.4	16.6
MQW	70/5000	100	21	5.5	34.5
MQW	75/4000	145	19	4.3	16.7

the narrower the wires, the shorter Δ_t and thus the higher E_{1D}. Moreover, the stronger intermixing achieved under the SiO_2 edges guarantees large lateral barrier heights (~ 130 meV). Strikingly, anisotropy is lost for wider wires, as in the case of the MQW structure, resulting in lower confinement energies due to flattening of the lateral potential. Such a behavior is consistent with aforementioned stress field calculations where more pronounced stress states, and thus expected anisotropy, were obtained for narrower openings.

The conduction and valence band profiles along the direction perpendicular to the wire axis calculated for an array of wires of width 20 nm and period 5 μm located on the SQW sample, have been represented in Figure 19. The calculated electron and hole levels are also indicated. The confinement along the tranversal direction x was considered independently from the confinement in the growth direction given by epitaxy. Several 1D confinement effects, such as heavy hole and light hole mixing in the valence band and excitonic effects [96], are not taken into account in our calculation.

The calculation predicts an energetic difference of roughly 20 meV between the first two electron-hole transitions. We notice that due to lateral interdiffusion, the potential is actually narrowed with respect to the initial 20 nm trench width. Indeed, the width of the lateral potential modulation at the position of the first electronic level is 14 nm, which is thus the QWRs effective width.

The shape of the two dimensional potential which confines carriers inside QWRs has been tentatively represented in Figure 20 for the conduction band, for the same array as in Figure 19. Figure 20(a) and Figure 20(b) are views of that same potential rotated 90° from each other. Far from the center of the wire, we observe the classical error function shape of the interdiffused QW along the z direction. When we get closer and closer to the wire center (moving towards the middle), the QW progressively experiences a reduced intermixing rate. At the wire center, the QW has been effectively protected (but not completely) from the interdiffusion and the shape along z tends to a square well potential. Passing through the wire region, we progressively find again the error function shape of the strongly interdiffused QW. This spatial modulation of the potential leads to the formation of a small valley under the

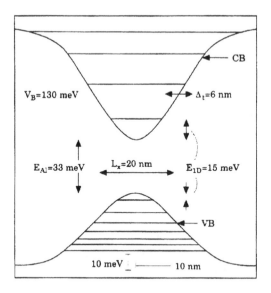

Fig. 19. Representation of the lateral potential modulation in the conduction band for an array of wires of width 20 nm and period 5 μm located on the SQW sample, with lateral interdiffusion length of 6 nm and barrier height 130 meV.

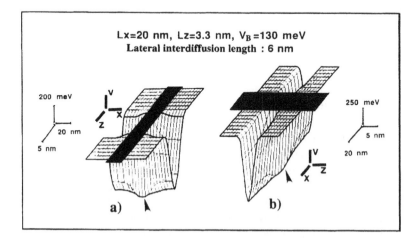

Fig. 20. Representation of the confining potential in the conduction band for a 3.3 nm-wide SQW (structure G), a 20 nm-wide trench, with lateral interdiffusion length of 6 nm and barrier height 130 meV. The z axis is oriented along the growth direction while the x axis is perpendicular to the direction of the wire. The black stripe represents the projection in the (x, z) plane of the regions located under the opening. The arrow points to the zone of 1D confinement.

wire center. This valley is constant along the x direction, parallel to the wire axis: it represents the 1D confinement zone. It is clear that the real potential is a complex function $V(x, z)$ which cannot be easily separated along the two directions of confinement. The calculation of the electronic levels for the conduction band and the valence band in such an intricate potential remains critical. However, according to the difference in dimensions between the QW thickness (3.3 nm) and the wire width (20 nm), we believe that our approach, which treats separately confinements along z and x, should give the correct order of magnitude of the lateral confinement effects. We point out here that since our technique spontaneously generates bordered QWRs, a slight decrease in the lateral potential amplitude should be represented on the potential when we moves away from the edges of the wire. However, this characteristic does not have a direct influence on the lateral barrier height which stays fixed by the potential created in the bordering regions. Moreover, the exact lateral extension of these stronger interdiffused region is not known.

In summary, we have reported in this section on the fabrication of double barrier QWRs by selective intermixing induced in GaAs/AlGaAs QW heterostructures by SiO_2 capping and annealing. Signatures of the confinement of carriers inside 1D structures have been observed in PL, PLE and linear polarization analysis. The qualitative interdiffusion model directly involving the effect of the stress field on point defect diffusion that we have proposed successfully accounts for the high lateral resolution achievable with this technique when narrow trenches are etched at relatively large periods (>750 nm).

3.7 Comparison with Quantum Wires Fabricated by Localized Ion Implantation

Ion implantation is a highly reproducible process which allows a precise control of the amount of injected defects and, in consequence, of the induced intermixing. In order to realize QWRs by ion implantation-induced interdiffusion and be able to probe their properties through optical spectroscopy, several requirements must be met. First of all, implantation has to be spatially localized and the lateral selectivity of the induced intermixing must be compatible with lateral carrier confinement. Next, the technique must be impurity-free and the choice of the implanted ion is thus restricted to inert species.

In AlGaAs/GaAs heterostructures, Ga^+ implantation was a ideal candidate and was investigated for this purpose by various groups [11,45–55]. Defect injection can be restricted to selected areas of a sample either through the use of masks deposited on the surface which completely stop incident ions [11,45–47,49,50–55,97], or by using a Ga^+ focused ion beam (FIB) [48]. In the broad beam case, nanometric regions can be effectively preserved from irradiation via a combination of high resolution electron beam lithography and metal lift-off at dimensions close to 20 nm. In the case of FIB, systems delivering a 10 nm spot size are available nowadays [98].

It appeared that high implanted Ga^+ doses were necessary to achieve optimal lateral resolution (minimization of active defect diffusion during annealing), a strong lateral barrier (large interdiffusion lengths provided by the numerous injected point defects) and a uniform intermixing in the implanted regions (overlap of the individual collision cascades). Moreover, the QWRs fabricated via Ga^+ implantation were typically realized on heterostructures containing QWs located very close to the surface. This condition is required in order to achieve very narrow wires for which the 1D quantum confinement effects are the more pronounced. Indeed, the deeper the region that must be intermixed, the higher the implantation energy that must be used. Unfortunately, the higher the energy, the higher the lateral defect dispersion and, thus, the poorer the lateral resolution. Under optimal conditions, complete intermixing can be obtained by Ga^+ implantation and typical lateral barrier heights of 100–200 meV are reproducibly achieved. The lateral interdiffusion lengths reported in the literature are as low as 8-10 nm. It is this high lateral selectivity that made people believe that, among selective interdiffusion techniques, ion implantation was the only one compatible with quantum carrier confinement.

As we have shown, high lateral barriers are also achievable with SiO_2 capping-induced intermixing. A precise control of the interdiffusion rate can be foreseen through a SIMS calibration of the amount of Ga pumped at fixed anneal temperature and SiO_2 thickness. A judicious choice of the geometric patterns should allow the fabrication of QWRs in a reproducible fashion as well. Moreover, we demonstrated that although our technique was solely based on vacancy diffusion processes, a lateral resolution as high as the one obtained by ion implantation, indeed better ($\Delta_l = 6$ nm), could be achieved due to diffusion anisotropy guided by the stress field. Like ion implantation depth dependence could become critical since lateral resolution is intrinsically linked to the stress field distribution imposed by the patterned SiO_2 layer, however, even on MQW structures containing QWs down to 120 nm, no such sensitivity could be observed. Calculated 1D confinement energies are also very close and relatively large (10–15 meV) for both techniques.

However, the true advantages of SiO_2 capping as compared to Ga^+ implantation lie in the quality of the fabricated nanostructures.

SiO_2 encapsulation-based interdiffusion gives rise naturally to bordered QWRs and guarantees excellent quality of the created lateral composition barriers. In constrast, conservation of electronic properties in strongly interdiffused QWs is never observed in the case of ion implantation. Due to the presence of unavoidable residual extended defects after annealing, the intermixed material is often highly degraded and full of non-radiative centers that quench PL intensity. Furthermore, the development of collision cascades during ion bombardment leads to a non-uniform interdiffusion, and although this statistical heterogeneous distribution of implantation defects can be smoothed by increasing the dose and the irradiation temperature, the intermixed material remains intrinsically disordered and rapidly looses its excitonic properties. In addition, the ultimate physical properties of the QWRs fabricated by ion implantation appear to be limited by the fluctuations of the confining potential along the wire axis which are the direct result of the spatial inhomogeneity of the intermixing due to the damage cascades.

We have given evidence for the high spatial homogeneity of SiO_2 encapsulation-induced intermixing and this strong asset made possible the PLE detection of reproducible features in the absorption edge which could be attributed to 1D subbands. Furthermore, to our knowledge, no effects of linear polarization anisotropy were ever reported in the literature for QWRs fabricated by ion implantation. The pronounced effect we observe leads us to believe that although confinement amplitudes are comparable in both competing techniques, regularity and barrier material quality are definitely superior via RTCVD SiO_2 capping and patterning. For these reasons, novel effects predicted by theory for QWRs but never clearly evidenced up to now in structures realized by other selective interdiffusion techniques could be observed.

Besides, the poor quality of the material intermixed by ion implantation renders the fabricated wires very difficult to exploit in active devices, such as QWR lasers for instance. On the other hand, the large minimal lateral barrier width necessary for the fabrication of QWRs by SiO_2 capping seems a significant obstacle for such applications where close-packed QWRs are necessary for gain to be achieved. All pattern geometries cannot be realized with this technique due to its high dependence upon stress field distribution. If, as we trust, this inconvenience can be overcome by a better control of the stress fields, the high quality of the QWRs fabricated by SiO_2 capping and patterning would then make them the ideal candidates for such sought-for devices.

4 SELECTIVE INTERDIFFUSION UNDER PATTERNED SIO_2/SI_3N_4 LAYERS

4.1 Purpose and Principle

Silicon nitride has long been used for passivation in III-V semiconductor device fabrication technology. More recently, Si_3N_4 capping of undoped GaAs/AlGaAs QW was found to effectively inhibit Al-Ga intermixing [26–28]. Naturally, Si_3N_4 was used in association with SiO_2 to obtain spatially selective intermixing of GaAs/AlGaAs QW heterostructures. This combination successfully resulted in the realization of a stripe-geometry

QW laser diode [26]. However, the various works published on the matter [26–28,99,100] indicated that the combined use of SiO_2 and Si_3N_4 could give quite different experimental results regarding intermixing rates and lateral extension, depending mainly on the patterning configuration and on the order of deposition of the layers (e.g. SiO_2 over Si_3N_4, or Si_3N_4 over SiO_2). In addition, like silicon dioxide, silicon nitride behavior was found to be dependent on the dielectric deposition technique used and the heterostructure composition. For instance, in several p-i-n structures silicon nitride was on the opposite found to enhance intermixing [99,100]. Although explanations have been proposed for specific behaviors, no global theory concerning the basic diffusion mechanisms involved has been reported. Moreover, the applicability of this technique to the fabrication of reduced size devices, e.g. laterally confined quantum structures, has never been investigated. The comparative work presented in this section provides a comprehensive explanation for the various behaviors reported in the literature, by highlighting the critical role played in vacancy diffusion by the stress distribution generated in the heterostructure by the capping layers [39,101,102].

Three different undoped QW heterostructures grown by molecular beam epitaxy (MBE) have been used for this study: structures C, D and G. We recall that structure C comprises a single 6 nm-thick GaAs QW, symmetrically bordered by two 1.5 nm-thick AlAs layers, and further embedded in thick $Al_{0.33}Ga_{0.67}As$ barriers, located at a depth of 30 nm from the surface. Structure D is a MQW containing ten identical uncoupled GaAs (5 nm) / AlAs (5 nm) QWs, embedded in 20 nm-thick upper and lower $Al_{0.33}Ga_{0.67}As$ barriers. And structure G contains a single 3.7 nm-thick GaAs QW also bordered by thin AlAs layers of thickness 1 nm and further embedded in thick $Al_{0.33}Ga_{0.67}As$ barriers, but the QW is located 60 nm below the surface. These heterostructures were capped by a thin capsule of GaAs to avoid any contact of reactive Al atoms with atmosphere and further deposited dielectric layers. As previously suggested [20,51], the effects of Ga-Al interdiffusion are enhanced for QWs bordered by thin AlAs layers, and become easier to detect by photoluminescence (PL) or by cross-sectional transmission electron microscopy (XTEM). Comparative experiments were undertaken on structure C for the case of patterned SiO_2 covered with Si_3N_4, and on structure G for the case of patterned Si_3N_4 covered with SiO_2.

Dielectric deposition was performed by RTCVD. As already mentioned, this unique technique involves the pyrolitic decomposition of reactive gases (SiH_4 and O_2 for SiO_2, SiH_4 and NH_3 for Si_3N_4) in an inert flowing N_2 atmosphere, accelerated by the rapid heating (700°C) of the sample holder by halogen lamps. The obtained SiO_2 layers present physical and compositional properties very close to those of thermally grown SiO_2, and Si_3N_4 layers show no sign of O incorporation. We showed in Section 2 that intermixing occurring under RTCVD SiO_2-capped and annealed GaAs/(Ga,Al)As heterostructures gave a higher reproducibility and higher uniformity than other SiO_2 deposition methods. In addition, we have also shown that the intermixing rate depends on the thickness of the SiO_2 capping layer and that optimized intermixing effects are obtained for 450 nm-thick SiO_2 films. Such a value was thus adopted for the present experiments, while Si_3N_4 layers of thickness 100 nm were systematically used for the complementary film. The dielectric layers obtained by RTCVD, using our working conditions, are under intrinsic stress. Measurements made at room temperature after deposition, give a high tensile stress of 2.26 GPa for Si_3N_4 thin films of thickness 100 nm, and a weaker compressive stress of -0.2 GPa for 450 nm-thick SiO_2 films.

Arrays of strips of width ranging from 60 nm to 2 μm and of different periods were patterned into the first deposited film using high resolution electron beam lithography (EBL) followed by reactive ion etching (RIE). EBL was performed on 200 nm-thick PMMA resist layers, and then after lift-off, metallic masks of 70 nm-thick Al were used for RIE pattern transfer. A (SF_6/CHF_3) gas mixture was used to obtain good etching anisotropy and a high selectivity between the dielectric layers and the semiconductor. Vertical edges are obtained when 100 nm-thick Si_3N_4 layers are processed, while a slope of 3° is observed by XTEM after etching of 450 nm-thick SiO_2 layers. After RIE, Al masks are easily removed by immersion in dilute NaOH, enabling subsequent deposition step to be performed. Thermal annealing was carried out at a temperature of 850°C for 30 minutes in a conventional furnace. During annealing the sample is sandwiched in between two GaAs clean samples and inserted in a graphite box filled with fine GaAs powder. Both constant temperature and As overpressure are thus guaranteed. On a same piece of the original MBE layer, different arrays of strips were fabricated on previously mesa-etched regions of 80 μm $\times$ 80 μm. This enables each array to be easily located and characterized independently. The mixing rate and its spatial distribution were characterized by low temperature PL and XTEM. In our PL experiments the exciting laser probe of roughly 40 μm diameter can be easily positioned on a specific mesa. XTEM specimen preparations were achieved using original highly localized techniques involving RIE and/or Focused Ion Beam (FIB) thinning [103,104] which enabled us to observe on a same specimen several arrays with different strip dimensions and periods. PL characterizations could thus be systematically compared to XTEM observations in order to get a better understanding of the intermixing process as a function of the structure of the patterned arrays.

4.2   Experimental Results: Patterned SiO_2 Coated with Si_3N_4

Figure 21 gives a schematic representation of the arrays patterned into the dielectric layers. Due to different intrinsic stress in the SiO_2 film and the Si_3N_4 film, such a patterning imposes a complex stress field on the near-surface region of the semiconductor. When the sample is cooled down

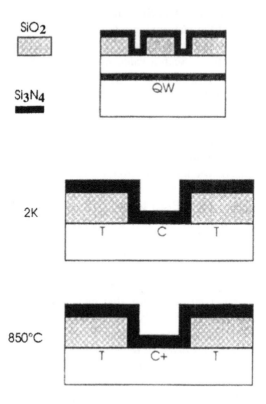

Fig. 21. Schematic representation of the arrays obtained when the SiO$_2$ film is deposited first, then patterned by Electron Beam Lithography and Reactive Ion Etching, and finally covered with the Si$_3$N$_4$ film. Also presented are the schematic stress distributions in the near surface region of the semiconductor induced by the dielectric layers, both at low temperature (2K) and at high temperature (850°C). Letters C and T denote compressive and tensile states respectively. Symbol C+ indicates a highly compressive region.

for PL experiments or annealed at high temperature for the interdiffusion process, this stress field is modified due to the thermal stress caused by the expansion coefficient mismatch between the dielectric films and GaAs. Let us recall that the thermal stress generated in a thin homogeneous film deposited on a thick substrate, during a temperature variation T, is given by equation (1). Table 1 listed values for the linear expansion coefficient of the dielectric materials and GaAs, as well as the elastic coefficients needed to calculate the thermal stress through equation (1). At low temperature RTCVD SiO$_2$ films are under compressive stress but due to a low expansion coefficient, it becomes tensile ($\sim$ +0.24 GPa) at the annealing temperature of 850°C. For RTCVD Si$_3$N$_4$ films, a different situation is obtained : films are under tensile stress at low temperature, as well as at the annealing temperature where a very high tensile stress is expected ($\sim$ +3.5 GPa). However, we do not have access to a direct measurement of these stress states at 2K and 850°C. Although the adhesion of the films does not appear to be modified by the thermal treatments, our simple calculation of the total stresses involved can only give a rough estimate since no chemical changes in the films are considered upon annealing.

Following patterning, the semiconductor surface will experience a non uniform stress field across the array of strips. Such a property has indeed been exploited in the past to fabricate QWRs by stress-induced lateral confinement [105,106]. In our case, the expected situations at 2K and 850°C are schematically represented in Figure 21. It is worth noting that due to its very high tensile stress, the Si_3N_4 film tends to impose the stress distribution to the whole structure. At low temperature, the semiconductor is under compressive stress below the Si_3N_4 strips, while the regions below the silica strips covered with Si_3N_4 are under tensile stress. We will see that this point will be confirmed by PL experiments. At high temperature, the situation is quite comparable : the highly stressed nitride film induces a compressive deformation of the semiconductor below the Si_3N_4 strips and by compressing the SiO_2 layer it drives the semiconductor under tension below the SiO_2 strips covered with Si_3N_4. The amplitude of the stress in the different regions depends strongly on the geometry of the patterned structures, and enhanced stress exists at the edges of the dielectric masks, as shown by previous calculations [93,107]. These calculations become rather complex in the case of two complementary deposited films, but the results presented in this paper can be qualitatively well understood on the basis of the simple description of the stress distribution presented in Figure 21.

Table 3 gives a summary of the PL results obtained on heterostructure C both before and after annealing. The energy position of the PL peak of the single QW is reported as a function of the structures patterned in the capping dielectric layers, for some areas also examined by XTEM. Figure 22 displays some of the most representative PL spectra. The as-grown structure C exhibits an intense low temperature PL signal at an energy of 1.626 eV corresponding to the QW emission, a PL line at 1.52 eV originating from the GaAs buffer layer, and a weak signal at an energy of 2.018 eV which is attributed to the thin AlAs layers bordering the QW and which forms quantum wells in the X valleys. Similar spectra are obtained on large areas uniformly capped with Si_3N_4, as observed in Figure 22(a), thus indicating that no degradation of the optical properties of the QW occurs after the deposition step.

However, in arrays, a splitting of the QW PL line into two peaks is observed after deposition and patterning. This splitting is due to the stress distribution induced in the QW by the capping layers. The influence of stress on the PL of GaAs/GaAlAs QWs is a well documented phenomenon [105,106]. In a tensile region the effective band gap energy of a QW is decreased, leading to a redshift of the PL peak, while on the opposite, a blueshift is induced

TABLE 3

Summary of the low temperature PL data obtained on heterostructure C for different processed regions containing arrays of SiO_2 strips coated with Si_3N_4. The table indicates the position of the observed PL peaks. When no comments are made, the PL intensity of the peaks is comparable to that of an as-grown sample. The data denoted by (*) correspond to nanometric arrays where the interdiffusion is weak with small modulations across the structures patterned in the dielectric caps.

Pattern geometry SiO_2 width/period (nm)	PL peak(s) energy (eV) before annealing	PL peak(s) energy (eV) after annealing
uniform (SiO_2 + Si_3N_4) bilayer	1.626	1.97 (very weak)
only uniform Si_3N_4	1.627	1.63
2200/3000	1.603/1.632	1.617 (very weak)/1.92 (weak)
2100/3000	1.607/1.63	1.617 (very weak)/1.92 (weak)
875/1200	1.620/1.625	1.617 (very weak)/1.9 (weak)
550/900*	1.613/1.625	1.710/1.84
375/600*	1.62/1.625	1.652/1.75 (weak)
70/120*	1.604/1.628	1.66

under compressive stress. The observation of two peaks before annealing is thus a confirmation that in these arrays, tensile regions under the SiO_2 strips alternate with compressive regions under the Si_3N_4 strips. This effect is particularly pronounced for wide SiO_2 and Si_3N_4 strips. Spectrum 22(b) shows an example of 2.1 μm-wide Si_3N_4-covered SiO_2 strips separated by 0.9 μm-wide Si_3N_4 strips, where the PL splitting reaches a value of 23 meV. When similar experiments are performed on arrays of trenches opened in the SiO_2 layer alone, no splitting of the PL peak is observed. It thus seems logical to conclude that it is the presence of the additional Si_3N_4 coating which induces a strong lateral stress gradient.

As can be seen in Table 3, regions uniformly capped with Si_3N_4 show very high stability against heat treatment. In these areas the PL intensity is not affected after annealing, which indicates that the RIE of the SiO_2 layer prior to Si_3N_4 deposition has not generated any defects in the structure. We thus confirm here the good encapsulating properties of uniform Si_3N_4 films for GaAs/GaAlAs structures reported in the literature [26–28], even in the case where the Si_3N_4 layer is deposited after several processing steps such as RIE.

The PL spectrum obtained after anneal on regions uniformly covered with a (SiO_2+Si_3N_4) bilayer is presented in Figure 22(c). We observe the GaAs buffer line at 1.52 eV, a low intensity and broad line around 1.97 eV which could correspond to the luminescence of the thick $Al_{0.33}Ga_{0.67}As$ barriers, and a small signal at 2.018 eV fitting well with the X luminescence of the thin AlAs layers bordering the QW. The quenching of the QW luminescence together with the observation of a line corresponding to the GaAlAs alloy,

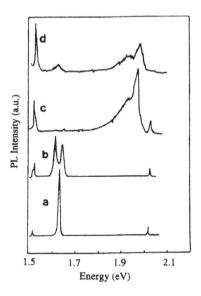

Fig. 22. Typical low temperature PL spectra obtained on structure C for different processed regions containing arrays of Si_3N_4 strips with adjacent $(SiO_2 + Si_3N_4)$ strips. (a) corresponds to a region uniformly capped with Si_3N_4 before annealing. (b) was obtained before annealing on an array of 2.1 μm-wide $(SiO_2 + Si_3N_4)$ strips with 0.9 μm-wide Si_3N_4 adjacent strips. (c) corresponds to an annealed region uniformly capped with a $(SiO_2 + Si_3N_4)$ bilayer. (d) was obtained after annealing on an array of 2.1 μm-wide $(SiO_2 + Si_3N_4)$ strips with 0.9 μm-wide Si_3N_4 adjacent strips.

could indicate the achievement of a complete mixing of the layers in these areas. However, this conclusion is not compatible with the observation of a line at 2.018 eV because complete mixing would also imply also the washing out of the thin AlAs bordering layers. In order to clarify this point, XTEM observations were carried out in these regions. Figure 23 presents a dark-field image of the QW structure capped with a $(SiO_2+Si_3N_4)$ bilayer.

We clearly observe the GaAs QW, visible as a dark stripe bordered by two bright layers corresponding to the thin AlAs layers. The observed contrast is very similar to the one obtained on the as-grown layer (not shown here). XTEM analysis thus clearly indicates that the interdiffusion has been inhibited in these regions. However, many defects are visible underneath the SiO_2/GaAs interface down to a depth of 15–20 nm. Their exact nature could not be determined by XTEM because of their large density. The presence of this high concentration of defects in the vicinity of the QW offers a possible explanation for the quenching of the QW PL. Moreover, the conservation of the thin AlAs layers at the QW interfaces, attested by XTEM, explains the origin of the PL signal observed at 2.018 eV. It is important to stress here

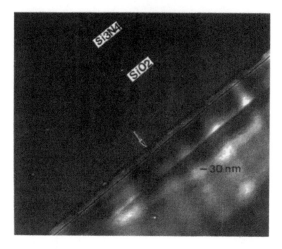

Fig. 23. XTEM (DF 200) micrograph obtained after annealing on a region uniformly capped with a (SiO$_2$ + Si$_3$N$_4$) bilayer. The vertical arrow points at the SiO$_2$/GaAs interface. The GaAs layer appears in black and the bordering thin AlAs layers appear almost white. Note the presence of a band of defects just below the SiO$_2$/GaAs interface.

that when a sample from heterostructure C is uniformly covered with SiO$_2$, without Si$_3$N$_4$ on top, and annealed in the same conditions, a blueshift of the QW PL peak of over 100 meV is observed without significant loss of intensity. In that configuration, interdiffusion occurs normally due to the generation of V$_{Ga}$ under the SiO$_2$/GaAs interface and no extended defects are formed during annealing, as previously reported in Section 2.

After annealing, the micrometric arrays of wide SiO$_2$ and wide Si$_3$N$_4$ strips exhibit two low intensity PL lines. Figure 22 (d) corresponds to the case of 2.1 μm-wide SiO$_2$ strips surrounded by 0.9 μm-wide Si$_3$N$_4$ strips. The very weak peak on the low energy side (1.62 eV) is slightly redshifted compared to the reference emission and thus corresponds to some regions protected from the interdiffusion. The broad peak on the high energy side exhibiting two maxima could correspond to highly interdiffused regions (quasi-complete mixing giving rise to emission around 1.92 eV) and to the Al$_{0.33}$Ga$_{0.67}$As alloy barriers (emission at 1.97 eV). In order to confirm this interpretation and to spatially localize these different emissions, XTEM observations were carried out on these arrays. Figure 24 presents a bright-field (BF) (a) and a dark-field (DF) (b) image of a 0.9 μm-wide trench etched through the SiO$_2$ layer and covered with the Si$_3$N$_4$ layer, which corresponds to the PL spectrum of Figure 22(d). The BF illumination enables us to image the two capping layers with an arrangement corresponding exactly to the situation pictured in Figure 21.

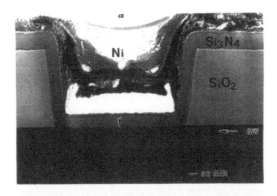

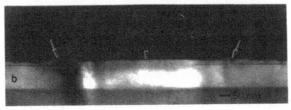

Fig. 24. XTEM micrographs obtained after annealing on a 0.9 μm-wide trench opened in the SiO$_2$ layer and covered with Si$_3$N$_4$. (a) is a bright-field image showing the different layers. The Ni layer visible on top of the structure is used for the thinning of the sample for XTEM observation. The horizontal arrow indicates the depth of the QW and the vertical one, the surface of the semiconductor. (b) is a dark-field (200) image showing the location of the interdiffused areas. The edges of the SiO$_2$-capped region are indicated by inclined arrows and the surface of the semiconductor, by a vertical arrow. The contrasts of the bordered QW disappear below the Si$_3$N$_4$ strip, attesting for a complete intermixing of the layers in this area.

On the DF image, the QW with its bordering thin AlAs layers is clearly seen under the SiO$_2$ strips covered with Si$_3$N$_4$. In these regions, as in the uniform areas covered with a (SiO$_2$+Si$_3$N$_4$) bilayer (see Figure 23) the interdiffusion is inhibited and a large concentration of defects is visible under the SiO$_2$/GaAs interface. In contrast, the QW disappears completely underneath the Si$_3$N$_4$ strips indicating a complete mixing. We also note that the Si$_3$N$_4$/GaAs interface is defect-free. The arrows on Figure 24(b) point at the position of the edges of the SiO$_2$ strips. We can see a transition region between the totally interdiffused areas and the protected ones. On a lateral distance extending over 200 nm on each side of the SiO$_2$ strips, a zone of intermediate mixing rate is observed. The width of the completely mixed area is thus reduced to 500 nm, as compared to the original 0.9 μm-wide Si$_3$N$_4$. The presence of this large transition region where the QW progressively changes from its original shape to a completely disordered layer thus explains the origin of the very broad peak observed systematically on the high energy side of the PL spectra of the arrays. The low intensity PL signal at 1.62 eV emanates from the non-interdiffused region located under the SiO_2 strips. The PL intensity is weak due to the large concentration of defects generated under the SiO_2/GaAs interface in these areas. Moreover, the PL peak position is slightly redshifted due to the tensile stress imposed under the SiO_2 strip, and also possibly to the presence of these defects.

XTEM analysis thus allows us to unambiguously state that the interdiffused regions are located under the Si_3N_4 strips and that the protected regions are in the SiO_2-capped areas. This surprising result is truly counterintuitive since we expected the protected regions to be in the Si_3N_4-capped areas because of the good encapsulating property of uniform nitride films. We can see here that PL used alone would have probably led us to a wrong conclusion concerning the location of the interdiffused regions.

In the case of the nanometric arrays of smaller pattern dimensions, the PL data of Table 3 (denoted by a *) indicate that intermediate situations are observed. No regions of complete mixing are generated and the difference in mixing rates between the two adjacent regions is attenuated. These results were confirmed by XTEM (data not presented here) which showed that the contrasts of the QW heterostructure were still visible. The modulations of the QW contrasts across those nanometric arrays were too small to be quantified through XTEM analysis.

In summary, coupled PL and XTEM experiments give the following indications for this specific configuration of dielectric layers where the SiO_2 film is deposited first. Strong lateral variations in the stress field are generated in the near surface region of the semiconductor due to the patterning of the SiO_2 film and the subsequent deposition of Si_3N_4 on top. After deposition and patterning, the induced stresses are sufficient to split the low temperature PL line of a QW located 30 nm below the surface. Large regions uniformly capped with Si_3N_4 are stable against heat treatment : neither interdiffusion nor crystalline defects are generated. In contrast, in large regions uniformly capped with a (SiO_2+Si_3N_4) bilayer, interdiffusion is inhibited due to the presence of the coating Si_3N_4 layer. However, a large density of defects are created under the SiO_2/GaAs interface which quench the QW PL signal. In arrays of wide Si_3N_4 strips with wide adjacent strips of SiO_2 covered with Si_3N_4, complete mixing of a 6 nm-wide QW occurs under the Si_3N_4 strips, while the interdiffusion is inhibited in the neighbouring regions covered with (SiO_2+Si_3N_4). A transition region as wide as 200 nm extending from the edge of the trench opened in the SiO_2 layer towards the Si_3N_4-capped region is observed. A large density of defects is formed under the SiO_2/GaAs interface, affecting considerably the PL of the non-interdiffused QW. In the arrays of nanometric dimensions, layer disordering is weaker and only small modulations are observed accross the different capped regions.

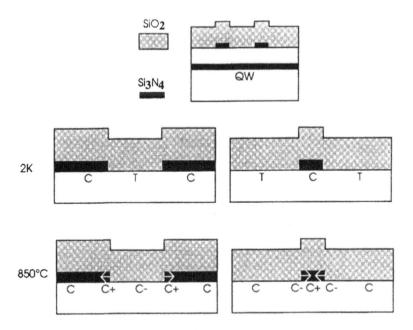

Fig. 25. Schematic representation of the arrays obtained when the Si_3N_4 film is deposited first, then patterned by Electron Beam Lithography and Reactive Ion Etching, and finally covered with the SiO_2 film. Also presented are the schematic stress distributions in the near surface region of the semiconductor induced by the dielectric layers, both at low temperature (2K) and at high temperature (850°C), for wide and narrow Si_3N_4 strips. Letters C and T denote compressive and tensile states respectively. Symbols C^+ and C^- indicate a highly (resp. weakly) compressive region. The white arrows symbolize the forces exerted on the semiconductor surface by the patterned Si_3N_4 structures.

4.3   Experimental Results: Patterned Si_3N_4 Coated with SiO_2

Figure 25 gives a schematic representation of the arrays patterned into the dielectric layers. In this experiment, the Si_3N_4 thin film is deposited first, patterned by EBL and subsequent RIE, and then covered with a thick SiO_2 layer. After patterning, due to the different intrinsic stresses in the deposited films, the surface region of the semiconductor experiences some stress which depends on the pattern geometry. The expected stress distribution for wide and narrow strips of Si_3N_4 at the two temperatures of interest for this study are sketched in Figure 5. At low temperature, the semiconductor surface is in a compressive state under the Si_3N_4 strips and in a tensile state in the SiO_2-capped regions. At high temperature, during annealing, both deposited films are under tensile stress, therefore no strong lateral stress gradient is expected in the near surface region of the semiconductor. The whole surface of the semiconductor is under compressive stress with

TABLE 4

Summary of the low temperature PL data obtained on heterostructure G for different processed regions containing arrays of Si_3N_4 strips covered with SiO_2. The table indicates the position of the observed PL peaks. When no comments are made, the PL intensity of the peaks is comparable to that of an as-grown sample, with no linewidth broadening.

Pattern geometry SiO_2 width/period (nm)	PL peak(s) energy (eV) before annealing	PL peak(s) energy (eV) after annealing
uniform (Si_3N_4 + SiO_2) bilayer	1.760	1.763
only uniform SiO_2	1.758	1.91 (weak)
1000/3000	1.758/1.762	1.617 (very weak)/1.92 (weak)
60/500	1.760	1.93 (weak and broad)
60/300	1.760	1.92 (weak and broad)
600/750	1.755/1.765	1.81/1.86 (weak and broad)
400/500	1.755/1.768	1.82/1.86 (weak and broad)
290/350	1.752/1.769	1.835/1.87 (weak and broad)

modulations at the edges of the structures which depend on the pattern geometry.

The experiments described in this section were obtained on the MBE structure G which contains a single 3.7 nm-wide GaAs QW. Arrays of narrow Si_3N_4 strips (<200 nm) with much larger periods ($\simeq 1$ μm) were patterned, together with inverse structures of wide Si_3N_4 strips (>200 nm) and smaller period. Table 4 gives a summary of the PL results obtained before and after annealing and Figure 26 presents the most representative PL spectra.

Before annealing, the QW PL line peak splits into two peaks only for the arrays of wide Si_3N_4 strips (>200 nm) surrounded by narrower SiO_2 strips (<150 nm) and for arrays of micrometric dimensions. In both cases, the splitting can reach a value of 17 meV (see Table 4). In contrast, for the arrays containing narrow Si_3N_4 strips (<200 nm) surrounded by wide SiO_2 strips, the PL peak is not split and corresponds to the emission obtained for large areas uniformly capped with Si_3N_4 (see Table 4). It thus appears that the splitting of the PL peak is related to a minimal surface coverage of the Si_3N_4 film which confirms that the nitride film imposes a higher stress on the structure than the SiO_2 film. This stress was indeed clearly visualized in XTEM on arrays performed on heterostructure D. Figure 27 is a DF XTEM micrograph obtained after anneal on an array of Si_3N_4 strips of width 200 nm and period 280 nm coated with a thick layer of SiO_2. The MQW is still visible but presents a very contrasted aspect: half circle-shaped halos are distinctly observed under each Si_3N_4 strip. We note the presence of a darker fringe surrounding a lighter region. Such contrasts can in no way be attributed to the presence of regions of different intermixing rate and, thus, of different Al concentration. Furthermore, we noticed that these contrasted halos were very

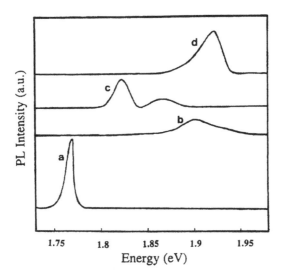

Fig. 26. Typical low temperature PL spectra obtained on structure G for different processed regions containing arrays of SiO_2 strips with adjacent ($Si_3N_4 + SiO_2$) strips. (a) corresponds to a region uniformly capped with a ($Si_3N_4 + SiO_2$) bilayer after annealing. (b) was obtained after annealing on a region uniformly capped with SiO_2. (c) corresponds to an array of 600 nm-wide ($Si_3N_4 + SiO_2$) strips with 150-nm wide SiO_2 adjacent strips after annealing. (d) was obtained after annealing on an array of 60 nm-wide ($Si_3N_4 + SiO_2$) strips with 240 nm-wide SiO_2 adjacent strips.

Fig. 27. XTEM (DF 200) micrograph obtained after anneal on an array of Si_3N_4 strips of width 200 nm and period 280 nm coated with SiO_2 realized on structure D. We clearly distinguish under each nitride strip a half circle-shaped halo showing a succession of light and dark fringes.

sensitive to illumination conditions, departure from Bragg angle, etc. Taking a closer look at these halos on other arrays [39], we could even distinguish double constrasts at the MQWs heterointerfaces where the darker ring of the halo passes through. The formation of this type of contrast is a signature of the presence of significant stress [108]. Since the (200) direction is perpendicular to the growth axis, we therefore visualize the deformation field along z, e.g. the distorsion between reticular planes parallel to the surface. The observed contrasts seem indeed to confirm that the nitride film imposes its stress to the SiO_2/Si_3N_4 ensemble.

The regions uniformly capped with a $(Si_3N_4+SiO_2)$ bilayer exhibit a good thermal stability. The PL spectrum of Figure 26(a) shows that the QW PL peak is blueshifted after annealing by less than 3 meV compared to a reference sample and does not suffer any broadening. We confirm once again the good encapsulating properties of uniform Si_3N_4 films even when covered with a thick SiO_2 layer.

The regions uniformly capped with SiO_2 after the complete etching of the Si_3N_4 layer are strongly interdiffused. The PL spectrum of Figure 26(b) shows that the QW PL peak is blueshifted by more than 140 meV up to an energy of 1.91 eV, which is slightly lower than the emission of the thick $Al_{0.33}Ga_{0.67}As$ alloy barriers. The emission of the interdiffused material is broadened and the PL intensity is lowered by a factor of 10 after interdiffusion. These results suggest a quasi-complete mixing accompanied by the formation of residual defects which are never observed when the homogeneous SiO_2 layer is directly deposited on the surface (with no pre-deposition of Si_3N_4 and RIE, as in Section 4.2). It thus seems that the defects generated during the RIE process can interact with the V_{Ga} generated during annealing under the SiO_2 film, leading to an interdiffusion process of lower quality. These conclusions are confirmed by XTEM observations. Figure 28 is a DF (200) XTEM micrograph of a large area uniformly capped with SiO_2. The horizontal arrow indicates the position of the original QW which is no longer visible due to the strong interdiffusion induced in the sample. Some defects can be seen under the $SiO_2/GaAs$ interface. Such defects are never observed when the SiO_2 layer is directly deposited onto the surface, and thus appear to be a consequence of the technological steps preceding SiO_2 deposition, mainly, the RIE of the Si_3N_4 thin film.

The situation after annealing varies as a function of the type of array considered. For arrays of wide Si_3N_4 strips (>260 nm), we observe an intense PL peak which is blueshifted compared to the reference QW PL emission (see Figure 26(c)). This blueshift increases for decreasing size of Si_3N_4 strips (60 meV for 400 nm-wide strips and 75 meV for 290 nm-wide strips, as can be seen in Table 4). Another weak emission is observed at higher energy (1.86 eV in Figure 26(c)). The position of this peak depends strongly on the pattern geometry. A very weak signal at an energy of 1.91 eV, very similar to the one obtained under large SiO_2-capped regions, is also sometimes detected. In order to fully interpret the origin of these emissions, XTEM analysis of this type of array was performed. Figure 29 is a typical DF (200) image of a 1 μm-wide Si_3N_4 strip from an array of period 3 μm. The QW turns out to have completely disappeared under the SiO_2-capped regions. In addition defects have been generated under the $SiO_2/GaAs$ interface, as observed in the regions uniformly capped with SiO_2 (see Figure 28). The very weak

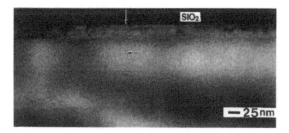

Fig. 28. XTEM (DF 200) micrograph obtained after annealing on a region uniformly capped with SiO_2. The vertical arrow points at the SiO_2/GaAs interface while the horizontal one indicates the depth position of the initial QW.

Fig. 29. XTEM (DF 200) micrograph obtained after annealing on a 1 μm-wide Si_3N_4 strip covered with SiO_2. The vertical arrow indicates the surface of the semiconductor and the inclined one points at the position of the QW. The contrasts of the bordered QW are still visible below the Si_3N_4 but disappear in the adjacent areas.

PL peak at an energy of 1.91 eV can thus been attributed to those highly interdiffused regions. In contrast, the bordered QW is perfectly visible under the Si_3N_4 strips as indicated by the inclined arrow in Figure 28. However, the interdiffusion presents a lateral extent of 150 nm under the Si_3N_4 strips. In this situation, where Si_3N_4 is deposited first and subsequently capped with SiO_2, the interdiffused regions are normally located under the SiO_2-capped areas. The interdiffusion is fairly isotropic, with a lateral interdiffusion length as large as 150 nm. The PL peak on the low energy side of the spectrum thus emanates from those partially protected regions located below the Si_3N_4 strips. Due to lateral interdiffusion effects, it is not surprising to observe an increasing blueshift of the QW PL peak when the size of the Si_3N_4 strips is reduced. Moreover, due to the high value of the lateral interdiffusion length (150 nm) the regions of the QW located under the Si_3N_4 cannot be completely protected from interdiffusion effects even for Si_3N_4 strips as wide as 400 nm.

Fig. 30. XTEM (DF 200) micrograph obtained after annealing on an array of 60 nm-wide Si$_3$N$_4$ strips covered with SiO$_2$ at a period of 1 μm. The vertical arrow indicates the surface of the semiconductor and the horizontal one points at the position of the QW. The contrasts of the bordered QW are not visible, attesting for a non selective interdiffusion.

The weak PL signal observed at an energy intermediate (at 1.86 eV in Figure 6c) between the emissions corresponding to the Si$_3$N$_4$-capped regions and the interdiffused SiO$_2$-capped regions, can possibly be attributed to this large transition region of intermediate mixing rate extending from the edge of the SiO$_2$ strips towards the Si$_3$N$_4$ strips.

For Si$_3$N$_4$ strips of nanometric dimension (<200 nm), only one PL peak at an energy around 1.92 eV can be observed (see Figure 26(d)). It thus seems that in this case the QW is uniformly interdiffused. This is confirmed by XTEM observation. Figure 30 presents a DF (200) image of an array of 60 nm-wide Si$_3$N$_4$ strips and period 1 μm. The horizontal arrow indicates the position of the original QW which has been completely interdiffused under the large SiO$_2$-capped regions as well as under the narrow Si$_3$N$_4$ strips. The regions capped with SiO$_2$ still exhibit some defects under the SiO$_2$/GaAs interface while the adjacent Si$_3$N$_4$/GaAs interface is defect-free. In this specific configuration of dielectric layers, the isotropic character of the induced interdiffusion leads to a quasi-complete interdiffusion of the QW under Si$_3$N$_4$ strips of dimension smaller than twice the lateral interdiffusion length.

In summary, coupled PL and XTEM experiments give the following indications for this configuration of dielectric layers where the Si$_3$N$_4$ film is deposited first. In this case, interdiffusion proceeds normally, i.e. the spatial localization of interdiffused regions under patterned structures can be properly deduced from the behavior observed under uniform capping layers. The Si$_3$N$_4$-capped regions are protected from the interdiffusion while the regions covered with SiO$_2$ are strongly mixed. The deposition of a thick SiO$_2$ layer on top of the Si$_3$N$_4$ thin film does not affect the inert encapsulating property of Si$_3$N$_4$. The interdiffusion process occurring in arrays of Si$_3$N$_4$ strips is isotropic with a lateral interdiffusion length of 150 nm. Defects are formed under the SiO_2/GaAs interface, likely resulting from the RIE process preceding the deposition of the SiO_2 film.

4.4   Vacancy Diffusion Under Stress

In all experiments reported in this paper, stimulation of thermal interdiffusion by SiO_2 capping can unambiguously be attributed to an impurity-free vacancy diffusion mechanism. In our structures, a GaAs cap layer was systematically grown by MBE on top of the semiconductor. Furthermore, the quality of the MBE growth guarantees the absence of any trace of Al close to the GaAs/SiO_2 interface. Moreover, careful XTEM observations of our deposited dielectric layers indicate the complete absence of any pinholes in the SiO_2 and Si_3N_4 films used for this study, when the layers are uniformly deposited as well as when patterned structures are built. For all these reasons, the observed interdiffusion in our samples is the consequence of the generation of an excess concentration of V_{Ga} in the semiconductor. The spatial distribution of the intermixing will therefore reflect the way by which V_{Ga} diffuse from the surface region where they are generated to the region of the QW heterostructure where they enhance Ga-Al interdiffusion. An interdiffusion model based on the elastic interaction existing between the created vacancies and the macroscopic stress field imposed on the semiconductor by the capping layer was proposed in Section 2.5. On the basis of this simple model, we now give an interpretation of the experimental results obtained in these SiO_2/Si_3N_4 patterning experiments.

4.4.1   Patterned silicon dioxide coated with silicon nitride

In large areas uniformly covered with a (SiO_2+Si_3N_4) bilayer disordering was found to be inhibited due to the Si_3N_4 top layer. Such an effect had been previously reported by Guido et al. [28], but in a completely different context where the enhanced interdiffusion mechanism was attributed to IILD due to the pinholes of the SiO_2 capping layer. The authors claimed that the role of the Si_3N_4 capping layer was to seal the SiO_2 pinholes from the influence of the arsenic ambient that would otherwise promote the penetration of Si inside the semiconductor. Such an interpretation cannot be retained to account for the present experiments since our dielectric layers show no pinholes. We believe that the observed inhibition of impurity-free layer disordering is mainly due to the different stress state imposed on the semiconductor surface when the SiO_2 layer is covered with Si_3N_4. Indeed, as can be seen in Figure 21, the surface of the semiconductor is in a tensile state at the annealing temperature for a (SiO_2 + Si_3N_4) capping layer. This effect is due to the top Si_3N_4 layer which, by compressing the SiO_2 layer, puts the semiconductor under tension. The effect of the Si_3N_4 top layer is thus to inverse the stress state in the semiconductor below the SiO_2/GaAs interface. The SIMS studies performed on different SiO_2 layers deposited by various techniques presented in Section 2 have shown that the pumping of Ga by SiO_2 could not be correlated to the stress state of the capping layer. We thus assume that the generation of V_{Ga} under the SiO_2/GaAs interface upon heating is fairly systematic and is not affected by the stress. Under $(SiO_2 + Si_3N_4)$ capping, V_{Ga} are injected in a tensile region which is not favourable to their penetration. Instead of distributing uniformly deep inside the semiconductor, like in the case of SiO_2 capping alone, they agglomerate rapidly in the near surface region of the semiconductor. The V_{Ga} trapped below the SiO_2/GaAs interface are no longer available to help in the interdiffusion process which is therefore inhibited. In other words, we think that the stress field has no influence on the Ga pumping by the SiO_2 layer but affects drastically the diffusion of the generated V_{Ga}. We attribute the extended defects observed by XTEM (see Figure 23) just below the SiO_2/GaAs interface to those agglomerated V_{Ga}.

For the micrometric arrays of wide Si_3N_4 strips with wide adjacent $(SiO_2 + Si_3N_4)$-capped regions, V_{Ga} are still injected under the SiO_2/GaAs interface only, because the nitride film is not porous to Al, Ga nor As. At the annealing temperature, the semiconductor is in a tensile stress under the SiO_2/GaAs interface. This tensile stress induces a rapid trapping of the V_{Ga} which inhibits layer disordering in the $(SiO_2 + Si_3N_4)$-capped regions. However, the adjacent regions capped by a thin Si_3N_4 film are under a high compressive stress. This lateral stress gradient attracts some of the V_{Ga}, initially generated below the SiO_2 layer, under the Si_3N_4 strips. The V_{Ga} diffusion is thus completely anisotropic: a driving force rapidly drains some vacancies laterally in the compressive regions where they can distribute quite uniformly in depth. A wind of vacancies blowing laterally from the SiO_2-capped regions towards the Si_3N_4-capped regions is generated in the semiconductor surface by the stress field gradient. After fast redistribution of the V_{Ga}, this stress is partially relaxed and a quasi-equilibrium state is reached in the structure. Interdiffusion can then proceed at a slower rate governed by the isotropic diffusion of the vacancies around their sites, which stimulates Ga-Al atomic exchanges at the heterointerfaces. Layer disordering thus unexpectedly occurs below the Si_3N_4 strips with some lateral interdiffusion length. The observation in XTEM of a transition region (of ~ 200 nm) between the non-interdiffused QW and the completely mixed QW, extending from the edges of the patterned SiO_2 layer towards the Si_3N_4 strips, is to our mind a signature of the lateral diffusion of the V_{Ga} under a transport force. Indeed, the vacancies are first drained laterally below the semiconductor surface towards the centre of the Si_3N_4 strips, and then distribute in the depth to accommodate the stress in the material. Finally, the interdiffusion process takes place with some lateral extent giving rise to the transition region of intermediate mixing rate. Clearly not all the vacancies generated below the

SiO_2/GaAs interface are drained towards the Si_3N_4-capped regions and many of them remain below the SiO_2/GaAs interface where they agglomerate as seen in XTEM (see Figure 24).

For the nanometric arrays of smaller dimensions, different situations slightly departing from the previous description can be encountered. The observed intermixing behavior depends strongly on the geometry of the patterns. When the (SiO_2 + Si_3N_4) strips are too narrow, the number of V_{Ga} generated is small and the induced interdiffusion below the adjacent Si_3N_4 strips is reduced. Moreover, due to the lower density of trapped V_{Ga}, the PL signals are generally higher. The lateral stress gradient is also highly dependent on the pattern geometry and the laterally-driven diffusion of the generated vacancies can be affected. This complex situation globally leads to a weak layer disordering with small modulations across the different capped areas of one array, especially when the sizes of the structures are comparable with the lateral interdiffusion length of the process (~ 200 nm), as observed through PL and XTEM analysis.

4.4.2   Patterned Silicon Nitride Coated with Silicon Dioxide

Regions uniformly capped with SiO_2 after the etching of the pre-deposited Si_3N_4 thin film are strongly intermixed. However, XTEM observations indicate the formation of some defects below the SiO_2/GaAs interface which are never observed when the SiO_2 layer is directly deposited on a clean surface. As can be seen in Figure 25, a compressive stress is generated at elevated temperature in the semiconductor below the SiO_2 film. The uniform depth distribution of the generated V_{Ga} can thus take place, and strong interdiffusion is induced. However, the RIE process preceding the SiO_2 deposition contributes to the formation of trapping centers for the V_{Ga} generated during the annealing under the SiO_2 film. As opposed to the case of a clean SiO_2/GaAs interface where the V_{Ga} generated by the pumping of Ga atoms by SiO_2 are free to quickly diffuse inside the semiconductor, due to RIE-induced point defects and roughness, some generated V_{Ga} are trapped below the SiO_2/GaAs interface, as can be seen by XTEM (see Figure 28) and are ineffective for the interdiffusion process. The result is a mixing of lower quality with a loss of PL signal and some lateral non-uniformity.

In the case of arrays of SiO_2 strips with adjacent (Si_3N_4+SiO_2) strips, the V_{Ga} are only generated in the regions directly covered with SiO_2. As can be seen in Figure 25, the surface of the semiconductor is under compressive stress during anneal, under the SiO_2 strips as well as in the adjacent regions. The fast diffusion of the vacancies towards depth can thus take place, but in this case there is no strong lateral stress gradient likely to drain laterally the V_{Ga}. In the absence of noticeable lateral stress gradient the V_{Ga} distribute in a quasi-isotropic fashion. After this fast redistribution of the V_{Ga}, a quasi-equilibrium state is reached and interdiffusion can proceed. According to this description, interdiffused regions are thus logically located below the SiO_2 strips where the V_{Ga} are initially formed. Moreover, it is not surprising to observe a lateral extent of the interdiffused region as large as 150 nm below the $(Si_3N_4+SiO_2)$ strips, due to the isotropic character of the V_{Ga} diffusion, related to the absence of noticeable lateral stress gradient at the annealing temperature.

4.4.3   Conclusions on intermixing selectivity

A qualitative model for the impurity-free intermixing of GaAs/GaAlAs QW structures induced by SiO_2/Si_3N_4 encapsulation has been proposed. Despite its simplicity, this model has the merit of underlining the major role played by stress on the diffusion of V_{Ga}, which are the point defects responsible for the enhancement of the thermal interdiffusion. Stress-enhanced Al-Ga interdiffusion has been previously reported [27] in Si_3N_4 strips patterned together with SiO_2, where dumbbell disordered regions were observed at the edges of the Si_3N_4 strips. However no clear interpretation of this result was proposed involving the influence of stress on defect diffusion. In our experiments, the layer disordering is unambiguously induced by an impurity-free mechanism due to the generation of excess V_{Ga}. The spatial distribution of the interdiffused regions thus intimately reflects the way these vacancies diffuse inside the semiconductor under the influence of stress. We show that although a tensile stress state in the SiO_2 layer is not essential for V_{Ga} creation, a high quality impurity-free intermixing under SiO_2 capping is only possible when the semiconductor surface is under compressive stress due to the elastic interaction of the V_{Ga} with the imposed stress field. Moreover, our experimental results support the idea that the localization of the disordered regions closely matches the regions under high compressive stress. When a strong lateral stress gradient is induced in the semiconductor, the V_{Ga} can be drained laterally from their site of creation towards the regions of highest compressive stress. This spectacular effect explains qualitatively the behavior observed in arrays of Si_3N_4 strips with adjacent $(SiO_2+Si_3N_4)$ strips where the disordering unexpectedly occurs below the Si_3N_4 strips. Our model gives the framework for understanding why opposite behaviors are observed when the SiO_2 film is patterned first and covered with Si_3N_4 and when the Si_3N_4 film is patterned first and then covered with SiO_2.

More quantitative predictions of our model are rather delicate. To begin with, a 3D calculation of the stress field under the patterned structures at the annealing temperature is required. Such types of calculations are now available stress-induced [93,107]. Then, the 3D resolution of the diffusion equation under a stress-induced driving force must be implemented. The input of this calculation is the gradient of the elastic interaction energy between the V_{Ga} and the stress field. This gradient obviously depends on the stress field distribution, but little is known in the literature on reasonable numerical values [78]. However, if we consider the homogeneous stress field in the nitride layer (3.5 GPa) applied to the typical volume occupied by a vacancy (roughly, the atomic volume), we find an energy around 0.4 eV. This crude estimation simply suggests that the effect of the elastic interaction on the V_{Ga} diffusion is far from being negligible. Moreover, under V_{Ga} injection, the stress field is modified due to its partial relaxation. Consequently, the gradient of the elastic interaction energy changes upon time until a quasi-stationary equilibrium regime is reached. Once the steady concentration of V_{Ga} is determined, the enhancement of interdiffusion can then be calculated by various ways [34,46]. The above inventory of the different steps required to fully simulate the induced disordering in the presence of patterned encapsulating layers, gives an idea of the difficulties involved in going beyond a simple qualitative description of the observed behavior.

In both patterning configurations, selective interdiffusion was demonstrated. However for both situations the lateral interdiffusion length was quite substantial (200 nm and 150 nm). We thus believe that this kind of combined patterning is not able to produce laterally-confined structures by selective interdiffusion.

5   CONCLUSION AND APPLICATION DOMAINS

The results we presented bring new light to the understanding of the mechanisms responsible for the enhancement of Ga-Al interdiffusion in SiO_2-capped GaAs/AlGaAs heterostructures during high temperature anneals. We show that higher density and purity SiO_2 films which can be obtained by RTCVD induce intense, homogeneous, and impurity-free intermixing of QWs with good reproducibility. In comparison, the other dielectric deposition techniques that we tested for this study (PECVD, UVCVD and RF-sputtering) give rise to spatial inhomogeneities and can lead to a significant degradation of the material's electronic properties. These drawbacks appear to be directly related to the poorer quality of the SiO_2/GaAs interface obtained via these processes. We have identified the main interdiffusion mechanism to be the quasi-instantaneous generation of excess Ga vacancies under the SiO_2/GaAs interface, resulting from the fast pumping of Ga atoms by the SiO_2 film during high temperature anneal, and their rapid diffusion inside the heterostructure where they can then promote

Ga-Al exchanges at the heterointerfaces. We also give evidence for the sponge-like behavior of the oxide layer, which quasi-instantly absorbs Ga up to its solubility limit, the latter increasing slightly with temperature.

In contrast with Ga ion implantation-enhanced intermixing, SiO_2 capping-induced intermixing does not give rise to extended residual defects which degrade the electronic properties, but instead offers great homogeneity both in depth (uniform interdiffusion of MQW structures is obtained) and in the lateral direction (QWs still show in their PL spectra a memory of their initial one-monolayer thickness variations). To account for all these unique properties, we propose a qualitative interdiffusion model based on the fast V_{Ga} diffusion towards depth driven by the thermal stress field imposed on the semiconductor by the SiO_2 film.

We demonstrated that our optimized RTCVD SiO_2 intermixing technique could be spatially localized on a length scale compatible with the lateral confinement of carriers into QWRs. Double barrier GaAs/AlGaAs QWRs are spontaneously fabricated with this process. For the first time using a selective interdiffusion technique, signatures of the lateral confinement of carriers inside quantum wires have been observed in PL, PLE and linear polarization analysis.

A qualitative interdiffusion model directly involving the effect of the stress field upon vacancy diffusion that successfully accounts for the high lateral resolution achievable with this technique when narrow trenches are etched at relatively large periods (>800 nm) was proposed. Evidence of this stress-dependent vacancy diffusion could also be observed in selective intermixing experiments carried out through the combined use of Si_3N_4 and SiO_2 encapsulation layers, and brings additional support to this simple model.

The high quality selective interdiffusion induced by SiO_2 capping make it an attractive alternative to widely-used ion implantion for various applications. Most particularly, SiO_2 encapsulation appears as a better candidate for the fabrication of such highly sought-after active devices as quantum wire lasers.

Moreover, our work clearly demonstrates that point defects such as V_{Ga} can be piloted by the stress field imposed on the semiconductor by patterned capping layers. This effect can be generalized to other kinds of point defects. For instance, large interstitial atoms can be drained towards tensile regions in order to accommodate the stress. By making it possible to intentionally move identified deleterious defects away from the active region of a device during annealing, this kind of effect opens new perspectives inside the area of diffusion and defect engineering and its application to the fabrication of optoelectronic devices.

Beyond these potential applications, SiO_2 capping-induced intermixing also turned out to be a unique experimental tool to investigate diffusion mechanisms and dynamics under stress in III-V compounds.

Acknowledgments

The authors gratefully thank M. Schneider and J-Y Marzin (L2M/CNRS) for help in the PL and SRPL experiments respectively, Y. Nissim and C. Mériadec (CNET-Bagneux) for RTCVD, E.V.K. Rao (CNET-Bagneux) for fruitful discussions, P. Krauz (CNET-Bagneux) for help during the annealing stage, M. Juhel (CNET-Bagneux) for SIMS analysis, and R. Planel (L2M/CNRS) for MBE growth.

References

1. *Opt. Quantum Electron.*, **23** (1991) Special Issue on Quantum Well Mixing for Optoelectronics.
2. J. Marsh, *Semicond. Sci. Technol.*, **8**, 1136 (1993) and references therein.
3. D.G. Deppe and N. Holonyak Jr., *J. Appl. Phys.*, **64**, R93 (1988) and references therein.
4. Y. Hirayama, Y. Suzuki, S. Tarucha and H. Okamoto, *Jpn. J. Appl. Phys.*, **24**, L516 (1985).
5. P. Gavrilovic, D.G. Deppe, K. Meehan, N. Holonyak, Jr., J.J. Coleman and R.D. Burnham, *Appl. Phys. Lett.*, **47**, 130 (1985).
6. Y. Hirayama, Y. Suzuki and H. Okamoto, *Jpn. J. Appl. Phys.*, **24**, 1498 (1985).
7. Y. Hirayama, Y. Suzuki and K. Okamoto, *Surface Science*, **174**, 98 (1986).
8. J. Cibert, P.M. Petroff, D.J. Werder, S.J. Pearton, A.C. Gossard and J.H. English, *Appl. Phys. Lett.*, **49**, 223 (1986).
9. J.R. Ralston, G.W. Wicks, L.F. Eastman, B.C. DeCooman and C.B. Carter, *J. Appl. Phys.*, **59**, 120 (1986).
10. K. Kash, B. Tell, P. Grabbe, E.A. Dobisz, H.G. Craighead and M.C. Tamargo, *J. Appl. Phys.*, **63**, 190 (1988).
11. J. Cibert, P.M. Petroff, D.J. Weider, S.J. Pearton, A.C. Gossard and J.H. English, *Appl. Phys. Lett.*, **49**, 1275 (1986).
12. G. Suzuki, Y. Hirayama and H. Okamoto, *Jpn. J. Appl. Phys.*, **25**, L912 (1986).
13. D. Kirillov, P. Ho and G.A Davis, *Appl. Phys. Lett*, **48**, 53 (1986).
14. D.J. Werder and S.J. Pearton, *J. Appl. Phys.*, **62**, 318 (1987).
15. P. Mei, T. Venkatesan, S.A. Schwarz, N.G. Stoffel, J.P. Harbison, D.L. Hart and L.A. Florez, *Appl. Phys. Lett.*, **52**, 1487 (1988).
16. B. Elman, Emil S. Koteles, P. Melman and C.A. Armiento, *J. Appl. Phys.*, **66**, 2104 (1989).
17. H. Sumida, H. Asahi, S. J. Yu, K. Asami, S. Gonda and H. Tanoue, *Appl. Phys. Lett.*, **54**, 520 (1989).
18. K.B. Kahen, D.L. Peterson and G. Rajeswaran, *J. Appl. Phys.*, **68**, 2087 (1990).
19. Y. Hirayama, *Jpn. J. Appl. Phys.*, **28**, L162 (1989).
20. F. Laruelle, Y.P. Hu, R. Simes, W. Robinson, J. Merz and P.M. Petroff, *Surface Science*, **228**, 306 (1990).
21. H. Leier, A. Forchel, G. Hrcher, J. Hommel, S. Bayer, H. Rothfritz, G. Weimann and W. Schlapp, *J. Appl. Phys.*, **67**, 1805 (1990).
22. C. Vieu, M. Schneider, R. Planel, H. Launois, B. Descouts and Y. Gao, *J. Appl. Phys.*, **70**, 1433 (1991).
23. E.V.K. Rao, M. Juhel, Ph. Krauz, Y. Gao and H. Thibierge, *Appl. Phys. Lett.*, **62**, 2096 (1993).
24. P.G. Piva, P.J. Poole, M. Buchanan, G. Champion, I. Templeton, G.C. Aers, R. Williams, Z.R. Wasilewski, E.S. Koteles and S. Charbonneau, *Appl. Phys. Lett.*, **65**, 621 (1994).
25. T. Miyazawa, H. Iwamura, O. Mikami and M. Naganuma, *Jpn. J. Appl. Phys.*, **28**, L1309 (1989).
26. D.G. Deppe, L.J. Guido, N. Holonyak Jr., J.J. Coleman and R. D. Burnham, *Appl. Phys. Lett.*, **49**, 510 (1986).
27. L.J. Guido, N. Holonyak Jr., K.C. Hsieh, R.W. Kaliski, W.E. Plano, R.D. Burnham, R.L. Thornton, J.E. Epler and T.L. Paoli, *J. Appl. Phys.*, **61**, 1372 (1987).

28. L.J. Guido, J.S. Major Jr., J.E. Baker, W.E. Plano, N. Holonyak Jr., K.C. Hsieh and R.D. Burnham, *J. Appl. Phys.*, **67**, 6813 (1990).
29. J.D. Ralston, S. OBrien, G.W. Wicks and L.F. Eastman, *Appl. Phys. Lett.*, **52**, 1511 (1988).
30. E.S. Koteles, B. Elman, R.P. Holmstrom, P. Melman, J.Y. Chi, X. Wen, J. Powers, D. Owens, S. Charbonneau and M.L.W. Thewalt, *Surperlatt. Microstruct.*, **5**, 321 (1989).
31. Y. Suzuki, H. Iwamura and O. Mikami, *Appl. Phys. Lett.*, **56**, 19 (1989).
32. E.S. Koteles, B. Elman, P. Melman, J.Y. Chi and C.A. Armiento, *Opt. Quantum. Electron.*, **23**, S779 (1991).
33. J.Y. Chi, X. Wen, E.S. Koteles and B. Elman, *Appl. Phys. Lett.*, **55**, 855 (1989).
34. K.B. Kahen, D.L. Peterson, G. Rajeswaran and D.J. Lawrence, *Appl. Phys. Lett.*, **55**, 651 (1989).
35. S.G. Ayling, J. Beauvais and J.H. Marsh, *Electron. Lett.*, **28**, 2241 (1992).
36. M. Ghisoni, P.J. Stevens, G. Parry and J.S. Roberts, *Opt. Quantum Electron.*, **23**, S915 (1991).
37. E.V.K. Rao, A. Hamoudi, Ph. Krauz, M. Juhel, and H. Thibierge, *Appl. Phys. Lett.*, **66**, 472 (1995).
38. A. Ppin, C.Vieu, M. Schneider, G. Ben Assayag, F.R. Ladan, R. Planel, H. Launois, Y. Nissim and M. Juhel, *Inst. Phys. Conf. Ser.*, **134**, 459 (1993).
39. A. Ppin, Ph.D. Thesis, Universit de Paris 6, (1995).
40. C. Francis, F.H. Julien, J.Y. Emery, R. Simes and L. Goldstein, *J. Appl. Phys.*, **75**, 3607 (1994).
41. R.W. Glew, A.T.R. Briggs, P.D. Greene and E.M. Alleh, Proceedings of the International Conference on InP and Related Compounds, 234 (1992).
42. J. Ralston, A. L. Moretti, R. K. Jain, and F. A. Chamber, *Appl. Phys. Lett.*,, **50**, 1817 (1987).
43. J. E. Epler, R. L. Thornton, and T. L. Paoli, *Appl. Phys. Lett.*,, **52**, 1371 (1988).
44. K. Brunner, U. Bockelmann, G. Abstreiter, M. Walther, G. Bhm, G. Trkle, and G. Weimann, *Phys. Rev. Lett.*, **69**, 3216 (1992).
45. J. Cibert, P.M. Petroff, G.J. Dolan, D.J. Werder, S.J. Pearton, A.C Gossard and J.H. English, *Superlattices and Microstructures*, **3**, 35 (1987).
46. J. Cibert and P.M. Petroff, *Phys. Rev.*, *B36*, 3243 (1987).
47. P.M. Petroff, J. Cibert, A.C. Gossard, G.J. Dolan and C.W. Tu, *J. Vac. Sci. Technol.*, *B5*, 1204 (1987).
48. Y. Hirayama, S. Tarucha, Y. Suzuki and H. Okamoto, *Phys. Rev.*, *B37*, 2774 (1988).
49. A. Forchel, H. Leier, B.E. Maile and R. Germann, in Festkrperprobleme (Advances in Solid State Physics), edited by U. Rssler (Vieweg, Braunschweig), **28**, 99 (1988).
50. H. Leier, B.E. Maile, A. Forchel, G. Weimann and W. Schlapp, *Mircrocircuit and Engineering*, **11**, 43 (1990).
51. C. Vieu, M. Schneider, D. Mailly, R. Planel, H. Launois, J.Y. Marzin and B. Descouts, *J. Appl. Phys.*, **70**, 1444 (1991).
52. C. Vieu, M. Schneider, G. Ben Assayag, R. Planel ,L. Birotheau, J.Y. Marzin and B. Descouts, *J. Appl. Phys.*, **71**, 5012 (1992).
53. C. Vieu, M. Schneider, D. Mailly, J.Y. Marzin and B. Descouts, *Microelectronic Engineering*, **15**, 23 (1990).
54. F.E. Prins, G. Lehr, M. Burkard, H. Schweizer, M.H. Pikuhn and G.W. Smith, *Appl. Phys. Lett.*, **62**, 1365 (1993).
55. F.E. Prins, G. Lehr, M. Burkard and H. Schweizer, *Nucl. Instr. Meth.*, *B80/81*, 827 (1993).
56. J.C. Lee, T.E. Schlesinger and T.F. Kuech, *J. Vac. Sci. Technol*, *B5*, 1187 (1987).
57. J. Bloch, U. Bockelmann and F. Laruelle, *Solid-State Electron.*, **37**, 529 (1994).
58. H. H. Andersen, *Appl. Phys.*, **18**, 131 (1979).
59. E.G. Bithell and W. Stobbs, *Phil. Mag.*, *A60*, 39 (1989).
60. F. Lebland, Ph.D. Thesis, Universit de Paris 7, (1993).
61. A. Pépin, C.Vieu, M. Schneider, E.V.K. Rao, H. Launois (submitted to *J. Vac. Sci. Technol B*).
62. T. Y. Tan and U. Gosele, *Appl. Phys. Lett.*, **52**, 1240 (1988).
63. D. G. Deppe., N. Jr. Holonyak, K. C. Hsieh, P. Gavrilovic, W. Stutius and J. Williams, *Appl. Phys. Lett.*, **51**, 581 (1987).

64. D. Bimberg, J. Christen, T. Fukunaga, H. Nakashima, D. E. Mars, and J. N. Miller, *J. Vac. Sci. Technol. B*, **5**, 1191 (1987).

65. R. F. Kopf, E. F. Schubert, T. D. Harris, and R. S. Becker, **58**, 631 (1991).

66. C. A. Warwick, W. Y. Jan, A. Ourmazd and T. D. Harris, **56**, 2666 (1990).

67. S. Y. Chang and G. L. Pearson, *Appl. Phys. Lett.*, **46**, 2986 (1975).

68. J. Gyulai, J. W. Mayer, I. V. Mitchell, and V. Rodriguez, *Appl. Phys. Lett.*, **17**, 332 (1970).

69. C. Vieu, *Defect and Diffusion Forum*, **119–120**, 127 (1995).

70. C. Vieu, A. Claverie, J. Faur, and J. Beauvillain, *Nucl. Instrum. Methods B*, **28**, 229 (1987).

71. H. Ryssel H. et I Ruge, *Ion implantation*, Wiley and sons Editors (1986).

72. S. T. Ahn, H. W. Kennel, J. D. Plummer, and W. A. Tiller, *J. Appl. Phys.*, **64**, 4914 (1988).

73. Corning Glass Works, Fused Silica, 7940 Data Sheets, Corning, N. Y. (1978).

74. P.J. Burkhars and R.F. Marvel, *J. Electrochem. Soc.*, **116**, 864 (1969).

75. C.E. Morosanu, *Thin Solid Films*, **65**, 208 (1980).

76. F.M. d'Heurle, *International Materials Reviews*, **34**, 53 (1989).

77. M.F. Doermer and W.D. Nix, *CRC Critical Reviews in Solid State and Material Sciences*, **14**, 225 (1988).

78. J. Philibert, *Diffusion et Transport de la Matire dans les Solides*, Les Editions de Physique, Les Ulis (1985).

79. P. Cusumano, B.S. Ooi, A. Saher Helmy, S.G. Ayling, A.C. Bryce, J.H. Marsh, B.Voegele, and M.J. Rose, *J. Appl. Phys.*, **81**, 2445 (1997).

80. M. Katayama, Y. Tokuda, N. Ando, Y. Inoue, A. Usami and T. Wada, *Appl. Phys. Lett.*, **54**, 2559 (1989).

81. P.J. Poole, S. Charbonneau, G.C. Aers, T.E. Jackman, M.Buchanan, M. Dion, R.D. Goldberg and I.V. Mitchell, *J. Appl. Phys.*, **78**, 2367 (1995).

82. W.P. Gillin, A.C. Kimber, D.J. Dunstan and R.P Webb, *J. Appl. Phys.*, **76**, 3367 (1994).

83. L.B. Allard, G.C.Aers, P.G. Piva, P.J. Poole, M.Buchanan, I.M. Templeton, T.E. Jackman, S. Charbonneau, U. Akano and I.V. Mitchell, *Appl. Phys. Lett.*, **64**, 2412 (1994).

84. H.H. Tan, J.S. Williams, C. Jagadish, P.T. Burke and M. Gal, *Appl. Phys. Lett.*, **68**, 2401 (1996).

85. Y. Arakawa and H. Sakaki, *Appl. Phys. Lett.*, **40**, 939 (1982).

86. E. Kapon, D.M. Hwang, and R. Bhat, *Phys. Rev. Lett.*, **63**, 430 (1989).

87. W. Wegscheider, L. N. Pfeiffer, M.M. Dignam, A. Pinczuk, K.W. West, S.L. McCall and R. Hull, *Phys. Rev. Lett.*, **71**, 4071 (1993).

88. T. Arakawa, M. Nishioka, Y. Nagamune and Y. Arakawa, *Appl. Phys. Lett.*, **64**, 220 (1994).

89. A. Pépin, C. Vieu, M. Schneider, G. Ben Assayag, R. Planel, J. Bloch, H. Launois, J.Y. Marzin, and Y. Nissim, *Surperlatt. and Microstruct.*, **18**, 229 (1995).

90. A. Pépin, C. Vieu, M. Schneider, G. Benassayag, R. Planel, J. Bloch, H. Launois, J.Y. Marzin and Y. Nissim, *Semiconductor heteroepitaxy: growth characterization and device applications*, B. Gil and R.L. Aulombard editors, 371 (1996).

91. A. Pépin, C. Vieu, M. Schneider, R. Planel, J. Bloch, G. Benassayag, H. Launois, J.Y. Marzin and Y. Nissim, *Appl. Phys. Lett.*, **69**, 61 (1996).

92. K. Kash, J.M. Worlock, M.D. Sturge, P. Grabbe, J.P. Harbison, A. Scherer and P.S.D. Lin, *Appl. Phys. Lett.*, **53**, 782 (1988).

93. S.M. Hu, *Appl. Phys. Lett.*, **32**, 5 (1978).

94. W.W. Liu and M. Fukuma, *J. Appl. Phys.*, **60**, 1555 (1988).

95. G. Danan, B. Etienne, F. Mollot, R. Planel, A.M. Jean-Louis, F. Alexandre, B. Jusserand, G. Le Roux, J.Y. Marzin, H. Savary and B. Sermage, *Phys. Rev. B*, **35**, 6207 (1987).

96. J.M. Luttinger, W. Kohn, *Phys. Rev.*, **97**, 869 (1955).

97. R.K. Kupka, Y. Chen, R. Planel and H. Launois, *Microelectron. Engin.*, **27**, 311 (1995).

98. J. Gierak, C. Vieu, H. Launois, G. Ben Assayag and A. Septier, *Appl. Phys. Lett.*, **70**, 2049, (1997).

99. H. Ribot, K.W. Lee, R.J. Simes, R.H. Yan and L.A. Coldren, *Appl. Phys. Lett.*, **55**, 672 (1989).

100. W. Choi, S. Lee, Y. Kim, D. Woo, S.K. Kim, S.H. Kim, J.I. Lee, K.N. Kang, J.H. Chu, S.K. Yu, J.C. Seo, D. Kim and K.Cho, *Appl. Phys. Lett.*, **67**, 3438 (1995).

101. A. Pépin, C. Vieu, M. Schneider, and H. Launois, Proceedings of the 9th IEEE Conference on Semiconducting and Semi-Insulating Materials Conference, 263, IEEE Publishers, Piscataway (1996).
102. A. Pépin, C. Vieu, M. Schneider, H. Launois, and Y. Nissim, *J. Vac. Sci. Technol. B*, **15**(1), 142 (1997).
103. G. Benassayag, C. Vieu, J. Gierak, H. Chaabane and A. Pépin, *J. Vac. Sci. Technol. B*, **11**, 531 (1993).
104. C. Vieu, A. Pépin, C. Vieu, G. Benassayag, J. Gierak and F.R. Ladan, *Inst. Phys. Conf. Ser.*, **134**, 385 (1993).
105. K. Kash, *Appl. Phys. Lett.*, **53**, 782 (1988).
106. Z. Xu and P. Petroff, *J. Appl. Phys.*, **69**, 6564 (1991).
107. P.A. Kirkby, P.R. Selway and L.D. Westbrook, *J. Appl. Phys.*, **50**, 4567 (1979).
108. D.D. Perovic, G.C. Wheatherly and D.C. Houghton, *Phil. Mag. A*, **64**, 1 (1993).

CHAPTER 11

Dependence of Dielectric Cap Quantum Well Disordering on the Characteristics of Dielectric Capping Film

W.J. CHOI

Photonics Research Center, Korea Institute of Science and Technology, Seoul 130–650, Korea

1 INTRODUCTION

The multiple quantum well(MQW) structure is a good cadidate to mono-lithically integrate lasers and waveguides for realizing photonic integrated circuits(PICs). For such an application, certain types of techniques are needed to define different bandgaps at the active and the passive part of the PICs. Quantum well disordering (QWD) teniques [1–13] have been used to locally disorder the MQW structure without any regrowth and/or selective growth step. Since the disordering of MQW results in a blue shift in bandgap and changes in refractive index near bandgap, these techniques can be used to fabricate lasers and waveguides monolithically with only one-step epitaxial growth [14–17].

There have been several techniques to selectively disorder III-V com-pound semiconductor QW structures such as impurity induced disordering [1,2], laser induced disordering [3], ion implantation disordering [4,5], and dielectric cap quantum well disordering [6–13] (DCQWD). Among these techniques, since DCQWD technique introduces much lower optical losses compared to the impurity induced disordering and ion implantation techniques which give high optical losses due to high number of defects and high doping concentrations introduced during the disordering process, this technique is thought to be better suited for fabricating high performance optical waveguide devices.

Recently, DCQWD technique has been used to fabricate laser diodes and photonic integrated circuits (PICs) [6,16,17], without any regrowth steps by disordering its structure selectively. Many dielectric films have been used to enhance QWD and to prevent QWD in DCQWD technique for the fabrication of LDs and PICs. SiO_2 [6,8] has been widely used as a capping layer to promote QWD, otherwise SiN [6], SrF_2 [9] and WN_x [10] have been used to prevent QWD in DCQWD technique. Since a SiO_2 capping layer induces a relatively larger blue shift than a SiN capping layer in a GaAs/AlGaAs QW system, SiO_2 is generally used to promote QWD while SiN is used as a mask to prevent QWD under the capped areas of QW [6]. The difference of degree of QWD between SiO_2 capped QWD and SiN capped QWD is thought to be a result of the difference Ga out-diffusion from GaAs into SiO_2 capping layer and into SiN capping layer during thermal annealing [18].

Although, for the fabrication of PICs with DCQWD technique, SiN capping layer has been proposed to be used as a mask to prevent QWD, it has been shown that SiN capping layer can enhance QWD in GaAs/AlGaAs QW system [6,11,12]. It also has been shown that the degree of QWD caused by the SiN capping layer is dependent on the film quality and/or the film growth process [11,12]. Therefore, if an optimum film growth condition could be found by varying the process condition, selective QWD can be achieved with the same dielectric capping material without surface degradations, which should be accompanied in the disordering process without an encapulation during thermal treatment.

There are various film growth techniques for DCQWD, such as chemical vapor deposition (CVD) technique [19], e-beam evaporation technique [20–22], sputtering technique [8], and plasma enhanced chemical vapor deposition(PECVD) technique [6,11,12], which can deposit SiN and/or SiO_2 films on a QW substrate. Among these technique, PECVD tehnique is a low-temperature process technique and offer films with a low damage, a high adhesion and lower stress than those grown by thermal CVD. The characteristics of film deposited by PECVD technique can be varied by varying process conditions, such as substrate temperature, the ratio of reactant gas and RF power [23]. In this study, PECVD-grown SiN was used to study the dependence of DCQWD on the characteristics of dielectric capping layer because it can promote QWD and its characteristics can be well controlled by controlling the growth condition.

In this article, the dependence of DCQWD on the characteristics of dielectric capping layers is mainly described, especially for GaAs/AlGaAs

MQW structure when PECVD-grown SiO_2 and SiN film are used as capping layers. The main topic of this article is described in section 3. The following section 2 describes the calculation of well shapes and quantized energy level of disordered QW structure. This kind of calculation is essential to compare the degree of QWD at each process condition. Setion 3 describes DCQWD by using PECVD grown dielectric film and its dependence on the characteristics of dielectric film, which is determined by film growing condition. In this article, such dependence is studied with PECVD grown SiN. To study the usefulness of DCQWD technique, DCQWD of low-temperature grown GaAs capped MQW structure was carried out and described in section 3.1. In section 4, the photogenerated carrier lifetime in disordered MQW structure by DCQWD technique is described. Finally the summary is presented in section 5.

2 ENERGY LEVEL CALCULATION OF INTERDIFFUSED QW STRUCTURE

Quantum well disordering has been explained by the interdiffusion of constituent atoms. In GaAs/AlGaAs QW system, QWD is thought to be induced by Al-Ga interdiffusion. Since Al diffusion from AlGaAs barrier into the GaAs well does not affect lattice parameters, the intermixed quantum well potential just follows the Al mole fraction. With an assumption of a constant concentration, the solution of Al diffusion can be expressed by the combination of error functions [9]. After a thermal treatment, the concentration of Al in the MQW region can be expressed [24] as

$$C(z, t) = C_o \left(1 - \frac{1}{2} \sum_{n=0}^{m-1} \left[\text{erf} \left(\frac{z - nL_w - nL_b}{2L_d} \right) \right. \right.$$
$$\left. \left. + \text{erf} \left(\frac{(n+1)L_w + nL_b - z}{2L_d} \right) \right] \right), \tag{1}$$

where erf is the error function, C_o is the initial concentration of Al, t is the diffusion time, L_b is the barrier width, $L_d (= \sqrt{Dt})$ is the diffusion length and D is the diffusion coefficient. By using this equation one can calculate the Al concentration and the corresponding intermixed quantum well potential profile for arbitrary different diffusion lengths. We obtained the Al concentration for different values of diffusion length. As the diffusion length increases, the shape of MQW is changed from square type to non-square type as shown in Figure 2. The Schrödinger equation is then solved numerically for quantized energy levels of electron and holes by using the intermixed quantum well potential profile. In our study, Schrödinger equation

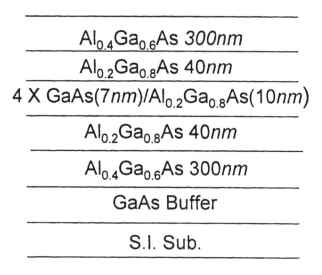

Fig. 1. MQW structure used for quantum well disordering.

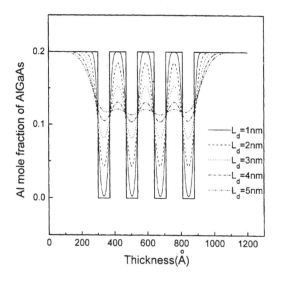

Fig. 2. Al mole fraction within MQW after Al-Ga interdiffusion.

TABLE 1
The material paramaters of $Al_x Ga_{1-x} As$.

Electron mass	$0.0667 + 0.0835x$
Heavy hole mass	$0.34 + 0.412x$
Bandgap (eV)	$1.424 + 1.247x$
Valence and conduction band offset ratio	40%:60%
Temperature dependence of bandgap	$1.519 - 5.405 \times 10^{-4} T^2 (204+T)^{-1}$

was solved by using finite difference method [9,24] for the calculation of subband energy in the interdiffused MQW.

Figure 1 shows the GaAs/AlGaAs MQW structure used in our study. As the Al-Ga interdiffusion increases, the well shape is deformed to non-square type accorrding to equation (1). Figure 2 shows the Al profile in the MQW structure after Al-Ga interdiffusion with finite interdiffusion lengths. As shown in Figure 2, the MQW structure is changed to bulk AlGaAs layer when the diffusion length becomes larger than 50 Å. For the calculation of transition energy shift due to Al-Ga interdiffusion in MQW region, at first, the lowest subband energies of electrons and heavy holes were calculated with and without Al-Ga interdiffusions. The transition energies of square MQW and interdiffused MQW can be calculated by considering the Al mole fraction at the bottom of both structures. The excitonic effect was ignored in this calculation. The material paramaters [26] used in the calculation of transition energy is summarized in Table 1.

Figure 3 shows the calculated transition energy shift as a function of diffusion length for our MQW structure. We used this kind of graph to get the diffusion length and corresponding diffusion coefficient from the transition energy shift obtained in the experiment. In our study, we assumed that the calculated transition energy reveals the QW peak energy obtained in PL measurement for the calculation of interdiffusion coefficient at each experiment. Total intermixing is occurred when transition energy shift is greater than 90 meV, i.e diffusion length is larger than 50 Å, in our MQW structure when considering the diffusion profiles in Figure 2.

In general, it is not easy to get exactly the same transition energy of designed square MQW from the PL measurement of epi-structure, because of difficulties in controlling growth system. If we ignore the excitonic effect on the transition energy obtained from PL measurement, this kind of difference may come from slight miscontrol of Al mole fraction of barrier and/or the well thickness during growth, though it depends on growth system. Since transition energy shifts by QWD depend on the composition of barrier and well width, it is very important to determine the exact parameters of MQW in order to get exact diffusion length. This kind of consideration is very

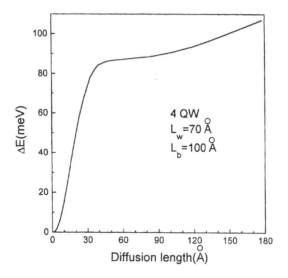

Fig. 3. Energy shift versus diffusion length.

important to determine the diffusion coefficient when short time thermal treatment, such as rapid thermal annealing (RTA), is used for QWD.

3 DCQWD BY USING PECVD GROWN DIELECTRIC FILM

3.1 DCQWD of Low Temperature Grown GaAs Capped QW Structure

The low-temperature grown (LT) GaAs has also been used as a vacancy source to promote QWD [28,29]. One can easily think of the combination of LT-GaAs capping layer and dielectric capping layer in vacancy induced QWD technique. In order to study this kind of combination, DCQWD technique was applied to LT-GaAs capped MQW structure.

The GaAs/AlGaAs MQW structures used in this study were grown by MBE on semi-insulating GaAs substrates without any intentional doping. Two samples were prepared for a comparson, whereas one with 100 nm thick normal temperature grown (NT) GaAs cap layer and the other with 100 nm thick low temperature grown (LT) GaAs cap layer on top of MQW structure. The schematic diagram of MQW structures is shown in Figure 4. All structures were grown at 600°C except for LT-GaAs epilayer, which was grown at 200°C. After a 300 nm thick SiO$_2$ was deposited on each sample by PECVD, RTA was performed at temperatures ranged from 900°C to 1000°C

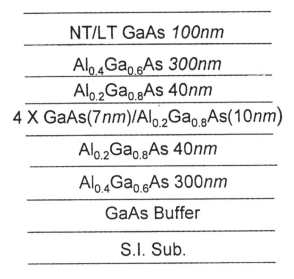

| NT/LT GaAs *100nm* |
| $Al_{0.4}Ga_{0.6}As$ *300nm* |
| $Al_{0.2}Ga_{0.8}As$ *40nm* |
| 4 X GaAs(*7nm*)/$Al_{0.2}Ga_{0.8}As$(*10nm*) |
| $Al_{0.2}Ga_{0.8}As$ *40nm* |
| $Al_{0.4}Ga_{0.6}As$ *300nm* |
| GaAs Buffer |
| S.I. Sub. |

Fig. 4. Sample structure of NT/LT GaAs capped MQW sample.

for 30 sec without any GaAs proximity cap under the Ar atmosphere. For a comparision, both samples were thermally treated without SiO_2 capping layer. In this case, GaAs proximity cap was used to protect sample surface. The samples were then characterized by PL measurements at $9°K$.

In all samples, it was observed that the full width at half maximum (FWHM) of the PL spectrum increases compared to that of the virgin sample. This may be due to the reduction of quantum confinement and the variation of quantum well width [13,27]. The PL peak shifts are summarized in Figure 5 with respect to RTA temperature. Without SiO_2 capping layer, poor surface was obtained because of desorption of As after RTA in our experimental condition, though GaAs proximity cap layer was used. Though there was no report about As desorption in the experiment of Tsang et al.'s [28], As desorption might be occured in our experiment because of higher RTA temperature than theirs. In our experiment desorption related defects might induce QWD. Though GaAs proximity cap could not protect desorption of constituent atoms from the sample surface perfectly, it could supply a certain number of constituent atoms to sample surface. Thus the total number of desorption related defects finally will be saturated at eqlibrium, which gave saturation behavior in energy shift for NT-GaAs capped case as seen in Figure 5. In contrast, energy shifts for LT-GaAs capped samples exhibit about 32 meV at all temperature as seen in Figure 5. It might be thought

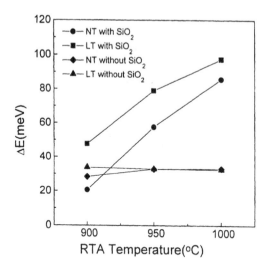

Fig. 5.   Energy shift versus RTA temperature for NT/LT GaAs capped samples with and without SiO₂ capping layer.

that the desorption can be easily occured in LT-GaAs cap layer than in NT-GaAs cap layer because of its poor crystal quality.

In contrast, with SiO₂ capping layer, very good surfaces were obtained after RTA treatment. Thus in SiO₂ capped case, QWD can be thought to be induced by the vacancies in LT-GaAs and/or SiO₂. As seen in Figure 5, SiO₂ capping layer enhances the QWD in all temperatures, compared to the energy shifts without SiO₂ capping layer for both LT-GaAs capped sample and NT-GaAs capped sample except for NT-GaAs capped sample at 900 ^{o}C. One can also see larger energy shifts for LT-GaAs capped samples than that for NT-GaAs capped samples, on which SiO₂ films were capped. As reported by Tsang et al. [28], LT-GaAs cap layer acts as a vacancy source. Therefore the total number of vacancies in LT-GaAs capped sample is larger than that of NT-GaAs capped sample when same thickness of SiO₂ capping was used. Since more vacancies induce larger energy shift [8,12], LT-GaAs capped sample exhibits larger energy shift at the same temperature when the same thickness of SiO₂ was capped.

The diffusion coefficients were calculated from the energy shifts by solving the potential profile induced by the interdiffusion and Schrödinger equation described in section 2. The temperature dependence of interdiffusion coefficient can be described by

$$D(T) = D_o \exp\left[\frac{-E_A}{K_B T}\right], \tag{2}$$

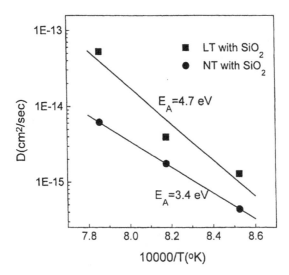

Fig. 6. Arhenius plot of interdiffusion coefficient for NT/LT GaAs capped samples with SiO$_2$ capping layer.

where E_A is the activation energy, K_B is Boltzman constant, T is the absolute temperature. Figure 6 shows the Arrhenius plots of interdiffusion coefficient for LT-GaAs capped samples and for the NT-GaAs capped samples, when SiO$_2$ capping layer was used. As clearly seen in Figure 6, the data of NT-GaAs capped sample were nicely fitted to give an activation energy of 3.4 eV. This activation energy of 3.4 eV is smaller than the value, 3.8 eV, reported by Ralston et al. [20], who used e-beam evaporated SiO$_2$ as a capping layer. As reported by Tsang et al. [28], the interdiffusion by LT-GaAs cap layer gave an activation energy of 4.04 eV. So PECVD grown SiO$_2$ capping film can enhance interdiffusion, compared to LT-GaAs cap layer as a vacancy source.

The interdiffusion coefficient of LT-GaAs capped sample is larger than that of NT-GaAs capped sample at each temperature. This can be explained by the difference of total number of vacancies in two samples as explained above. Moreover, LT-GaAs capped sample has thinner top layer on MQW structure than NT-GaAs capped sample because LT-GaAs cap layer acts as a vacancy source. The activation energy from LT-GaAs capped samples with SiO$_2$ capping is 4.7 eV. Although LT-GaAs capped samples with SiO$_2$ capping has two vacancy sources, vacancy indiffusion might depend on the LT-GaAs cap layer. Thus 4.7 eV of activation energy may largely reveal the characteristics of LT-GaAs cap layer as a vacancy source. In this case, the activation energy of 4.7 eV is lager than 4.04 eV reported by Tsang et al. [28]. It is thought that this may come from the different LT-GaAs growth condition which determines the quality of LT-GaAs [31–33]. Therefore this larger activation energy of LT-GaAs capped sample also conforms that PECVD grown SiO$_2$ capping film enhance interdiffusion when compared to LT-GaAs cap layer as a vacancy source.

The interdiffusion coefficient of NT-GaAs capped sample with SiO$_2$ capping is nearly the same at all temperatures when comparing our results with the results of Tsang et al.'s [28] by using their temperature dependence of interdiffusion coefficient, $D(T)$. The thicknesses of the top layer on QW are 440 nm in our structure and 25 nm in Tsang et al.'s. Therefore one can easily conclude that our 300 nm thick SiO$_2$ vacancy source induces larger intermixing than their 100 nm thick LT-GaAs vacancy source. Moreover the interdiffusion coefficient of LT-GaAs capped samples with SiO$_2$ capping is nearly an order of magnitude higher than Tsang et al.'s.

As described in detail, SiO$_2$ capping layer not only preserves sample surface fairly well but also induces QWD. In addition, larger energy shift was obtained when SiO$_2$ capping layer was used over the LT-GaAs capped sample. Since thicker LT-GaAs layer also induces larger energy shift [29], this combined technique can be used to get selective QWD which is very useful for the fabrication of optoelectronic devices.

3.2   Dependence on the Film Growing Condition

Dielectric capping layer acts as a vacancy source to induce the outdiffusion of constituent atoms in DCQWD technique. Thus the degree of QWD depends on the film quality which is dependent on growth conditions. The properties of SiN film prepared by PECVD method depend on the growth parameters including reactant gas ratio, gas flow rate, background pressure, substrate temperature, RF power and geometric configuration of the reactor chamber [34–37]. Thus PECVD-grown SiN is one of the good cadidates to study the dependence of DCQWD on the characteristics of dielectric capping layer. In order to find which growth parameters affect DCQWD, several SiN films were grown on MQW structures by varying PECVD growth parameters such as growth time, RF power and gas flow rates.

We used a GaAs/AlGaAs MQW laser structure which was grown by metal organic chemical vapor deposition (MOCVD) method on a Si-doped n$^+$ GaAs substrate. The vertical structure of the MQW substrate is shown in Figure 7. SiN capping layer was deposited by the PECVD technique for capping layers [23]. We used dilute silane (5% SiH$_4$ in N$_2$) and NH$_3$ (99.999%) as reactant gases. During the plasma deposition process, the flow rate of SiH$_4$ and NH$_3$ were 40.5 sccm and 24.7 sccm, respectively. N$_2$ (99.999%) gas was used as a buffer and the total pressure is maintained at 0.9 Torr. Temperature of the substrate was maintained at 300°C. SiN films

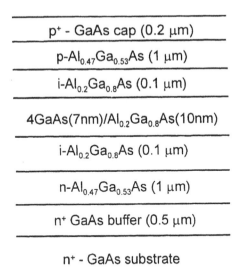

| p⁺ - GaAs cap (0.2 μm) |

Fig. 7. Epitaxial structure of GaAs/AlGaAs MQW LD substrate.

were grown at 30 watts RF power for 3, 6, 9 and 12 min., which gave SiN thicknesses of 30, 59, 85 and 123 nm, respectively. The thickness of a SiN film was measured using an ellipsometer (Gaertner, L117). The measured growth rate and refractive index of the SiN film are approximately 10 nm/min. and 1.91, respectively. Thermal treatment of the SiN capped MQW samples was accomplished simultaneously by rapid thermal annealing (RTA) at 850°C for 35 sec. in Ar atmosphere with a heating rate of 65°C/sec. Disordering of the MQW sample was observed by PL measurement at 20 K after removing the SiN film.

PL spectra, at 20 K, of the MQW samples which were PECVD SiN capped and RTA treated are shown in Figure 8. As shown in Figure 8, a virgin sample exhibits a QW peak at 789.2 nm. QW peak shifts are 14.0 meV, 36.0 meV, 48.6 meV and 62.5 meV for the samples whose SiN capping layers were grown for 3, 6, 9 and 12 min., respectively. Figure 9 shows QW peak shifts as a function of SiN thickness. QW peak shift becomes larger with SiN thickness as shown in Figure 8 and 9. Such a result is similar to the result on SiO_2 cap disordering reported by Chi et al. [8].

There seems to be a linear relationship between the thickness of SiN capping layer and the diffusion length calculated from the energy shift as shown in Figure 9. The nonlinear relation between the energy shift and the thickness of SiN capping layer is a result of nonlinear relation between the energy shift and the diffusion length as shown in Figure 3, though it seems

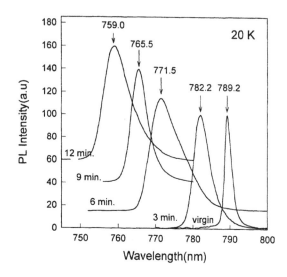

Fig. 8. PL spectra of MQW LD structure disordered with different thick SiN capping layers. Each time represents growing time.

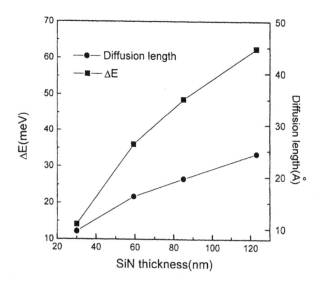

Fig. 9. Energy shift and diffusion length as a function of SiN capping layer thickness.

to show a linear relation within finite diffusion length range less than 30 Å. Since vacancy indiffusion from dielectric capping layer is thought to cause QWD in DCQWD technique [8,28], the diffusion length depends on the total vacancies in dielectric capping layer. Therefore diffusion length can be a measure of the total number of vacancies in a dielectric film. One can easily see, in Figure 9, that SiN films whose thickness are thicker than 60 nm have a constant vacancy density. However, 30 nm thick SiN film has a relatively low vacancy density. As widely studied [41,42], PECVD SiN film is not homogeneous along its depth and has a Si-rich bottom layer. The thickness of Si-rich bottom layer near substrate reaches up to 30 nm [43]. Thus a 30 nm thick SiN film is largely affected by the Si-rich layer, though this effect can be averaged out in thicker SiN films.

For a comparison, a 150 nm thick PECVD SiO_2 capping layer was used for dielectric cap quantum well disordering. The SiO_2 capping layer was also deposited with PECVD method at 30 watt RF condition using SiH_4 and N_2O as reactant gases with an N_2 buffer. The RTA condition for this SiO_2 capped sample is the same as that for SiN capped samples. Figure 10 shows the PL spectrum of this sample, measured at 14 K. As shown in Figure 10, the SiO_2 capped sample exhibits a QW peak at 760.0 nm. Since the temperature difference of 6 K in the PL measurements result in approximately 0.5 meV of energy difference [38], PL spectrum measured at 14 K can be compared to the PL data in Figure 8. Considering a 6 K temperature difference, there is 59.8 meV (29.2 nm) of QW peak shift.

SiN capping layer has been used to prevent QWD in GaAs/AlGaAs QW system [6] for device applications, because its effect on QW peak shift was smaller than a SiO_2 capping layer. However, as shown in Figure 9 and 10, there is a smaller QW peak shift for a SiO_2 capped sample than for a SiN capped sample, even though the SiO_2 capping layer is thicker than SiN capping layer. Therefore, one can conclude that it is impossible for a PECVD SiN capping layer to be used as a mask to prevent QWD, otherwise, it should be used to enhance QWD as reported by Beauvais et al. [13]. Beauvais et al. [13] also reported that a PECVD SiN cap disordering could cause QW peak shift. In their experiment, they used a 5 nm thick single QW inserted between two 2 nm thick AlAs layers which were used to enhance disordering. Their experimental results showed that a thicker $Al_{0.4}Ga_{0.8}As$ barrier caused a reduction in QW peak shift. Especially for a QW substrate with an 1.0 μm thick $Al_{0.4}Ga_{0.8}As$ barrier, they reported a QW peak shift of 43 meV and 38 meV for a 200 nm thick SiO_2 capped sample and for a 240 nm thick SiN capped sample, repectively, thermally treated with RTA at 950°C for 30 sec. It has been reported by many authors [6,8,13] that, for dielectric cap disordering, a higher RTA temperature, thicker capping layer thickness, and narrower QW width in its QW structure cause larger QW peak shift. Therefore it can be concluded that the effect on QW peak shift is more profound for

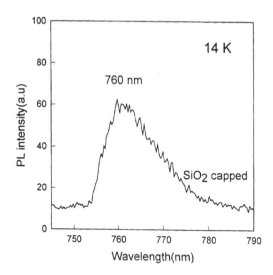

Fig. 10. PL spectrum of MQW LD structure disordered with SiO$_2$ capping layer.

our result than for Beauvais et al.'s because of the 1.3 μm thick barrier on QW, a 100 K lower RTA temperature, and a 117 nm thinner SiN capping layer used for our experiment. Moreover, they inserted 2 nm thick AlAs thin layers between the QW and the barrier to enhance disordering and their SiN capping layer exhibited a refractive index of 1.76 which indicated the existence of oxygens in the SiN layer, which was reported to enhance the Ga atom outdiffusion [39] that is thought to be a main reason of QW disordering [8]. Since the refractive index of our SiN capping layer, 1.91, it indicates that this SiN film is well grown to form a stoichiometric Si$_3$N$_4$ whose refractive index is 1.94 [34], our SiN capping layer may not have oxides to enhance disordering.

Table 2 shows the calculated interdiffusion coefficient. Comparing this result to that of section 3.1 for SiO$_2$ capping layer, this result shows the enhancement of QWD. In the substrate of section 3.1, there is no any doping, however, in this experiment, the MQW LD structure employed Zn (5×10^{17} cm^{-3}) as a p-type dopant in the upper cladding layer. Therefore it might be Zn induced disordering reported by Nagai et al. [40] who used Zn in p-type cladding layer as a disordering source. Though our Zn concentration is much lower than that of their's ($1 \times 10^{18} - 4 \times 10^{18}$), there might be some effect of Zn impurity induced disordering in our results because of relatively high thermal treatment. Though there is an enhancement of QWD by Zn induced disordering, as seen in Figures 8, 9 and 10, the effect of dielectric cappings on the QWD holds its validity.

TABLE 2
Parameters of DCQWD with different thick SiN capping layers grown by PECVD.

Growth time (min.)	SiN thickness (nm)	ΔE (meV)	L_d (Å)	D (cm^2/sec)
3	30.0	14.0	9.5	2.6×10^{-16}
6	59.3	36.0	16.2	7.5×10^{-16}
9	85.1	48.6	19.6	1.1×10^{-15}
12	123.0	62.5	24.4	1.7×10^{-15}

In order to study the dependence of film growth condition on DCQWD, RF power was also varied during the SiN film growth. We used the same reactant gases such as dilute SiH_4 (5% in N_2) and NH_3. During the SiN growth, the ratio of partial pressure (P_{NH3}/P_{SiH4}) was kept as 1, and total pressure was kept as 0.9 Torr by adding N_2 gas and the substrate temperature was kept at 300 °C. SiN films were grown with various RF powers (30 watts, 60 watts and 90 watts) on the MQW LD structure shown in Figure 7. Disordering of samples was also accomplished by RTA at 850 °C for 35 seconds in Ar atmosphere with a heating rate of 65 °C/sec. Disordering of MQW samples was observed by PL measurement at 12 K after removing the SiN film.

Figure 11 shows PL spectra of the MQW samples before and after RTA treatment. After RTA treatment, relative to the PL peak before RTA treatment, the PL peaks of SiN capped MQW sample were blue shifted by 10.6 nm (30 watts), 15.4 nm (60 watts), and 29.8 nm (90 watts), respectively. As shown in Figure 11, the energy shift becomes larger as the RF power became stronger. This result seems to show that the higher RF power condition of film growth by PECVD is much more effective for disordering the MQW structure. However, as discussed above, the consideration of film thickness should be followed for a relative comparison.

As discussed above, one can get normalized vacancy density of a dielectric film by normalizing diffusion constant with a film thickness. As shown in Table 3, all thicknesses are thick enough to ignore the effect of the Si-rich interface between the sample and the SiN film. Thus diffusion length can be used as a measure to compare the film qualities determined by different RF powers. As shown in Table 3, relative vacancy density is 3 times larger at 90 W RF power condition than those at lower RF power conditions. Therefore one can conclude that SiN film grown at high RF power condition induces large QWD and the RF power can be a good process parameter for DCQWD when PECVD SiN is used as a capping layer.

In order to find the dependences of DCQWD on the SiN film growth condition, reactant gas ratio was also changed during SiN growth. NH_3 gas flow rates were changed from 0 sccm to 40 sccm at fixed SiH_4 gas flow rate

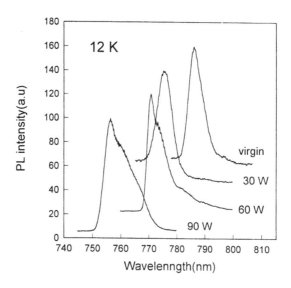

Fig. 11. PL spectra of disordered MQW LD structure with SiN capping layer grown at different RF power.

of 20 sccm. During the growth, 30 W RF power was applied and the base pressure was kept at 0.9 Torr by adding N_2 gas. In this case, we also used diluted SiH_4 (5% in Nitrogen). The growth temperature was 300 °C. Two MQW samples were used as shown in Figure 4. These MQW samples have no any intentional doping. After SiN growth on the MQW sample, thermal treatment of the samples was accomplished by RTA at 950°C for 30 sec.

The properties of SiN capping layer on the MQW samples were characterized before RTA treatment by an ellipsometer, a surface profiler, Auger electron spectroscopic method and an elastic recoil detection (ERD) method of Rutherford Back Scattering (RBS). As shown in Figure 12, the thickness of SiN film decreases with the increase of the NH_3 flow rate from 270 nm

TABLE 3
Parameters of DCQWD with SiN capping layers grown at different RF powers.

RF power (watt)	SiN thickness (nm)	ΔE (meV)	L_d (Å)	Relative vacancy density (%)
30	95	21.7	11.8	100
60	132	31.2	14.7	89
90	64	61.9	24.3	306

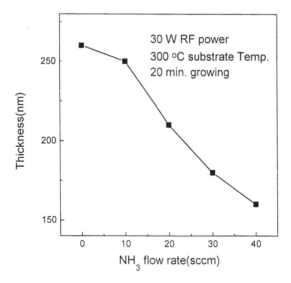

Fig. 12. Thicknesses of SiN films grown with different NH_3 flow rate.

to 160 nm. The refractive index of SiN film also decreases slightly with the increase of the NH_3 flow rate from 1.85 to 1.83. These refractive indices indicate that these SiN films are N-rich films. The data from Auger electron sepectroscopy also showed that the atomic percent of Nitrogen of the SiN film is almost 67 % for all SiN films as seen in Figure 13. As reported by Dun et al.'s [35], N_2 gas can be used as a Nitrogen source of the SiN film grown by PECVD method. Since N_2 gas was used not only as a carrier gas but also as a Nitrogen source in our experimental condition, all SiN films could exhibit nearly same composition. This N-rich SiN might be the result of the lack of SiH_4 in our experimental condition.

However, thought all SiN films exhibit the same composition, the data from ERD reveals different characteristics in Hydrogen content. The Hydrogen content of the film increases with the increase of the NH_3 flow rate as shown in Figure 14. This result is well coincident with the results of Dun et al.'s [35] who used SiH_4/N_2 gas mixture to lower Hydrogen content in the SiN film.

Figure 15 shows PL spectra of NT-GaAs capped samples after RTA treatment at 9 K. As one can see, the amount of blue shift increases with NH_3 flow rate. In this case, one may consider the thickness of SiN capping layer to compare the QWD to each other. However, as shown in Figure 12, the thickness of SiN film decreases with NH_3 flow rate. As discussed above, relative degree of QWD can be compared with the relative vacancy density. Table 4 shows relative vacancy density at each NH_3 flow rate. As shown in

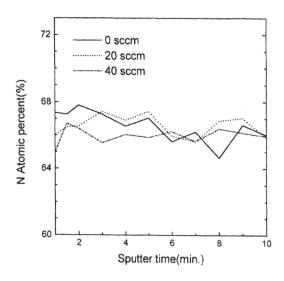

Fig. 13. Nitrogen atomic percent in SiN film grown with different NH_3 flow rate.

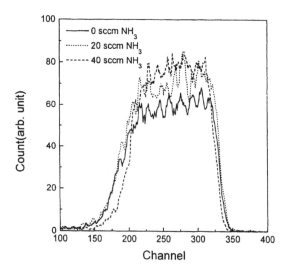

Fig. 14. ERD data of Hydrogen in SiN films grown with different NH_3 flow rate.

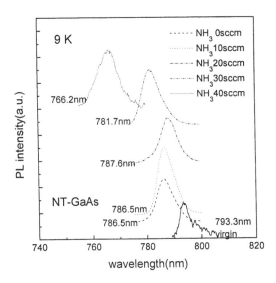

Fig. 15. PL spectra of NT-GaAs capped MQW structure disordered with SiN capping layer grown with different NH3 flow rate.

Table 4, relative vacancy density increases with NH3 flow rate. Therefore one can conclude that SiN film grown at higher NH3 flow rate have larger vacancy density to cause larger QWD.

The maximum energy shift is 55.2 meV for the sample capped with SiN grown at 40 sccm NH3 flow rate. This value is comparable to SiO2 cappled sample which is also RTA treated at 950°C for 30 sec. As seen in Figure 5, 300 nm thick SiO2 capped sample exhibited energy shift of 57.6 meV. When considering the thickness of dielectric capping layer, SiO2 capping layer seems to have lower vacancy density than SiN film grown at 40 sccm NH3 flow rate.

TABLE 4
Parameters of DCQWD with SiN capping layers grown with different NH3 flow rate.

NH3 flow rate (sccm)	SiN Thickness (nm)	ΔE (meV)	L_d (Å)	Relative vacancy density (%)
0	260	13.5	9.3	100
10	250	13.5	9.3	104
20	210	11.3	8.4	112
30	180	23.2	12.5	194
40	160	55.3	22.1	386

As shown in Figure 14, SiN film grown at higher NH_3 flow rate exhibited higher Hydrogen content. As reported by Kapoor et al. [44], SiN film with high Hydrogen concentration exhibits large trap density. Therefore higher Hydrogen content is thought to be a source of larger vacancy density of SiN film grown at higher NH_3 flow rate. Since SiH_4 was fixed during SiN growing, the increase of Hydrogen content in SiN film is a result of increase of NH_3 flow rate and hence increased Hydrogen in SiN has N-H bond configuration. The decrease of refractive index with NH_3 flow rate may also be a result of the increase of N-H bond in SiN film [23].

It is noteworthy that one can get the maximum energy difference of 42 meV by simply changing NH_3 flow rate as seen Figure 15 and Table 4. This is large enough to get selective disordered MQW structure spatially for the fabrication of optoelectronic devices monolithically. Moreover, this technique may reduce the problem caused by the difference of thermal expansion coefficient between different dielectric capping layers used for selective QWD. Therefore this technique can be widely used for the integration of optolectronic devices.

4. CARRIER LIFETIMES IN DISORDERED QW STRUCTURE

Since ultrafast scattering and relaxation processes of carriers in semicon-ductors affect the electronic and optical performances of LD's and PIC's, especially in high speed operation, it is important to understand the carrier dynamics of disordered QW structures. QWD is thought to be induced by the vacancy indiffusion which creates defects in QW structure. Thus it is easily thought that there should be changes in carrier dynamics in disordered QW structures compared to virgin structure and also such changes depend on the degree of QWD. We investigated the carrier lifetimes of dielectric cap disordered QW structures with different degrees of QWD.

We employed a typical GaAs/AlGaAs MQW LD structure shown in Figure 7. This structure was grown by MOCVD method. The substrate structure represents a typical p-i-n type separate confinement heterostructure (SCH). The substrate was divided into pieces, on which SiN capping layers were deposited by PECVD method. Four samples with SiN thicknesses of 30, 59, 85 and 123 nm were prepared. These samples are refered to as S1, S2, S3 and S4, respectively, and S0 represents the virgin MQW sample. These SiN capped MQW samples were treated by RTA at 850°C for 35 s in Ar gas environment with a heating rate of 65°C/s. Disordering of the MQW samples was observed by PL spectra at 20 K after removing the SiN film.

Carrier lifetimes were measured at 20 K by using a time correlated single photon counting (TCSPC) system [48] equipped with a micro-channel plate photomutiplier tube. A cavity dumped dye laser employing Rhodamine

6G as a gain dye and DODCI as a saturable absorber was used as an excitation source. The output pulses had 1–2 ps in pulsewidth at 3.8 MHz dumping rate. The combination of a constant fraction discriminator, a delay, a time-to-amplitude converter and a multichannel analyzer was utilized to obtain the temporal profile of luminescence decay. The instrumental response of our TCSPC system is typically 50 ps (**fwhm**), which gives about 10 ps time resolution through a deconvolution technique. The laser intensity was controlled by various neutral density filters to reduce the exciton-exciton or other multiple carrier interactions in the quantum well. The excitation pulses at 585 nm were employed to excite electrons above the energy gap of the $Al_{0.4}Ga_{0.6}As$ cladding layer. Therefore the PL decay profiles obtained with this excitation wavelength reveal the characteristics of the carrier dynamics not only in the QW region but also in the SCH region. The excited sheet carrier concentration was estimated to be in the order of 10^{10} cm^{-2}.

PL spectra of the MQW samples at 20 K are shown in Figure 16. As shown in Figure 16, a virgin sample exhibits a QW peak at 789.2 nm. The blue shift of this PL peak increases as the SiN capping layer thinkness increases, indicating that a thicker SiN capping layer causes a more disordering of the QW relative to SiO_2 capping layer [8] as described in section 3.2. There is also a broadening of the PL peaks for the samples capped with SiN film from 4.1 meV in **fwhm** of the virgin sample to 10.3 (S1), 20.1 (S2), 10.8 (S3) and 17.6 meV (S4), respectively. Though the broadening of PL peaks after QW disordering is well known [6,13,16], there seems to be no correlation between the QW peak shift and the broadening of its PL peak in our case. Although it is difficult to draw a quantitative relationship between the PL peak broadening and the degree of QWD, the induced defects by QWD seem to contribute to the boadening of PL peaks by reducing carrier lifetimes.

Figure 17 shows PL decay profiles of a virgin sample and dielectric cap disordered samples. From these PL decays, photo-generated carrier lifetimes were obtained and summerized in Table 5. As seen in Table 5, there is a decrease in carrier lifetimes and an increase in QW PL peak shift as the SiN capping layer thickness increases. The dielectric cap annealing of compound semiconductors induces defects in the QW region after RTA treatment of the QW samples [18]. As pointed out by many researchers [45–47], the carrier lifetime decreases due to these defects which can trap carriers efficiently and relax the momentum conservation in the radiative transition [46]. The thicker SiN capping layer introduces more defects and enhances the interdiffusion of Ga and Al through RTA in the QW region [8].

It is also noteworthy that QWD induced disordering and defects mainly contribute as nonradiative quenching sites for PL decay processes. This indicates that the excitation-relaxation process is dispersive due to the energetic disorder in MQW semiconductors. Such inhomogeneous systems are characteristic for inhomogeneous line broadening effect. Thus, the

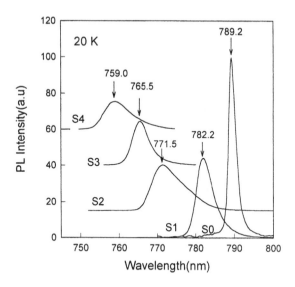

Fig. 16.   PL spectra of a virgin and dielectric cap disordered samples.

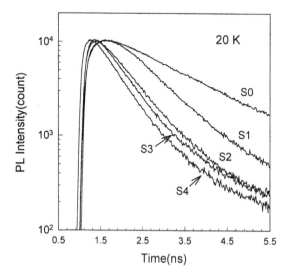

Fig. 17.   PL decay profiles of a virgin sample and dielectric cap disordered sample.

TABLE 5
Sample characteristics and photoluminescence lifetimes.

sample code	SiN thickness (nm)	ΔE (meV)	Rise time (ns)	Decay time (ns)
S0	–	–	0.21	1.92
S1	30	14.0	0.37	0.81
S2	59	36.0	0.14	0.71
S3	85	48.6	0.12	0.62
S4	123	62.5	0.11	0.57

QWD induced defects in MQW's decrease the internal quantum efficiency providing nonradiative and dispersive quenching sites for radiative carrier relaxation processes in MQW semiconductor. Thus, as shown in Figure 16, the PL peak broadening increases and its intensity decreases as QWD increases.

There is also a decrease in rise time along with an increase in QWD. The rise times for S0, S1, and S2 are slightly longer than reported value of about 120 ps at 20 K [48]. The reported rise time in the results of Jeong et al.'s [48] represents the carrier dynamics in QW region with a thin (30 nm thick) barrier layer, indicating the relaxation time of photo-generated carriers in the QW region only. On the other hand, in our experiment the excitation photon energy was high enough to excite the carriers in the relatively thick $Al_{0.4}Ga_{0.6}As$ cladding layer as well as in the QW and $Al_{0.2}Ga_{0.8}As$ barrier layer. Considering the 1.2 μm thick top layers (GaAs and $Al_{0.4}Ga_{0.6}As$ cladding layer) on the top of QW region, most of carriers are generated in these regions. This indicates that the rise times in Table 5 are the sum of the relaxation time inside the QW region and the drift time of photo-generated carriers in the $Al_{0.4}Ga_{0.6}As$ cladding layer and $Al_{0.2}Ga_{0.8}As$ barrier layer into the QW region. Electrons in the lowest lying states in these cladding and barrier layers scatter into virtual conduction-band states in QW regions, and then cool down to the edge of the conduction-band. The scattering is caused by all possible "mixing" potentials, which might be potential fluctuations due to the QWD process. Probably, the increased interface scattering of photo-generated carriers with increased defects caused by QWD might be the dominant contribution for the shortening of rise times. However, this explanation may lose its validity for the lowest disordered sample, S1, because of its relatively long rise time. Since this increased rise time can not be explained in detail.

The dynamic behaviors of LD's are often characterized with a relaxation oscillation frequency, which determines the frequency response of LD direct modulation and its high-frequency roll-off [50]. The relaxation oscillation frequency and 3-dB frequency of LD direct modulation are given, respectively as [51],

$$\Omega = \left[\frac{1 + \Gamma v_g a n_o \tau_p}{\tau_e \tau_p} \left(\frac{I}{I_{th}} - 1 \right) \right]^{1/2}, \tag{3}$$

$$f_{3-dB} = \frac{1}{2\pi} \left[\Omega^2 - \Gamma^2 + 2(\Omega^2(\Omega^2 + \Gamma^2) + \Gamma^4)^{1/2} \right]^{1/2}, \tag{4}$$

where Γ is a optical confinement factor, v_g is a group velocity of optical waves in a cavity, a is a gain coefficient, n_o is a transparent carrier density, τ_e is a carrier lifetime, τ_p is a photon lifetime in a cavity, I is an injection current and I_{th} is a threshold current. Thus, for a high speed modulation of LD, the carrier lifetime should be reduced.

Since the relaxation oscillation frequency limits the modulation frequency, n-type or p-type doping has been introduced in the MQW active region to enhance the characteristics of the relaxation oscillation frequency for high speed MQW LD's [49,50,52]. As represented in equation 3, the relaxation oscillation frequency of LD is inversely proportional to the square root of the carrier lifetime. As seen in Table 5, a shorter decay time of photo-generated carrier can be achieved by increasing the capping layer thickness. Therefore, a DCQWD process could be used to enhance the direct modulation speed of LD's as well as to improve lightwave communication system performance such as mode partition noise, eye opening, and power penalty [53]. Although a SiN capping layer has been known to induce relatively less disordering than a SiO$_2$ capping layer [6], the experimental results shown in Table 5 reveal that the relaxation oscillation frequency can be increased by a factor of 2 for 123 nm thick SiN capping layer.

The transport mechanism of carriers in a QW structure which includes active QW medium and its transport medium such as the SCH region in our LD structure may affect the modulation characteristics of LD's and modulators. Though the rise time (i.e. sum of relaxation time and drift time in our case as explained above) is shorter than the decay time, it is not negligible. Especially for situations where n-type doping [49] or p-type doping [50,52] is introduced in the QW region to get shorter lifetimes for high speed LD modulation, the drift time becomes comparable to the decay time [50] and it would limit the modulation speed of LD's. Therefore, in addition to the decrease in decay time, the reduction of the drift time with an increase in disordering makes it possible to expect further increase in the modulation bandwidth of LD, though it would require higher threshold current because of its nonradiative recombination centers.

A fundamental limit to the speed of MQW optical modulators is the dynamics of photogenerated carriers within the QW structure. For optical modulators with high speed and low power dissipation, defects have been introduced in the QW region by means of a lattice mismatched structure [54] and proton implantation [47]. For high speed and low power dissipation, fast recovery time and low photo-current generation are requried. As pointed out by Wang et al. [54], defects in the QW region can cause carrier trapping or nonradiative recombinations, which result in fast recovery time and low photo-current in an optical modulator. Therefore the defects introduced by DCQWD can result in a decrease in the recovery time and photo-current. Furthermore, since the DCQWD technique introduces defects not only in the QW region but also in the SCH region, this technique would result in further suppressed photo-current as in the case of proton implantation [47].

5 SUMMARY

In summary, dependence of DCQWD on the characteristics of dielectric capping layer was presented especially for GaAs/AlGaAs MQW structure. In vacancy induced QWD technique, the degree of QWD is dependent on the total number of vacancies and the vacancy denity in vacancy source. DCQWD technique was applied to LT/NT-GaAs capped MQW structure by using PECVD-grown SiO_2 capping layer. This technique fairly preserved the surface of MQW substrate after high temperature thermal treatment. The activation energy of interdiffusion due to vacancy in SiO_2 capping layer was smaller than that due to vacancy in LT-GaAs. It was found that the larger number of vacancy caused the larger QWD. The result from this kind of combined technique shows the possiblity that DCQWD technique can be used with other QWD techniques to give various benifits. Normalized vacancy density was introduced for the semi-quantitative comparison of degrees of QWD due to different characteristics of dielectric capping layer. It was shown that PECVD-grown SiN could promote QWD as much as SiO_2 capping layer. It was also shown that the vacancy in SiN capping film could be controlled by controlling the film growth conditions, such as RF power, NH_3 flow rate and growth time. SiN growth conditions of higher RF power, higher NH_3 flow rate caused larger QWD. The thicker SiN film also caused the larger QWD at fixed growth condition. These results show that one can achieve selectively disordered MQW structure, for applications of PICs, with SiN capping layers grown at different growth condition, although many experimental approaches have used different materials to get selectively disordered QW structure.

The carrier lifetime was measured for a SiN cap disordered GaAs/AlGaAs MQW structure to study the usefulness of DCQWD technique. The carrier lifetime was dependent on the degree of disordering. It was found that the carrier lifetime decreases along with disordering which increases with the thickness of the SiN capping layer for the DCQWD technique. This behavior can be explained in terms of the enhanced carrier trapping and relaxation of momentum conservation due to increased defects by QWD. This result indicates that the DCQWD technique can be used to enhance the high speed performances of laser diodes and modulators.

Acknowledgments

This work has been made possible by many people in Korea Institute of Science and Technology. Above all, I greatly thank to K. N. Kang for his continuous support and encouragement. I thank to J. I. Lee, S. H. Kim and S. K. Kim who gave me critical discussions and supports. I also thank to S. Lee, D. H. Woo, Y. Kim, I. K. Han, H. J. Kim for their fruitful discussion and help. Special thank to J. Zhang in Chinese Academic of Science for his collaboration in the calculation of diffused QW. I thank to J. H. Chu and D. Kim in Korea Research Institute of Standards and Science for their collaboration. I am greatly indebted to professor K. Cho in Sogang University who give me continuous supports and encouragements.

References

1. R.L. Thornton, W.J. Bosby and T.L. Paoli, *J. Lightwave Technol.*, **LT-6**, 786 (1988).
2. F. Julien, P.D. Swanson, M.A. Emanuel, D.G. Deppe, T.A. DeTemple, J.J. Coleman and N. Holonyak, Jr., *Appl. Phys. Lett.*, **50**, 866, (1987).
3. C.J. McLean, A. McKee, G. Lullo, A.C. Bryce, R.M. De La Rue, J.H. March, *Electron. Lett.*, **31**, 1285 (1995).
4. K.K. Anderson, J.P. Donnelly, C.A. Wang, J.D. Woodhouse and H.A. Haus, *Appl. Phys. Lett.*, **53**, 1632 (1988).
5. J.E. Zucker, B. Tell, K.L. Jones, M.D. Divino, K.F. Brown-Goebeler, C.H. Joyner, B.I. Miller and M.G. Young, *Appl. Phys. Lett.*, **60**, 3036 (1992).
6. H. Ribot, K.W. Lee, R.J. Simes, R.H. Yan and L.A. Coldren, *Appl. Phys. Lett.*, **55**, 672 (1989).
7. Y. Suzuki, H. Iwamura, T. Miyazawa and O. Mikami, *IEEE J. Quantum Electron.*, **30**, 1794 (1994).
8. J.Y. Chi, X. Wen, E.S. Koteles and B. Elman, *Appl. Phys. Lett.*, **55**, 855 (1989).
9. I. Gontijo, T. Krauss, J.H. Marsh and R.M. De La Rue, *IEEE J. of Quantum Electron.*, **QE-30**, 1189 (1994).
10. E.L. Allen, C.J. Pass, M.D. Deal, J.D. Plummer and V.F.K. Chia, *Appl. Phys. Lett.*, **59**, 3252 (1991).
11. W.J. Choi, J.I. Lee, I.K. Han, K.N. Kang, Y. Kim, H.L. Park and K. Cho, *J. Mat. Sci. Lett.*, **13**, 326 (1994).
12. W.J. Choi, S. Lee, Y. Kim, S.K. Kim, J.I. Lee, K.N. Kang, N. Park, H.L. Park and K. Cho, *J. Mat. Sci. Lett.*, **14**, 1433 (1995).
13. J. Beauvais, J.H. Marsh, A.H. Kean, A.C. Bryce and C. Button, *Electron. Lett.*, **28**, 1670 (1992).
14. A. Ramdane, P. Krauz, E.V.K. Rao, A. Hamoudi, A. Ougazzaden, D. Robin, A. Gloukhian and M. Carre, *IEEE Photon. Technol. Lett.*, **PTL-7**, 1016 (1995).
15. D. Hofsteter, H.P. Zappe, J.E. Elper and P. Riel, *IEEE Photon. Technol. Lett.*, **PTL-7**, 1022 (1995).

16. T. Miyazawa, H. Iwamura, and M. Naganuma, *IEEE Photon. Technol. Lett.*, **PTL-3**, 421 (1991).
17. J. Beauvais, G.S. Ayling, and J.H. Marsh, *IEEE Photon. Technol. Lett.*, **PTL-4**, 372 (1993).
18. M. Kuzuhara, T. Nozaki, and T. Kamejima, *J. Appl. Phys.*, **66**, 5833 (1989).
19. D.G. Deppe, L.J. Guido, N. Holonyak, Jr., K.C. Hsieh, R.D. Burnham, R.L. Thornton and T.L. Paoli, *Appl. Phys. Lett.*, **49**, 510 (1986).
20. J.D. Ralston, S. O'Brien, G.W. Wicks and L.F. Eastman, *Appl. Phys. Lett.*, **52**, 1511 (1988).
21. S. O'Brien, J.R. Shealy, D.P. Bour, L. Elbaum and J.Y. Chi, *Appl. Phys. Lett.*, **56**, 1365 (1990).
22. J.D. Ralston, W.J. Schaff, D.P. Bour and L.F. Eastman, *Appl. Phys. Lett.*, **54**, 534 (1989).
23. I.K. Han, Y.J. Lee, J.W. Jo, J.I. Lee and K.N. Kang, *Appl. Surf. Sci.*, **48/49**, 104 (1991).
24. W.J. Choi, S. Lee, J. Zhang, Y. Kim, S.K. Kim, J.I. Lee, K.N. Kang and K. Cho, *Jpn. J. Appl. Phys.*, **34**, L418 (1995).
25. W.J. Choi, S. Lee, Y. Kim, D. Woo, S.H. Kim, J.I. Lee, K.N. Kang, J.H. Chu, S.K. Yu, J.C. Seo and D. Kim, *Appl. Phys. Lett.*, **67**, 3438 (1995).
26. S. Adachi, *J. Appl. Phys.*, **58**, R1 (1985).
27. E.H. Li and B.L. Beiss, *IEEE J. Quantum Electron.*, **29**, 311 (1993).
28. J.S. Tsang, C.P. Lee, S.H. Lee, K.L. Tsai, H.R. Chen, *J. Appl. Phys.*, **77**, 4302 (1995).
29. J.S. Tsang, C.P. Lee, J.C. Fan, S.H. Lee, K.L. Tsai, *Jpn. J. Appl. Phys.*, **34**, 1089 (1995).
30. T.Y. Tan and U. Gosele, *Appl. Phys. Lett.*, **52**, 1240 (1988).
31. M. Kaminska, Z.L. Eber, E.R. Weber and T. George, *Appl. Phys. Lett.*, **56**, 1881 (1989).
32. M. Kaminska, E.R. Weber, Z.L. Eber and R.J. Leon, *J. Vac. Sci. Technol. B*, **4**, 710 (1989).
33. R.P. Mirin, J.P. Ibbeston, U.K. Mishra and A.C. Gossard, *Appl. Phys. Lett.*, **65**, 2335 (1994).
34. A.K. Sinha, H.J. levinstein, T.E. Smith, G. Quintana and S.E. Haszako, *J. Electrochem. Soc.*, **125**, 601 (1978).
35. H. Dun, P. Pan, F.R. White and R.W. Douse, *J. Electrochem. Soc.*, **128**, 1555 (1981).
36. G.M. Samuelson and K.M. Mar, *J. Electrochem. Soc.*, **129**, 1773 (1982).
37. W.A.P. Claassen, W.G.J.N. Valkenburg, M.F.C. willemsen, W.M. v.d. Wijgert, *J. Electrochem. Soc.*, **132**, 893 (1985).
38. H.C. Casey, Jr. and M.B. Panish, Heterostructure Lasers (Academic, New York, 1978), Part A, p. 189.
39. M. Kuzuhara, T. Nozaki, and T. Kamejima, *J. Appl. Phys.*, **66**, 5833 (1989).
40. Y. Nagai, N. Shigihara, S. Karakida, S. Kakimoto, M. Otsubo and K. Ikeda, *IEEE J. Quantum Electron.*, **QE-31**, 1364 (1995).
41. I.K. Han, Y.J. Lee, J.I. Lee, K.N. Kang and S.Y. Kim, *J. Mat. Sci. Lett.*, **11** 1689 (1992).
42. S.V. Nguyen, *J. Vac. Sci. Technol.*, **B 4**, 1159 (1986).
43. S.V. Nguyen and P.H. Pan, *Appl. Phys. Lett.*, **45**, 134 (1984).
44. Vikram J. Kapoor, Robert S. Bailey and Herman J. Stein, *J. Vac. Sci. Technol.*, **A 1**, 660 (1983).
45. M.B. Johnson, T.C. McGill, and N.G. Paulter, *Appl. Phys. Lett.*, **54**, 2424 (1989).
46. M. Kasu, T. Yamamoto, S. Noda, and A. Sasaki, *Appl. Phys. Lett.*, **59**, 800 (1991).
47. T.K. Woodward, B. Tell, W.H. Knox, and J.B. Stark, *Appl. Phys. Lett.*, **60**, 742 (1992).
48. H. Jeong, I.-J. Lee, J.-C. Seo, M. Lee, D. Kim, S.-J. Lee, S.-H. Park and U. Kim, *Solid State Commun.*, **85**, 111 (1993).
49. J.N. Sweester, T.J. Dunn, L. Waxer, I.A. Walmsley, S.M. Shank, and G.W. Wicks, *Appl. Phys. Lett.*, **25**, 3461 (1993).
50. J.D. Ralston, S. Weisser, I. Esquivias, E.C. Larkins, J. Rosenzweig, P.J. Tasker, and J. Fleissner, *IEEE J. Quantum Electron.*, **QE-29**, 1648 (1993).
51. G.P. Agrawal, and N.K. Dutta, Semiconductor Lasers (Van Nostrand Reinhold, New York, 1993), p. 258-283.
52. K. Uomi, T. Mishima, and N. Chinone, *Appl. Phys. Lett.*, **51** 78 (1987).
53. W-H. Cheng, *IEEE Photon. Technol. Lett.*, **PTL-6**, 355 (1994).
54. H. Wang, P. LiKamWa, M. Ghisoni, G. Parry, P. N. Stavrinou, C. Robert, and A. Miller, *IEEE Photon. Technol. Lett.*, **PTL-7**, 173 (1995).

CHAPTER 12

Selective Area Disordering of Quantum Wells for Integrated All-Optical Devices

PATRICK LIKAMWA

Center for Research and Education in Optics and Lasers (CREOL) and the Dept. of Electrical and Computer Engineering, The University of Central Florida Orlando, FL 32816-2700, USA

1 INTRODUCTION

A key requirement for the fabrication of highly functional integrated optoelectronic devices is the ability to area-selectively bandgap engineer the optical property of the material. One attractive option is to etch off islands of semiconductor material and epitaxially re-grow [1] another semiconductor with a slightly different bandgap energy. This method of integrating optoelectronic devices requires extensive growth capabilities and the growth conditions are extremely critical in the realization of high quality structures with reproducible results. On the other hand, if multiple quantum wells (MQWs) are employed in the design of the devices then an alternative method of area-selective post-growth disordering of the quantum wells can be used to achieve different bandgap materials monolithically on a single substrate. Disordering of MQWs, or quantum well intermixing, takes place when the different constituent atoms interdiffuse between the well and barrier regions, resulting in diffuse quantum well interfaces and graded potential barriers. Controlled interdiffusion of barrier and well atoms across the heterointerface leads to a blue-shift of the band-edge of the MQW material as a result of the re-shaping of the quantum well potential profile. Consequently a spatially selective QW intermixing process can be employed to produce different band-gap regions on the same wafer with correspondingly different optical properties such as absorption spectrum and refractive index. Hence this procedure enables the fabrication of optoelectronic components for optical communication (e.g. modulators, all-optical switching devices, waveguides and photodetectors). MQW's can be selectively disordered in a variety of ways, such as impurity induced disordering [2], laser induced disordering [3,4], ion-implantation-enhanced interdiffusion [5,6] and vacancy induced disordering [7,8]. There is a wide range of uses of this area-selective bandgap engineering. This chapter concentrates on the realization of optically integrated all-optical structures on a single semiconductor wafer. Of the processes mentioned earlier, the impurity-free vacancy induced disordering (IFVD) appear to have the best characteristics. Since this latter method does not require the incorporation of impurity dopants, waveguides fabricated in such disordered materials do not experience attenuation due to free-carrier absorption.

Quantum well structures also have an important nonlinear optical property that can be exploited to produce all-optical switches. This nonlinearity which is resonant with the band-gap, arises from the saturation of the excitons [9,10] and Moss-Burstein [11] shift of the band-gap energy due to photo-generated free carriers. It manifests itself in that the material's optical transparency and its refractive index are dependent on the intensity of light passing through it. The intensity dependent nonlinear refractive index causes an alteration in the net phase change of the propagating optical wave. In a waveguide, a high optical intensity can be maintained over the length of the device (several hundreds to thousands of wavelengths long) so that a significant nonlinear phase change can be accumulated. Various guided-wave devices can transform the nonlinear change in phase to produce either an all-optical switching of intensities between two ports or an all-optical modulation of the transmitted intensity. The devices that have been realized in this work employ disordered passive waveguide sections to effectively control the switching mechanism. However, an extension of this work could lead to integrated all-optical devices that can exhibit high functionality.

2   BANDGAP RESONANT NONLINEAR OPTICAL PROPERTIES OF MULTIPLE QUANTUM WELL STRUCTURES

In a QW, a restriction on the motion of carriers in the direction normal to the interface is imposed. As a result, only discrete levels of electron and hole energy states are allowed in the conduction and the valence bands, respectively. In a MQW structure, the wells and barriers are repeated a number of times to realize a periodic structure. In order for the MQW heterostructure to retain the properties of the SQW, the barrier layer needs to be thick enough to prevent the carrier's wave function in one QW from having substantial amplitude overlap with the neighboring QWs. Thus, carriers confined in adjacent quantum wells do not interact. The reduced dimensionality of the electron and the hole changes the band structure and the exciton binding energy of the heterostructure and leads to changes in its linear and nonlinear physical properties such as the absorption spectrum, effective refractive index, etc.

One of the important nonlinear optical properties of QW is the saturation of exciton resonances [9,10]. An exciton is an electron-hole pair state that is created by the absorption of a photon. The energy needed to free the electron-hole pair is called the binding energy. In QWs, the binding energy increases due to the restriction of the exciton orbit in one direction , thus increasing the exciton oscillator strength. When the input optical intensity is increased, a large population of excitons is created. Within several hundreds of femtoseconds, these excitons ionize rapidly into free electrons and holes. At high optical intensities, saturation of the exciton absorption occurs owing to two principal effects. Firstly, each exciton is associated with an electron-hole pair which when freed occupy a quantum level in the conduction band and the valence band, respectively. Increasing the population of excitons effectively fills the quasi-quantum levels due to Pauli's exclusion principle and leads to "phase-space filling" and therefore, excitons cannot be produced beyond a certain density. Secondly, as excitons are created, they readily ionize into free electron-hole pairs. These cause "Coulombic screening" and make the creation of further excitons harder. Exciton saturation results in a decrease of the exciton absorption peak, which in turn gives rise to changes in refractive index. Unfortunately, the nonlinear effect due to saturation of the exciton resonances saturates at relatively low intensities and leads to insufficient phase changes for optical switching. A further increase in excitation power beyond the exciton saturation level results in the creation of more free electrons that fills in the Γ band. Thus the effective band gap is increased, and a change of refractive index takes place. This important nonlinear effect due to band-filling is enhanced in MQW structures due to the effect of the reduced dimensionality on the density of states function. There are other nonlinear mechanisms in MQW structures that are transient and require coherent interactions. Examples of such mechanisms include the nonlinear absorption and refraction due to two-photon absorption, the Kerr nonlinearity that can be tapped at photon energies below the two-photon bandgap and the optical (a.c.) Stark effect that is an interaction of the electric field of the optical wave with the exciton orbit.

3   DISORDERING OF QUANTUM WELLS BY ZINC DIFFUSION

Previous reports [12] suggested that in impurity induced disordering (IID), the impurities induce the disordering process through the generation of free carriers that increase the equilibrium number of vacancies at the annealing temperatures. The models proposed to describe the IID process are based on the fact that the species diffuse via interstitial-substitutional mechanisms [13,14]. For the case of Zn as the impurity, two different mechanisms have been proposed for the incorporation of interstitial donors on the substitutional acceptor site. Assuming that the native defects are either neutral or singly ionized, group III self-diffusion rate, D_{III} is given by [15];

$$D_{III} = f_1'' P_{As_4}^{1/4} \left(D_{V_{III}^x} + D_{V_{III}} \exp \left\{ \frac{E_F - E_A}{k_B T} \right\} \right)$$
$$+ f_2'' P_{As_4}^{-1/4} \left(D_{I_{III}^x} + D_{I_{III}} \exp \left\{ \frac{E_D - E_F}{k_B T} \right\} \right)$$

where P_{As_4} is the partial pressure of As$_4$, f_1'' and f_2'' are functions which depend on the crystal structure, and E_A and E_D are the vacancy acceptor energy level and the interstitial donor level, respectively. $D_{V_{III}^x}$, $D_{V_{III}}$, $D_{I_{III}^x}$, and $D_{I_{III}}$ are the diffusion coefficients of the neutral and charged group III vacancies, and neutral and charged group III interstitials, respectively. It can be seen from the previous equation that the rate of IID depends on As overpressure and the type of doping. In p-type QW structures, intermixing takes place under As-deficient conditions, and in n-type QWs it takes place under As-rich condition because the concentration of group III interstitials and group III vacancies is raised in the respective cases. Thus the role of impurities in the intermixing is to change the Fermi level in the material and, consequently, raise the point defect concentration.

The samples used in this work were grown by molecular beam epitaxy (MBE) on undoped substrates and most of them were designed so that the MQW regions would form the core of a single mode slab waveguide with AlGaAs cladding layers on either side.

The zinc diffusion experiments were initially carried out in a semi-sealed ceramic box placed inside an open diffusion furnace with flowing high purity nitrogen gas. The samples were cleaved into pieces of 2×2 mm^2 and then cleaned by boiling in acetone and methanol. Each sample was placed together with 0.1 g of pure zinc inside a covered ceramic boat within a quartz annealing furnace. The flow of high purity nitrogen was maintained at 4SCFH (standard cubic feet per hour) 15 minutes prior to the start of the annealing and was only turned off when the temperature of the sample was less than 150°C. Depending on the temperature/time combinations complete or partial intermixing of the quantum wells would be encountered. Moreover, if the diffusion of Zn is not carried out for long enough duration, the quantum wells at the bottom of the stack remain un-intermixed. The effects of the zinc diffusion on the disordering of the quantum wells were studied by a combination of photoluminescence (PL), secondary ion mass spectroscopy (SIMS) and Auger electron spectroscopy (AES). A 15 mW HeNe laser emitting at 633nm wavelength was used to excite the photoluminescence, and the emission spectra were captured on a quarter-meter monochromator in combination with a linear array detector and optical multichannel analyzer. Figure 1a shows the photoluminescence of one of the as-grown samples. The main peak at 838 nm was due to direct recombination of the heavy-hole exciton while the "hump" at 825 nm was emission from the light-hole exciton recombination. Figure 1c shows the photoluminescence of the sample that had been subjected to zinc diffusion at a temperature of 510°C for 60 mins. It is observed that the PL spectrum is broadened somewhat with a clear shift to much shorter wavelengths. The peak of the PL in this case was at 737 nm. We therefore assume that the quantum wells have been uniformly disordered. Figure 1b shows the PL emission from a sample that was subjected to zinc diffusion at 490°C for 60 mins. The emission spectrum in this case was extremely broad and there was no clear evidence of a single peak intensity. This is attributed to the fact that non-uniform disordering of the quantum wells occurred and the zinc doping concentration which followed an erf diffusion profile was not high enough in the deeper quantum wells to induce disordering there.

A SIMS measurement was performed on that particular sample to detect the concentration of zinc and also monitor the fluctuations in the concentration of aluminum atoms. The SIMS profile shown in Figure 2b, indicate that the Al mole fraction is constant for the first 0.3 μm below the surface and begin to oscillate as the measurement progresses deeper into the sample with the largest oscillations occurring furthest from the surface. The zinc concentration profile indicates a steady drop from 10^{18} cm^{-3} to below 10^{17} cm^{-3}. By correlating the two profiles for Zn and Al, it is apparent that the Zn concentration needs to exceed a level of 5×10^{17} cm^{-3} for intermixing to take place.

A comparative SIMS measurements done on an as-grown sample is shown in Figure 2a. As expected, the Al mole fraction is high in the barrier layers and low in the quantum well layers (ideally should be zero but is limited by

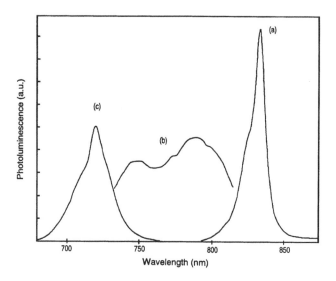

Fig. 1. Photoluminescence spectra of MQW samples (a) as-grown (b) subjected to zinc diffusion @490°C for 60 mins. (c) subjected to zinc diffusion @510°C for 60 mins.

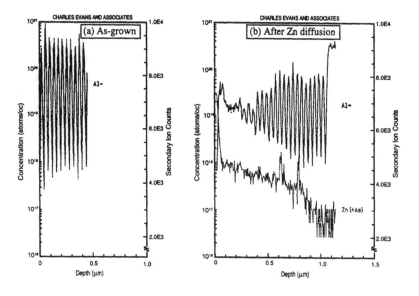

Fig. 2. SIMS measurements of MQW sample (a) before and (b) after zinc diffusion.

the depth resolution of the equipment) and the magnitude of the modulation remains nearly constant with a slow degradation as the probing is carrier out further into the sample depth.

It is anticipated that the position of the diffusion front leads the intermixing edge very slightly. Consequently, it is required that the diffusion depth should exceed the total thickness of the quantum well layer and the zinc concentration must exceed the minimum level of 5×10^{17} cm^{-3}, so as to achieve uniform intermixing. This theory would predict that for a given sample there is a minimum time-temperature condition that leads to a uniformly disordered MQW as evidenced by a well-defined shifted PL spectrum. From diffusion theory, the diffusion depth, x is given by,

$$x_d \propto \sqrt{D_s t}$$

where t is the duration of the diffusion and D_s is the surface diffusivity that is proportional to the Zn overpressure.

i.e.
$$D_s \propto \text{Zn surface pressure}$$
$$\propto (T - T_o)^2$$

where T is the temperature of the annealing and To is a critical temperature which depends on the melting point of zinc. Therefore,

$$x_d \propto (T - T_o)\sqrt{t}$$

From the distribution of samples that were annealed at various temperatures for different times, a graph of the minimum time v/s temperature required for complete intermixing as shown in Figure 3, was obtained. The fit to these data was obtained by using,

$$T = T_o + \frac{A}{\sqrt{t - t_o}}$$

where A is an arbitrary constant and t_o is a fitting parameter.

The graph represents the minimum conditions for uniform intermixing.

A ridge waveguide was delineated in the zinc IID layer by wet chemical etching through a photoresist mask defined through contact photolithography. Light from a cw Ti:sapphire laser tuned to a wavelength of 890 nm was then launched into it using the endfire coupling technique. The attenuation coefficient of the waveguide was measured by a sequential cleaving method and was found to be prohibitively high (>200 dB/cm). Even after much care in optimizing the zinc diffusion to produce uniform intermixing with the absolute minimum doping concentration, the waveguide loss was measured to be 30 dB/cm. Although the free carrier absorption is the most serious problem associated with IID, another problem that is a consequence of using electrically active dopants is the reduced electrical isolation (or decreased resistivity). Consequently, disordering of MQW using zinc diffusion is deemed to be inappropriate for producing sufficiently low-loss waveguides for optoelectronic interconnects.

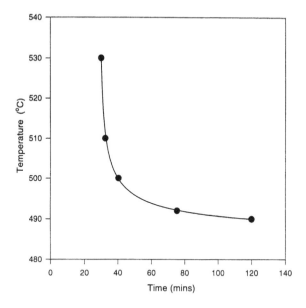

Fig. 3.   Minumum temperature-time combinations for uniform disordering.

4   VACANCY INDUCED DISORDERING BY OXIDATION AND ANNEALING

The QW intermixing in impurity-free vacancy disordering (IFVD) is also caused by the interdiffusion of Ga and Al atoms across the barrier-well interface. The diffusion coefficient (D) of Al in an undoped as grown Al-As-GaAs heterostructures is of the order of 10^{-19} cm^2/s^{-1} at 850°C [16]. However, D is also dramatically enhanced when the sample is annealed after extra vacancies are created at the surface of the sample. Different methods have been proposed and demonstrated to create surface vacancies. In this section a process of surface oxidation followed by etching is used to create the surface vacancies. Subsequent high temperature anneals are used to diffuse the vacancies into the MQW structure resulting in disordering of the QWs by interdiffusion of Al and Ga with the vacancies.

The MQW layer used in this section consisted of 38 periods of 70 Å thickness GaAs layers alternating with 70 Å wide Al$_{0.3}$Ga$_{0.7}$As layers. The MQW waveguide core was clad on either side by a 0.5 μm thick Al$_{0.3}$Ga$_{0.7}$As layer on the top and a 2 μm thick Al$_{0.3}$Ga$_{0.7}$As layer on the bottom. The structure was deliberately not capped by a thin protective layer of GaAs so as to expose the uppermost AlGaAs surface for easier oxidation. The wafer was first cleaved into small pieces measuring about 3×4 mm^2. The samples were cleaned by boiling twice in acetone and then twice in methanol.

Several sets of samples were prepared under different controlled conditions. One set of samples, (A), were left in their as-grown state. The second set of samples, (B), was etched in concentrated hydrofluoric (HF) acid for 15 minutes. This etched the native oxide only, resulting in the formation of Al and/or Ga vacancies on the surface of the sample. A third set of samples, (C), was oxidized by heating in air for 30 minutes at 600°C before etching in HF in order to achieve a larger population of vacancies. A fourth set of samples, (D), was oxidized by heating, but not etched in order to test whether the structures could be protected from disordering.

Some non-uniform disordering over the sample area was freqeuntly observed in our experiments. In particular, areas near the edges showed a larger shift of the emission wavelength than the middle part. To overcome this problem, we first made grooves on one set of samples. The grooves were 20 μm wide and 1.5 μm deep with a separation of 200 μm. This set of samples, (C$'$) was oxidized by heating and etched by HF similar to sample C.

All of the samples were finally annealed at a temperature of 615°C for 3 hours in a tube furnace with a continuous flow of purified N$_2$ gas. A piece of freshly cleaned GaAs substrate was used to cover the samples in order to minimize As loss during the annealing stage. The N$_2$ gas flow was started 5 minutes prior to turning on the furnace heating element, and the diffusion process was timed from when the temperature reached 50°C below the set point. After the required duration, the furnace was turned off and cooled down rapidly by blowing a stream of air onto the outside of the tube. The N$_2$ gas was turned off when the furnace temperature dropped below 70°C.

The samples were then characterized by measurements of the room temperature photoluminescence spectra. The PL emission from the as-grown samples, A, (Figure 4a) depicts a peak at 843 nm due to the heavy-hole, (HH), exciton with a shoulder at 835 nm due to light hole, (LH), exciton. The second sample, B, which had its naturally occurring oxide etched off before being subjected to the 3 hours annealing, exhibited evidence of partial disordering. Figure 4b shows that the peak of the PL emission from this sample has shifted from 843nm to 820 nm. In the case of sample C, which was thermally oxidized and etched before annealing, the PL emission was blue-shifted by 40 nm as shown in Figure 4c. On the other hand, sample D did not show any wavelength shift. In this case, the PL spectrum was identical to the as-grown sample, Figure 4a. Therefore, for this sample, with the HF etching step missed out, there was no evidence of disordering although the sample had been oxidized for 30 mins. at 600°C. The LH exciton shoulder was not observed for the most shifted PL spectrum, however the width of the two shifted PL spectra remained almost the same as that of the as-grown sample.

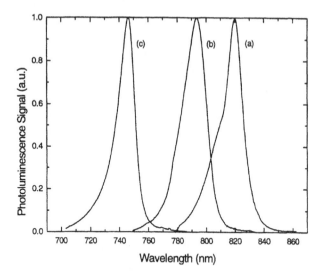

Fig. 4. Room temperature photoluminescence spectra of the MQW samples (a) from as-grown layer (b) from the sample that was oxidized, etched and annealed for 2.5 h (c) from the sample that was oxidized, etched and annealed for 3 h.

It should be noted that the PL spectra taken from the center of the disordered samples were shifted by a lesser extent (~ 10 nm) than the spectra emitted from the top perimeter surface (within 1mm of the edges). The spectra shown in Figure 4, were taken from the edge regions of the disordered samples. However, we found that the sample C′ that had a grooved pattern etched on the surface could be disordered uniformly across the whole sample area. The PL spectrum from this sample overlaid exactly on top of the PL spectrum shown in Figure 1c, for sample C.

Strip-loaded, single-mode waveguides were then produced on the intermixed waveguides in order to assess optical propagation losses in the disordered samples. These were fabricated by defining 4 μm wide strips and etching off 0.45 μm from the cladding layer over the rest of the sample surface. Optical propagation loss measurements at wavelengths below the band gap energy were made on strip-loaded waveguides fabricated in the most disordered samples, C′, with grooves. The waveguide attenuation was measured by scanning the fringes of the Fabry-Perot created by the cleaved end facets of the waveguide [17]. The beam from a narrow bandwidth (<1.5 GHz) CW Ti:sapphire laser was chopped and launched into the waveguide and the output was captured by a photo-detector connected to a lock-in amplifier. The amplified signal was recorded on a chart recorder. The thermo-optic effect in the waveguide was employed to scan the Fabry-Perot

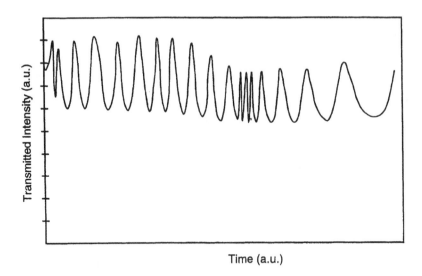

Time (a.u.)

Fig. 5. Thermally scanned Fabry-Perot interferogram for the measurement of waveguide attenuation.

fringes by gentle heating and the optical loss was deduced from the measured visibility (maximum to minimum intensity) of the interferogram. An example of an interferogram obtained from a 1mm long sample is shown in Figure 5.

Since the two facets of the waveguide form a Fabry-Perot cavity, the transmission of the waveguide is given by the Airy function;

$$\frac{I_{out}}{I_{\epsilon}} = \frac{(1-R)^2 \exp(-\alpha L)}{1 + R^2 \exp(-2\alpha L) - 2R \exp(-\alpha L) \cos(4\pi n L/\lambda)}$$

where λ is the wavelength of the probing light, R is the power reflection coefficient of each facet, L, n and α are respectively the length, refractive index and attenuation constant of the waveguide.

The ratio of the maximum to minimum transmission is therefore,

$$V = \frac{1 + R \exp(-\alpha L)}{1 - R \exp(-\alpha L)}$$

which gives the attenuation constant (in units of inverse length) as,

$$\alpha = -\frac{1}{L} \ln \left(\frac{1}{R} \frac{\sqrt{V}-1}{\sqrt{V}+1} \right)$$

From the visibility of the Fabry-Perot fringes the waveguide loss was measured to be 15 dB/cm at a wavelength of 860 nm.

Hence, partial disordering of the quantum well structure was achieved by allowing or enhancing oxidation of the surface of the sample and then removing this oxide by etching in concentrated HF acid just prior to annealing the sample at 615°C for 3 hours. It is believed that by etching off the surface oxide, a thin layer of Ga or Al vacancies is created. These vacancies are driven into the MQW region by thermal annealing thereby resulting in intermixing of Ga and Al atoms as these randomly exchange lattice positions with the vacancies. This mechanism fits the model proposed by Longini [18] which suggested a diffusion mechanism based on group III vacancies. Theoretically, therefore, this technique should be able to disorder any III-V MQW system, such as InGaAs/GaAs, InP/InGaAs, etc.

The results also show that if the surface oxide is not etched by HF, the MQW structure underneath that region remains unchanged after an annealing cycle of 3 hours at 615°C. This is a very important property that could allow devices with different optical band-gaps to be fabricated on the same sample for optoelectronic integrated circuit applications. It was also found that this technique is very reproducible. It is not yet understood why the edge parts of the samples exhibited more shift of emission wavelength than the middle part. This makes disordering of larger size samples ($> 3 \times 4$ mm^2) non-uniform. However, by etching a pattern of grooved lines on the sample prior to the disordering process, edges that seem to enhance the disordering process are effectively provided to the whole sample. This results in uniform disordering across the sample.

Since no impurity is required for this process, optical waveguides fabricated on the disordered sample does not suffer loss due to free carriers. However, the measured propagation losses are still relatively high. The sample surfaces were observed to be rougher after annealing which accounts for the waveguide losses in the annealed samples. This appears to be related to either the loss of As, or non-uniform oxidization caused by the O$_2$ or H$_2$O in the N$_2$ gas.

5 VACANCY INDUCED DISORDERING BY SPIN-ON SILICA FILM CAPPING AND ANNEALING

Another method of producing the surface vacancies is by depositing a SiO$_2$ layer on top of the sample [19,20] by either chemical vapor deposition or electron beam evaporation. The silicon dioxide capping layer promotes outdiffusion of Ga atoms into the cap layer, thus generating group III vacancies at the surface. High temperature rapid thermal annealing causes these vacancies to interdiffused with Ga/Al atoms in the well and barrier layers of the MQW region and alters the composition profile across the quantum well. The initially square well compositional profile is changed to a gradually graded profile.

To model the interdiffusion of Al and Ga atoms in a single QW, Fick's equation is solved assuming a constant diffusion coefficient, D, that is independent of the initial Al composition of the barrier layers

$$\frac{\partial x(z, t)}{\partial t} = D\frac{\partial^2 x}{\partial z^2}$$

where $x(z, t)$ is the Al concentration at position z after annealing for t seconds. The Al composition varies across an intermixed quantum well centered at $z = 0$ is given by

$$x(z) = x_o\left[1 + \frac{1}{2}\text{erf}\left(\frac{z - L/2}{2\sqrt{Dt}}\right) - \frac{1}{2}\text{erf}\left(\frac{z + L/2}{2\sqrt{Dt}}\right)\right]$$

where x_0 is the initial Al composition of the barrier, L is the initial well width, D is the diffusion, t is the annealing time, and erf is the error function. As a result of the alteration of this potential profile, the quantized energy levels inside the QW are increased. This increase is dependent on the initial width of the QW, the annealing time, the initial concentration of the Al atoms, x_0, and the diffusion coefficient which depends of the number of vacancies at the surface.

The use of SiO_2 may result in complications that occur due to chemical reaction between the dielectric cap and the Al-containing semiconductor alloy resulting in freeing of Si atoms that act as IID species

$$3SiO_2 + 4Al \rightarrow 2Al_2O_3 + 3Si$$

In the case of GaAs/AlGaAs, this complication can be avoided by capping the sample by a thin layer of GaAs (~ 0.1 μm).

The MQW waveguide structure used in this section is similar to the one used in the previous section except that the new sample was capped by a thin (~ 1000 Å) layer of GaAs on the surface. In order to achieve controlled intermixing between the GaAs quantum wells and the AlGaAs barriers, the surface of the sample was coated with a thin film of "spin-on-glass" which is a commercially available solution of glass (SiO_x) forming compound. The thin film was deposited by spinning the liquid on the sample surface at a speed of 3000 rpm for 30 seconds, this resulted in a film thickness of 230 nm. The film was then cured at 400°C for 30 minutes in a constant flow of ultra high purity air (78% nitrogen and 22% oxygen). Localized compositional disordering was then induced by photolithographic definition and removal of the SiO_x film in selected regions followed by rapid thermally annealing. The rapid heating of the sample was carried out in a flowing nitrogen atmosphere in an AG Associates Heatpulse 210 rapid thermal annealer. The sample is placed face down on a mechanical grade undoped GaAs substrate to protect the face of the sample and to provide overpressure of As so as to minimize its desorption. It takes seven seconds to reach a steady state temperature of 1050°C, and about 14 seconds to cool down to 500°C.

The silicon oxide cap at the surface promotes out-diffusion of Ga atoms into the cap layer, thus generating group III vacancies at the surface of the sample. The intermixing is caused by the inter-diffusion of these vacancies with Ga and Al atoms, with a subsequent diffusion of the vacancies into the MQW region. There have been previous reports on the use of SiO_2 to enhance the intermixing of group III-V multiquantum well structures [18,20,21]. Rapid thermal annealing of a sample that is encapsulated by an SiO_x cap layer deposited on the surface by either chemical vapor deposition or electron beam evaporation also results in a significant blue shift of the bandedge. The magnitude of the blue shift is a function of the thickness of the dielectric film, and the annealing time and temperature. The main difference between this work and the previously reported work is the simplicity of the technique that we used to generate the vacancies and enhance the impurity free vacancy diffusion in the heterostructure.

In order to characterize our method of IFVD, several samples measuring 3×4 mm^2 of the MQW structure described above were first cleaned with solvents, and then coated with the spin-on glass and cured at 400°C for 30 minutes in a nitrogen/oxygen atmosphere. Conventional photolithography and photoresist masking was then used to etch off the SiO_x layer from the surface of one half of each sample by immersing them in a buffered oxide etchant (BOE) (1:9 HF:H_2O) for 30 seconds. The samples were then annealed separately for 20 seconds at temperatures of 960, 970, 980, and 1000°C respectively.

The degree of disordering was then measured by comparing the room temperature photoluminescence (PL) and/or the absorption spectra of the SiO_x-capped region with the uncapped region, and also with an as-grown sample. As the Al atoms diffuse into the quantum well region, the abruptness of the interfaces is destroyed and the subband energies move apart, resulting in an increase in the $n = 1$ electron to heavy-hole transition energy. Figure 6 shows the measured room temperature photoluminescence of (a) an as-grown sample, (b) an annealed sample with no film on its surface, and (c) an annealed silicon oxide film coated sample. The thermal annealing was carried out at 980°C for 20 seconds. The heavy hole (hh) and light hole (lh) exciton transitions are clearly resolved in the case of the as-grown and the uncapped sample. The measurement indicates that the MQW intermixing achieved by this method, results in a difference of 37 nm of blue shifting between the effective band edge of the coated and the uncoated samples.

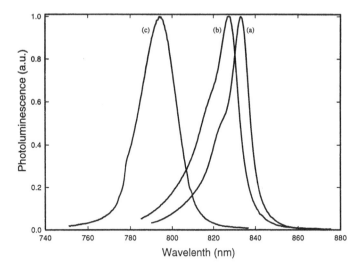

Fig. 6. Room temperature photoluminescence spectra of the MQW samples: a) from the as-grown sample; b) from the annealed sample with no film on its surface, and c) from the annealed SiO_x-capped sample.

The measured PL shift (in meV) as a function of annealing temperature is shown in Figure 7. It is clear that as the temperature is increased, the difference between the PL shifts of the encapsulated regions and the unencapsulated ones increases. However, the difference in PL shifts is not the only desired attribute, because a minimal shift of the PL from the unencapsulated regions is also an important consideration. From our experimental data plotted in Figure 7, it is evident that the optimum temperature at which the PL shift of the covered samples is substantial yet that of the uncovered remains small, is 980°C. It should be noted that there are applications in which some disordering of the "protected" MQW region is beneficial to the device performance. For example, in the devices studied in this work and described later on, the slight disordering in the switching region resulted in a faster recovery time of the devices.

5.1 Theoretical Modeling

As discussed earlier, the Al-Ga interdiffusion across the heterointerfaces has been modeled using an error function profile [16]. For a quantum well centered at $z = 0$, the Al composition varies across the well as;

$$x(z) = x_o \left[1 + \frac{1}{2}\text{erf}\left(\frac{z - W/2}{2\sqrt{Dt}}\right) - \frac{1}{2}\text{erf}\left(\frac{z + W/2}{2\sqrt{Dt}}\right) \right]$$

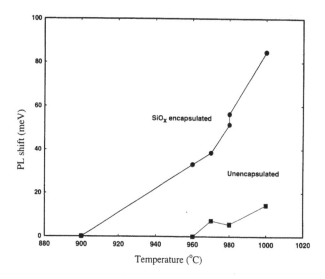

Fig. 7. Plot of PL energy shift as a fimction of anneal temperature (anneal time was 20s in all cases).

where x_0 is the initial Al concentration in the region surrounding the QW, W is the initial well width, D is the diffusion coefficient assuming isotropic Al-Ga interdiffusion, t is the annealing time, and erf( ) is the error function. To find the diffusion coefficient at a given annealing temperature, we assume that the quantum well closest to the surface contributes to the part of the PL at its peak wavelength. We also assume that there is no coupling between the neighboring quantum wells, this assumption is valid because the barrier separating the wells is thick enough to prevent any noticeable interaction between the wave functions of the eigenstates inside the QWs. Thus, we can safely assume that the calculated e1-hh1 transition energy of a single well corresponds to that of our MQW structure. In the calculations of the electron and heavy-hole subband energies, the following parameters were used: the effective masses of the electron and the hh were considered to be 0.067 m and 0.48 m, respectively (where m is the free electron mass). The bandgap energy as a function of the Al composition, $x(z)$, was approximated by $E_g(\text{eV}) = 1.43 + 1.44x(z)$ and the energy offset ratio of the conduction to the valence band was taken to be 65:35. The $n = 1$ e-hh transition wavelength for the 80 Å QW was calculated to be 834.6 nm, this value agrees with the measured PL of the structure which was found to peak at $\lambda = 835$ nm. The values of the diffusion coefficient that correspond to the blue shift at the annealing temperatures were calculated by an iterative procedure. For an

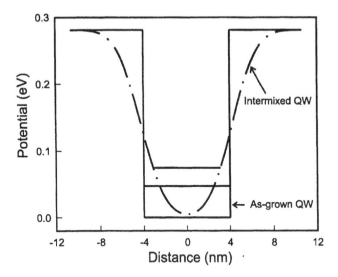

Fig. 8.   Effect of disordering on the effective bandgap energy of the quantum well.

assumed value of D, the e1-hh1 transition energy of the intermixed QW with
the corresponding error function potential barriers (Figure 8), was calculated
using the transmission matrix method [22] and compared to the measured
value until they are less than 0.5 meV apart. In Figure 9 we show the calculated
values of D plotted as a function of $\left(\frac{1}{k_b T}\right)$ for both the capped and the
uncapped samples. A curve fit of the data gives a single activation energy of
$E_a \approx 4.4$ eV. This value is in good agreement with the results reported in
the literature [22,23]. The fact that almost identical activation energies were
obtained for both curves implies that the dissimilarity in disordering behavior
between the SiO_x capped and uncapped sample is due to the difference in the
number of vacancies generated at the surface of the samples. From Figure 9,
it is evident that the diffusion coefficient for the SiO_x capped sample is one
order of magnitude greater than that for the uncapped samples.

Single-mode ridge waveguides were fabricated from as-grown and intermixed samples and the waveguiding losses were measured using a
thermally scanned Fabry-Perot transmission technique described above. The
measurements indicated that the additional losses due to the vacancy induced
disordering were around 8–10 dB/cm.

While the losses are still too high for the technique to be applicable to the
integration of many devices, a simple demonstration of the application of this
disordering technique has been performed with the realization of integrated
all-optical devices described in the next section.

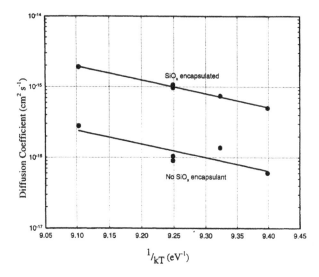

Fig. 9. Plot of the deduced values of the diffusion coefficient vs. 1/kT for both the capped and the uncapped samples.

6   GUIDED WAVE OPTICAL DEVICES

Resonant nonlinearities in MQW structures have emerged as promising effects that can be utilized to achieve all-optical switching in integrated waveguide devices. However, the integrated devices need to operate at wavelengths close to the band edge where the nonlinearity is high. This results in substantial losses due to the linear absorption that is present in all sections of the device of interest. Since the change in the refractive index is only needed in the active switching region, it is beneficial to limit the absorption of photons to that region only. Hence area-selective disordering can be used to blue-shift the absorption edge of the passive sections with a minimal effect on that of the switching section.

6.1   The Zero-Gap Nonlinear Directional Coupler

The first device described in this chapter is the zero-gap nonlinear directional coupler (NLDC) which consists of a single mode input waveguide, a dual-mode coupling section and two single-mode output waveguides (Figure 10). This device is a variant of the directional coupler proposed by Jensen [23]. The light is launched into the input waveguide and as it reaches the dual mode section it excites both the lowest symmetric and the lowest asymmetric

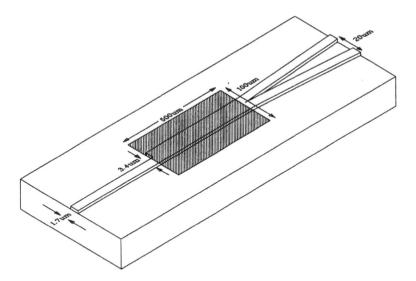

Fig. 10. Schematic drawing of integrated devices nonlinear zero-gap coupler switch. The shaded regions represent the non-intermixed areas, all other areas are intermixed.

modes. These modes propagate with different phase constants along that central section to the branching region. If the length of the dual mode section is equal to the critical coupling length, then owing to destructive interference between the modes, power is completely transferred into the output waveguide that is "crossed" to the input waveguide [24]. If the length of the central region extends beyond the critical coupling length, the light will be coupled back to the adjacent output waveguide that is "straight-through" to the input waveguide. By the proper choice of the length of the dual-mode section, we can achieve the coupling of a desired optical power from one waveguide to the other. Since the waveguide coupler is designed with a specific critical coupling length at a certain refractive index, then changing the refractive index results in changing L_c. Thus, the output port of the light can be switched from one port to the other. This change can be achieved optically by using the band-filling optical nonlinearity of the MQW structure. The phase difference can be controlled by the intensity of the input light, which affects the refractive index of the switching region resulting in an all-optical switch.

Except for the mode-beating section, all regions of the MQWs were intermixed to blue shift the band-edge away from the device operating wavelength. After the silicon oxide was spun on the surface, the film was cured as described above. Then using a contact mask photolithographic process to define windows measuring $100 \times 500 \ \mu m^2$ in positive photoresist, a buffered oxide etch (BOE) was used to remove the oxide film within the exposed windows. After removing the photoresist, the sample was subjected to rapid thermal annealing at $980°C$ for 20 seconds to diffuse the vacancies into the MQW region. Thus apart from the windowed sections, the whole sample was intermixed. The zero-gap directional coupler was defined using a second step photolithography and aligned carefully so that the double-mode waveguide sections were exactly within the non-intermixed windows. Using a wet chemical etch consisting of $H_3PO_4:H_2O_2:H_2O$ solution in the ratios of $1:1:10$, a 350 nm thickness of the cladding layer is etched off to form cladding ridges that were 3.4 μm wide in the double-mode switching sections and 1.7 μm wide in the single mode waveguide branching sections. The length of the mode beating section was 500mm and the half-angle of the output capture waveguides was $0.8°$. Finally the GaAs substrate was polished down to 100 μm thickness and the sample was cleaved to a total length of 1.7 mm assuring good facets. Figure 11 shows the simulation of a ridge waveguide that was solved by a beam propagation method.

6.2   The Integrated Mach-Zehnder Interferometer

Figure 12 shows a schematic drawing of the integrated Mach-Zehnder optical modulator, which is essentially two of the Y-junctions connected back to back. It contains a 500 μm long active switching region of non-intermixed MQW in one arm, while the other arm and the rest of the structure were made of disordered MQW. The same process described for the above was also used in the fabrication of this device except that in this case the defined windows were $20 \times 500 \ \mu m^2$. The angle between the branching output arms was $1.6°$ and after the sample was cleaved, the length of the input guide was 450 μm and the device length was 1.5 mm. PL measurements showed a distinctive difference between the peak wavelengths of the luminescence from the windowed (non-intermixed) and the intermixed regions. This indicates that the technique is highly area-selective. The absorption spectra for the intermixed and non-intermixed samples shown in Figure 13, indicate that the absorption spectrum is bodily shifted with only a slight degradation in the excitonic resonances.

Under low intensity conditions, the device is basically a balanced symmetric Mach-Zehnder interferometer [25]. The signal beam is split equally at the first Y-junction and the two waves propagate along the separate arms of the interferometer before getting recombined at the second Y-junction. With no control beam present, the two waves recombine in phase and hence the signal beam is reconstructed at the output and the device has a high optical transmission. An optical control pulse injected at the input is also

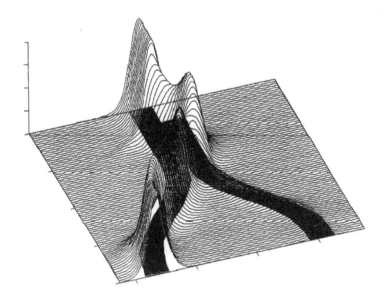

Fig. 11. BPM simulations of the zero-gap directional coupler.

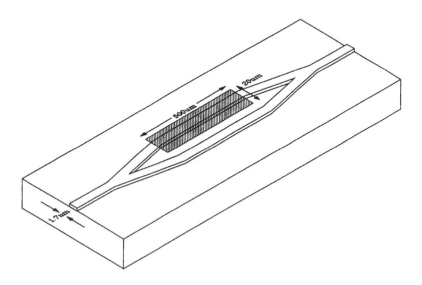

Fig. 12. Drawing of symmetric nonlinear Mach Zehnder modulator. The shaded regions represent the non-intermixed areas while all other areas are intermixed.

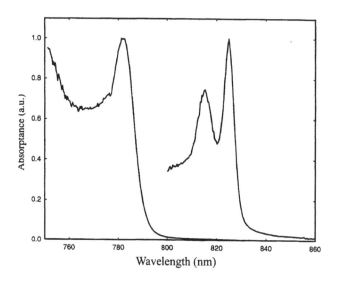

Fig. 13.   Absorption spectra of disordered and non-disordered regions.

split into two parts at the first Y-junction. Only the arm that contains the
non-intermixed MQW will experience a nonlinear refractive index change.
Consequently, only the signal wave propagating through the interferometer
arm containing the non-disordered MQW will sense a nonlinear change in
phase. When the two signal waves recombine at the second Y-junction, the
phase difference leads to an incomplete reconstruction of the optical signal. If
a nonlinear phase change of π is induced then destructive interference occurs
and the signal is totally switched into radiation modes and the transmission of
the device is turned off. Hence the device functions as an optically controlled
modulator that could be useful for optical demultiplexing applications.

6.3   Measurement of All-Optical Switching

To characterize the switching devices we used a pump-probe experimental
set-up [26] shown in Figure 14, with a mode-locked Ti-sapphire laser as the
light source. The laser wavelength was tunable in the range from 800 nm to
900 nm with a pulse width of 150 fs. In this pump-probe experiment, the laser
beam was split into a strong pump-beam and a weaker signal beam (probe)
that was modulated into rectangular pulses using a mechanical chopper. A
half-wave plate was used to rotate the polarization of the control beam so
that it was polarized in the horizontal direction, while the signal beam was
polarized in the vertical direction. The temporal delay between the two pulses

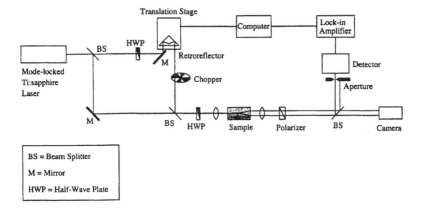

Fig. 14. Optical set-up for time resolved measurements.

was adjusted by changing the path length traversed by the signal beam. This was accomplished using a retro-reflector mounted on a computer-controlled delay stage. The two beams were recombined colinearly and focused into the input waveguide using a ×45 microscope objective lens with a numerical aperture of 0.65, resulting in an estimated coupling efficiency of 10%. To observe the output beam on a video monitor, another lens was used to image the output facets of the waveguide onto a CCD camera. The control beam was filtered out by the use of a cross-polarizer placed at the output of the device. A Si photodetector connected to a lock-in amplifier was used to detect the mechanically chopped signal beam from either output port of the device. Figure 15 shows the transmission of the signal in both output ports of the NLDC as a function of the delay between the control and the signal beams. The initial 63:37 split ratio of the output ports was switched to 30:70 with an exponential recovery time constant of around 300 ps. The total power through the two output ports remained constant and hence indicated that negligible nonlinear absorption took place.

The same optical setup was used to measure the switching performance of the integrated Mach-Zehnder device [27]. Figure 16 shows the output power being modulated by the injection of a strong optical control pulse. The device has a switch contrast of 15 to 1 and it too exhibited an exponential recovery time constant of about 300 ps. The effect of increasing the control beam intensity on the switching characteristics of the device is shown in Figure 17. As the intensity of the control beam is increased, the transmission increases until a certain intensity level is reached when the transmission starts to decrease with increasing intensity. This agrees well with the predicted sinusoidal[2] dependence of the output signal on the control beam intensity.

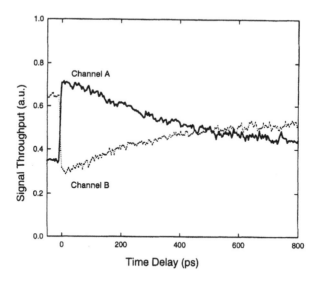

Fig. 15. Switching of the transmission of the signal pulses through the output channels A and B as a function of the time delay with the control pulses.

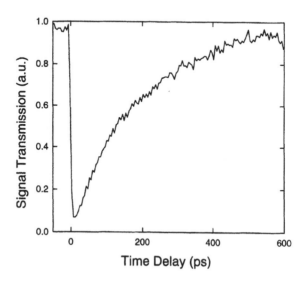

Fig. 16. Probe transmission in an integrated Mach-Zehnder interferometer as a function of the delay between the signal and the control pulses.

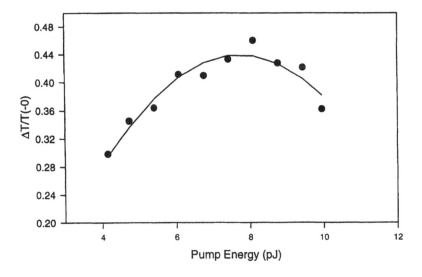

Fig. 17. Change in transmission (before and 60 ps after switching) as a function of the energy of the optical control pulse.

7 PROSPECTS FOR FUTURE INTEGRATED ALL-OPTICAL CIRCUITS

We have developed an inexpensive and reliable process for the area-selective disordering of MQW structures. The scattering losses due to the vacancy induced disordering was measured to be less than 10 dB/cm and work is currently underway to further improve on the waveguiding quality of the disordered MQWs. This technique has been applied to the fabrication of two integrated optical devices. In both devices, the mechanism for the switching is the nonlinear refractive index that is caused by photo-generated carriers. Since this mechanism entails absorption of some of the control beam, it is hence very important that the optical absorption be confined to the active sections only. Selective area disordering is shown to be very effective at defining regions of different bandgap energies. Hence it can be ensured that the energy of the control laser beam is too low in comparison to the bandgap energy of the passive regions to be absorbed and the free carriers are only created in the non-intermixed active sections. The controlled selective area intermixing of MQW structures will potentially play a significant role in the advancement of photonic integrated circuits.

Acknowledgments

Much of the data reported in this article were results obtained by Dr. Shawn Shi and Dr. Ayman Kanan as part of their graduate research work. The semiconductor quantum well structures were grown by Dr. Jagadeesh Pamulapati and Dr. Paul Cooke through a collaboration with Dr. Mitra Dutta while they were at the Army Research Laboratory in Ft. Monmouth, New Jersey. The SIMS measurements were provided by Charles Evans and Associates through a collaboration with Dr. John Zavada. Furthermore the author wishes to acknowledge the helpful discussions and encouragement from Professor Peter Robson and Professor Alan Miller.

References

1. T. Roehr, H. Kratzer, G. Boehm, W. Klein, G. Traenkle and G. Weimann, "Direct molecular beam epitaxial growth of low-dimensional structures on reactive ion etched surfaces", *J. of Crystal Growth*, **150**, 1, 306–310, 1995.
2. W.D. Laidig, N. Holonyak Jr., M.D. Camras, K. Hess, J.J. Coleman, P. D. Dapkus and J. Bardeen, "Disorder of an AlAs-GaAs superlattice by impurity diffusion", *Appl. Phys. Lett.*, **38**, 776, 1981.
3. D. Kirillov, J.L. Merz, P.D. Dapkus, J.J. Coleman, "Laser beam heating and transformation of a GaAs-AlAs multiple-quantum-well structure", *J. Appl. Phys.*, **55**, 1105, 1984.
4. J.E. Epler, R.D. Burnham, R.L. Thornton, T.L. Paoli and M.C. Bashaw, "Laser induced disordering of GaAs-AlGaAs superlattice and incorporation of Si impurity, " *Appl. Phys. Lett.*, **49**, 1447, 1986.
5. Y. Hirayama, Y. Suzuki, and H. Okamoto, , "Ion-species dependence of interdiffusion in ion-implanted GaAs-AlAs superlattices", *Jpn. J. Appl. Phys.*, **24**, 1498 (1985).
6. J. Cibert, P.M. Petroff, D.J. Werder, S.J. Pearton, A.C. Gossard and E. English, "Kinetics of implantation enhanced interdiffusion of Ga and AL GaAs-Ga$_x$Al$_{1-x}$As interfaces", *Appl. Phys. Lett.*, **49**, 223, 1986.
7. E.S. Koteles, B. Elman, P. Melman, J.Y. Chi and C.A. Armiento, "Quantum well shape modification using vacancy generation and rapid thermal annealing", *Optic. Quantum Electron.*, **23**, S779, 1991.
8. D.G. Deppe, L.J. Guido, N. Holonyak, Jr., K.C. Hsieh, R.D. Burnham, R.L. Thornton and T.L. Paoli, "Stripe-geometry quantum well heterostructure Al$_x$Ga$_{1-x}$As lasers defined by defect diffusion", *Appl. Phys. Lett.*, **49**, 510 (1986).
9. D.S. Chemla and D.A.B Miller, "Room-Temperature excitonic nonlinear-optical effects in semiconductor quantum-well structures," *J. Opt. Soc. Am. B*, **2**, 1155 (1985)
10. D.S. Chelma, D.A.B Miller, P.W. Smith, A.C. Gossard, and W. Weigmann, "Room Temperature Excitonic nonlinear absorption and Refraction in GaAs/AlGaAs multiple quantum well structures," *IEEE J. Quant. Electron.*, **QE-20**, 3, 265 (1984)
11. T.S. Moss, *Phys. Status Solidi (b)*, **101**, 555, 1980.
12. F.H. Julien, M..A. Bradley, E.V.K Rao, M. Razeghi, L. Goldstein, "InGaAs(P)/InP MQW mixing by Zn diffusion, Ge and S implantation for photoelectronic applications," *Optic. Quant. Electron.*, **23**, 847 (1991).
13. B. Tuck, "Atomic diffusion in III-V semiconductors," Hilger, Philadelphia (1988)
14. Pavesi, D. Araujo, N.H. Ky, J.D. Ganiere, F.K. Reinhart. P.A. Buffat, and G. Burri, "Zinc diffusion in GaAs and zinc-induced disordering of GaAs/AlGaAs multiple quantum wells: a multitechnique study," *Optical and Quant. Electronics*, **23**, S789 (1991).
15. J.H. Marsh, "Quantum well intermixing," Semiconductor science and technology, 8, 1136 (1993) and the references cited therein.
16. T.E. Schlesinger, and T. Kuech, "Determination of the interdiffusion of Al and Ga in undoped (Al, Ga) As/GaAs superlattice", *Appl. Phys. Lett.*, **49**, 519 (1986).

17. R.G. Walker, "Simple and accurate loss measurement technique for semiconductor optical waveguides", *Electron. Lett.*, **21**, 581, 1985.
18. R.L. Longini, "Rapid zinc diffusion in gallium arsenide", *Solid-St. Electron.*, **5**, 127, 1962.
19. I. Gontijo, T. Kraus, J.H. Marsh, and R.M. De La Rue, "Postgrowth control of GaAs/AlGaAs quantum well shapes by impurity-free vacancy diffusion", *J. of Quantum Elect.*, **30**(5), 1189 (1994) and references therein.
20. J.D. Ralston, S. O'Brien, G.W. Wicks, and L.F. Eastman, "Room-temperature exciton transitions in partially intermixed GaAs/AlGaAs superlattices," *Appl. Phys. Lett.*, **52**, 1511 (1988).
21. Ghisoni, P.J. Stevens, G. Parry, and J.S. Roberts, "Post growth tailoring of the optical properties of GaAs/AlGaAs quantum well structures", *Optics and Quantum Elect.*, **23** (1991) S915-S924.
22. Y. Zebda and A. M. Kan'an, "Resonant tunneling current calculations using the transmission matrix method", *J. of Appl. Phys.*, **72** (1992) 559.
23. S.M. Jensen, "The nonlinear coherent coupler", *IEEE J. of Quantum Electronics*, **18**, 1580, 1982.
24. Y. Silberberg, G.I. Stegeman, "Nonlinear coupling of waveguide modes", *Appl. Phys. Lett.*, **50**, 1562, 1987.
25. See for example G.I. Stegeman, and A. Miller, "Physics of all-optical switching devices" . In Photonic switching (ed. J. Midwinter) vol. 1, 81–146. Academic press.
26. P. Li Kam Wa, A. Miller, J.S. Roberts, P.N. Robson, "130ps recovery of all-optical switching in a GaAs multi-quantum well directional coupler", *Appl. Phys. Lett.*, **58**(19), 2055, 1991.
27. A.M. Kan'an, P. LiKamWa, Mitra Dutta, J. Pamulapati, "Area-Selective Disordering of Multiple Quantum Well Structures and its Applications to All-Optical Devices", *J. Appl. Phys.*, **80**(6), 3179–83, 1996

CHAPTER 13

Polarization-dependent Refractive-Index Change Induced by Superlattice Disordering and Its Applications

YASUHIRO SUZUKI

NTT Photonics Laboratories

1 INTRODUCTION

Disordering in a superlattice (SL) occurs when the constituent atoms interdiffuse and the interfaces between quantum wells and barriers become fuzzy. With 100% disordering, the interface can disappear. Disordering gives rise to some interesting and useful properties — for example, a change in the energy gap and polarization-dependent changes in the refractive index. These changes can be locally induced on the wafer, and in most cases, the procedure for inducing them does not involve any complicated processes. This technique therefore seems very promising for the fabrication of semiconductor optical integrated circuits [1]–[12].

This chapter focuses on the polarization-dependent refractive-index changes induced by SL disordering and their applications to devices in GaAs/AlGaAs and InGaAs/InP systems.

The signs of refractive-index change with SL disordering are different for TE and TM modes. By using these polarization-dependent refractive-index changes, we can achieve a polarization mode-selective channel waveguide, or a polarization mode filter that passes only one polarization mode light (TE or TM light) [6–9]. Moreover, a polarization mode splitter, which divides the TE and TM polarization modes, can be achieved by utilizing mode-selective waveguides [12]. These polarization control devices are expected to play an important role in the semiconductor monolithic functional devices for future fiber transmission systems that will exploit polarization diversity.

In this chapter, first, the disordering methods that are suitable for polarization control devices using polarization-dependent refractive-index changes are described. Second, polarization-dependent refractive-index changes are explained based on experimental results. Next, polarization control devices such as a mode filter and a mode splitter are introduced as applications of polarization-dependent refractive-index changes. Finally, calculation models for polarization-dependent refractive-index change with SL disordering are discussed.

2 DISORDERING TECHNIQUES FOR PASSIVE FUNCTIONAL DEVICES — IMPURITY-FREE SELECTIVE DISORDERING

2.1 SiO$_2$/Si$_3$N$_4$ Cap Annealing

Two main methods have already been developed to induce disordering in selected areas of SLs. The first introduces impurities like Zn or Si, whose diffusion enhances the interdiffusion of the constituent atoms [13–18]. The other is impurity-free disordering [19–20].

Typical methods for the former type are the diffusion or implantation of Zn, Si and so on. In these techniques, diffused impurities induce the interdiffusion of constituent atoms in SLs, so that they are able to disorder. In disordering using Si atoms in GaAs/AlGaAs systems, the number of atoms necessary for this is $1 - 5 \times 10^{18}$ cm^{-3} [21,22]. When impurities are introduced, however, they cause the free carrier absorption to increase. The loss of disordered waveguide formed by Si implantation at the long wavelength side of the bandgap of undisordered superlattices is 3–5 cm^{-1} [22]. This value should be decreased for guided-wave devices. Implantations of B or F have been investigated in an attempt to reduce free carrier absorption and allow the fabrication of low-loss waveguides [23].

Impurity-free disordering methods, in contrast, are suitable for making waveguides because no carriers are generated. Figure 1 shows an example of the process involved in impurity-free disordering methods using rapid thermal annealing (RTA). An SiO$_2$ or Si$_3$N$_4$ film is deposited on the top of an SL structure by chemical vapor deposition, and patterned by a photolithographic technique. After that, rapid thermal annealing is done in an H$_2$ atmosphere. Only the region just under the cap becomes disordered. Disordering can be confirmed by observing photoluminescence (PL) peak shifts.

Details of impurity-free disordering using RTA in GaAs/AlGaAs and InGaAs/InP systems will be discussed in the following sections.

2.2 GaAs/AlGaAs Systems

In GaAs/AlGaAs systems, impurity free disordering is accomplished by SiO$_2$ cap rapid thermal annealing [20]. Ga vacancies are considered to contribute to disordering in this method. There is the possibility that waveguide loss will increase due to p or n type defects formed by Ga vacancies. The loss of waveguides formed this method, however, is 5–6 dB/cm which is small enough for guided-wave devices [7].

We will now discuss an example of RTA disordering in GaAs/AlGaAs systems. Undoped GaAs/AlAs SL structures are grown on (100) semi-insulating GaAs substrates by molecular beam epitaxy (MBE). Next, an approximately 200-nm-thick layer of SiO$_2$ is deposited on the SL. Then the SiO$_2$ is patterned by conventional photolithography and reactive ion etching followed by RTA at 950°C for 30s in an H$_2$ atmosphere at a heating rate of 30°C/s. This RTA partially disorders the SL structure under the SiO$_2$. Deppe et al. showed that impurities like Si do not diffuse from the SiO$_2$ cap [19]. The width of the disordered region was found to be wider than the SiO$_2$ pattern itself. This will be discussed in Section 4.1.1.

The disordering can be checked using PL measurements. Figure 2 shows an example of PL spectra in the disordering of SLs by SiO$_2$ RTA cap annealing [7]. The as-grown SL structure consists of 100 periods of 8-nm-thick GaAs

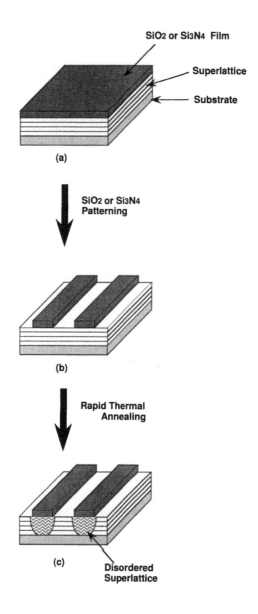

Fig. 1. Disordering process by cap annealing.

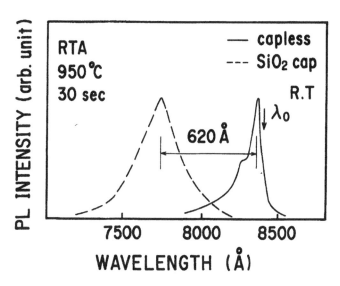

Fig. 2. PL spectra of SLs and disordered SLs by SiO₂ cap annealing. λ_0 is the PL peak position of the as-grown SL structure. The solid line is for the SiO₂ capless region after RTA, and the broken line is for the SiO₂ cap region after RTA.

well layers alternated with 8-nm-thick AlAs barrier layers. λ_0 is the PL peak position for the as-grown SL structure. The solid curve represents the PL spectrum of the SiO₂ capless region after RTA. The PL spectrum of the capless region still has a shoulder due to electron-light hole transitions. This indicates that the characteristics of the SL do not change significantly even after RTA in the capless region. The broken curve represents the SiO₂ cap region after RTA. The PL peak position of the cap region shifts by 62 nm toward the shorter wavelength from that of the capless region. This indicates that the cap region is disordered because of the RTA.

Figure 3 shows an example of measured propagation loss for slab waveguides fabricated using a disordered SL. The slab waveguides consist of the disordered SL (the as-grown SL structure has 100 periods of 8-nm-thick GaAs wells and 8-nm-thick AlAs barriers) and an $Al_xGa_{1-x}As$ ($x = 0.7$) cladding layer between the disordered SL and the substrate. A wavelength of 835 nm was used, which is the PL peak position of the non-disordered region. The solid line is the propagation loss of a disordered waveguide made by RTA with an SiO₂ cap. It is as low as 4–5 dB/cm. The broken line is the propagation loss of a waveguide with the impurity-induced disordering reported by Thornton et al. [3]. It is ten times greater than that of the channel waveguides fabricated using RTA with an SiO₂ cap. This shows that the disordered regions formed by annealing with a SiO₂ cap do not induce free

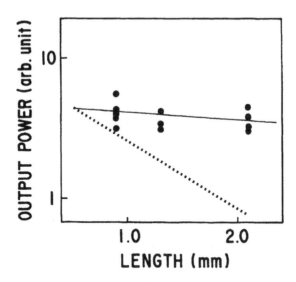

Fig. 3. Propagation loss of disordered SL waveguides. The wavelength used in the measurements was 0.835 μm, which was close to the absorption edge of the non-disordered SL. The solid line is the propagation loss of a disordered region by annealing with an SiO_2 cap, and the broken line is that of a waveguide with impurity-induced disordering, reported by Thronton et al. [3].

carrier absorption and so they can be used as low-loss waveguides even at the PL peak wavelength of the as-grown SL.

2.3   InGaAs/InP Systems

For InGaAs/InP systems, there has been the report that SiO_2 cap rapid thermal annealing is not very effective and various improvements are needed [24].

Si_3N_4 cap annealing is one useful technique in disordering InGaAs/InP SLs and it is impurity-free and suitable for passive waveguided devices [5,9].

We will now discuss an example of disordering induced by Si_3N_4 cap annealing [5]. An undoped InGaAs/InP SL structure is grown on a semi-insulating InP substrate by gas source MBE. After a 150-nm-thick Si_3N_4 film is deposited and patterned, the structure is rapidly annealed once for 15 sec at 800°C in an H_2 atmosphere. This annealing partially disorders the region under the cap but does not significantly affect the characteristics of other regions. The lateral spread of disordering was estimated to be 1–2 μm as described in Section 4.1.2 [7].

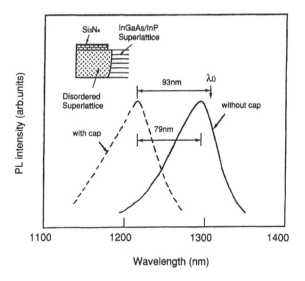

Fig. 4. PL spectra in disordered SLs. λ_0 is the PL peak position of the as-grown SL with a band-gap of 1.31 μm.
Inset: Schematic representation of the disordering method. After RTA, only the region under the 150 nm-thick Si_3N_4 cap is disordered.

The degree of disordering in InGaAs/InP systems can be determined by measuring PL spectra in the same manner as in GaAs/AlGaAs systems. Figure 4 shows typical measured PL spectra. The as-grown SL structure consists of 125 periods of 2.8-nm-thick lattice-matched InGaAs well layers and 4.6-nm-thick InP barrier layers. The PL peak from the region under the Si_3N_4 cap was strongly blue-shifted because the energy level shift occurred due to the transformation from an initially square well compositional profile to a gradually graded profile.

The typical loss of waveguide formed by disordered SLs in InGaAs/InP systems is almost the same as that in GaAs/AlAs systems; about 4–5 dB/cm. Although the mechanism for rapid thermal disordering in InGaAs/InP systems is not clear, low loss disordered waveguides can be obtained by this disordering method.

3   POLARIZATION-DEPENDENT REFRACTIVE-INDEX CHANGE INDUCED BY SUPERLATTICE DISORDERING

Kapon et al. [6], have shown the following relationship exists among the refractive indices of the ordered SLs in the TE and TM modes or n(TE), and n(TM) and that of completely disordered SLs or n_{CD};

$$n(TE) > n_{CD} > n(TM) \tag{1}$$

For partial disordering of SLs, polarization-dependent refractive-index changes [3,5,6] in the transparence region have the following relationship in GaAs/AlGaAs and InGaAs/InP SLs:

$$n\,(TE) > n_D(TE) > n_D(TM) > n(TM), \tag{2}$$

where $n_D(TE)$ and $n_D(TM)$ are the refractive indices of partially disordered SLs in the TE and TM modes, respectively. This relationship has been confirmed experimentally [6,25].

The refractive index must be discussed based on the band structure of materials. However, when considering the refractive indices of MQW structures at the transparent wavelength apart from exciton peaks, we can discuss the refractive indices for MQW using the thin film approximation (the Root Mean Square (R.M.S.) model) [26–30]. For polarization control devices using polarization-dependent refractive-index changes, operating wavelengths are apart from the bandgap of the devices. Therefore, thin film approximation can be used.

In terms of the thin film approximation, relationship (2) is considered to result from the following mechanism. The interdiffusion of the constituent atoms smoothes the interfaces between wells and barriers, which changes the composition profile, causing the refractive index profile to change. In the transparence region, the exciton effects are thought to contribute very little to refractive-index changes [26].

The variation in the refractive index is an important parameter to consider when designing optical devices. We can roughly estimate the amount of refractive-index changes based on PL peak shift due to disordering as Figure 2 shows. The degree of disordering can be roughly estimated by using the amount of PL peak shift. It seems that the difference in the refractive indices of non-disordered SLs for the TE and TM modes ($|n(TE) - n(TM)|$) decreases along with disordering and reaches zero at complete disordering. The measured PL peak shift is 62 nm in Figure 2, so the degree of intermixing is estimated to be about 15% determined by considering the PL peak wavelength of completely disordered SLs. Accordingly, the difference between the indices, $\Delta n(TE) = |n(TE) - n_D(TE)|$, $\Delta n(TM) - |n_D(TM) - n(TM)|$ are roughly estimated to be 4×10^{-3}.

The amount of refractive-index changes induced by disordering can be more directly obtained by measuring the wavelength responses of the waveguide with Bragg gratings [25].

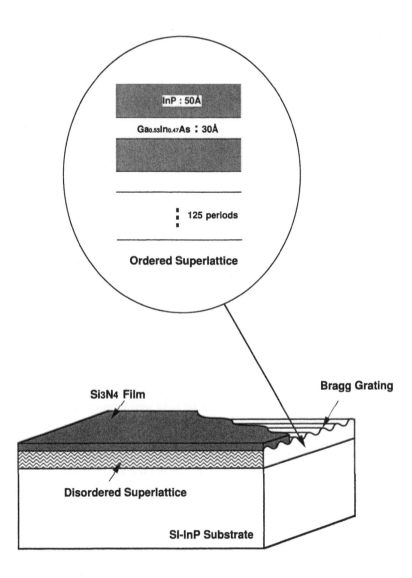

Fig. 5. Structures of the InGaAs/InP ordered SL and the disordered SL waveguide with Bragg gratings.

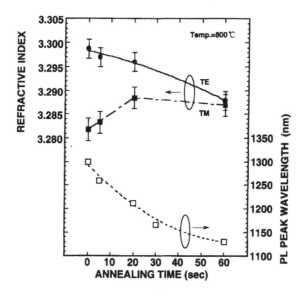

Fig. 6. Refractive-index change and PL peak in InGaAs/InP SL.

Figures 5 and 6 show an example of estimating refractive-index changes
using the waveguide with Bragg gratings. As Figure 5 shows, the SL structure
was grown on an SI-InP substrate by gas source MBE. The SL structure
consisted of 125 periods of 3-nm-thick InGaAs well layers and 5-nm-thick
InP barrier layers.

The PL peak wavelength before disordering was 1300 nm. The thickness of
the SL was 1.0 μm and this layer structure worked as a transparent waveguide
in the 1.55 μm-wavelength region. A Bragg grating was fabricated on the
SL surface of the waveguide to act as a band-stop wavelength filter. The
effective index of the waveguide was obtained from its wavelength response.
A grating pitch of 231.9 nm was chosen so that the designed Bragg wavelength
was inside the 1.48 to 1.55 μm range. The grating depth and waveguide
length were 50 nm and 1–2 μm, respectively. The structural parameters were
designed for the waveguide to support a single mode.

By using the measured effective index of the waveguide, the refractive
index of the disordered SLs was calculated.

Figure 6 shows the annealing time dependence of refractive indices and
PL peak wavelength. As the annealing time increased, the refractive index
for the TE mode decreased and that for the TM mode increased. At the same
time, the PL peak wavelength decreased, and this shift indicates that SL
disordering occurred. The amount of disordering was found to increase with

annealing time, as indicated by the PL peak shift. The experimental results on the refractive-index changes agree with relationship (2). The refractive-index changes for an annealing time of 60s are -1.1×10^{-2} for the TE mode and 5×10^{-3} for the TM mode at a wavelength of 1.55 μm. As the degree of disordering increases, the refractive indices of the TE and TM modes tend to approach the same value. The values are expected to coincide in the extreme case of 100% disordering. This is because with 100% disordering, material anisotropy disappears and uniform properties as from a bulk crystal appear.

These polarization-dependent refractive-index changes can be explained by the microscopic refractive index profile changes in the disordered SLs [30], which can be estimated from the refractive indices of alloy materials and diffusion theory. The overall refractive-index change can be estimated from the refractive index profile by extending the R.M.S. model. This will be discussed in detail in Section 5.

4   DEVICE APPLICATION

4.1   Mode-Selective Channel Waveguides / Mode Filters

Based on relationship (2) in Section 3, channel waveguides are expected to be fabricated whose lateral confinement is dependent on the optical polarization, i.e., TE and TM modes. The device structure for mode- selective channel waveguides is shown in Figure 7.

In Figure 7(a), the core and cladding regions of the channel waveguide are disordered SL and SL, respectively. The refractive index of the core regions is higher than that of the cladding region only for the TM mode. Therefore, this structure becomes a TM-mode-selective waveguide. In Figure 7(b), the structure is just the reverse, so it works as a TE-mode-selective waveguide.

4.1.1   *Characteristics of mode-selective channel waveguides —*
GaAs/AlGaAs systems

By using disordered SLs with SiO_2 cap RTA, we can obtain mode-selective waveguides in GaAs/AlGaAs systems. An example of characteristics of the waveguides is described as follows [7]. Figures 8(a) and (b) show observed near-field patterns for the TM-mode and the TE-mode waveguides, respectively. A laser beam at a wavelength of 850 nm was endfire-coupled into the waveguides. The SL structure consists of 62 periods of 9.4-nm-thick GaAs well layers and 5.8-nm-thick AlAs barrier layers, and $Al_x Ga_{1-x} As$ ($x = 0.43$) is used as a cladding layer. Both the disordered and the non-disordered SLs are transparent for this wavelength. The length of the waveguide is about 1 μm. In Figure 8(a), the SiO_2 stripe is 3 μm wide.

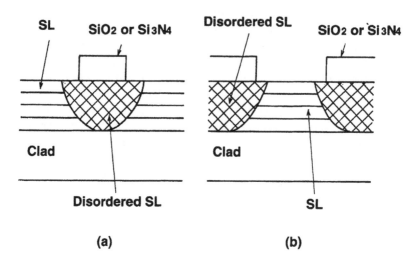

Fig. 7. Schematic configuration of fabricated channel waveguides. (a) TM mode-selective waveguide. (b) TE mode-selective waveguide.

It was confirmed that the TM mode light was guided along the channel in multimode and that the TE mode light was not laterally confined. On the other hand, in Figure 8(b), the TE mode light incident on the TE waveguide (with a capless region width of 15 μm) was guided in single mode whereas the TM mode light was not laterally guided. These pictures clearly indicate that mode selectivity is functioning in these channel waveguides.

In the TM waveguide in Figure 8(a), the optical confinement is about 15 μm wide, which is 10–12 μm wider than the width of the SiO$_2$ stripe. For the TE waveguide in Figure 8(b), the optical confinement is about 3 μm wide, which is much narrower than the capless region which is 15 μm wide. This is due to the lateral spread of SL disordering which extends beyond the SiO$_2$ cap width.

In GaAs/AlAs disordered SL mode-selective channel waveguides, it was difficult to fabricate single-mode channel waveguides because of the large lateral spread of disordering (about 6 μm). Especially for TM-mode-selective waveguides whose channel core region had to be formed by disordered SL, single-mode waveguides could not be achieved. However, a small lateral spread of disordering in GaAs/AlGaAs SL has been reported by Chi et al. [31]. A large lateral spread of disordering in GaAs/AlAs mode selective waveguides may not be inherent, although further investigation is necessary.

It was found that the photoelastic effect of SiO$_2$ in the SL contributes little to waveguiding using disordering because the near-field distributions are very different from those for waveguides based on photoelastic effects [32].

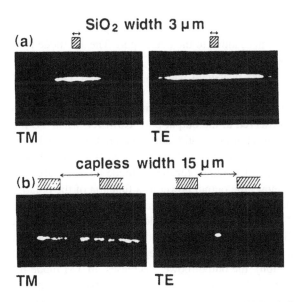

TM   TE

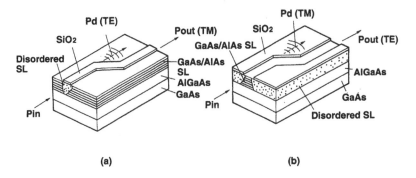

Fig. 8. Observed near-field patterns and intensity profiles. TM and TE polarized lights are launched into (a) TM mode-selective waveguides and (b) TE mode- selective waveguides, respectively.

Fig. 9. Schematic diagrams of mode filters. (a) TM mode filter. (b) TE mode filter.

By using mode-selective channel waveguides, a mode filter can be achieved. Figures 9(a) and (b) are schematic diagrams of mode filters.

These filters have bending mode-selective channel waveguides to prevent unwanted light from going into the output port. The bending angle is 1 degree and the whole device is about 5 μm long. In the TM mode filter (Fig. 9(a)), the channel core is formed with a disordered SL, while the cladding layer is formed with an SL. TM light is guided along the channel to the output port (Pout (TM)); at the bend, however, TE mode light diffuses into the substrate (Pd(TE)) and is not guided to the output port. The TE mode filter functions in a similar way.

Figure 10 shows near-field patterns for TE and TM mode filters. The wavelength used for the measurements was 1.15 μm. In the TM mode filter (Figure 10(a)), when TM mode light is introduced into the input port, guided light can be observed at the output port. However, when TE mode light is coupled into the input port, guided light can not be observed at the output port, and only some scattered light is produced at the output side.

Similar results are obtained for the TE mode filter (Figure 10(b)).

Figure 11 shows the output power as a function of the polarization angle of the incident light. The TE polarization corresponds to $\theta = 0$ degree and the TM polarization to $\theta = 90$ degrees. The maximum output power was normalized to zero dB for both filters. For TE mode filters, as the polarization angle increases from 0 degree (TE polarization) to 90 degrees (TM polarization), the output power decreases. In TM mode filters, as the polarization angle decreases from 90 degrees to 0 degree, the output power decreases. These results clearly indicate that these waveguides can function as TE/TM polarization filters. An extinction ratio of about 14 dB was attained for both filters.

4.1.2   Characteristics of mode-selective channel waveguides — InGaAs/InP systems

TE and TM mode-selective waveguides in InGaAs/InP systems can also be fabricated using the same waveguide structure described in Section 4.1.1. An example of mode-selective waveguides in InGaAs/InP systems is described as follows [9]. The SL structure and the fabrication process are the same as those in Section 2.3. The near-field patterns for the TE and TM waveguides at a wavelength of 1.52 μm are shown in Figs. 12(a) and 12(b), respectively. Both the disordered SL and the non-disordered SL are transparent at this wavelength. The length of the waveguides is about 1.5 μm. In Figure 12(a), the width of the Si_3N_4 capless region is 8 μm. Only the TE mode light was guided along the channel in a single mode. The TM mode light was not laterally confined. In Figure 12(b), the Si_3N_4 stripe is 5 μm wide. The result is just the opposite: the TM mode light coupled into the TM waveguide is guided in a single mode, whereas the TE mode light is not laterally guided.

In InGaAs/InP SL, both TE and TM mode- selective channel waveguides can guide single-mode-light because the amount of lateral spread of disordering is small (around 1 μm) compared to that in GaAs/AlAs systems.

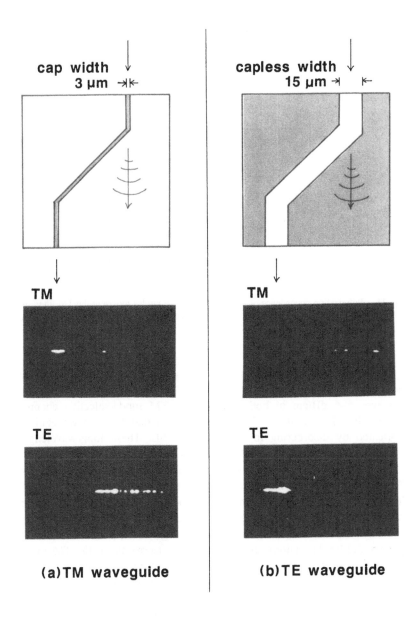

cap width
3 μm

capless width
15 μm

TM

TM

TE

TE

(a)TM waveguide

(b)TE waveguide

Fig. 10. Observed near-field patterns of mode filters. (a) TM mode filter. (b) TE mode filter.

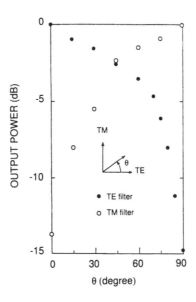

Fig. 11. Polarization dependence of output power. Open circles represent the TM filter and closed circles represent the TE filter. Zero degree in the polarization angle corresponds to the TE mode and 90 degrees to the TM mode.

The propagation loss was measured at a wavelength of 1.52 μm by the cut back method for disorder-fabricated channel waveguides. The propagation loss was 5–6 dB/cm in both the TE and TM mode-selective channel waveguides. This value is almost the same as that for slab waveguides fabricated by disordering in a GaAs/AlAs SL. These long-wavelength channel waveguides can be used as low-loss waveguides.

The polarization dependence of the output power is shown in Figure 13. The TE and TM polarizations correspond to $\theta = 0$ degree and $\theta = 90$ degrees, respectively. The maximum output power was normalized to zero dB for both selective waveguides. For the TE mode-selective channel waveguide, as the polarization angle increases from 0 degree (TE polarization) to 90 degrees (TM polarization), the output power decreases. In the TM mode waveguides, the function is just the reverse. As Figure 13 shows, the extinction ratio was found to be about 13 dB in both mode waveguides. However, this extinction ratio can probably be improved by introducing a bent waveguide structure to diffuse unwanted polarization light into the substrate from the waveguide core as discussed in Section 4.1.1.

The amount of refractive-index changes induced by disordering (n(TE)-n_D(TE) and n_D(TM)-n(TM)) was estimated by using PL peak shifts and the wavelength response of the waveguide with Bragg grating in Section 3.

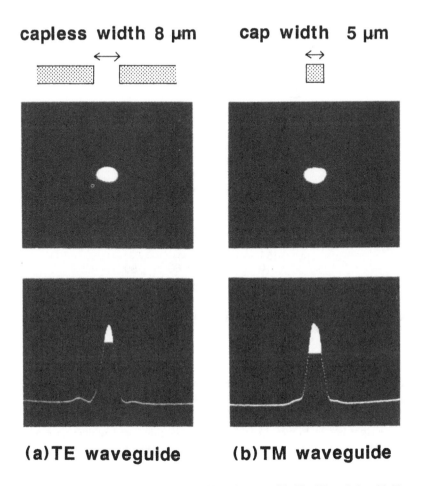

Fig. 12. Observed near-field patterns. (a) TE mode waveguide. The TE mode is guided in a single mode. The width of the Si_3N_4 capless region is 8 μm. (b) TM mode waveguide. The TM mode is guided in a single mode. The width of the Si_3N_4 cap region is 5 μm.

The index changes can also be estimated by observing the near-field patterns of TE and TM mode-selective waveguides which have various waveguide widths. Figure 14 shows an example of experimental data. The abscissa denotes the widths of Si_3N_4 capless widths and cap widths, and the ordinate the amount of the variation in indices induced by disordering. For TE waveguides, the capless region widths ranged from 4 to 14 μm and for TM waveguides, the cap region widths ranged from 3–7 μm. The observed mode patterns are very dependent on the Si_3N_4 capless and cap widths for TE and TM waveguides, respectively. The two bars correspond to the experimental

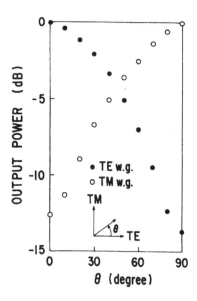

Fig. 13. Polarization dependence of output power. The closed circles are for the TE waveguide, and the open circles for the TM waveguide.

transition regions between single mode and multi-mode guiding. Figure 15 shows device structures and examples of near-field patterns for TE- and TM-mode-selective waveguides with various channel widths.

The right bar in Figure 14 is for TE waveguides, and widths narrower than 10 μm; a single mode can be observed. On the other hand, for the waveguides with widths wider than 10 μm, multimode is observed (see Figure 15(a)). Comparing the experimental results with the single mode conditions calculated by the effective index method (the solid curve in Figure 14), the amount of variation in the indices between the disordered regions and the nondisordered regions is estimated to be 1×10^{-3}.

In TM-mode-selective waveguides, the guided mode changed from single mode to multimode in the 7 μm-wide waveguide (see Figures 14 and 15(b)). This corresponds to a refractive-index change of 2×10^{-3}. These index change values are almost the same as those for the annealing time of 15 s in Figure 6.

The characteristics of mode-selective waveguides were measured with Si_3N_4 film coated. The refractive-index changes due to stress field induced by Si_3N_4 film can be neglected because they are much smaller than those due to disordering [9,32].

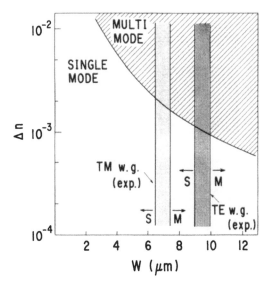

Fig. 14.   Variation in the refractive index. The abscissa represents the widths of the waveguides and the ordinate the amount of variation in the indices. The two bars correspond to the transition regions of the guided mode; the one at right is for the TE mode and the left one is for the TM mode. The solid curve is the boundary on single-mode propagation as calculated by the effective index method.

4.2   Mode Splitter

A semiconductor guided-wave polarization mode splitter that divides the TE and TM polarization modes as well as the mode filter is expected to be one of the key devices in future fiber transmission systems that will exploit polarization diversity. Several kinds of mode splitters have already been reported [10,11]. A directional coupler that has one waveguide with a metal cladding works as a polarization mode splitter because of its asymmetric structure [10]. Another kind of directional-coupler-type mode splitter uses both current and voltage to control TE and TM polarizations independently [11]. To adjust coupling length, however, these devices require tuning electrodes or high accuracy in fabrication, or both.

Here, we will discuss a mode splitter using polarization-dependent refractive-index changes induced by SL disordering [12], which does not require any electric control schemes and complicated in fabrication.

4.2.1   Device Structure

The device consists of input, output, and mode-splitting regions (Figure 16). The most important is the mode-splitting region, where disordered and

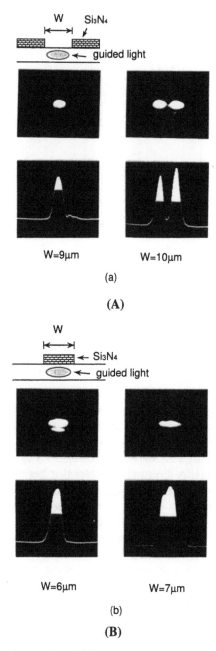

Fig. 15. Device structures and near field patterns of mode-selective waveguides with various channel widths. (A) TE mode-selective waveguide. (B) TM mode-selective waveguide.

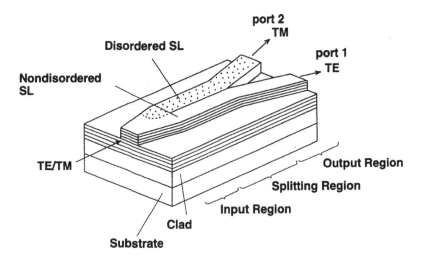

Fig. 16. Structure of a mode splitter. The device consists of input, output, and mode-splitting regions. In the mode-splitting region, disordered and non-disordered waveguides are formed adjacent to each other

non-disordered waveguides are formed adjacent to each other. Figure 17 shows a cross-section of this region. The channel waveguide has a rib structure, and the core is composed of both disordered and nondisordered SLs. The TE mode tends to propagate in the nondisordered core region because, as Figure 17(b) shows, the refractive index for TE mode light is higher there. The TE mode light therefore can be guided to the nondisordered output port. For the TM mode, the relationship of the refractive indices is just the reverse, so the TM mode light can be guided from the disordered core region to the output port.

How effective splitting would be for the TE mode can be calculated by using the beam propagation method (BPM). The amplitude and the spatial distribution changes of the field in the guided wave can be obtained using this method [33,34]. Figure 18 shows the calculated results for a device with a total length of 800 μm. The angle between the branches is 1 degree, and the core is 10 μm wide. It is assumed that n(TE) = 3.337, the refractive index of the non-disordered region, and that $\Delta n = n(\text{TE})\text{-}n_D(\text{TE}) = 1.2 \times 10^{-3}$, the difference between the nondisordered and the disordered regions. Almost all the TE-mode light shifts over to the non-disordered region.

4.2.2 *Typical experimental results*

The SL structure for the mode splitter is the same as that used for the mode-selective waveguides in Section 4.1.2. A typical structure for the mode

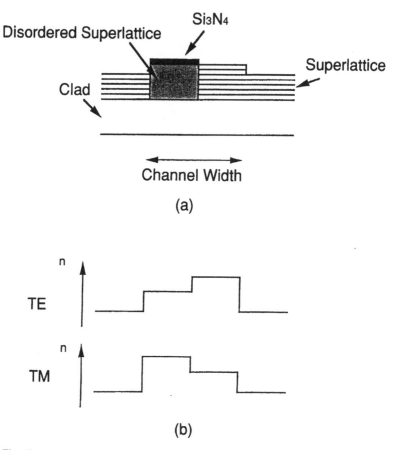

Fig. 17. (a) Cross-section of the splitting region. (b) Refractive index profiles for TE and TM modes.

splitter is as follows. The mode-splitting region and the tapered input and the output regions have 25-nm-high rib structures. The splitting region is 1000 μm-long and both the disordered and the nondisordered core regions are 7 μm-wide. The total device is 3 mm-long. The propagation loss of channel waveguides is 5–6 dB/cm, and the excess loss in the branch region is 1–2 dB [9].

The fabrication of the device is almost the same as that of mode-selective waveguides in InGaAs/InP systems. The ribbed structure of the waveguides, including the input and the output regions, is fabricated by wet etching.

Typical near-field patterns and intensity profiles of the mode split-ter at a wavelength of 1.52 μm are shown in Figure 19(a) and 19(b).

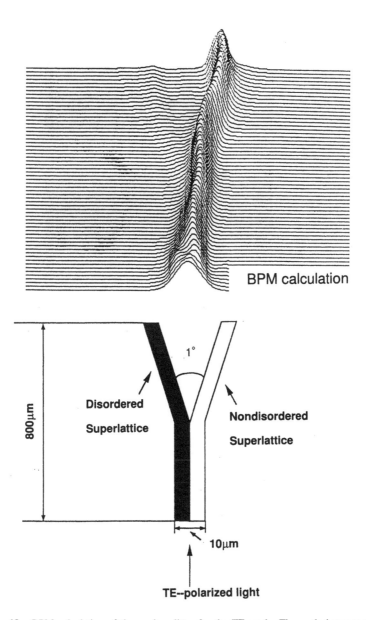

Fig. 18. BPM calculation of the mode splitter for the TE mode. The angle between the branches in the calculation model was 1 degree, and the core was 10 μm wide. We assumed that $n(\text{TE}) = 3.337$, the refractive index of the non-disordered region, and that $\Delta n = n(\text{TE}) - n_D(\text{TE}) = 1.2 \times 10^{-3}$, the difference between the nondisordered and the disordered regions.

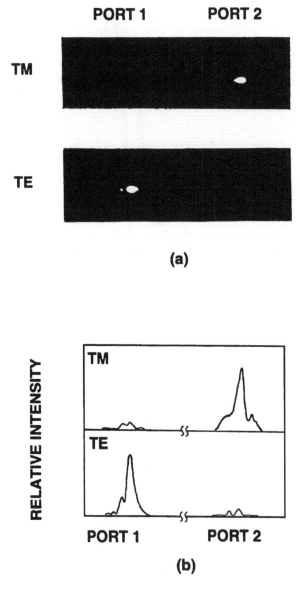

Fig. 19. (a) Near-field patterns and (b) intensity profiles of the mode splitter at a wavelength of 1.52 μm.

A 45-degree-polarized incident light is endfire-coupled into the input port. The output light is observed through a polarizer. The TE-mode light is guided along the non-disordered region to one output port and the TM-mode light is guided along the disordered region to the other output port. These results clearly indicate that mode-splitting works. The crosstalk is about −10 dB (Figure 19(b)).

Various semiconductor mode splitters have already been reported, but their bandwidth is limited and their fabrication tolerance is low because they all use directional-couplers as a key component. The device described here, on the other hand, is based on a different operation principle, and its bandwidth should be fairly large and its fabrication tolerance should be fairly high. In addition, no electrical control scheme, such as current injection and voltage application, is necessary.

5 CALCULATION MODELS FOR POLARIZATION-DEPENDENT REFRACTIVE-INDEX CHANGE

5.1 Method of calculation

The relationship of refractive-index changes can be explained by a model based on refractive index profiles by considering the interdiffusion of SL interfaces. The refractive indices of an SL dependent on polarization modes in the transparency wavelength region are given by

$$n^2(\text{TE}) = \frac{n_W^2 L_W + n_B^2 L_B}{L_W + L_B} \tag{3a}$$

$$n^2(\text{TM}) = \frac{L_W + L_B}{\frac{L_W}{n_W^2} + \frac{L_B}{n_B^2}} \tag{3b}$$

where L_W and L_B are the thicknesses of the well and the barrier, respectively, and n_W and n_B are the corresponding refractive indices of the wells and the barriers (Figure 20(a)) [27–29]. These equations are given by the Root Mean Square (R.M.S.) model.

After disordering, the interdiffusion of constituent atoms at the interfaces of well layers and barrier layers generates an intermediate layer where the composition changes gradually (Figure 20(b)). This composition profile change induces the refractive index profile to change.

By extending Eqs. 3a and 3b, the refractive indices for TE and TM modes are given by

$$n_D^2(\text{TE}) = \frac{1}{T} \int_0^T n^2(z)dz \tag{4a}$$

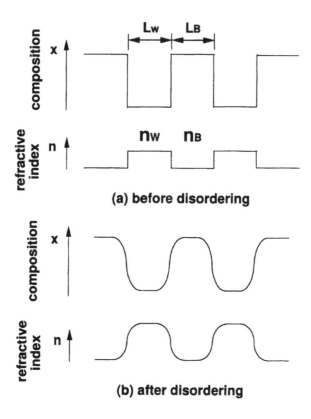

Fig. 20. Composition profile and refractive index profile. L_w and L_B are the thicknesses of the well and the barrier, and n_W and n_B are the refractive indices of the well and the barrier.

$$n_D^2(\text{TM}) = \frac{T}{\int_0^T \frac{dz}{n^2(z)}} \tag{4b}$$

where $n(z)$ and T are the refractive index profile and the film thickness of the SL period, respectively. The details are provided in the appendix. We call these equations (4a and 4b) the extended R.M.S. model.

The refractive-index changes are also induced by the change of the bandgap along with the composition profile change accompanied by disordering. Here, we will focus on the refractive-index changes in the transparent wavelength region which is far enough away from the bandgap. The energy difference between the wavelength considered and the bandgap wavelength (the detuning energy) is more than 0.1 eV. Referring to the detuning dependence of refractive- index change, due to bandgap change induced by Quantum Confined Stark Effect, the amount of refractive-index change, due to bandgap change in the above transparent wavelength region, can be estimated to be at most in the order of 10^{-4}. This value is sufficiently small compared to the amount of refractive-index change due to disordering that is discussed here. The refractive-index change due to disordering is considered to be mainly induced by composition profile change. Therefore, we can neglect the effect of refractive-index changes due to bandgap change accompanied by disordering.

Polarization-dependent refractive-index changes are calculated as follows. First, the composition profile is calculated using the diffusion equation of constituent atoms. Second, the composition profile is transformed into a relationship between the refractive indices and mole fractions of ternary or quaternary materials. Finally, the polarization-dependent refractive-index changes can be determined using the extended R.M.S. model (Eq. (4)).

5.2   GaAs/AlGaAs Systems

In the GaAs/AlGaAs system, only group III atoms diffuse into well and barrier layers when disordering. The diffusion constant has been reported to be the same in the wells and barriers [35,36]. The Al composition profile after disordering can be calculated based on the simple diffusion equation that follows:

$$\frac{\partial C}{\partial t} = D \frac{\partial^2 C}{\partial x^2} \tag{5}$$

where $C(x, t)$ is the Al concentration and $D(= D_W = D_B)$ is the diffusion constant in wells (D_W) and barriers (D_B). The solution is

$$C(x, t) = \frac{1}{2\sqrt{\pi Dt}} \int_{-\infty}^{\infty} f(\lambda) \exp\left(-\frac{(\lambda - x)^2}{4Dt}\right) d\lambda \tag{6}$$

where the initial condition is

$$t = 0, \quad c(x, 0) = f(x), \quad -\infty < x < \infty \tag{7}$$

The diffusion length is defined as $\Delta = \sqrt{Dt}$ which is related to the degree of disordering. Figure 21(a) shows examples of calculated Al composition profiles of ordered and partially disordered SLs. The ordered SL structure has 9.4-nm-thick GaAs wells and 5.6-nm-thick AlAs barriers. The Al atom diffusion length Δ is 0.5 nm. From this Al composition profile, we can determine the refractive index profile using the relationship between refractive indices and Al mole fractions obtained by Casey et al. [37]. Figure 21(b) shows the refractive index profile at a diffusion length of 0.5 nm.

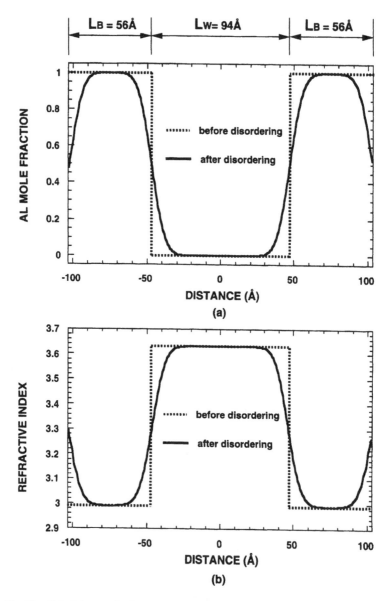

Fig. 21. Calculation results for (a) the Al composition profile and (b) the refractive index profile in ordered and disordered SLs. The ordered SL structure has 9.4 nm-thick GaAs wells and 5.6 nm-thick AlAs barriers. The diffusion length is 0.5 nm.

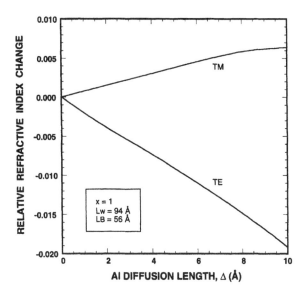

Fig. 22. Relative change in refractive index as a function of aluminum atom diffusion length at a wavelength of 0.85 μm. The ordered SL structure has 9.4 nm-thick GaAs wells and 5.6 nm-thick AlAs barriers. The aluminum mole fraction is x.

Using the extended R.M.S. model (Eq. (4)), we can calculate the refractive-index changes. Figure 22 shows the dependence of relative refractive index changes for TE and TM modes (n_D(TE)-n(TE) and n_D(TM)-n(TM)) at a wavelength of 0.85mm on Al atom diffusion length. The SL has 9.4-nm-thick GaAs well layers and 5.6-nm-thick AlAs (Al mole fraction $x = 1$) barrier layers. As the diffusion length becomes longer (the degree of disordering increases), Δn(TE) $= n_D$(TE) $- n$(TE) decreases and Δn(TM) $= n_D$(TM)$-n$(TM) increases. The refractive-index changes range from $10^{-3} - 10^{-2}$. These results agree well with the experimental results discussed in Section 3. By comparing calculated and experimental results, it can be estimated that an annealing time of 30s yields a degree of disordering that corresponds to a diffusion length of around 0.5 nm.

The refractive-index changes are dependent on the thickness of the well and barrier. Figure 23 shows the relative change in refractive index (Δn(TE) and Δn(TM)) as functions of well width at a diffusion length of 0.5 nm. The barrier thickness is 5.6 nm and the Al mole fraction of the barriers (x) equals one. As the well thickness increases, the relative change in refractive indices for both TE and TM modes decreases (in absolute terms). The TE mode yields larger (absolute terms) relative refractive-index changes than the TM mode. Larger relative changes in refractive index are obtained with narrow

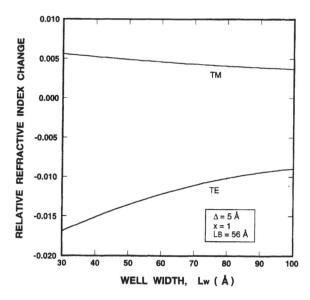

Fig. 23. Relative change in refractive index as a function of well width at a diffusion length of 0.5 nm.

well widths. At a well width of 3 nm, the relative refractive-index change for the TE mode is −0.017 and that for the TM mode is 0.0055. Figure 24 shows the relative refractive-index changes as functions of barrier width at a diffusion length of 0.5 nm. The well thickness is 9.4 nm. Almost the same relationship is apparent.

Figure 25 shows the relative change in refractive index as functions of the Al mole fraction of the barrier at a diffusion length of 0.5 nm. As the Al mole fraction increases (barrier height increases), the relative change in refractive index increases for both TE and TM modes; the TE mode has a the larger increase (in absolute terms).

The results in Figures 23, 24 and 25 indicate that narrow wells, as well as high and narrow barriers, yield large changes in the refractive index; these are useful in fabricating polarization-control waveguide devices.

5.3   InGaAs/InP Systems

In InGaAs/InP systems, the diffusion of group V atoms plays an important role in disordering. Here, it is assumed that (i) only group V atoms contribute to disordering [38] and that (ii) the diffusion constants are different in wells and barriers [39]. With these assumptions, the simple error function fails to develop the composition profile of this system. Instead, we can calculate the

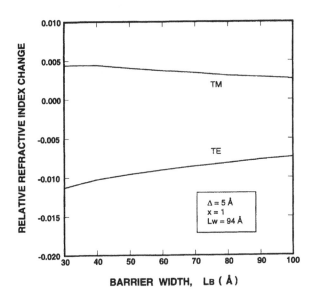

Fig. 24. Relative change in refractive index as a function of barrier width at a diffusion length of 0.5 nm.

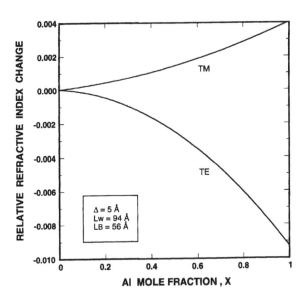

Fig. 25. Relative change in refractive index as a function of the aluminum mole fraction of the barrier at a diffusion length of 0.5 nm.

composition profile directly from the diffusion equation and then determine the refractive index profile using the following formula which gives the refractive index for $Ga_x In_{1-x} As_y P_{1-y}$ [40]:

$$n^2 = 1 + \frac{E_d}{E_0} + \frac{E_d}{E_0^3} E^2 + \frac{\eta}{\pi} E^4 \ln \left(\frac{2E_0^2 - E_g^2 - E^2}{E_g^2 - E^2} \right) \tag{8}$$

where

$$\eta = \frac{\pi E_d}{2 E_0^3 (E_0^2 - E_g^2)} \tag{9}$$

for photon energy $E (= h\nu)$ and with

$$E_0 = 0.595x^2(1 - y) + 1.62xy - 1.891y - 0.524x + 3.391$$
$$E_d = (12.36x - 12.71)y + 7.54x + 28.91,$$
$$E_g = 1.35 + 0.668x - 1.17y - 0.758x^2 + 0.18y^2 - 0.069xy$$
$$- 0.322x^2y + 0.03xy^2.$$

The lattice-matching condition to InP is $y = 2.197x$.

Figure 26 shows examples of relative changes in refractive indices for TE and TM modes as functions of P atom diffusion length in barriers at a wavelength of 1.55 μm. The SL consists of 3-nm-thick InGaAs well layers and 5-nm-thick InP barrier layers. This SL has the same structure as the mode splitter [12]. In this figure, $\gamma (= D_W / D_B)$ is the ratio of diffusion constant in the well ($= D_W$) to that in the barrier ($= D_B$).

In the InGaAs/InP system, the refractive-index changes are strongly dependent on the diffusion constant ratio γ. With a simple diffusion constant ($\gamma = 1$), the index change that accompanies the increase in diffusion length (an increase in the degree of disordering) does not agree with the experimental results. The calculated results indicate that the signs of the relative changes in refractive indices for TE and TM modes are minus whereas the signs recorded experimentally are plus and minus (see Figure 6). At $\gamma = 0.7$, the signs agree.

Figure 27 shows the relative change in refractive index as functions of the diffusion constant ratio at a diffusion length of 0.8 nm (barriers). The hatched area indicates that the signs of relative refractive-index changes, TE and TM modes, are opposite to one another. This agrees with the experimental results in Figure 6. Taking into consideration the sign and the change in the refractive indices, a diffusion length of 0.8 nm and a diffusion constant ratio of around 0.74 agree with the experimental results for an annealing time of 60s. For this case, the changes in refractive indices are about -1.1×10^{-2} for TE and 5×10^{-3} for TM modes.

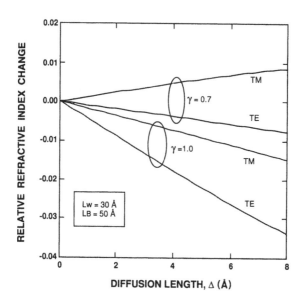

Fig. 26. Relative change in refractive index as a function of P atom diffusion length in barriers at a wavelength of 1.55 μm. γ is the diffusion constant ratio (D_W/D_B).

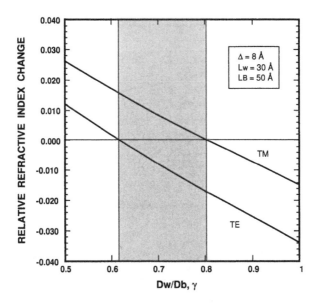

Fig. 27. Relative change in refractive index as a function of diffusion constant ratio at a diffusion length of 0.8 nm (barriers).

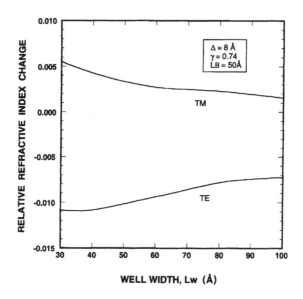

Fig. 28. Relative change in refractive index as a function of well width at a diffusion length of 0.8 nm (barriers).

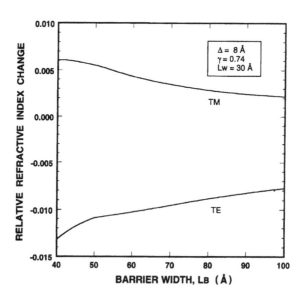

Fig. 29. Relative change in refractive index as a function of barrier widths at a diffusion length of 0.8 nm (barriers).

The relative change in refractive index in the InGaAs/InP systems is also dependent on well and barrier thickness. Figure 28 shows the relative change in refractive index as a function of well widths at a diffusion length of 0.8 nm (barriers). The barrier thickness is 5 nm. As the well thickness increases, the absolute value of the change in refractive index decreases for both TE ($|n_D(\text{TE}) - n(\text{TE})|$) and TM ($|n_D(\text{TM}) - n(\text{TM})|$) modes. Figure 29 is a corresponding plot for barrier width; the diffusion length is 0.8 nm (barriers). The well thickness is 3 nm. Figures 28 and 29 indicate the basic relationship; the relative change in refractive index decreases with well or barrier thickness. This is true for both systems.

6 SUMMARY

This chapter described the polarization-dependent refractive-index changes induced by SL disordering and its application in devices such as polarization mode filters and mode splitters in GaAs/AlGaAs and InGaAs/InP systems. The refractive-index changes were explained based on the extended R.M.S. model considering the interdiffusion of SL interfaces. By using SL disordering, polarization-dependent index changes can be created in selected areas. This means that the polarization control devices described here using polarization-dependent refractive-index changes can easily be integrated with other semiconductor optical functional devices such as lasers, detectors and so on. Such monolithic devices are expected to play an important role in future photonic networks.

Appendix

Refractive indices of thin film materials are introduced intuitively based on Maxwell's equation [41]. Figure 30(a) shows the model of 2-layer thin film materials. t_i is the thickness of the i material ($i = 1, 2$) and ε_i is its dielectric constant ($i = 1, 2$). First, we consider the case where the electric vector is perpendicular to the interfaces (TM mode). According to the continuous conditions of Maxwell's equation, the normal component of the electric displacement must be continuous across the material interface. Therefore the electric displacement must have the same value D in both materials. If E_i is the corresponding electric field,

$$E_i = \frac{D}{\varepsilon_i} (i = 1, 2). \tag{10}$$

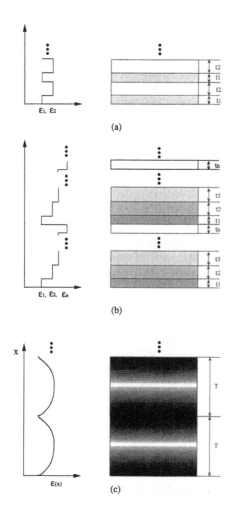

Fig. 30. Calculation model of refractive indices for thin film materials. (a) Two-layer material; (b) n-layer material; (c) Material whose dielectric constant changes continuously with period of T.

The mean field E averaged over the total volume is

$$E = \frac{t_1 E_1 + t_2 E_2}{t_1 + t_2} = \frac{t_1 \frac{D}{\varepsilon_1} + t_2 \frac{D}{\varepsilon_2}}{t_1 + t_2} \tag{11}$$

The effective dielectric constant ε_{TM} is

$$\varepsilon_{TM} = \frac{D}{E} = \frac{1}{\frac{\frac{t_1}{\varepsilon_1} + \frac{t_2}{\varepsilon_2}}{t_1 + t_2}} = \frac{t_1 + t_2}{\frac{t_1}{\varepsilon_1} + \frac{t_2}{\varepsilon_2}} \tag{12}$$

Using the relationship between the dielectric constant ε and refractive index n described by

$$\varepsilon = n^2, \tag{13}$$

we can obtain the refractive index relationship based on the R.M.S. model given by

$$n_{TM}^2 = \frac{t_1 + t_2}{\frac{t_1}{n_1^2} + \frac{t_2}{n_2^2}} \tag{14}$$

where n_{TM} is the averaged refractive index of thin film materials and $n_i (i = 1, 2)$ is the refractive index of the material.

Next, we consider the case where the electric vector is parallel to the interfaces (TE mode). According to the continuous conditions of Maxwell's equation, the tangential component of the electric vector is continuous across the interface. Therefore, the electric field must have the same value E in both materials. If D_1 and D_2 are the corresponding electric displacements,

$$D_i = \varepsilon_i E (i = 1, 2). \tag{15}$$

The mean electric displacement D is

$$D = \frac{t_1 D_1 + t_2 D_2}{t_1 + t_2} = \frac{t_1 \varepsilon_1 E + t_2 \varepsilon_2 E}{t_1 + t_2}. \tag{16}$$

Considering the effective dielectric constant $\varepsilon_{TE} (= D/E)$, we can obtain the averaged refractive index n_{TE} in the TE mode as shown by

$$n_{TE}^2 = \frac{t_1 n_1^2 + t_2 n_2^2}{t_1 + t_2}. \tag{17}$$

In the case of n-layer materials in Figure 30(b), the averaged refractive indices are given by

$$n_{TM}^2 = \frac{\sum_{i=1}^{n} t_1}{\sum_{i=1}^{n} \frac{t_i}{n_i^2}} \tag{18-1}$$

$$n_{TE}^2 = \frac{\sum_{i=1}^{n} t_1 n_i^2}{\sum_{i=1}^{n} t_i}. \tag{18-2}$$

When the refractive index n changes continuously (Figure 30(c)), the averaged refractive indices are given by

$$n_{TM}^2 = \frac{T}{\int_0^T \frac{dx}{n(x)^2}} \tag{19-1}$$

$$n_{TE}^2 = \frac{\int_0^T n(x)^2 dx}{T}, \tag{19-2}$$

where $n(x)$ is the refractive index profile and T is the period of the thin film material.

References

1. H. Nakajima, S. Semura, T. Ohta, Y. Uchida, H. Saito, T. Fukuzawa, T. Kuroda, and K.L.I. Kobayashi, *IEEE J. Quantum Electron.* **21**, 629–633, 1985.
2. A. Furuya, M. Makiuchi, and O. Wada, *Electron. Lett.* **24**, 1282–1283, 1988.
3. R.L. Thornton, J.E. Epler, and T.L. Paoli, *Appl. Phys. Lett.* **51**, 1983–1985, 1987.
4. J. Werner, E. Kapon, N.G. Stoffel, E. Colas, S.A. Schwarz, C.L. Schwartz, and N. Andreadakis, *Appl. Phys. Lett.* **55**, 540–542, 1989.
5. T. Miyazawa, H. Iwamura, and M. Naganuma, *IEEE Photon. Technol. Lett.* **3**, 421–423, 1991.
6. E. Kapon, N.G. Stoffel, E.A. Dobisz, and R. Bhat, *Appl. Phys. Lett.* **52**, 351–353, 1988.
7. Y. Suzuki, H. Iwamura, and O. Mikami, *Appl. Phys. Lett.* **56**, 19–20, 1990.
8. Y. Suzuki, H. Iwamura, and O. Mikami, *IEEE Photon. Technol. Lett.* **2**, 818–819, 1990.
9. Y. Suzuki, H. Iwamura, T. Miyazawa, and O. Mikami, *Appl. Phys. Lett.* **57**, 2745–2747, 1990.
10. P. Albrecht, M. Hamacher, H. Heidrich, H. -P. Nolting, and C. M. Weinert, *IEEE Photon. Technol. Lett.* **2**, 114–115, 1990.
11. H. Yanagawa, H. Mak, K. Ueki, and Y. Kamata, *ECOC'90 Technical Digest*, 125–128, 1990.
12. Y. Suzuki, H. Iwamura, T. Miyazawa, and O. Mikami, *IEEE J. Quantum Electron.*, **30**, 1794–1800, 1994.
13. W.D. Ladig, N. Holonyak, Jr., M.D. Camras, K. Hess, J.J. Coleman, P.D. Dapuks, and J. Bardeen, *Appl. Phys. Lett.*, **38**, 776–778, 1981.
14. R.W. Kliski, P. Gavrilovic, K. Meehan, J. Gavrilovic, K.C. Hsieh, G.S. Jackson, N. Holonyak, Jr., and J.J. Coleman, *J. Appl. Phys.* **58**, 101–107, 1985.
15. E.V.K. Rao, H. Thibierge, F. Brillouet, F. Alexandre, and R. Azoulay, *Appl. Phys. Lett.* **46**, 867–869, 1985.
16. P. Gavrilovic, D.G. Deppe, K. Meehan, N. Holonyak. Jr., J.J. Coleman, and R.D. Burnham, *Appl. Phys. Lett.* **47**, 130–132, 1985.
17. B. Tell, B.C. Johnson, J.L. Zyskind, J.M. Brown, J.W. Sulhoff, K.F. Brown-Goebeler, B.I. Miller, and U. Koren, *Appl. Phys. Lett.* **52**, 1428–1430, 1988.
18. S.A. Schwartz, P. Mei, T. Venkatesan, R. Bhat, D.M. Hwang, C.L. Schwartz, M. Koza, L. Nazar, and B.J. Skromme, *Appl. Phys. Lett.* **53**, 1051–1053, 1988.
19. D.G. Deppe, L.J. Guido, N. Holonyak, Jr., K.C. Hsieh, R.D. Burnham, R.L. Thornton, and T.L. Paoli, *Appl. Phys. Lett.* **49**, 510–512, 1988.
20. J.D. Ralston, S. O'Brien, G.W. Wicks, and L.F. Eastman, *Appl. Phys. Lett.* **52**, 1511–1513, 1988.
21. P. Mei, H.W. Yoon, T. Venkatesan, S.A. Schwartz, and J.P. Harbison, *Appl. Phys. Lett.* **50**, 1823–1825, 1987.
22. J. Werner, E. Kapon, A.C. Von Lehmen, R. Bhat, E. Colas, N.G. Stoffel, and S.A. Scwartz, *Appl. Phys. Lett.*, **53**, 1693–1695, 1988.

23. J.H. Marsh, S.I. Hansen, A.C. Bryce, and R.M. De La Rue, Optical and Quantum Electron. **23, p. S941–, 1991.**
24. E.V.K. Rao, A. Hamoudi, Ph. Krauz, M. Juhel, and H. Thibierge, *Appl. Phys. Lett.*, **66**, 472–474, 1995.
25. A. Wakatsuki, H. Iwamura, Y. Suzuki, T. Miyazawa, and O. Mikami., *IEEE Photon. Technol. Lett.* **3**, 905–907, 1991.
26. G.J. Sonek, J.M. Ballantyne, Y.J. Chen, G.M. Carter, S.W. Brown, E.S. Koteles, and J.P. Salerno, *IEEE J. Quantum Electron.* **22**, 1015–1018, 1986.
27. J.P. van der Ziel and A.C. Gossard, *J. Appl. Phys.* **49**, 2919–2921, 1978.
28. S. Ohke, T. Umeda, and Y. Cho, *Opt. Commun.*, **56**, 235–239, 1985.
29. G.M. Alman, L. A. Molter, H. Shen, and M. Dutta, *IEEE J. Quantum Electron.* **28**, 650–657, 1992.
30. Y. Suzuki, H. Iwamura, T. Miyazawa, A. Wakatsuki, and O. Mikami, *IEEE J. Quantum Electron.* **32**, 1922–1931, 1996.
31. J.Y. Chi, X. Wen, Emil S. Koteles, and B. Elman, *Appl. Phys. Lett.* **55**, 855–857, 1989.
32. P.A. Kirkby, P.R. Selway, and L.D. Westbrook, *J. Appl. Phys.*, **50**, 4567–4579, 1979.
33. L. Thylen, *Opt. Quantum Electron.*, **15**, 433–439, 1983.
34. M.D. Feit and J.A. Fleck, Jr., *J. Opt. Soc. Am.*, **A7**, 73–79, 1990.
35. J. Cibert, P.M. Petroff, D.J. Werder, S.J. Pearton, A.C. Gossard, and J.H. English, *Appl. Phys. Lett.*, **49**, 223–225, 1986.
36. T.E. Schlesinger and T. Kuech, *Appl. Phys. Lett.*, **49**, 519–520, 1986.
37. H.C. Casey, Jr. and M.B. Panish, Academic Press, San Diego, 1978.
38. K. Nakashima, Y. Kawaguchi, Y. Kamamura, H. Asahi, and Y. Imamura, *Japan. J. Appl. Phys.*, **26**, L1620–L1622, 1987.
39. K. Mukai, M. Sugawara, and S. Yamazaki, *J. Crystal Growth*, **115**, 433–438, 1991.
40. K. Utaka, Y. Suematsu, K. Kobayashi and H. Kawanishi, *Jpn. J. Appl. Phys.*, **19**, L137–L140, February, 1980.
41. M. Born and E. Wolf, Principles of Optics, 6th edn. Oxford: Pergamon Press, pp. 705–708, 1987.

CHAPTER 14

Broadspectrum InGaAs/InP Quantum Well Infrared Photodetector via Quantum Well Intermixing

DEEPAK SENGUPTA*, YIA-CHUNG CHANG AND GREG STILLMAN

Department of Electrical & Computer Engineering and Department of Physics, University of Illinois at Urbana-Champaign, 1406 W. Green Street, Urbana, IL 61801, USA

Abstract

We have demonstrated red shifting and broadening of the wavelength response of a bound-to-continuum ultra-thin p-type InGaAs/InP quantum well infrared photodetector (QWIP) after growth via quantum well intermixing. A substantial bandgap blue shift, as much as 292.5 meV at 900°C have been measured and the value of the bandgap shift can be controlled by the anneal time. Compared to the as-grown detectors, the peak spectral response of the intermixed detector was shifted to longer wavelengths without any major degradation in the responsivity characteristics. In general, the overall performance of the intermixed QWIP has not dropped significantly, with the spectral broadening taken into account. Thus, the post-growth control of the

---

*Presently with the Jet Propulsion Laboratory, Pasadena, CA 91109, USA.

quantum well composition profiles offers unique opportunities to fine tune various aspects of a photodetector's response. Theoretical modeling of the intermixing effect on the energy levels is performed based on the effective bond-orbital method and obtain a very good fit to the photoluminescence data.

KEY WORDS: red shift, peak response, multiple quantum well, dark current characteristics, absolute response, quantum efficiency, quantum well intermixing, and rapid thermal annealing.

1 INTRODUCTION

Since the first proposal of the semiconductor multiple quantum wells (MQW's) by Esaki and Tsu [1], there have been extensive theoretical and experimental studies on this subject. To date, lattice-matched MQW's including GaAs-$Al_xGa_{1-x}As$, InAsGaSb, $In_xGa_{1-x}As$-InP, $Ga_{1-x}In_xAs$-$Al_{1-x}Ga_xAs$ can be grown by molecular beam epitaxy (MBE) with nearly perfect interface quality [2–5]. Good quality strained-layer superlattices [6] have recently been successfully grown. These include Ga(As,P), Ga(As,Sb), (Al,Ga)Sb, (In,Ga,Al)As, and Si-Ge systems [7–9]. Detailed information about the electronic, optical, and transport properties of semiconductor MQW's has been accumulated via various experimental techniques [10–14]. Theoretically, many sophisticated methods have been used to calculate the electronic band structures and optical properties of these systems. These include psuedopotential [15,16] tight-binding [17–19], effective mass [20–22], k.p [23], wannier-orbital [24], and bond-orbital [25] methods. With all of these efforts, fundamental physical phenomena such as excitonic excitation, electron and hole tunneling, impurity states, quantum stark effect, lattice vibrations, electron-phonon coupling, inter-subband transitions, and quasi-two dimensional plasma are more or less understood. With ample knowledge about the fundamental physical phenomena available to us, we have reached the stage where the combination of experimental characterization with detailed theoretical analysis can now provide meaningful engineering design of quantum well devices that will find immediate application. Some of the more successful applications include quantum-well lasers [26–28], optical switches [29–31], photodetectors [32–34], and optical modulators [35–37].

In recent years, considerable interest has focused on the fabrication and characterization of III-V based quantum well infrared photodetector. [38]. Modern photoconductive quantum well infrared photodetectors, with responses at various wavelengths from 5–12 μm, were first demonstrated by B.F. Levine et al. in 1987 [39]. A schematic of the type I quantum well structure is shown in Figure 1. In most QWIP devices, a signal is generated

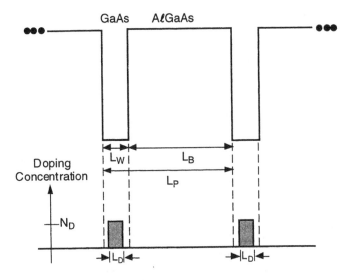

L_W = Well width (40Å)

L_B = Barrier width (300Å)

L_P = Width of one period of the structure (340Å)

L_D = Width of intentionally doped region in the quantum well (center 20Å of quantum well)

N_D = Doping concentration intentionally introduced in the quantum well over a region of width L_D (1.0 x 10^{18}/cm³)

Fig. 1. Schematic drawing of the type I quantum well structure. The parameters for the standard structure are indicated above.

when a quantum well absorbs infrared radiation. This excites electrons from the ground state to an excited state from which the excited electron is collected by an applied field, resulting in photocurrent. The use of these materials rather than conventional mercury cadmium telluride(MCT) detector is based upon two advantages:

First, these detectors are fabricated from GaAs/Al$_x$Ga$_{1-x}$As quantum well superlattices, a significantly less expensive, more uniformly grown and more abundant material than MCT. Second, standard high yield GaAs integrated circuit fabrication techniques may be used, paving the way for advanced sensor arrays and their cost effective manufacture with existing infrastructures. The price paid for the easy fabrication of QWIPs is that they have lower quantum efficiency and higher dark current than ideal narrow gap detectors such as MCT [40]. The many fabrication problems in MCT make it generally unsuitable for a large array technology or monolithic integration. Here is where QWIPs find their niche. The GaAs/Al$_x$Ga$_{1-x}$As QWIP is easier to fabricate, less expensive to grow, and non uniform across large area two-dimensional (2-D) arrays at low cost. Long-wavelength imaging systems such as the 15 μm cut-off 128 $\times$ 128 QWIP FPA camera, 256 $\times$ 256 Portable QWIP FPA camera, 9 μm cut off 640 $\times$ 436 QWIP FPA camera, and 256 $\times$ 256 Palm size QWIP FPA camera have been demonstrated by the Jet Propulsion Laboratory and achieved an excellent imagery with a noise equivalent differential temperature of 30 mK [41]. Superlattice detectors are therefore extremely versatile due to the nearly limitless possibilities of bandgap engineering (three fundamental structures: bound-to-bound, bound-to-continuum, and bound-to-miniband). Besides the successful GaAs/Al$_x$Ga$_{1-x}$As superlattice, other III-V materials have indicated promise as LWIR detectors. Incremental engineering advancements are quickly changing the perspective on QWIP, towards one of great hope, as its successor to MCT in LWIR array technology. Cost issues affect the development and function commercialization of such technology.

In spite of the successful developments of intersubband GaAs/Al$_x$Ga$_{1-x}$As QWIP technology, only little effort has focused on broadspectrum / multi-color detection using the InGaAs/InP material system [42]. The InGaAs/InP material system [43] has the advantage of binary InP barriers which have the potential for improved barriers quality and transport. This implies a lower dark current relative to the GaAs/Al$_x$Ga$_{1-x}$As material system which has ternary barrier employing aluminium. Also due to the high quality InP barriers, lattice matched InGaAs/InP ternary may have superior properties compared to GaAs/AlGaAs QWIPs. Apart from the excitation of electrons, using holes in the InGaAs/InP system offers three potential advantages: The first is a reduction in the dark current. The larger mass of the holes in the InGaAs/InP system relative to the mass of electrons, leads to lower tunneling probability and therefore lower dark currents. The second advantage is that a broader spectral response can be achieved. In the plane of the well, the energy-momentum relation for holes is not a single parabola, as in the case of electrons, and therefore absorption from one subband to another does not occur at the same transition energy for all hole moments. This results in a broadening of the absorption spectrum which is beneficial for broad spectral imaging. Finally, most of the GaAs/AlGaAs bandgap discontinuity is in the conduction band, whereas in InGaAs/InP p-QWIP intersubband absorption can occur at a much shorter wavelength.

Modifying the layer (i.e., quantum well and/or barrier width) thickness will change the absorption wavelengths of the raw material. In this manner, one can grow detectors which are tuned for absorption at various wavelengths. This method, however, relies on a separately grown superlattice, possibly on different substrates for each different absorption peak wavelength desired. An alternative method to achieving variation in absorption peaks which can easily be employed is the technique known as impurity-free vacancy disordering. In this case, a dielectric cap layer is deposited on top of the structure followed by a rapid thermal anneal process [44–47]. Such tunability is of potential interest for infrared detector applications as it would facilitate the fine-tuning of the peak detector response of as-grown structures to a desired operating wavelength. In addition, rapid thermal annealing can be used to broadened spectral responses, a direct result of the linear super position of adjacent absorption maxima and as well as multiple-color detections by shifting each detector by different amount.

In this article, we describe the growth and characterization of 4.55 μm p-doped InGaAs/InP QWIPs and the optical and device studies of the effect of intermixing on an ultra-thin (~ 10 Å) p-type InGaAs/InP multiple quantum well structures using SiO_2 caping. Finally, theoretical calculations based on the effective band-orbital method is performed to compare with the experimental results. We first present the details of our experiments and measurements procedures. Next, a brief discussion of our theoretical calculations is conducted, and the model is compared to experimental data.

2 GROWTH AND MATERIAL CHARACTERIZATION OF AN ULTRA-THIN P-TYPE INGAAS/INP QWIP

Epitaxial growth was performed in a modified Perkins-Elmer 430P gas source molecular beam epitaxy (GSMBE) system [48]. The growth chamber was equipped with a 5000 ℓ/s cryopump and a 200 ℓ/s turbo molecular pump. Cracked AsH_3 and PH_3 (>99%) were used as the group V sources, while elemental solid sources in effusion cells were used for the group III sources. Cracked AsH_3 and PH_3 were injected separately from two independent low-pressure crackers equipped. Switching of ASH_3 and PH_3 was controlled by the fast vent/run valves in front of each injector. Details of the MBE system and cracker design are reported elsewhere [48]. p-type InGaAs/InP QWIP structure were grown on semi-insulating (001) InP substrates at 500°C. InP was grown at a rate of 0.59 μm/hr with a PH_3 flow of 2–5 sccm, and lattice-matched $Ga_{0.47}In_{0.53}As$ was grown at 1.05 μm/hr with an AsH_3 flow of 2.5 sccm. Growth chamber pressure with hydride gas flowing was 2×10^{-6} torr as measured by a cold cathode gauge. The pressure reached the gauge baseline value of $<10^{-7}$ torr approximately 7 seconds after hydride gas flow was switched from the growth chamber into the vent lines. The p-type structure shown in Figure 2 consisted of 30 periods of 10 Å Be center

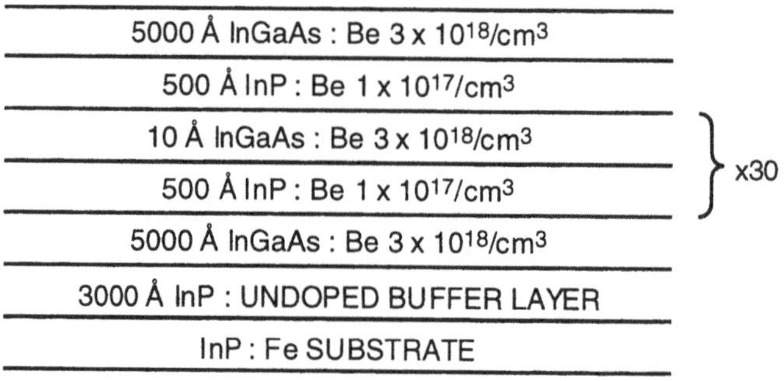

5000 Å InGaAs : Be 3 x 10^{18}/cm³
500 Å InP : Be 1 x 10^{17}/cm³
10 Å InGaAs : Be 3 x 10^{18}/cm³
500 Å InP : Be 1 x 10^{17}/cm³
5000 Å InGaAs : Be 3 x 10^{18}/cm³
3000 Å InP : UNDOPED BUFFER LAYER
InP : Fe SUBSTRATE

$\left. \right\}$ x30

Fig. 2. Structure of InGaAs/InP p-type QWIP.

doped (3×10^{18} cm^{-3}) In$_{0.53}$Ga$_{0.47}$As quantum wells (QWs) and 500 Å Be doped (1×10^{17} cm^{-3}) InP barriers sandwiched between 5000 Å Be doped (3×10^{18} cm^{-3}) In$_{0.53}$Ga$_{0.47}$As contacts on InP substrate. The material quality of the p-type InGaAs/InP quantum wells were investigated using double crystal X-ray diffraction and cross-sectional scanning tunneling microscopy. In the case of p-type InGaAs/InP QWIPs, for each type of interface, normal (InGaAs-on-InP) and inverted (InP-on-InGaAs), a different source switching scheme was used [48]. A two-step procedure was implemented for InGaAs-on-InP interface. Growth of InP was terminated by closing the In shutter and switching the PH$_3$ flow from the growth chamber to a vent line. Thus the InP surface remained free of any flux for fixed period of time, T_P seconds. Then AsH$_3$ was introduced into the growth chamber exposing the InP surface to As$_2$ overpressure for T_{As} seconds before the growth of InGaAs, providing longer P$_2$ pumping time to remove the phosphorous while preserving the InP surface from degradation. Growth was resumed after the completion of a second time period by opening the Ga and In shutters. Growth of InGaAs was terminated by closing the In and Ga shutters. Simultaneously, the flow of AsH$_3$ was directed from the growth chamber to the vent line. Thus, no flux or overpressure was applied to the InGaAs surface to T_{As} seconds. During this time, residual arsine was pumped from the growth chamber. Growth was resumed after the completion of the T_{As} interval by opening the in shutter and flowing PH$_3$ into the growth chamber. Various combinations of the three times $T_P/T_{As}/T_{As}$ were applied to the growth of the samples, and the optimized times 7/10/20 were established based on the photoluminescence and DCXRD measurements.

DCXRD measurements were carried out to characterize the crystalline quality of the p-type InGaAs/InP superlattice layer (SL). A Phillips double-crystal X-ray diffractometer was used to measure the diffraction patterns.

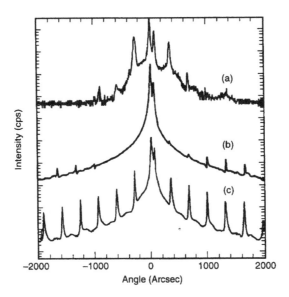

Fig. 3. High resolution double-crystal x-ray diffraction rocking curves of a 30-period InGaAs/InP QWIP structure. (a) Experimental rocking curves of the p-type QWIP structure (b) Simulation assuming abrupt interface for both InGaAs-on-InP and InP-on-InGaAs interfaces; (c) Simulation assuming a monolayer of $In_{0.5}As_{0.5}P$ inserted at each inverted interface.

Each of the diffraction patterns were measured in a $\theta - 2\theta$ mode with a fine collimation of the incident and diffracted X-rays in order to track the evolution of its superlattice harmonics over several degrees. Figure 3 (a) shows the experimented and simulated (004) CuKα1 rocking curves of the p-QWIP structure with 30 periods of InGaAs/InP SL (10 Å InGaAs well and 500 Å InP barrier). The superlattice satellites are observed up to the 4th order, and the average FWHM of the experimental satellite peaks is 31 arcsec., demonstrating excellent material quality. The simulation in Figure 3(b) assumed abrupt interfaces for both InGaAs-on-InP and InP-on-InGaAs interfaces, and that in Figure 3(c) assumes a monolayer of $InAs_{0.5}P_{0.5}$ inserted at the upper interfaces of the InGaAs layer. Clearly, a transitional monolayer in the group V sublattice gives a much better agreement than the abrupt interface does between experiment and simulation, suggesting residual As_2 incorporation in the InP layers and formation of strained monolayers at the interfaces. In this case, a period consists of three layers: 10.5 Å $In_{0.475}Ga_{0.525}As$, 3.04 Å $InAs_{0.5}P_{0.5}$ and 564.3 Å InP and to atomic layer schematic can be expressed as In-P-In-P*-GaIn-As-GaInAs-GaIn-As*-In*AsP*-In-P-In-P.

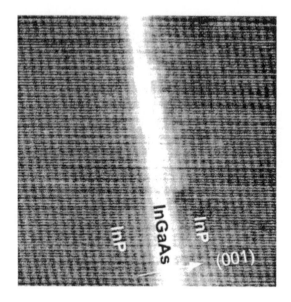

Fig. 4. Cross-sectional scanning tunneling microscopy (XSTM) of a p-type InGaAs/InP QWIP with about 4 to 5 monolayers within quantum well.

Figure 4 shows cross-sectional scanning tunneling microscope (XSTM) image of InP/InGaAs p-QWIP structure with ultra thin quantum wells [49]. The InP barrier appears darker than the adjacent InGaAs well region because of its valence band offset. The InGaAs regions also have a mottled appearance due to alloy fluctuations. A marked difference in roughness between the normal and inverted interface was observed. As shown in Figure 4, the normal interface is extremely sharp, with the intermixing within a monolayer in the [001] direction. The roughness observed at the inverted interface can be attributed to:

1. Alloy fluctuations within the InGaAs.
2. The exchange of group V species. Due to the large chemical binding energy difference between As and P compounds, the exchange and diffusion of group V species is expected during the growth interruption. Because of larger binding energy (GaP > GaAs), the exchange and diffusion of P into InGaAs is likely and the diffusion rate of P is typically much higher than that of As. DCXRD measurements on InGaAs/InP (GaAs/InGaP) interfaces suggest partial exchange of As with P within the first 3–4 InGaAs (GaAs) monolayers during exposure to a P_2 beam.

3. The residual As_2 incorporated into the InP layers. Since As_2 has a larger incorporation coefficient than P_2, even a little amount of residual As_2 in the growth chamber will cause the carryover of As into InP layer, forming InAs(P) layer. As shown in Figure 4, higher density of As and scattered As atoms were observed at the inverted interface and in the first 5–6 InP monolayers, respectively. In addition to providing direct evidence of well-defined quantum wells, barriers, and interfaces, the XSTM results allowed independent determination of the thicknesses for later comparison to results derived from DCXRD rocking curve fits calculated using dynamical theory.

3 RED-SHIFTING AND BROADENING VIA QUANTUM WELL INTERMIXING

The bandgap tuning of multi-quantum well structures requires changing the bandgap spatially across the wafer. One way to reach the goal is to use the technique of intermixing via rapid thermal annealing. This process modifies the geometry and composition of QWs and leads to a blue shifted bandgap. The post-growth wavelength shifting of QWIPs by rapid thermal annealing is accomplished by dielectric encapsulation of the QWIP and exposing it to a high temperature for a short period of time. The quantum well is changed from a square well with a sharp interface, to an error-function shaped well with a corresponding change in the confined energy levels. The redistributed profile makes the quantum well thinner at the bottom than to originally grown well, shown in Figure 5(a).

This modification of the well width shifts the absorption to longer wavelengths as the energy level rises in the well. Dielectric film stress, transferred to the surface of the substrate, influences the amount of diffusion induced wavelength shifting which will occur during annealing. Deposition may take place via chemical vapor deposition (CVD), plasma enhanced chemical vapor deposition (PECVD), or sputtering. Substrate temperature, reactant gas composition, and for PECVD, microwave power applied to the plasma will change the film characteristics. Films such as silicon dioxide and silicon nitride, with their myriad of compositions and mixtures should be adequate to provide a wide intermixing of the interface in bit layer chambers. The annealing operation may be performed via rapid thermal annealing (RTA). Once the anneal is complete, the dielectric cap layer may be stripped and the wafers prepared for subsequent detector processing. Broad-spectrum/multi-color detection makes use of the variation in wavelength shift for different dielectric induced stress Figures 5(b), 5(c) and 5(d) illustrates that the broad spectrum response is a direct result of the

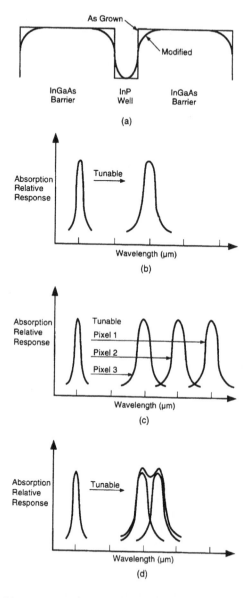

Fig. 5. (a) Modification of quantum well profile due to QW-intermixing. (b) Schematic representation of shifting effect and its utility for response detector tuning. (c) Multi-color detection by shifting each pixel a different amount. (d) Schematic representation of a broadband detector made possible through the linear superposition of the relative absorption response of two neighboring peaks.

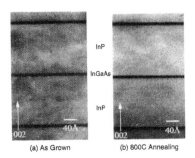

(a) As Grown   (b) 800C Annealing

Fig. 6. Cross-sectional TEM of the (a) as-grown, and (b) RTA (800°C for 30s) MQW structures.

linear super position of two adjacent absorption maxima. In all studies, prior to annealing the samples were first degreased in trichlorethane, acetone, and methanol followed by a light surface etch using NH_4OH. Then, a 1500 Å SiO_2 encapsulant was deposited by plasma enhanced chemical vapor deposition. Rapid thermal annealing was performed in an AET RTA reactor with 10 sccm of N_2 flowing. The temperature was stabilized at 200°C prior to the high temperature annealing.

The cross-sectioned TEM micrographs are shown in Figure 6 for (a) an as-grown MQW and (b) an RTA MQW annealed at 800°C for 30 seconds. No defects or dislocations are observed for both as-grown and annealed MQW regions.

In Figure 7, the 6K PL spectrum of as-grown and annealed InGaAs/InP QWIP samples are shown. A blue shift (~ 24.0 meV @ 700°C, ~ 130.2 meV @ 800°C and ~ 292.5 meV @ 900°C) of photoluminescence peak was observed. This is expected, as the grown state is higher and the effective barrier height lower for the annealed QWIP than the as-grown QWIP. In addition to the blue shift, the annealed QWIP samples also exhibit a reduction in the peak luminescent intensity and is attributed to the out-diffusion of the p-type (Be) dopants from the well and is strongly dependent on the amount of disordering during the annealing process. Rapid thermal annealing at 900°C leads to a blue shift of ~ 292.5 meV and almost complete alloying. In Figure 8, the energy shift of the luminescence line is displayed against the corresponding annealing temperature for both as-grown and rapid thermally annealed MQW investigated. The change of the emission energy is very pronounced. It is ~ 10 Å quantum well, a small narrowing of the effective width causes a significant blue-shift due to the well width dependence of the first subband energies. In Figure 9, 6K PL spectrum of the as-grown and annealed (800°C) InGaAs/InP QWIP samples at 20, 40, and 60 sec are shown.

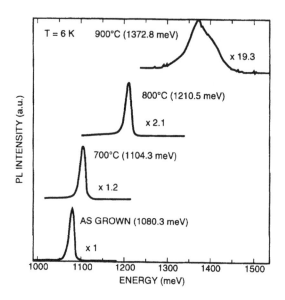

Fig. 7. Photoluminescence spectra at 6K of the as-grown and RTA (700°C, 800°C, and 900°C for 30s) MQW structures.

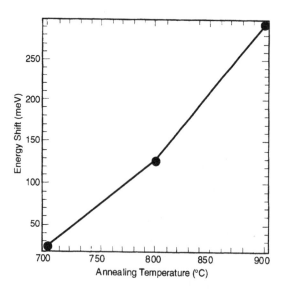

Fig. 8. Energy shift of the photoluminescence peak versus RTA (700°C, 800°C, and 900°C for 30s) MQW structures.

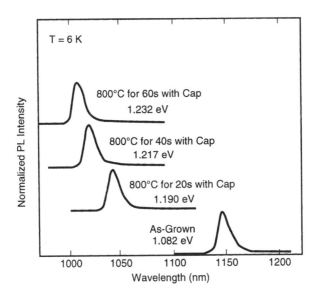

Fig. 9. 6K photoluminescence spectra of the as-grown and RTA treatment (800°C for 20, 40, and 60s).

A blue shift ($\sim$ 108 meV for 20 sec, $\sim$ 135 meV for 40 sec and $\sim$ 150 meV for 60 sec) of the photoluminescence peak was observed. Figures 10 and 11 show the integrated PL intensities and half width of the emission line versus the annealing temperatures. Up to an annealing temperature of 800°C, the half width of the luminescence spectrum remain approximately 19.3 to 19.6 meV and results in no significant degradation of the PL linewidth up to 800°C rapid thermal annealing.

In summary, we have achieved strong and homogeneous intermixing in InGaAs/InP MQWs, using SiO_2 capping and subsequently annealing. No significant degradation of optical properties was observed up to 800°C. A substantial PL blue shift, as much as 130.2 meV, was found in the structure and the value of the shift can be controlled by its anneal time. In addition, beyond 800°C annealed QWIP structure also exhibits a reduction in peak luminescent intensity, which may be due to the overall broadening of the peak response as well as any defects the annealing process might have introduced.

4 DEVICE FABRICATION AND CHARACTERIZATION

QWIPs were fabricated into 200 μm diameter mesa diodes by chemically etching and evaporating Ti/Pt (p-type) contacts. Devices were polished with

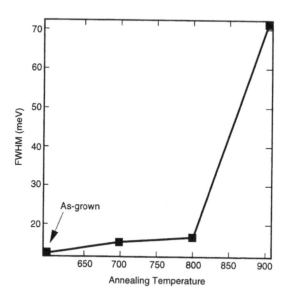

Fig. 10. 6K photoluminescence linewidth of the as-grown and RTA (700°C, 800°C, and 900°C for 30s) MQW structures.

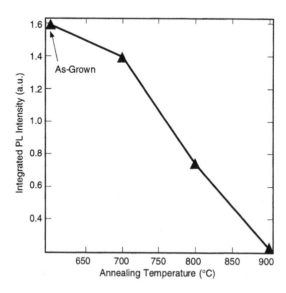

Fig. 11. 6K photoluminescence integrated PL intensities of the as-grown and RTA (700°C, 800°C, and 900°C for 30s) MQW structures.

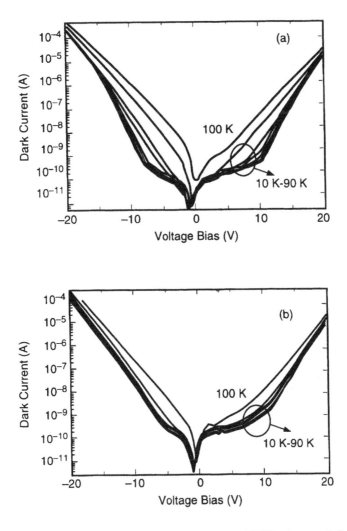

Fig. 12. Dark current versus bias for temperatures from 10 to 100 K for the p-type InGaAs/InP QWs with (a) full cold shielding and (b) 2p Sr 300 K background.

45° bevels (to accommodate selection rules) and indium bonded to a copper heat sink. The I-V measurements were done with a Keithley source measure unit and a two-stage closed cycle refrigerator. The sample was fully cold shielded. The detector response spectrum was taken with a Bomem Fourier Transform Spectrometer. The responsivity calibration was made using a 1000 K blackbody source and a narrow band pass filter.

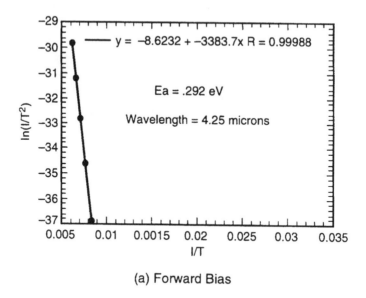

(a) Forward Bias

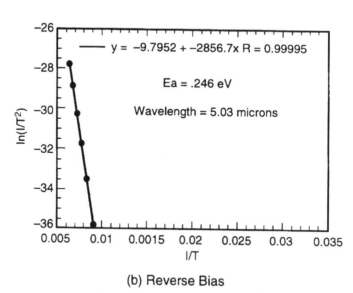

(b) Reverse Bias

Fig. 13. Arrhenius plot of a p-type InGaAs/InP QWIP under (a) forward and (b) reverse biases. The activation energies correspond to wavelengths of 4.25 and 5 μm, respectively.

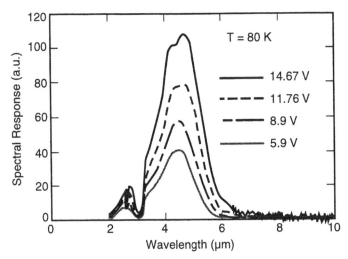

Fig. 14. Bias dependence of the spectral response for the as-grown QWIP at 80 K.

The temperature dependent current-voltage curves both with full cold shielding and with 2π Sr 300 K background of the p-type InGaAs/InP QWIP are shown in Figures 12(a) and (b), respectively. The dark current was observed to be insensitive to temperature changes up to 60 K. The dark current decreased significantly as the temperature was lowered due to the decrease of thermionic emission of carriers from the QWs at higher temperatures. It is also worth noting that for these QWIPs, the positive bias direction (defined mesa top positive) is the low dark current direction. That is, for both cases, the lowest dark current direction is composed of carriers flowing into the substrate.

In Figures 13(a,b), on a log scale, we plot the normalized dark current (Id/T) measured at a bias of 50 mV, against the inverse temperature for the sample with peak response of 4.55 μm for the p-type InGaAs/InP. The excellent fit over several orders of magnitude in current yields activation energy E_a corresponding to wavelength 4.25 μm and 5.0 μm for the p-type InGaAs/InP. This is in good agreement with the optically measured peak response. This demonstrates that there is negligible tunneling contribution to the dark current, and thus the barriers are nearly ideal in their current blocking.

The bias dependence of the 45° incidence responsivity is shown in Figure 14 for the p-type InGaAs/InP. An increase in response is observed with bias voltage and the peak wavelength shifts slightly with bias. The increase in response with bias is attributed to the step-like increase in the number of QWs included in the high field bias regime.

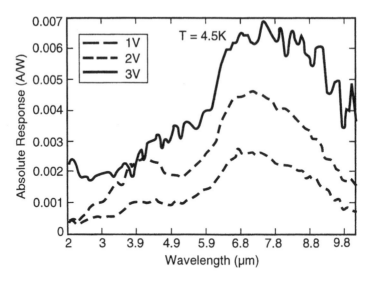

Fig. 15.   Bias dependence of the spectral response measured for the RTA QWIP at 4.5 K.

Figures 14 & 15 show the photoresponse for both the as-grown and RTA detector with bias. The peak response wavelengths measured are ~4.55 μm (80 K and 5.9 V) for the as-grown detector and ~7.0 μm (4.5 K and 4 V) for the RTA detector. Since the ground state is higher and the effective barrier height is lower in the RTA detector than in the as-grown detector, the peak response wavelength of the RTA detector experiences a long wavelength shift. The peak absolute responses are calculated to be ~2.5 mA/W (80 K and 5.9 V) for the as-grown detector and ~2.0 mA/W (4.5 K and 4 V) for the RTA detector from the blackbody and relative spectral response measurements. The peak responsivity of the RTA detector is of a similar magnitude compared to the as-grown detector. Although the broadened absorption spectrum of the RTA detector can result in a reduced spectral response, we believe that the dominant reduction is a consequence of the out-diffusion of the Be dopant from the well and the increased dark current from the RTA detector structure. The small reduction in the response may still be acceptable for the focal plane array detector applications.

5   COMPARISONS WITH THEORY

Theoretical modeling of the intermixing effect on the energy levels is performed based on the effective bond-orbital method (EBOM). This method casts the k.p formalism into a local-orbital formalism so that the complicated boundary conditions in heterostructures can be easily managed. Details of this method is given in Ref. [50]. This method has been used to calculate the electronic structures and optical responses of many kinds of multiple quantum wells (MQW's), including GaAs-$Al_xGa_{1-x}As$ [50], InAs-GaSb [50], HgTe-CdTe [51], and GaInAs-InP [52] with good success. Here we are interested in the intermixing effect in $In_{0.53}Ga_{0.47}As/InP$ MQWs. Both Group III (Ga) and Group V (P) interdiffusion will be considered. Prior to the interdiffusion, the system is strain-free. After the interdiffusion, there is a weak strain distribution. Furthermore, with the interdiffusion, the system turns into a InGaAsP quaternary compound with compositional modulation. The strain distribution is related to the lattice mismatch by

$$\varepsilon_\|(z) = (a(z) - a_{InP})/a_{InP}$$

$$\varepsilon_{xx} = \varepsilon_{yy} = \varepsilon_\|, \quad \varepsilon_{zz} = -(2C_{12}/C_{11})\varepsilon_\|,$$

where C_{12} and C_{11} are the elastic constants which are position dependent. The effect of strain on the Hamiltonian will be modeled according to the Bir-Picus theory [53]. The strain Hamiltonian takes a diagonal form [in the EBOM basis labeled by $(J, M) = (1/2, 1/2),)(1/2, -1/2), (3/2, 3/2), (3/2, 1/2), (3/2, -1/2), (3/2, -3/2)]$ with the diagonal matrix elements given by

$$\Delta V_C, \Delta V_c, -\Delta V_H + D, -\Delta V_H - D, -\Delta V_H - D, \quad \text{and} \quad -\Delta V_H + D,$$

where

$$\Delta V_C = 2c_1(1 - C_{12}/C_{11})\varepsilon_\|$$

$$\Delta V_H = 2(a_1 + a_2)(1 - C_{12}/C_{11})\varepsilon_\|$$

and

$$D - b(2C_{12}/C_{11} + 1)\varepsilon_\|$$

The deformation potentials $(c_1 + a_1 + a_2)$ and b for InP, GaAs, and InAs can be found in Ref. [54]. For the alloys, we simply use the linear interpolation.

The input band parameters used in our modeling are listed in Table 1. Here E_v and E_c are the zone-center energies for the valence band and conduction band, respectively. Note that we use the experimental bandgap of bulk $In_{0.53}Ga_{0.47}As$ as the input parameter rather than the linear interpolation of band gaps of InAs and GaAs, since the bandgap bowing effect is significant for InGaAs alloy.

The model system for the as-grown sample is a superlattice with each period consisting of 4 monolayers (11A) of $In_{0.53}Ga_{0.47}As$, one monolayer of $InAs_{0.5}P_{0.5}$ and 169 monolayers (497A) of InP. The presence of one monolayer of $InAs_{0.5}P_{0.5}$ is suggested by the DCXRD analysis. Including this $InAs_{0.5}P_{0.5}$ monolayer at one interface between $In_{0.53}Ga_{0.47}As$ and InP in our model also makes the resulting energy gap (1.102 eV) in better agreement with the PL measurement (1.082 eV). Without this $InAs_{0.5}P_{0.5}$ monolayer, the energy gap obtained in our model would have been 0.05 eV higher. The difference of 20 meV between the theoretical band gap predicted here and the PL peak position can be attributed to the excitonic effect (which accounts for approximately 10 meV) and the effects due to difference in the realistic geometry and the ideal geometry used here. With intermixing, the system turns into a $In_{1-x}Ga_xAs_{1-y}P_y$ quaternary superlattice with both x and y being functions of z (the coordinate in growth direction). Since x is close to 0.47 in the well region and 0 in the barrier region, we obtain the EBOM parameters for $In_{1-x}Ga_xAs_{1-y}P_y$ by taking the linear interpolation between the corresponding EBOM parameters for $In_{0.53}Ga_{0.47}As_{1-y}P_y$ and $InAs_{1-y}P_y$ via the following relation

$$V(In_{1-x}Ga_xAs_{1-y}P_y) = (x/0.47)V(In_{0.53}Ga_{0.47}As_{1-y}P_y)$$
$$+ (1 - x/0.47)V(InAs_{1-y}P_y),$$

where $V(In_{0.53}Ga_{0.47}As_{1-y}P_y)=(1-y)V(In_{0.53}Ga_{0.47}As)+yV(In_{0.53}Ga_{0.47}P)$ and

$$V(InAs_{1-y}P_y) = (1-y)V(InAs)+yV(InP).$$

The composition (x or y) distribution is modeled by a simple diffusion theory [55]. For Group III (Ga) interdiffusion, we obtain

$$x = 0.47\{erf[(L/2 + z)/2\sqrt{D_1t}] + erf[(L/2 - z)/2\sqrt{D_1t}]\}/2$$

and for Group V (P) interdiffusion, we have

$$y = 1 - \{erf[(L/2 + z)/2\sqrt{D_2t}] + erf[(L/2 - z)/2\sqrt{D_2t}]\}/$$
$$2 - 0.5\{erf[(a_1/2 + z_1)/2\sqrt{D_2t}] + erf[(a_1/2 - z_1)/2\sqrt{D_2t}]\}/2,$$

where D_1 and D_2 are the diffusion constants for describing Ga and P interdiffusions, respectively and t is the annealing time. L is the width of the quantum well and z is distance measured from the center of the well. The second term in the expression for $y(z)$ describes the interdiffusion due to the one monolayer $InAs_{0.5}P_{0.5}$ at the interface. In this term, a_1 denotes the thickness of the monolayer and z_1 is the distance measured from the center of the $InAs_{0.5}P_{0.5}$ monolayer. The diffusion constant is determined by fitting the calculated band gap energies to the experimental results. With $D_1 = 7.03 \times 10^{-17}$ cm²/s and $D_2 = 3.92 \times 10^{-17}$ cm²/s we obtain a very good fit to the PL data as shown in Figure 16.

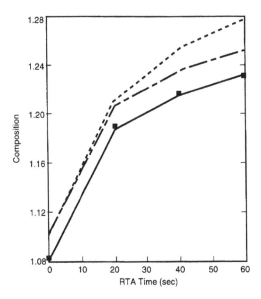

Fig. 16. Calculated bandgap as a function of the annealing time at 800°C. Dashed curve: with both P and Ga interdiffusion. Dotted curve: with P interdiffusion only. Filled squares: 6K photoluminescence data. The second solid line is just the first dashed curve down-shifted by 20 meV to match the experiment.

In Figure 16, the dotted line shows the calculated band gap as a function of the annealing time with Group V (P) interdiffusion alone and the dashed line shows the results with both Group III and Group V interdiffusions. To compare with the PL data, we have rigidly shifted the calculated band gap by 20 meV downward so that the calculated results for as-grown sample agree with the PL data. We found that the Group III (Ga) interdiffusion has much smaller effect on the band gap shift compared with the Group V (P) interdiffusion. This also means that the diffusion constant for Group III interdiffusion obtained by fitting the PL data is subject to a much larger error. With the diffusion constants used, the resulting composition distributions for $x(z)$ (solid curves) and $y(z)$ (dashed curves) for different annealing times ($t = 20$, 40, and 60 seconds) are shown in Figure 17. Note that $y(z)$ is not symmetric with respect to z because of the contribution from the monolayer $InAs_{0.5}P_{0.5}$ at the interface.

6 CONCLUSION

In conclusion, we have demonstrated that RTA can be employed to both shift the operating wavelength and to broaden the response of an ultra-thin p-type

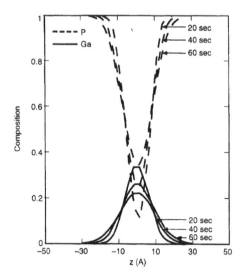

Fig. 17. P (dashed line) and Ga (solid line) compositions versus z (the coordinate in growth direction).

InGaAs/InP quantum well infrared photodetector following intermixing of the well and barrier layers during rapid thermal annealing. The use of RTA changes the well profile of a QWIP and peak wavelength, but the reduced responsivity indicates that this technique is limited for sensitive IR detectors. Recent advances in growth, complimented by innovative structures (random gratings and reflector layers) should offset any degradation in performance. This makes feasible integration of multiple-colored pixels.

Acknowledgment

Research described in this paper was performed by the Center for Space Microelectronics Technology, Jet Propulsion Laboratory, California Institute of Technology, Pasadena, CA 91109 and by the Microelectronics Laboratory, University of Illinois at Urbana-Champaign, Urbana, IL 61801. The authors would like to acknowledge Prof. N. Holonyak Jr., and the members of the Microelectronics Laboratory at the University of Illinois at Urbana-Champaign and also Drs. S.D. Gunapala, S.V. Bandara, H.C. Liu, and K.K. Choi for their advice and technical assistance. We would also like to acknowledge B. Payne (UIUC) and D. Cuda (JPL) for help with the manuscript preparation. One of the authors (D.K.S.) acknowledges the fellowship awarded by the National Aademy of Sciences-National Research Council.

References

1. L. Esaki and R. Tsu, *IBM J. Res. Develop.*, **14**, 61 (1970).
2. A.C. Gossard, P.M. Petroff, W. Wiegman, R. Dingle, and A. Savage, *Appl. Phys. Lett.*, **29**, 323 (1976).
3. E.E. Mendez, L.L. Change, C.A. Chang, L.F. Alexander, and L. Esaki, *Surf. Sci.*, **142**, 215 (1984).
4. Y.D. Galeuchel, P. Roentgen, and V. Graf, *Appl. Phys. Lett.*, **53**, 2638 (1988).
5. K.Y. Cheng, A.Y. Cho, T.J. Drummond, and H. Morkoc, *Appl. Phys. Lett.*, **40**, 147 (1982).
6. G.C. Osbourn, *J. Appl. Phys.*, **53**, 1586 (1982); G.C. Osbourn, R.M. Riefield, and P.L. Gourley, *Appl. Phys. Lett.*, **41**, 172 (1982).
7. T.P. Pearsall, F.H. Pollak, and J.C. Bean, *Bull. Am. Phys. Soc.*, **30**, 266 (1985).
8. A. Orchel, U. Cebulla, G.T. Pankle, H. Kroemer, S. Subbana, and G. Griffiths, *Surf. Sci.*, **174**, 143 (1986).
9. G. Abstreiter, H. Brugger, T. Wolf, H. Jorke, and H.J. Herzog, *Phys. Rev. Lett.*, **54**, 2441 (1985).
10. J.N. Schulman and T.C. McGill, *Appl. Phys. Lett.*, **34**, 663 (1979).
11. J.P. Faurie, A. Millon, and J. Piaguet, *Appl. Phys. Lett.*, **41**, 713 (1982).
12. X.C. Zhang, S.K. Chang, A.V. Nurmikko, L.A. Kolodziejski, R.L. Gunshor, and S. Datta, *Phys. Rev.*, **31**, 4056 (1985).
13. L.A. Kolodziejski, R.L. Gunshor, T.C. Bonsett, R. Venkatasubramanian, S. Datta, R.B. Bylsma, W.M. Becker, and N. Otsuka, *Appl. Phys. Lett.*, **47**, 169 (1985).
14. G.H. Dohler, H. Kunzel, D. Oiego, K. Ploog, P. Ruden, H.J. Stolz, and G. Abstriter, *Phys. Rev. Lett.*, **47**, 864, (1981).
15. E. Caruthes and P.J. Lin-Chung, *Phys. Rev. Lett.*, **39**, 1543 (1977).
16. D. Ninno, K.B. Wong, M.A. Gell, and M. Jaros, *Phys. Rev. B*, **32**, 2700 (1985).
17. N.N. Schulman and T.C. McGill, *Phys. Rev. Lett.*, **39**, 1680 (1977).
18. J.N. Schulman and Y.C. Chang, *Phys. Rev. B*, **24**, 4445 (1981); *Phys. Rev. B*, **27**, 2346 (1983).
19. Y.T. Lu and L.J. Sham, *Phys. Rev. B*, **40**, 5567 (1989).
20. L.L. Chang and L. Esaki, *Surf. Sci.*, **98**, 70 (1980).
21. G. Bastard, *Phys. Rev.*, **25**, 7584 (1982).
22. G.D. Sanders and Y.C. Chang, *Phys. Rev. B*, **31**, 6892 (1985).
23. C. Mailhiot, D.L. Smith, and T.C. McGill, *J. Vac. Sci. Technol.*, **B2**(3), 371 (1984).
24. D.Z.-Y. Ting and Y.C. Chang, *Phys. Rev. B*, **36**, 4357 (1987).
25. Y.C. Chang, *Phys. Rev. B*, **37**, 8215 (1988).
26. D. Ahn, S.L. Chuang and Y.C. Chang, *J. Appl. Phys.*, **64**, 4056 (1988).
27. D.A. B. Miller, D.S. Chemla, T.C. Damen, A.C. Gossard, W. Wiegmann, T.H. Wood, and C.A. Burrus, *Appl. Phys. Lett.*, **45**, 13 (1984).
28. D.S. Chemla and D.A.B. Miller, *J. Opt. Soc. Am.*, **B2**, 1155 (1985).
29. S. Schmitt-Rink and C. Ell, *J. Lumimn.*, **30**, 585 (1985).
30. B.F. Levine, G. Hasnain, C.G. Bethea, and N. Chand, *Appl. Phys. Lett.*, **54**, 27044 (1989).
31. B.F. Levine, *J. Appl. Phys.*, **74**, R1 (1993).
32. K.K. Choi, B.F. Levine, C.G. Betha, J. Walker, and R.J. Malik, *Appl. Phys. Lett.*, **50**, 1814 (1987).
33. T.H. Wood, R.W. Tkach, and A.R. Chraplyvy, *Appl. Phys. Lett.*, **50**, 798 (1987).
34. S. Weiner, D.A.B. Miller, and D.S. Chemla, *Appl. Phys. Lett.*, **50**, 842 (1987).
35. E. Efron, T.Y. Hsu, J.N. Schulman, W.Y. Wu, I. Rouse, I.J. D'Haenens, and Y.C. Chang, *Proceedings of the SPIE Conference on Quantum Well and Superlattice Physics*, Vol. **792**, p. 197 (1987).
36. D.L. Smith, T.C. McGill, and J.N. Schulman, *Appl. Phys. Lett.*, **43**, 180 (1983).
37. G.D. Sanders and Y.C. Chang, *Phys. Rev.*, **B35**, 1300 (1987).
38. S.D. Gunapala and S.V. Bandara, *Physics of Thin Films*, edited by M.H. Francombe, and J.L. Vossen, Vol. 21, pp. 113–237, Academic Press, N.Y., (1995).
39. B. Levine, *J. Appl. Phys.*, **74**, R1 (1993).
40. S.A. Lyon, *Surf. Sci.*, eightb 228, 508, (1990).

41. S.D. Gunapala, J.S. Park, G. Sarusi, T.L. Lin, J.K. Liu, P.D. Maller, R.E. Muller, C.A. Shott, and T. Hoeltz, *IEEE Trans on Electron Devices*, Vol. 44, No. 1, Jan 1997.
42. H.C. Liu, *Optical Engineering*, **33**(J), 1961–1967 (1994).
43. S.D. Gunapala, B.F. Levine, D. Ritter, R. Hamm and M.R. Danish, *Appl. Phys. Lett.*, **58**, 202–4 (1991).
44. J.D. Ralston, M. Ramsteiner, B. Dischler, M. Maier, P. Koidl, and D.J. As, *J. Appl. Phys.*, **70**, 2195 (1991).
45. D.G. Deppe and N. Holonyak, Jr., *J. Appl. Phys.*, **64**, R93 (1988).
46. J. Beauvais, J.H. Marsh, A.H. Kean, A.C. Boyle, B. Garrett, and R.W. Celew, *Electron. Lett.*, **28**, 1670 (1992).
47. C.J. McLean, J.H. Marsh, R.M. DeLaRue, A.C. Boyle, B. Garrett, and R.W. Glen, *Electron Lett.*, **28**, 1117 (1992).
48. S.L. Jackson, J.N. Ballargeon, A.P. Curtis, X. Liu, J.E. Burkes, J.I. Malin, K.C. Hseih, S.G. Bishop, K.Y. Cheng, and G.E. Stillman, *J. Val. Sci. Technol.*, **B11**, 1045 (1993).
49. W. Wu, S. Skala, J.R. Tucker, J.W. Lyding, A. Seabaugh, E.A. Beam III, and D. Joranovic, *J. Vac. Sci. Technol.*, **A13**, b02 (1995).
50. Y.C. Chang, *Phys. Rev.*, **B37**, 8215 (1988).
51. Y.C. Chang, J. Cheung, A. Chiou, and M. Khoshnerisan, *J. Appl. Phys.*, **66**, 829–834 (1989).
52. M.P. Houng and Y.C. Chang, *J. Appl. Phys.*, **65**, 3092 (1989).
53. G.L. Bir and G. E. Pikus, *Symmetry and Static Induced Effects in Semiconductors* (Halsted, United Kingdom, 1974).
54. S. Adachi and C. Hamaguchi, *Phys. Rev.*, **B19**, 938 (1979).
55. E.H. Li and W.C.H. Choy, *J. Appl. Phys.*, **82**, 3861 (1997).

CHAPTER 15

Diffused Quantum Well Modulators

WALLACE C.H. CHOY^a AND E. HERBERT LI^b

^{a} *School of Electronic Engineering, Information Technology and Mathematics, University of Surrey, Guildford, Surrey, GU2 5XH, UK*
^{b} *Department of Electrical and Electronic Engineering, University of Hong Kong, Pokfulam Road, Hong Kong*

1 INTRODUCTION

Optical communication systems offer the potential for rapid transfer of parallel data from one system to another without electromagnetic interference problems and with minimum signal dispersion (commonly observed in high-bit-rate electrical interconnects). In these systems, one of important processes is to add information onto optical carrier. One of the common ways to encode a signal to an optical beam is to use modulation devices and switches such as a self-electro-optic effect device (SEED) [1] which can produce bi-stable states in the large bit-carry communication system and a high speed signal processing.

Advanced semiconductor growth techniques such as molecular beam epitaxy (MBE) and metal organic chemical vapor deposition (MOCVD) create thin layers (down to 10 Å) of high quality III-V semiconductor materials and therefore permit the construction of lattice matched (e.g. AlGaAs/GaAs) and strained (e.g. InGaAs/InP) heterostructures known as quantum wells (QWs) [2]. These QWs can be engineered at atomic level, giving rise to new physical mechanisms to create some new devices. The excitonic electroabsorption, highlighting important features of QW for modulation devices, was demonstrated in practice in 1983 [3], by applying a reverse bias field perpendicular to a stack of QWs structure. Since then, efforts have been put into investigating excitonic electroabsorption and it applications in optical modulation provided by the quantum-confined Stark effect (QCSE) of QW structures [4–6].

In 1984 and 1985, transverse and waveguide type QW modulators were produced respectively [6,7]. The excitonic electroabsorption of QW is used to produce absorption change and thus optical intensity modulation by alternative reverse bias and unbias of the QW structures. Apart from electroabsorption change, we can also obtain electro-refractive index change at the same time. (The relation between these two changes is explained by the Kramers-Krönig relation [8].) The refractive index change can be used to produce a phase modulation [9–12] to optical signals. These two optical changes provided by QW structures are normally larger than that of the conventional bulk LiNbO$_3$ and III-V semiconductor modulator and hence increase the modulation strength. (Note that the bulk electroabsorption effect may be similarly used for better optical bandwidth. However, this is at the cost of operation voltage and electrical bandwidth.) As a consequence, these external QW modulation components to laser sources play an important role in the communication system for modulating the intensity, phase and polarization of the optical signals. Moreover, external modulators can function as a remedy to reduce penalties of direct modulated laser, that are, chirping and modulation speed subject to the direct modulation. The widest 3dB bandwidth of a directly modulated laser reported to date is 37GHz [13]. However, an ultra-high-speed (50GHz) QWs electroabsorptive (EA) modulator integrated with waveguides for 40 Gbit/s optical modulation has been developed [14]. This modulator is therefore considered as a potential candidate in developing a high-speed and low-chirp modulation [15]. In this chapter, we will focus on waveguide and transverse (vertical) type EA modulator.

Figure 1 is a schematic view of the transverse modulator in which an optical beam is modulated vertically. Together with its compatibility with the technology already used in vertical cavity lasers [16] and detectors [17], three dimensional optoelectronic integration can be realized in the future. This device has the advantage of being polarization independent with respect

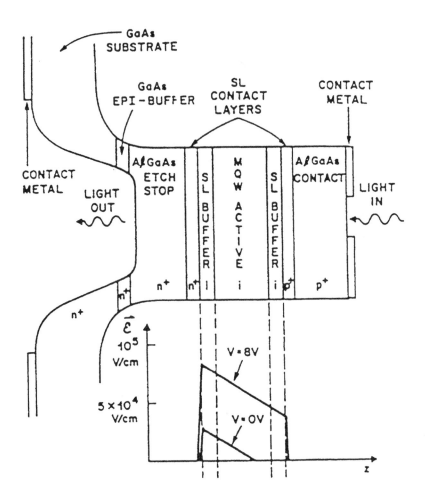

Fig. 1.   Schematic view of sample used in MQW absorption experiments. The quantum wells, in the "MQW Active", are in the center of a p-i-n device, which is operated back-biased. The electric filed profile calculated in the depletion approximation at two different applied voltage, is shown in the lower half of the figure. Light propagates perpendicular to the MQW layers. [20]

to the electrooptic materials [18]. It can also be used in multimode and single-mode systems. However, it has a limitation of short active region. This can be tackled by introducing quarter-wavelength Bragg reflectors on the top and bottom of the QW cavity structure to form Fabry Perot modulator [19], These reflectors increase the effective modulation interaction length and thus enhance the modulation performance. Another way to increase the effective interaction length of the device is to change the propagation direction of

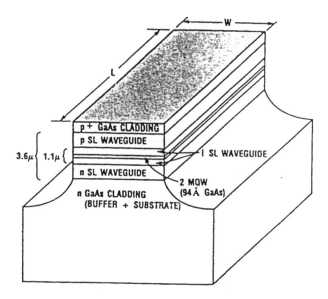

Fig. 2.   Schematic view of waveguide configuration modulator. "SL" stands for "superlattice". Two MQW's 9.4 nm thick were placed in the center of a p-i-n device. Light was confined in the plane of the layers by a waveguide. The device active interaction length (L) is 150 μm and device width (W) of 40 μm. [20]

light beam by 90° i.e. lying in the plane of the QWs layers, and to have a slab optical waveguide added to the structure to confine the light along the length of the device. This is the so-called waveguide (longitudinal) type modulator as illustrated in Figure 2, where light is confined on the plane of QW by a waveguide. In this structure, the prescribed geometric limitation between the thickness of the QWs region and the optical interaction length is no longer a problem. The interaction length can be freely adjusted to a length necessary for producing the desired ON-OFF ratio (contrast ratio) [9]. In addition, since the layers of QW can be reduced in the device, the electric field is more homogenous in the active region. However, this type of QW device presents both TE and TM modes polarization [20]. The optical electric field perpendicular to the growth direction of the QW (TE mode) has different optical properties to the electric field parallel to the growth direction of the QW (TM mode).

With the ability to control disordering in III-V semiconductor materials by using disordering technologies such as impurity-induced disordering [21], impurity-free vacancy diffusion [22], photoabsorption-induced disordering (PAID) [23] as well as impurity free interdiffusion using anodization followed by rapid thermal annealing [24] — the as-grown QW material can be modified to become a diffused quantum well (DFQW). The bandstructure and the optical parameters (such as absorption coefficient and refractive index) of the material can have post-grown modification. We can use this process as an off-the-shelf technology to modify the optical properties and to localize selected areas of the QW structures for photonic and optoelectronic integration. In order to monolithically integrate a modulator to a laser and a waveguide on the same wafer by using the technology of intermixing, DFQW modulators have been investigated since 1989 [25] in which DFQW has been developed as the active cavity to modulate a light beam. In 1988, it was reported that properties of superlattice structure could be improved by a suitable annealing [26] and in 1996–7, the intensity PL spectra of quantum wire and QW were enhanced after the interdiffusion [24,27], i.e. the optical properties of superlattice or QWs and thus the modulation performance can be improved. Moreover, due to the modification of the QW bandstructure in intermixing, the transition energy and thus the operation wavelength (λ_{op}) of the modulators can be tuned [25]. In theory, it has been shown that the modulation performance such as the absorption change of DFQW modulator can be also improved as compared with the unannealed QW one [28].

In this chapter, we will introduce the modulation parameters of EA modulators, which are the tools to indicate the modulation performance of the devices. We will then discuss two types of EA modulators including the waveguide type DFQW modulator and Fabry Perot (FP) vertical reflection type DFQW modulator in terms of the modulation parameters. For the waveguide type modulator, we also discuss the remedies of polarization dependent optical properties. The integration of DFQW modulators to other optical components such as lasers and waveguides is finally addressed.

2 MODULATION PARAMETERS

There are several important parameters used to characterize a waveguide type EA modulator: optical confinement, modulation efficiency, chirping parameter, bandwidth, insertion loss and electrical bias. [12,20]. For a DFQW EA modulator, the adjustment of λ_{op} and the modification of quantum confined Stark shift (QCSS) due to interdiffusion are also used to examine the performance of the modulator [28].

Optical confinement factor Γ is defined as:

$$\Gamma = \frac{\underset{\substack{\text{the depletion region of} \\ \text{the waveguide device}}}{\int \varphi(x, y)\varphi^*(x, y)dA}}{\underset{\substack{\text{the entire cover range} \\ \text{of a guide field}}}{\int \varphi(x, y)^*(x, y)dA}}, \tag{1}$$

where $\varphi(x, y)$ is the guiding optical field, dA is a small but finite area normal to the optical field at (x, y). Since the extent of the interdiffusion in the QW region is inhomogeneous throughout the cross section of a modulator, Γ is dependent on the geometrical directions x and y. From this equation, we can observed that the Γ parameter indicates the portion of the optical power which overlaps the depletion region of the p-i-n device structure. The depletion region consists of a QW active region and part of the two (top and bottom) cladding layers, within the entire device structure. Therefore, an efficient modulator requires a large value of Γ.

The effective absorption coefficient, Γ_{eff}, is given by :

$$\alpha_{\text{eff}} = \frac{\underset{\substack{\text{the wells within} \\ \text{the active region}}}{\int} \alpha(x, y)\varphi(x, y)\varphi^*(x, y)dA}{\underset{\substack{\text{the entire range of} \\ \text{the guiding field}}}{\int} \varphi(x, y)\varphi^*(x, y)dA}, \tag{2}$$

where $\alpha(x, y)$ is the absorption coefficient of the QW structure, Equation (2) shows that α_{eff} is determined by the fraction of the optical field intensity $\varphi(x, y)\varphi^*(x, y)$ within the wells of the active region.

Modulation efficiency, η, established the fraction of optical intensity change in the modulator is defined as

$$\eta = \frac{I_{ON} - I_{OFF}}{I_{ON}}(\times 100\%) \tag{3}$$

where I_{ON} is the intensity of ON-state transmitted light, and I_{OFF} is the intensity of OFF-state light. Usually, the modulation depth is described in decibels, using the term ON-OFF ratio or contrast ratio (CR) [29]:

$$CR = 10\log\frac{I_{ON}}{I_{OFF}} = 10\log\frac{\exp(-\alpha_{\text{eff}}(ON)l)}{\exp(-\alpha_{\text{eff}}(OFF)l)} \tag{4}$$

where $\alpha_{\text{eff}}(ON)$ and $\alpha_{\text{eff}}(OFF)$ are the effective absorption at the ON-state and the OFF-state, respectively, and l is modulation interaction length of the active cavity.

Modulation bandwidth $\Delta\nu$ is defined by the frequencies where the modulation depth or contrast ratio reduces to half of its maximum value. If the switching time τ is defined instead of a frequency bandwidth, then the equivalent bandwidth is

$$\Delta\nu = \frac{2\pi}{\tau} \text{ Hz} \tag{5}$$

where τ is the 10- to 90-percent rise time. $\Delta\nu$ can be also defined by the device capacitance, C, and load resistance, R, of the devices' operation circuit as

$$\Delta\nu = \frac{1}{2\pi RC} \tag{6}$$

Insertion loss L in decibels describes the fraction of intensity lost induced by the insertion of the modulator to the propagation medium of the optical beam. The definition is

$$L = 10 \log \frac{I_0}{I_{ON}} = 10 \log(\exp(\alpha_{\text{eff}}(ON)l)) \tag{7}$$

where I_0 is the transmitted intensity of the system when the modulator is not in the beam. L can be expressed in terms of l and the effective absorption coefficient of MQW at ON-state.

Both α and refractive index, n, change under bias and does the chirping parameter [30]. The static chirping parameter β_{mod} of modulator is given by:

$$\beta_{\text{mod}} = \frac{4\pi \, \Delta n_{\text{eff}}}{\lambda_{op} \, \Delta \alpha_{\text{eff}}} \tag{8}$$

where Δn_{eff} and $\Delta \alpha_{\text{eff}}$ are the change of the effective refractive index and effective absorption coefficient due to the applied electric field respectively. β_{mod} can be considered as the fraction of phase modulation strength (due to Δn_{eff}) and intensity modulation strength (due to $\Delta \alpha_{\text{eff}}$). It can be observed that, for pure and perfect electroabsorption modulation, β_{mod} equals zero.

The adjustment of λ_{op} is considered as the change of λ_{op} of the DFQW modulator from that of the as-grown QW one. Similarly, the modification of QCSS due to an applied electric field is defined as the deviation of QCSS of the DFQW modulator from that of the as-grown structure. The electrical bias required for modulation is also a crucial parameter which indicates the power consumption of the device. This comsumption is due to a power dissipation in the load resistance of the drive circuit.

For a Fabry Perot DFQW reflection modulator [28,31], apart from the chirping parameter, insertion loss, adjustment of λ_{op} and modification of QCSS, the modulator also be characterized by using reflectance (R_{tot}) of the entire FP modulator, change of the reflectance (ΔR_{tot}), and finesse.

R_{tot} is given by

$$R_{tot} = \left| \frac{\sqrt{R_T} - \sqrt{R_B} e^{-\alpha l + 2\sqrt{-1}(\delta l)}}{1 - \sqrt{R_T}\sqrt{R_B} e^{-\alpha l + 2\sqrt{-1}(\delta l)}} \right|^2 \tag{9}$$

where R_T and R_B is the reflectance of the top and bottom mirror respectively. δ is the phase change per unit length of cavity, i.e. $\delta = 2\pi n_c / \lambda$, where n_c is the refractive index of the cavity and λ is the photon wavelength. l is the modulation interaction length and, for a FP modulator, it equals the length of the active cavity.

From equation (9), ΔR_{tot} can be calculated. This calculation can be simplified by using the following three assumptions: 1) the change of λ_{op} due to an applied electric field = 0 implies ignoring the shift of the FP mode due to the change of n_c; 2) the OFF state $R_{tot} = 0$ implies the FP modulator operates at an OFF-state on-resonance mode perfectly while holding the impedance and phase-matching conditions [28]; 3) R_T and R_B are both <1 and remain constant under a bias, which is a reasonable assumption since the λ_{op} used is far from the bandgap of the bulk materials used in the Bragg reflectors, and hence, the change of their refractive index can be neglected. Under these assumptions, R_{tot} can be expressed as

$$\Delta R_{tot} = \left| \frac{\sqrt{R_T} - \sqrt{R_T}e^{-\Delta\alpha l}}{1 - R_T e^{-\Delta\alpha l}} \right|^2 \tag{10}$$

For a comprehensive analysis, other parameters developed from ΔR_{tot} such as $\Delta R_{tot}/R_{tot}$ (min) and $\Delta R_{tot}/V$ are usually used where R_{tot} (min) is the OFF state reflectance and V is the reverse bias.

The finesse defined as the ratio of free-space FP mode spacing to full-width-half-maximum (FWHM) of the mode by assuming a uniform refractive index is given as

$$\text{Finesse} = \left| \frac{\pi (R_T R_B)^{1/4} e^{-\frac{1}{2}\alpha l)}}{1 - \sqrt{R_T}\sqrt{R_B}e^{\alpha l}} \right|^2 \tag{11}$$

The built-in voltage and applied voltage of a p-i-n multi-layer FP modulator is determined by using Poisson's equation. The electrostatic potential V inside the depletion region is expressed as

$$V = V(\text{build-in}) - V(\text{applied}) = \Psi(-X_n) - \Psi(X_p), \tag{12}$$

where $\Psi(-X_n)$ and $\Psi(X_p)$ are the potentials at $-X_n$ and X_p, respectively, which are the coordinates (zero reference taken at the i-p interface) of the depleted end points in the n-doped and p-doped reflectors respectively. The $\Psi(-X_n)$ and $\Psi(X_p)$ considered in the Poisson's equation are expressed in terms of the depletion thickness X_m and the carrier density of the multi-layered materials in the stacked structure. The integrated form of the Poisson's equation is expressed as a quadratic equation for X_m as shown below:

$$\left(\frac{N_D}{2\varepsilon_i^T}a^2 + \frac{N_A}{2\varepsilon_j^B}g^2 \right) X_m^2 + \left(\frac{N_D}{\varepsilon_i^T}da - \frac{N_A}{\varepsilon_j^B}dg - \zeta \right) \times$$
$$X_m + \left[\frac{N_D}{2\varepsilon_i^T} + \frac{N_A}{2\varepsilon_j^B} \right](b_j l)^2 + \Xi - \frac{V}{q} = 0, \tag{13}$$

where $a = \frac{N_A}{N_A+N_D}$, $d = \left(\frac{N_D-N_c}{N_D+N_A}\right)(l)$, $c = \frac{N_D}{N_A+N_D}$, $g = \frac{N_D}{N_D+N_A}$

$$\zeta = N_A S_2^B + N_D a S_2^T + l\left(\frac{N_D a}{\varepsilon_1^T} - \frac{N_A c}{\varepsilon_c}\right),$$

$$\Xi = N_D\left(\frac{S_1^T}{2} - S_2^T d\right) + N_A\left(\frac{S_1^B}{2} + S_2^B d\right) + \frac{(l)^2}{2}\left[\frac{N_D}{\varepsilon_1^T} - \frac{N_c}{\varepsilon_c}\right]$$

$$-dl\left[\frac{qN_D}{\varepsilon_1^T} + \frac{qN_A}{\varepsilon_c}\right].$$

$$S_1^T = \sum_{i=2}^{Y}\left[\frac{1}{\varepsilon_i^T} - \frac{1}{\varepsilon_{i-1}^T}\right](t_{i-1})^2, \qquad S_2^T = \sum_{i=2}^{T}\left[-\frac{1}{\varepsilon_{i-1}^T} + \frac{1}{\varepsilon_i^T}\right](t_{i-1})$$

$$S_1^B = \sum_{j=2}^{Z}\left[\frac{1}{\varepsilon_j^B} - \frac{1}{\varepsilon_{j-1}^B}\right](b_{j-1})^2, \qquad S_2^B = \sum_{j=2}^{Z}\left[-\frac{1}{\varepsilon_1^B} + \frac{1}{\varepsilon_{j-1}^B}\right](b_{j-1})$$

ε_i^T and ε_j^B are the x-independent terms of the dielectric constant in the i-layer and j-layer of the top and bottom reflector respectively. N_A, N_c and N_D are the doping concentrations of the p-i-n structure, q is the carrier charge, t_i and b_j are the depletion width extensions of the layer i (top reflector) and layer j (bottom reflector) respectively due to the applied field. $i = Y$ and $j = Z$ are the outermost layers of the top and bottom reflectors that the depletion range will extend to. The applied voltage which is equivalent to the expected electric field in the depletion layer is then be calculated from equation (13). Therefore, the voltage swing of the modulator is determined from equation (12).

For a FP modulator with DFQW as the active region, interdiffusion will modify the optical properties of the active region and thus the performance of the modulator. The relation between the absorption coefficient and ΔR_{tot} is obtained by taking derivative with respect to $\exp(\Delta\alpha l)$ in equation (10) and considering $exp(\Delta\alpha l)$ is greater than 1 that

$$\frac{d(\sqrt{\Delta R_{tot}})}{d(\exp(\Delta\alpha l))} = \pm\frac{\sqrt{R_T}(1 - R_T)}{(1 - R_T \exp(\Delta\alpha l))^2}. \tag{14}$$

For the sign convention, the positive relation holds in the domain of $1 < \exp(\Delta\alpha l) < 1/R_T$, while the negative relation holds when $\exp(\Delta\alpha l) > 1/R_T$. From relation (14), it is observed $\exp(\Delta\alpha l)$ increases with $|\Delta R_{tot}|$, i.e. as $\Delta\alpha_c$ increases, $|\Delta R_{tot}|$ increases.

Generally, a high performance absorption modulator requires a large $\Delta\alpha_{eff}$ and thus a high CR, a low Δn_{eff}, i.e. a low β_{mod}, and a low α_{loss}. In order to obtain a large $\Delta\alpha_{eff}$ the absorption modulator is selected to operate at the wavelength of the exciton peak of the first electron (C1) and first heavy hole (HH1) in the presence of an applied electric field and α_{loss} is the α_{eff} of the QW structure at this wavelength without an applied field.

3 WAVEGUIDE TYPE DFQW ELECTRO-ABSORPTIVE MODULATOR

3.1 Devices Structure

A QW modulator is fabricated as a p-i-n structure where light is coupled into the intrinsic guiding region (core) including an absorbing (active) region of QW. The growth direction of QW is normal to the propagation direction of the traveling optical wave. The p- and n- type cladding layers are served as ohmic contacts which enable an electric field to be applied to the active region. This modulator operates based on QCSE which functions as shifting the exciton peak with an applied voltage. Thus, at the photon wavelength of the biased fundamental heavy hole exciton peak, which is usually selected as λ_{op} of a modulator, the optical loss coefficient of the structure can be varied by orders of magnitude.

Different types of III-V semiconductor DFQW materials have been used to produce the modulation such as GaAs-based DFQWs of AlGaAs/GaAs [25,32] and InP-based DFQWs of InGaAsP/InP [33,34] and InAlGaAs/InGaAs DFQW [35]. The GaAs-based DFQW modulator has the advantages that AlGaAs/GaAs QW are easily and routinely fabricated with features of sharp exciton peak and large change in absorption coefficient caused by reverse bias. The InP-based modulators are intensely investigated because it can operate at 1.33 μm to 1.55 μm wavelength range at which fiber optical communication systems will likely operate as a result of the low loss and minimum dispersion characteristics exhibited by optical fibers in this wavelength range.

There are several interdiffusion methods (otherwise known as intermixing methods) among the materials systems. AlGaAs/GaAs DFQW electroabsorptive modulator can be developed by using the impurity-free vacancy diffusion (IFVD) method [25,32,36]. This technique produces large blue shifts of the band edge (and thus λ_{op} can be adjusted) while a clearly definable room-temperature excitonic characteristic is obtained [37]. Moreover, the maintenance of electrical properties, especially QCSE, is very promising [25]. These are the characteristics that make the quantum wells structures so useful from a modulation device viewpoint. For InP-based materials, In-

GaAsP quaternary sublattices DFQW electroabsorption modulator has been produced by using photoabsorption induced disordering (PAID) [33,38] and carrier-free cap-annealing technique based on phosphorous-doped silicon-dioxide encapsulation (SiO:P) [34,39].

3.2 Characteristics of the Modulator

3.2.1 AlGaAs/GaAs DFQW

The modulation characteristics of AlGaAs/GaAs DFQW modulators including λ_{op}, QCSE, exciton linewidth, and absorption coefficient under applied field have been reported [25,32]. The λ_{op} (in eV) of the modulator blue shifts, i.e. λ_{op} can be adjusted when interdiffusion proceeds because Al sublattice diffuses into the well, the diffused potential profile moves upwards and hence the energy of transition increases [28]. Under an applied electric field, the energy of eigenstate maintains higher than that of the as-grown structure so that λ_{op} blue shifts. The extent of interdiffusion is indicated by diffusion length L_d, defined as $(Dt)^{1/2}$, where D is the material and temperature dependent diffusion coefficient and t is the annealing time [40]. Consequently, since λ_{op} of an electro-absorption modulator is set at the biased exciton peak of the transition energy between C1 and HH1 (C1-HH1), the adjusted amount of λ_{op} depends on not only the diffusion temperature, diffusion time and thus L_d, but also QCSE of the material structure.

Due to the grading and lowering of the potential barriers in partially intermixed QWs, QCSE reduces in DFQW as compared to its as-grown QWs structure. However, subject to different annealing times, the reduction of QCSE is not significant as shown in Figure 3. Moreover, for a proper comparison, the diffused well should be compared with its 'equivalent' square well, i.e. a well width (L_z) which gives the same zero-field first electron and first heavy hole position [32]. The theoretical results show that QCSS of DFQW should be greater than that obtained for the 'equivalent' square QW as shown in Figure 4. In this figure, a 85 Å GaAs wells and 60 Å $Al_{0.31}Ga_{0.69}As$ barriers DFQWs are compared to the 'equivalent' square QW with $L_z = 70$ Å.

The exciton peaks with and without the applied field of DFQWs are weaker and broader than that of the as-grown QWs. Due to a reduction of quantum confinement in graded and lowered DFQW potential profiles, the exciton broadening increases with particle tunneling in DFQW [25]. Figure 5 and 6 show the absorption spectra of the as-grown QWs and DFQWs [25]. We can observe that the exciton of DFQW remains clearly resolved for electro-absorption modulation. The figures also indicate that the as-grown p-n junction is retained. These features are important for device applications. Under an applied field, the exciton absorption coefficient of DFQW reduces

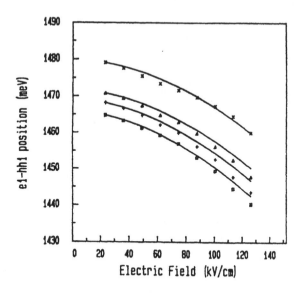

Fig. 3. The e1-hh1 position as a function of electric field for the as-grown (□) and three annealed MQW samples by using IFVD. The full curves represent the calculated theoretical Stark shifts. Annealing times 90s (+), 120s (Δ) and 150s (×). [32]

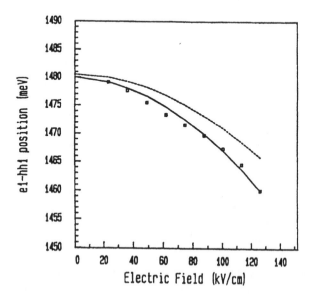

Fig. 4. Comparison of the Stark for a diffused (error function) well (theory (——)) and experimental (□) and the 'equivalent' square well (theory (-----)). [32]

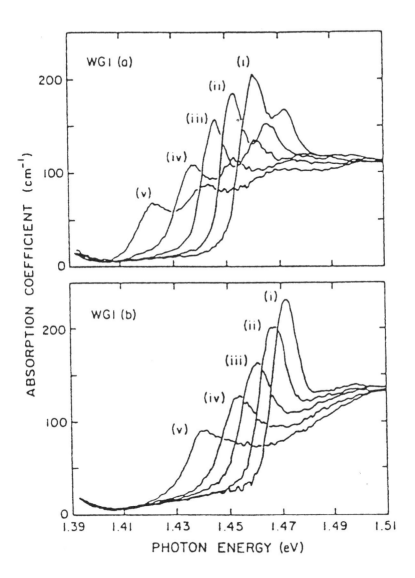

Fig. 5. Absorption spectra of WG1 — as-grown QW — for (a) TE polarization and (b) TM polarization as a function of applied voltage (i) 0, (ii) −7.5 V, (iii) −11 V, (iv) −14 V, and (v) −18 V. [25]

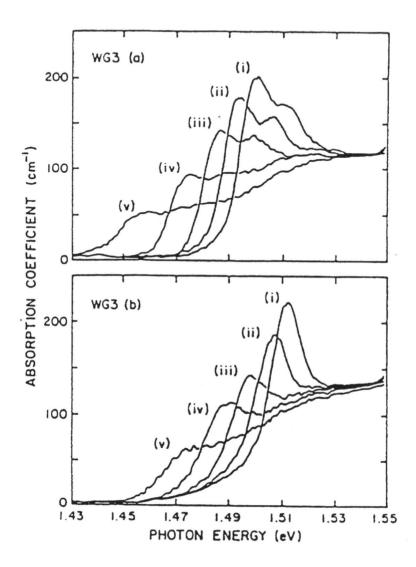

Fig. 6. Absorption spectra of WG3 — (annealing temperature 985°C, annealing time 28s RTA) — for (a) TE polarization and (b) TM polarization as a function of applied voltage (i) 0, (ii) −7.5 V, (iii) −11 V, (iv) −14 V, and (v) −18 V. [25]

faster than that of the as-grown structure because the reduction of oscillator strength is more rapid for DFQW than for the unannealed QW [32]. This causes a decrement in modulation depth. However, the use of shallow DFQW has been suggested as a solution for increasing the saturation intensity in electro-absorption modulators. Shallow DFQW can also prevent hole pin-up in the active region, allowing a higher cutoff frequency at bias voltages and an improvement in quantum efficiency [41]. Recently, another technology of interdiffusion using anodization followed by rapid thermal annealing has demonstrated that the linewidth broadening of DFQW PL spectra do not change significantly and the PL intensity of DFQW enhances simultaneously [27]. In other words, the optical properties which is an important factor for optical modulators can be improved by using interdiffusion. This is in a good agreement with theoretical studies [42].

A theoretical study of a wide range of AlGaAs/GaAs DFQW structures has been carried out to explore the effects of interdiffusion in this material system [42]. They are combinations of different Al concentrations ($x = 0.2$, 0.3 and 0.4) and well widths ($L_z = 8$, 10 and 12 nm). The extension of interdiffusion are from $L_d = 1$ nm to 4 nm. Since an EA modulator operates at the wavelength of the respective biased band-edge of C1-HH1 exciton peak [43], it is important to understand C1 and HH1 eigenstate properties and their effect on the modulation performance including the absorption change $\Delta\alpha$, Stark Shift, α_{loss}, λ_{op} and the required bias. Since the linewidth broadening of DFQW PL has no significant change by using the interdiffusion method of anodization [24,27]. The broadening factor is considered to be constant in the calculation of absorption coefficient. The Lorentzian broadening factors half-width-half-maximum (HWHM) are extracted from experimental data [44,45]. The HWHM of HH and LH are considered to be the same and have a value of 3meV which is averaged from the values taken from ref. [44,45]. Although the broadening factor increases with an applied electric field [32], its relation is still not well known, and has been considered as constant in the theoretical study.

A Interdiffusion contribution to a QW confinement

In an interdiffused QW, the confinement profile is graded in nature with no well-defined well width and well depth. Therefore, the following analysis will be based on a changeable well potential depth (under no bias), an effective width of QW at the first eigenstate energy and its wavefunction. Generally, both a deeper well depth and a narrower well width enhance the confinement of carriers in QW.

In the typical unbiased QW structure which is $Al_{0.3}Ga_{0.7}As/GaAs$ 100 Å/ 100 Å QW, the effective width reduces by 40% for $L_d \leq 2$ nm, whilst

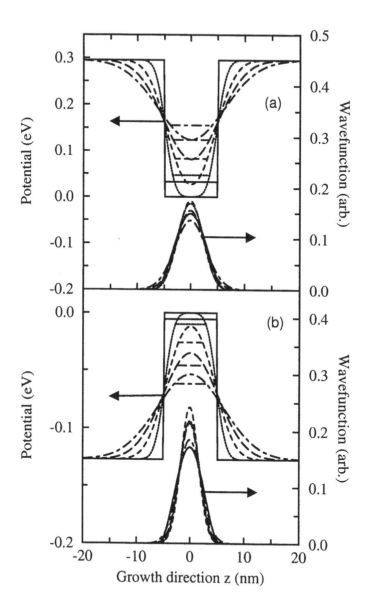

Fig. 7. The potential profile, wavefunction of first eigenstate, and the states' L_z of a 100 Å/100 Å $Al_{0.3}Ga_{0.7}As$/GaAs DFQW. (a) conduction band, and (b) valence band. $L_d = 0$ (solid line), $L_d = 1$ nm (dot line), $L_d = 2$ nm (dash line), $L_d = 3$ nm (long dash line) and $L_d = 4$ nm (dotted-dash line). [28]

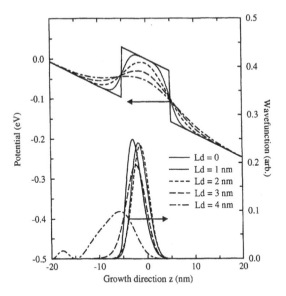

Fig. 8. The heavy hole (HH) potential profile and wavefunction of first HH eigenstate of 100 Å/100 Å$Al_{0.3}Ga_{0.7}As$/GaAs DFQWs with F=60 kV/cm. $L_d = 0$ (solid line), $L_d = 1$ nm (dot line), $L_d = 2$ nm (dash line), $L_d = 3$ nm (long dash line) and $L_d = 4$ nm (dotted-dash line). [42]

the potential depth remains unchanged and slightly reduces by 10% for $L_d = 1$ nm and for $L_d = 2$ nm respectively, as shown in Figure 7. The wavefunctions of both C1 and HH1 are therefore stronger confined with a higher wavefunction peak in the DFQW as compared to that of the as-grown structure ($L_d = 0$). As the interdiffusion proceeds further (3 nm $\leq L_d \leq$ 4 nm), the well depth becomes much lower and the wavefunctions of C1 and HH1 broaden out with their peaks reducing back to (or lower than) the as-grown case. The confinement only improves in the initial stages of interdiffusion ($L_d \leq 2$ nm).

When the DFQW is tilted by an applied electric field, an interesting feature can be obtained as shown in Figure 8. For the case of $L_d \leq 2$ nm, the shift of wavefunctions weakens as compared to that of the as-grown structure because of the enhanced confinement, while the wavefunction of the extensive interdiffused QW ($L_d = 4$ nm) shifts substantially and partly tunnel out. Consequently, the Stark shift of wavefunction can be tailored by using the QW interdiffusion.

The increase of x from 0.2 to 0.4 (i.e. the increase of the QW barrier potential) improves the confinement in the as-grown QW. Its effect on the interdiffusion is best illustrated by the DFQWs with $L_d = 2$ nm under an

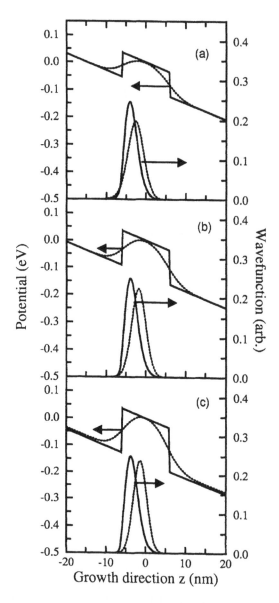

Fig. 9. As-grown Al concentration effect: the heavy hole (HH) potential profile and wavefunction of first HH eigenstate of the $Al_x Ga_{1-x} As$/GaAs DFQW with $L_z = 12$ nm and F = 90 kV/cm, Solid line and represents $L_d = 0$ (rectangular QW), and dot line represents $L_d = 2$ nm (a) the case of $x = 0.2$; (b) $x = 0.3$, and (c) $x = 0.4$. [42]

applied field of 60kV/cm as shown in Figure 9(a) to (c). From these figures, the entire HH1 wavefunctions of the tilted DFQW move towards the center of the well as compared to the HH1 wavefunctions of the as-grown structure. Its peak amplitude, although slightly less than the as-grown ones, also increases when x increases from 0.2 to 0.4. This is contributed by the deepening of the potential depth and narrowing of the effective width due to a different magnitude of interdiffusion to the increasing x content. For the HH valence band, the potential depth increases from 0.085 eV to 0.155 eV and the effective width reduces from 5.54 nm to 4.66nm, as x increases. These mean that the biased first eigenstates are better confined with increasing x content in the DFQWs.

More importantly, the biased eigenstates stay closer to the center of the DFQW as x content increases. This is in contrary to the as-grown QW case in general. The consequence of this is a much improved overlapping between the electron and hole states under bias and will give rise to better modulation performance because the modulator is operated at the 1S HH biased exciton peak.

A similar feature of the HH1 wavefunction can also be found in $L_d = 2$ nm as the DFQW as-grown well width increases from $L_z = 8$ nm to $L_z = 12$ nm, as shown in Figure 10. The wavefunctions stay closer to the central of QW as L_z increases, although the amplitude of the HH1 wavefunction of the DFQW reduces in the widening of the as-grown well width. It is interesting to note that interdiffusion offers a special modification to the confinement of the biased first eigenstates. Generally, when well width increases, their confinement and overlapping reduce. However, as shown in Figure 10, the confinement of HH1 wavefunction in the tilted DFQWs illustrates that the centralization of the eigenstates enhances with the increase of the as-grown well width. This enhancement in the DFQW can be explained by the increasing of the potential depth, which increases from 0.103 eV to 0.121 eV as L_z increases from 8 nm to 12 nm. The effective well width also increases from 5.68 nm to 6.65 nm at the same time. This implies that the potential depth is the dominant factor contributing to the centralization. Although the effects of the two initial conditions, i.e. the value of x and the as-grown well-width, on the C1 confinements and its wavefunctions of DFQWs have not been discussed here, the results show that their effects on C1 are the same as those on HH1. As a consequence, interdiffusion can recover the poor performance in wide well-width QW which is used as an active region to produce a large Stark Shift and a low α_{loss} in the EA-modulator [46].

B Interdiffusion contributions to modulation properties

DFQWs under different strengths of the applied electric field and the use of QW interdiffusion to adjust λ_{op} have been investigated [42]; the tunability

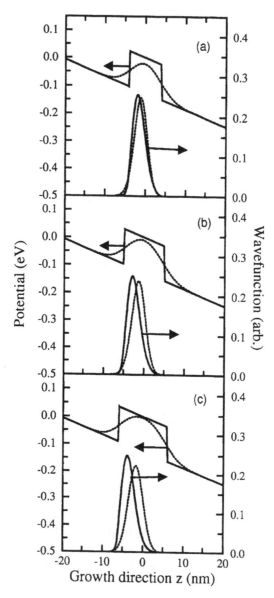

Fig. 10. As-grown well-width effect: the heavy hole (HH) potential profile and wavefunction of first HH eigenstate of the $Al_{0.3}Ga_{0.7}As/GaAs$ DFQW with F=90 kV/cm, Solid line and represents = 0 (rectangular QW), and dot line represents $L_d = 2$ nm (a) the case of $L_z = 8$ nm, (b) $L_z = 10$ nm, and (c) $L_z = 12$ nm. [42]

of λ_{op} is defined as the difference of λ_{op} of a DFQW to its corresponding as-grown QW. The OFF-state and ON-state absorption coefficient are defined as the absorption coefficient of the biased 1S HH exciton absorption peak and of the unbiased rising absorption edge at the wavelength of the biased exciton peak, respectively. α_{loss} is considered as the OFF-state absorption coefficient.

(i) Fixed as-grown QW conditions

The absorption spectra of the typical QW structure ($Al_{0.3}Ga_{0.7}As$/GaAs 100 Å/100 Å QW) under different extents of interdiffusion ($0 \leq L_d \leq 4$ nm) are shown in Figure 11(a). The DFQW 1S HH exciton absorption peaks remain fairly constant with interdiffusion. However, with increasing interdiffusion, the 1S HH exciton peak wavelength blue-shifts at a different rate. Under an applied electric field and thus producing the quantum confined Stark effect, the exciton absorption peak of the DFQWs red-shifts and its amplitude reduces, such as for the case of DFQW with $L_d = 2$ nm shown in Figure 11(b).

$\Delta\alpha$, i.e. the difference between the ON- and OFF- states absorption at different applied fields of this DFQW, is extracted to generate the dash line in Figure 12(b). Similarly, $\Delta\alpha$ of other L_d's cases are extracted to form the other lines in Figure 12(b). It should be noted that, due to an enhance confinement in cases of $L_d = 1$ nm and 2 nm, $\Delta\alpha$ at different applied fields from $F = 50$ kV/cm to $F = 130$ kV/cm, i.e. dot line and dash line of Figure 12(b) respectively, are greater than that of the as-growth case (solid line). When interdiffusion increases to $L_d \geq 3$ nm, its $\Delta\alpha$ at different applied fields reduces. The $\Delta\alpha$ of $L_d = 3$ nm terminates at $F = 110$ kV/cm because of their 1S HH exciton tunnels out. For even more extensive interdiffusion ($L_d = 4$ nm), it HH1 wavefunction is weakly bounded even under a small applied field of $F = 60$ kV/cm, as can be seen in Figure 8, and its 1S HH exciton tunnels out at $F \geq 90$ kV/cm. Its $\Delta\alpha$ can only be shown up to $F = 80$ kV/cm. For developing a large electro-absorption modulation, the use of DFQW, for example with $L_d = 1$ nm at $F = 110$ kV/cm, can produce an improved (two-fold) 4.6 dB CR per μm (propagation length) as compared to the as-grown QW of 2.3 dB at this applied field.

An important intrinsic parameter for the EA modulation is the quantum confined Stark shift, this is because a large shift provides a larger ON/OFF ratio. As listed in Table 1 (column 4), the Stark shift reduces with interdiffusion until $L_d = 2$ nm because of the enhanced confinement. When $L_d > 2$ nm, the Stark shift increases due to the relaxation of confinement as shown in Figure 8. The Stark Shift is also dominated by the valence band since the potential barrier of valence band is shorter than that of conduction band.

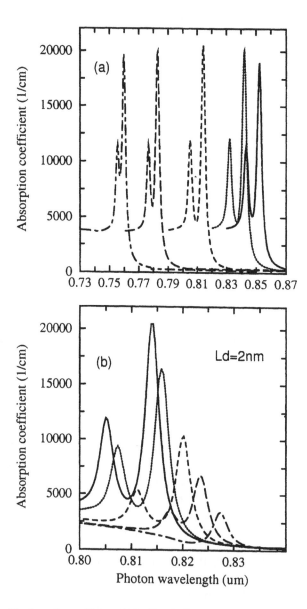

Fig. 11. The absorption coefficient of $Al_{0.3}Ga_{0.7}As$/GaAs DFQW. (a) the absorptive spectra of the unbiased DFQW with L_d from 0 to 4 nm stepped by 1 nm; $L_d = 0$ (solid line); $L_d = 1$ nm (dot line); $L_d = 2$ nm (dash line); $L_d = 3$ nm (long dash line); and $L_d = 4$ nm (dotted-dash line). (b) the absorptive spectra of the biased DFQW with $L_d = 2$ nm, F=0 (solid line), F=50 kV/cm (dot line), F=90 kV/cm (dash line), F=110 kV/cm (long dash line), and F=130 kV/cm (dotted-dash line). [42]

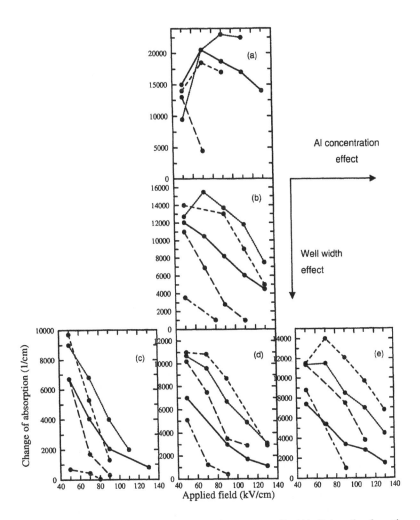

Fig. 12. The effects of Al concentration x and as-grown well width (L_z) on the absorption change of $Al_x Ga_{1-x} As/GaAs$ DFQWs. (a) $x = 0.3$ and $L_z = 8$ nm, (b) $x = 0.3$ and $L_z = 10$ nm, (c) $x = 0.2$ and $L_z = 12$ nm, (d) $x = 0.3$ and $L_z = 12$ nm, and (e) $x = 0.4$ and $L_z = 12$ nm, in different L_d from 0 to 4 nm stepped by 1 nm. $L_d = 0$ (solid line), $L_d = 1$ nm (dot line), $L_d = 2$ nm (dash line), $L_d = 3$ nm (long dash line) and $L_d = 4$ nm (dotted dash line). [42]

α_{loss} depends on the tail of rising unbiased absorption exciton edge and is inversely proportional to the Stark shift. The increase (although small) of unbiased absorption coefficient due to initial interdiffusion, as shown in Figure 11(a), and the reduction of Stark shift, see Table 1, these make the α_{loss} increase for $L_d \leq 2$ nm as shown in Table 2. However, α_{loss} reduces at

TABLE 1

The Stark shift and the normalised shift, i.e. which is denoted as $SS_{Ld \neq 0}/SS_{Ld=0}$, of DFQWs with $L_d = 0$ to 4 nm stepped by 1 nm in an applied field F = 90 kV/cm. [42]

L_d (nm)	$x = 0.2$, $L_z = 12$ nm		$x = 0.3$, $L_z = 8$ nm		$x = 0.3$, $L_z = 10$ nm		$x = 0.3$, $L_z = 12$ nm		$x = 0.4$, $L_z = 12$ nm	
	Shift (meV)	$SS_{Ld \neq 0}/$ $SS_{Ld=0}$	Shift (meV)	$SS_{Ld \neq 0}/$ $SS_{Ld=0}$	Shift (meV)	$SS_{Ld \neq 0}/$ $SS_{Ld=0}$	Shift (meV)	$SS_{Ld \neq 0}/$ $SS_{Ld=0}$	Shift (meV)	$SS_{Ld \neq 0}/$ $SS_{Ld=0}$
0	32.9	1	11.1	1	20.2	1	30.4	1	28.1	1
1	25.5	0.76	6.5	0.59	12.2	0.60	19.4	0.64	17.0	0.60
2	21.5	0.65	10.0	0.90	11.1	0.55	14.8	0.49	12.7	0.45
3	–	–	–	–	18.2	0.90	15.3	0.50	15.9	0.57
4	–	–	–	–	–	–	–	–	26.3	0.94

TABLE 2

The residence loss of DFQWs that the operation wavelength is selected at bias exciton peak (F = 90 kV/cm) and the normal ON-state is at F = 0. [42]

	absorption loss (cm^{-1})								
	$x = 0.2$			$x = 0.3$			$x = 0.4$		
L_d (nm)	L_z = 8 nm	L_z = 10 nm	L_z = 12 nm	L_z = 8 nm	L_z = 10 nm	L_z = 12 nm	L_z = 8 nm	L_z = 10 nm	L_z = 12 nm
0	2457	990	606	3161	1169	657	3695	1278	730
1	3630	1487	715	6653	2354	1000	10228	3171	1230
2	2500	1275	892	3787	2558	1464	6586	4048	2243
3	–	–	–	–	1250	1100	2065	1820	1483
4	–	–	–	–	–	–	–	–	779

L_d = 3 nm due to an increasing Stark shift. For a more extensive interdiffusion, the 1S HH exciton tunnels out beyond F = 90 kV/cm, therefore, a slightly lower field (F = 80 V/cm) is used. Under this field strength, α_{loss} is only ≈ 580 cm^{-1} whilst Stark Shift is 24 meV. It should be noted that the α_{loss} considered here is purely the material absorption coefficient. In the case of a practical device structure, the optical confinement is usually < 1 [47]. Take the optical confinement equal to 0.5 as an example, the α_{loss} of the extensive interdiffused QW (L_d = 4 nm) reduces to 290 cm^{-1}.

Another important modulation parameter is the required bias (voltage) to make the ON/OFF modulation. For the case of extensive interdiffusion, since the confinement becomes effectively weaker, the DFQW offers a higher tunneling rate [48], Equivalently, a lower voltage is required [49] by using the DFQW as compared to the as-grown QW. This higher switching rate combined with low α_{loss} (580 cm^{-1} at F = 80 kV/cm) can be very useful in high speed EA-modulators, while the lowered voltage feature can be used to

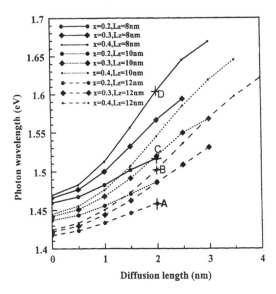

Fig. 13. The operation wavelengths of AlGaAs/GaAs DFQW in different L_d from 0 to 4 nm stepped by 1 nm. The operation wavelengths are at the 1S HH exciton peak biased by applied field F = 90 kV/cm. The solid-lines, dotted-lines and dashed-lines are the cases L_z = 8, 10, 12 nm respectively. For a fixed diffusion, i.e. L_d = 2 nm, the points A and C represent the wavelength at L_z = 12 nm and 8 nm, respectively, for x = 0.2. Likewise, the points B and D correspond to the cases for x = 0.4. [42]

satisfy the low bias requirement of switching arrays from by self-electro-optic effect device (SEED) [50].

An interesting feature of DFQW is in its adjustability of λ_{op}. A summary of λ_{op} working at $F = 90\,\text{kV/cm}$ as a function of L_d (from 0 to 4 nm), x (from 0.2 to 0.4), and L_z (from 8 nm to 12 nm) are shown in Figure 13. Due to the diffused potential profile moves upwards with increasing L_d, for the cases of $F \geq 0$, the energy of eigenstate in general increases with respect to that of the as-grown structure for cases of $F \geq 0$, and thus λ_{op} blue shifts. For the typical case which is denoted by ($\blacklozenge \ldots \blacklozenge$) in the figure, the λ_{op} blue-shifts from 1.442 to 1.565 eV when L_d increases from 0 to 4 nm. This blue shift property of λ_{op} is nonlinear in nature that the tuning of λ_{op} is wider in the intermediate interdiffusion range (1 nm $\leq L_d \leq$ 2.5 nm) than those in the initial stage ($L_d < 1$ nm) and final stage ($L_d > 2.5$ nm). In practice, the precise control of the operation wavelength by a fine adjustment of L_d can be realized when the annealing temperature reduces to around 900°C [51]. For instance, an annealing time of 10 seconds implies $L_d = \sqrt{[(2.9 \times 10^{-17}\,\text{cm}^2/\text{s})\,(10\text{s})]} = 1.7$ Å. Therefore, the required L_d for a targeted wavelength discussed here is experimentally achievable.

(ii) Effect of as-grown Al concentration variation in barrier

When x increases from 0.2 to 0.4 and L_z is fixed at 12 nm, $\Delta\alpha$ under different applied fields and for all interdiffusion extends gradually and rises up due to an enhancement of the quantum confinement, as shown from Figure 12(c) to (e). The increasing DFQW $\Delta\alpha$ over its corresponding as-grown QW enhances from the case of $x = 0.2$ to $x = 0.4$. As a consequence, when Al concentration increases, a longer range of interdiffusion exists with an enhanced $\Delta\alpha$, and an increased magnitude of $\Delta\alpha$. However, it should be noted that, for all cases of x, $\Delta\alpha$ terminates at a lower applied electric field when L_d increases, see Figure 12(c) to (e). Generally, the maximum allowed applied field (voltage) of DFQWs increases when x increases.

The variation of the Stark shift, it variation of DFQWs due to the increase of Al concentration is illustrated by its normalized value, i.e. $\frac{\text{Stark shift}(L_d \neq 0)}{\text{Stark shift}(L_d = 0)}$. The normalized Stark shift of the DFQWs reduces, see Table 1 (column 2, 4, 6), and thus α_{loss} of the DFQWs increases, see Table 2, when x increases. Adjustability of λ_{op} of DFQWs also widens with the increase of x as shown in Figure 13. By comparing the adjustable range of λ_{op} of the DFQW (from $L_d = 0$ to $L_d = 2$ nm), when x increases through 0.2, 0.3, and then 0.4, it increases from 35.4 meV to 64.3 meV and 78.1 meV, respectively.

(iii) Effect of as-grown well width variation

The effect of the as-grown well width on $\Delta\alpha$ of DFQWs has been summarized as shown in Figure 12(a), (b) and (d) [42]. It is found that for almost all the interdiffusion cases and $L_d = 0$ (the as-grown structure), $\Delta\alpha$ on the whole reduces with increasing L_z. However, the range of the interdiffusion extent with an enhanced $\Delta\alpha$ (over that of $L_d = 0$ case) widens when the as-grown well width increases. This is because the confinement of QW with a wider as-grown well width improves with interdiffusion, In addition, interdiffusion can produce a recovery to the reduction of $\Delta\alpha$ due to increasing well width. For instance, when L_z of an as-grown QW increases from 8 nm to 10 nm, its $\Delta\alpha$ at $F = 110$ kV/cm reduces from 17000 cm^{-1} to 6000 cm^{-1}, however, the $\Delta\alpha$ of the latter QW can be improved to 12000 cm^{-1} by interdiffusion to $L_d = 1$ nm. Consequently, there are two benefits in using interdiffusion in the wide well-width QW structure; first, the range of an enhanced $\Delta\alpha$ increases when L_z increases, second, interdiffusion can, to a certain extent, compensate the degradation caused by a widened well width on $\Delta\alpha$.

As L_z increases, the amount of Stark Shift of all the interdiffusion cases increases, as shown in Table 1 (the middle three columns). The α_{loss} reduces as shown in Table 2, Moreover, the adjustability of λ_{op} decreases as the as-grown well width widens. On the other hand, since a narrower as-grown

TABLE 3

The normalized $\Delta\alpha$, i.e. $\Delta\alpha_{Ld\neq0}/\Delta\alpha_{Ld=0}$, of DFQW with respect to the corresponding rectangular QWs for different L_d cases (from 1 nm to 4 nm stepped by 1 nm) under applied field $F = 90$ kV/cm. [42]

| x = 0.2 | $(\Delta\alpha_{Ld\neq0})/\Delta\alpha_{Ld=0})$ | | | |
	$L_d = 1$ nm	$L_d = 2$ nm	$L_d = 3$ nm	$L_d = 4$ nm
$L_z = 8$ nm	1.17	0.53	–	–
$L_z = 10$ nm	1.37	0.73	–	–
$L_z = 12$ nm	1.86	0.65	–	–
x = 0.3				
$L_z = 8$ nm	1.23	0.96	–	–
$L_z = 10$ nm	1.76	1.70	0.41	–
$L_z = 12$ nm	2.22	2.91	1.16	–
x = 0.4				
Lz = 8 nm	1.19	1.15	0.09	–
Lz = 10 nm	1.80	1.90	0.91	–
Lz = 12 nm	2.46	3.58	2.18	0.34

width QW will have a more rapid rate of reduction in the diffused potential depth (thus weaken the confinement), the applied voltage for a required level of ON/OFF in the extensively interdiffused QW is expected to be lowered by reducing the well width, see Figure 12(d) to Figure 12(b) to Figure 12(a).

C Combined effects of the structural initial conditions to DFQW

(a) In general, an increase of the both initial conditions including the Al concentration and well width widens the interdiffusion range with an improved $\Delta\alpha$, as shown in Figure 12. An illustrative example is shown in Table 3 which lists the normalized $\Delta\alpha$ ($\Delta\alpha_{Ld\neq0}/\Delta\alpha_{Ld=0}$) of DFQWs under bias $F = 90$ kV/cm. By comparing the enhanced $\Delta\alpha$ caused by increasing the Al concentration with that due to the increase of the well width, it can be observed that the former one is more obvious and dominating. Take DFQW with $L_d = 1$ nm as an example and use the ratio of different normalized $\Delta\alpha$ cases for comparison. The increment of x from 0.2 to 0.4 provides an enhancement to the effect of widening L_z on the DFQW from $1.86/1.17 = 1.59$ to $2.46/1.19 = 2.07$. While the widening of L_z from 8 nm to 12 nm, however, brings a weaker enhancement to the effect of increasing x on the DFQW from $1.19/1.17 = 1.05$ to $2.46/1.86 = 1.32$. This means that the effect of interdiffusion on $\Delta\alpha$ is intensified by an increasing Al concentration of the QW in the study ranges considered here for the two initial conditions. This is in coherence with the analysis of section A of 3.2.1, the dominant factor of an enhanced confinement and thus an improved $\Delta\alpha$ [52] is the diffused potential depth rather than the diffused well width.

(b) The variations of α_{loss} and Stark shift in the DFQWs, caused by a change of the Al concentration and as-grown well width, are inversely related to each other; the former one increases when Al concentration increases and L_z decreases, and vice verse for the later one. Of the two effects, the variation of α_{loss} and Stark shift strongly depends on the Al concentration and as-grown well width, respectively. This can be seen by comparing the ratio of the normalized values as used in (a).

(c) An enlargement of the Al concentration and a reduction of L_z can widen the adjustable range of λ_{op}. With the initial condition of x increased from 0.2 to 0.4, the variation of λ_{op} of a DFQW (at $L_d = 2$ nm and $F = 90$ kV/cm) between $L_Z = 8$ nm and 12 nm will be modified from a bandwidth of AC to BD, see Figure 13. Their relative ratio, i.e. BD/AC, is 1.75 which shows the effect of increment in x. With the initial condition of L_z reduced from 12 nm to 8 nm, the modification of λ_{op} of the DFQW due to an increment in x from 0.2 to 0.4 will change from the width AB to CD. The influence of reduction in L_z produces a relative ratio (i.e. CD/AB) of 1.99. Therefore, it can be concluded that the adjustability of λ_{op} through interdiffusion is more sensitively dependent on the as-grown well width.

(d) One of the interesting features of the DFQW is a lowered required voltage swing when the interdiffusion is moderately extensive (the relevant value of L_d depends on the Al concentration and L_z). The maximum allowed applied voltage increases with the increase of both x and L_z.

D Advantages of employing DFQWs as the active region material

Based on the four features discussed in the section C of 3.2.1, for general applications, the $Al_xGa_{1-x}As/GaAs$ with x between 0.3 and 0.4 and L_z between 10 nm and 12 nm should be employed to develop DFQW EA modulation devices for general applications. Within these ranges, the magnitudes of $\Delta\alpha$, the extension of L_d with improved $\Delta\alpha$ and the adjustability of λ_{op} are at their best.

A wide well-width rectangular QW to be used as the active-region material can provide a large Stark Shift and a low α_{loss} in the EA modulator [46]. With interdiffusion, the magnitude of $\Delta\alpha$ can be further improved in this wide-well-width QW system. The range of interdiffusion with an enhanced $\Delta\alpha$ widens with increasing L_Z. On the other hand, the low-Al concentration in rectangular QWs are attractive instruments for developing SEED devices [46] since it provides a fast tunneling rate. However, these QWs suffer from low $\Delta\alpha$. Interdiffusion can then be a remedy to restore the $\Delta\alpha$ as demonstrated in Figure 12.

The only drawbacks of the DFQW for the above two devices seem to have a narrow Stark shift and a large α_{loss}. However, for an accurate comparison, a DFQW should be compared with an "equivalent" rectangular QW [53] having the same amount of the Stark shift. In this case, the α_{loss} of DFQW is found to be less. In order to develop an equivalent-quantum well with the same $\Delta\alpha$ as that of DFQW, a narrower well width or a deeper potential depth is required. This will reduce the Stark shift and the tunneling sensitivity [46], thus increasing the loss as compared to the DFQW. On the other hand, in order to provide a similar Stark shift to the DFQW structures, a wider well width and shallower well depth in the equivalent-QW are required. In this case, the $\Delta\alpha$ of the equivalent-QW reduces. This means that an equivalent-QW cannot attain both the $\Delta\alpha$ and Stark shift (and thus residual loss) of a DFQW at the same time. As a whole, the improvement of DFQW in terms of confinement and tunneling sensitivity cannot be replaced by a single rectangular QW. This is also one of the advanced feature in using DFQW for the EA-modulator.

The required bias of the modulator can also be reduced by introducing an extensively interdiffused QW structure, thus found applications in high-speed modulators. DFQW can provide an adjustability of λ_{op}, therefore, a single substrate QW structure is only required with a selective interdiffusion process to develop EA-modulators which operate at different λ_{op}'s. A multiple bandgap integrated structure can be realized for multi-colors or wide bandwidth applications. The yield of fabrication may also be improved by using interdiffusion as a post-growth wavelength correction technique.

3.2.2 InGaAs(P)/InP DFQW

For InP-based DFQW, it can also provide adjustability of α_{op} and modulation depth. For instance, a five periods of 85 Å $In_{0.53}Ga_{0.47}As$ well and 120 Å $In_{0.76}Ga_{0.24}As_{0.51}P_{0.49}$ barrier DFQW provide 80 nm, 95 nm and 120 nm bandgap shift which depends on the irradiation conditions during PAID [33]. The modulation depth of the samples reduces, as shown in Figure 14. At the optimum modulation wavelength of 1522 um for the sample with 120 nm shift, CR with value of 20 dB is obtained when the bias voltage is varied between +0.5 V to −2.5 V. The CR increases to 27 dB by using a sample with less shift of 80 nm. In another quaternary InGaAsP DFQW modulator [34], the transition energy of C1-HH1 blue shifts with a value of 80 meV. CR of the modulator is 15 dB under −6 V bias. Consequently, although the modulation depth of the InP-based DFQW modulator reduces with interdiffusion, the value is large enough for modulation application. In addition, the adjustability

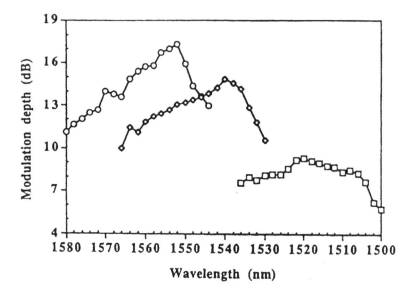

Fig. 14. Modulation depth when bias voltage was varied between +0.5 V and −1 V plotted as a function of wavelength for three samples which had undergone different bandgap shifts: —○— shifts 80 nm, —◇— shifts 950 nm, and —□— shifts 120 nm. [33]

of λ_{op} enables successful development of a single substrate QW structure to operate at different photon wavelengths by a selective interdiffusion process. The InP-based DFQW also provides an extra property of strain over lattice matched GaAs based materials. Two types of strain including compressive and tensile are possible in InGaAsP DFQW to offer blue-shift or red-shift in the exciton absorption peaks. The engineering of these types of strain can act as another degree of freedom to modify the modulation properties.

Through a theoretical study, the effects of strain on the interdiffused InGaAsP/InP potential profile, implies the quantum confinement, and thus λ_{op} and QCSE can be considered in detail [43]. For a InGaAsP/InP material system, interdiffusion can be experimentally controlled as group III sublattices dominant by implanting dopant of Zn [54] or group V sublattices dominant by implanting dopant of phosphorus [55].

The transition energy, λ_{op} and QCSE strongly depend on the quantum confinement, therefore the interdiffused potential profile and subband wave-functions are to be first discussed. As Group III interdiffusion proceeds, the well shape is modified due to a large compressive strain created in the well layer near the interface, thus "miniwells" are generated at the interfaces [56], see Figure 15 for $In_{0.61}Ga_{0.39}As_{0.84}P_{0.16}/InP$ DFQW. All the confined electron states are located above the miniwells for interdiffusion

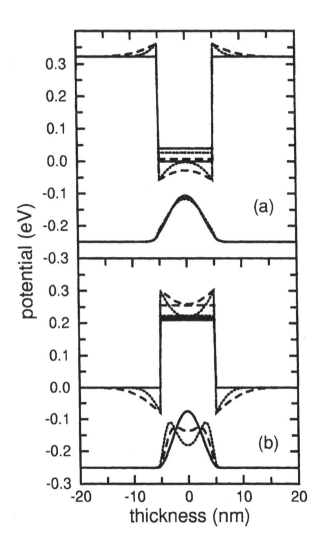

Fig. 15. (a) The conduction band potential profiles, and first electron C1 wavefunction square of the In$_x$Ga$_{1-x}$As$_y$P$_{1-y}$/InP Group III diffused QWs with L$_z$ = 10 nm and L$_d$ = 0 (solid line), L$_d$ = 1.5 nm (dot line), and L$_d$ = 3 nm (dash line) where 1 − y = 0.16 and under zero bias. The material system is lattice-matched before any interdiffusion. (b) The heavy hole HH potential profiles, and first heavy hole HH1 wavefunction square of the same DFQWs. [43]

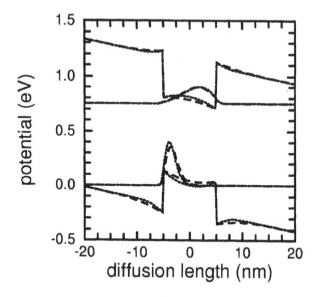

Fig. 16. The biased (F = 100 kV/cm) C and HH profiles as well as C1 and HH1 wavefunction square of the DFQW with $L_d = 1.5$ nm (dot line) and 3 nm (dash line). [43]

up to $L_d = 3$ nm, hence the wavefunction of C1 is mainly localized at the center of the well, see Figure 15(a). On the other hand, the HH1 wavefunction is localized in the miniwells with a double peaks shape. As $L_d \rightarrow 1.5$ nm, the miniwells become sharper and deeper, where the splitting of double peaked HH1 becomes more pronounced, see Figure 15(b). As interdiffusion proceeds further (1.5 nm < $L_d \leq 3$ nm), the amount of In atoms penetration into the well center is large enough to create a more uniform alloy distribution to even out the strain energy. As a result, the HH potential profile becomes much flatter where the miniwells are diminishing. This forces the HH1 wavefunction to be pushed out of the miniwells and re-localized at the center of the well, as can be seen in Figure 15(b). The eigenstates under an applied electric field are very similar in trend to the case of an unstrained interdiffused QW or square QW, see Figure 16.

The potential profiles of the QWs can also be modified by Group V interdiffusion, see Figure 17 for $In_{0.61}Ga_{0.39}As_{0.84}P_{0.16}/InP$ DFQW. The diffused quaternary material modifies the confinement profile in the central well layer, at the same time, two extra triangular wells are generated in the two barriers near the interfaces [43]. Consequently, a "three-wells" potential profile is produced. When an electric field is applied to the Group V interdiffused QW structure, an interesting feature can be obtained.

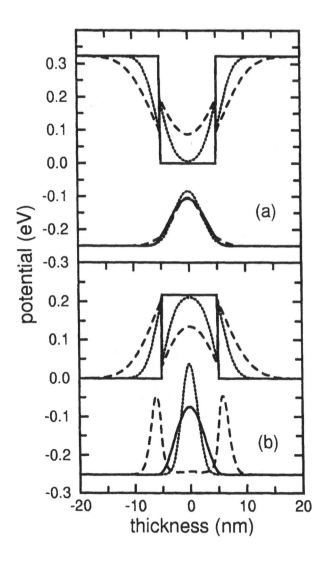

Fig. 17. The Group V interdiffusion for the $In_xGa_{1-x}As_yP_{1-y}$/InP QWs with $L_z = 10$ nm and $L_d = 0$ (solid line), $L_d = 1.5$ nm (dot line), and $L_d = 3$ nm (dash line) where $1 - y = 0.16$ and under zero bias. The material system is lattice-matched before any interdiffusion. (a) Conduction band. (b) Valence band. [43]

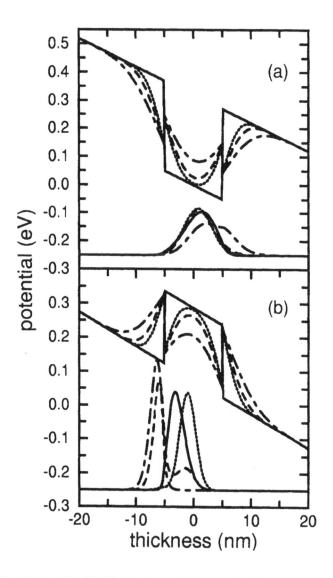

Fig. 18. The biased (F = 100 kV/cm) for the case in Figure 17. The diffusion lengths are $L_d = 0$ (solid line), $L_d = 1.5$ nm (dot line), $L_d = 2$ nm (dash line), and $L_d = 3$ nm (dot-dash line). (a) Conduction band. (b) Valence band. [43]

As interdiffusion initially proceeds, see Figure 18, the bottom of the central well maintains at a level as deep as that of the rectangular QW while the lower part of the central well becomes narrower. These contribute to a stronger confinement and a movement of the wavefunction to the well center (i.e. $z = 0$). In other words, in a biased well structure and as a result of diffusion, the HH1 wavefunction localized from a position originally near the interface (when $L_d = 0$ nm) moves towards the well center, while the C1 wavefunction remains roughly localized at the well center, as shown in Figure 18. As interdiffusion proceeds further to $L_d = 2$ nm, more P penetrates into the center of the well layer and thus causes the well depth to reduce. It should be noted that at this time, however, the HH potential depth of the two side-triangular-wells remains the same. Therefore, the HH1 wavefunction re-distributes to a strong localization in the triangular well while it only weakly localizes in the central well (dash line). In all cases considered here, the C1 wavefunction remains confined at the well center (in fact, this feature maintains until $L_d = 3$ nm). When L_d increases to 3 nm, since the energy of the HH1 state is below the minimum potential of the well center, this state is confined in the triangular well and thus the wavefunction (dotted dash line) is completely localized into the triangular well. All these variations will significantly modify the interband transition energy, λ_{op} and QCSE of the DFQW under Group V interdiffusion.

For the interband transition energy and λ_{op}, Figure 19(a) shows the C1-HH1 transition energy (under zero bias) as a function of Group III diffusion length L_d within a range of P concentrations in the well layer. This figure can serve as a guideline (without considering the modulation properties) for selecting a particular initial as-grown composition that best suits a desired tunable λ_{op}. In the case of P in well = 0.10 which terminates at $L_d = 1$ nm. This is because, at that point, the top width of the diffused well has reached the critical layer thickness. In order to have a meaningful analysis of the effect of interdiffusion on the optical properties, the P content can only be taken here as low as 0.16 for L_d to have an extension up to 4 nm.

In the case of Group V interdiffusion, the C1-HH1 (solid lines) and C1-LH1 (dash lines) transition energies generally blue-shift (contrast to Group III interdiffusion) but to a lesser extent when P concentration increases. Since tensile strain is produced in the well layer during interdiffusion [43], the HH1 and LH1 cross each other at a particular Group V interdiffusion extent. At the crossing, polarization (TE and TM) insensitivity photonic materials can be obtained. A plot of the transition energies are shown in Figure 19(b) for a range of concentrations. For the case of P in well = 0.10, the blue shift terminates at $L_d = 2.5$ nm since after which, dislocation will be generated.

The magnitude of QCSS is an important parameter for modulation. For the Group III interdiffusion, as listed in Table 4, the QCSS increases with L_d until diffusion becomes extensive; it is dominated by the variation of

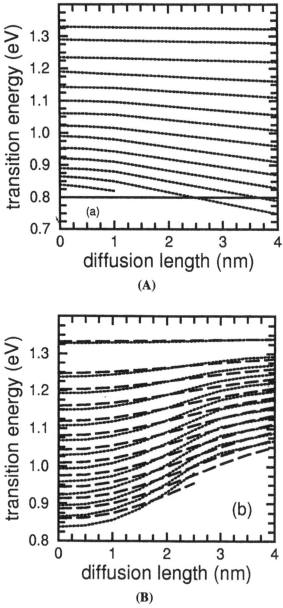

(A)

(B)

Fig. 19. (a) The C1-HH1 transition energies in different extent of group III one phase interdiffusion with $L_z = 10$ nm and L_d ranged from 0 to 4nm. The considered material system is $In_xGa_{1-x}As_yP_{1-y}$/InP with $P_{1-y} = 0.10, 0.16, 0.23, 0.30, 0.36, 0.43, 0.49, 0.56, 0.63, 0.69, 0.76, 0.82, 0.89,$ and 0.96 from lowest to highest dotted lines. (b) The C1-HH1 (dot lines) and C1-LH1 (long dashed line) transition energy for the same materials in (a) under different extent of group V one phase interdiffusion ranged from $L_d = 0$ to 4 nm. [43]

TABLE 4

The quantum confined Stark shift of $In_{0.61}Ga_{0.39}As_{0.84}P_{0.16}$/InP DFQW with L_z of 10 nm [43].

L_d (nm)	Stark Shift (meV)	
	Group III	Group V
0	21	21
1	33	9
1.5	34	20
2	33	18
3	28	26

TABLE 5

The shift of C1 and HH1 state, and Stark Shift of the DFQW with L_z of 10 nm and as-grown P = 0.164 in well [43].

L_d (nm)	shift of C1 (meV)	shift of HH1 (meV)	Stark Shift (meV)
0	5	15	21
1.5	6	28	34
3	5	23	28

the HH1 state in the miniwells. As L_d increases to 1.5 nm (under a bias of 100 kV/cm), this HH1 state is attracted into the bottom of the deepened left-hand side miniwell, as shown in Figure 16. This creates a large shift of the HH1 state, see Table 5, and thus the bandedge shifts accordingly. When $L_d > 1.5$ nm, the depth of the entire well increases while the miniwell shape flattens out; this reduces the shift of HH1 state and contributes to a reduced QCSS. Under the Group V interdiffusion, QCSS of the low-P material system reduces when $L_d \rightarrow 1.5$ nm and then increases when $L_d \rightarrow 2$ nm. This is due to a pronounced quantum confinement in the beginning, followed by a weakened confinement.

3.3 Polarization Insensitive Modulation

As discussed in section 1, a waveguide type modulator exhibits a drawback of strong polarization dependence on the incoming optical TE and TM fields. Polarization insensitivity of both AlGaAs/GaAs and InGaAsP/InP QW structure have been intensively investigated recently.

For lattice matched AlGaAs/GaAs QW structure, one of the ways to produce polarization insensitivity is to develop a parabolic-shaped confinement in the finite depth QW structure (hereafter denote as PQW) [57].

The reason for this is that QCSS of the fundamental transitions are insensitive to HH and LH effective masses in the PQW, thereby resulting in an identical energy of QCSS [58]. This feature gives the PQW an ability to modulate light with a transverse magnetic (TM) polarization in a similar degree to that with a transverse electric (TE) polarization. Currently, the selection of an appropriate DFQW and the chronological steps to determine the ON-state and OFF-state of an interdiffusion induced parabolic-like AlGaAs/GaAs QW electro-absorption modulator has been proposed for producing polarization insensitivity [59].

There are two features of the PQW that can be used to assess whether the potential profile of the DFQW has reached a parabolic-like shape. They are (i) an equally spaced eigenstates, and (ii) an equal QCSS energy of the C1-HH1 and C1-LH1 transitions as well as the shift of their 1S exciton transition energies.

As shown in Figure 20, the upper half of the DFQW potential profile will widen-up from the as-grown rectangular QW structure (dotted-dash line) as L_d increases. At the same time, the lower half of the QW width will narrow down to match that of the PQW. With the appropriate extent of interdiffusion, i.e. $L_d = 2$ nm (solid line), the diffused profile can become very close to the PQW profile (dash line). Consequently, to achieve the parabolic-like DFQWs, a narrower as-grown rectangular well width L_z than that of the targeted PQW is required.

The polarization insensitivity considerably depends on the structural parameters of the DFQWs such as L_z and Al content of the as-grown rectangular QW as well as the extent of interdiffusion, indicated by L_d. Under interdiffusion, the interfaces between the barrier and well of lattice-matched AlGaAs/GaAs QW system are modified from abrupt to graded gradually [28]. Consequently, the energy of bound states will change and thus the energy difference of adjacent bound states i and j, $\Delta E_{ij} = E_i$ (energy of bound state i) $-\Delta E_j$ (energy of bound state j), is modified. For $Al_{0.32}Ga_{0.68}As/GaAs$ DFQW with $L_z = 14$ nm, the deviation of ΔE_{ij} from ΔE_{12} to ΔE_{54} of conduction band and HH valence band gradually reduces at the beginning of interdiffusion, i.e. the confinement profile becomes more parabolic-like in shape. The mean deviation of ΔE_{ij} becomes minimum at $L_d = 2$ nm and $L_d \approx 1.5$ nm for HH and electron, respectively, as shown in Figure 21. Upon further increasing of L_d, the DFQW is over-interdiffused so that the mean deviation increases.

QCSS due to an applied electric field of the $Al_{0.32}Ga_{0.68}As/GaAs$ DFQW will also be modified by interdiffusion and thus the difference between QCSS of C1-HH1 and C1-LH1 is varied as shown in dashed line of Figure 21 where the demonstrated applied field is $F = 50$ kV/cm. At $L_d = 2$ nm, the difference reduces to minimum. The minimum of the mean deviation of HH bound state and the minimum of the QCSS difference indicate that the DFQW

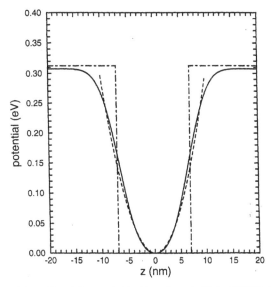

Fig. 20. Potential profile of the parabolic $Al_{0.32}Ga_{0.68}As$/GaAs DFQW (solid line) with $L_z = 14$ nm and $L_d = 2.0$ nm, and its corresponding as-grown QW (dashed-dot line) with $L_d = 0$, and the ideal PQW (dash line) $x = 0.3$, $L = 20$ nm. [59]

with $L_d = 2$ nm is most parabolic-like in shape. As shown in Figure 21, both the minimum of the QCSS difference and the mean deviation of HH bound state locate at the same L_d when the minimum mean deviation of electron bound state positions at the slightly shorter L_d. This implies that the variation of QCSS is dominated by changes in HH valence band, This feature can also be obtained in an interdiffused strained InGaAsP/InP QW system [43]. It is also worthy to note that in another report of non-square AlGaAs/GaAs QW, the modeling fitted experimental result shows that the modification of interband transition energies of unbiased QW system is dominated by the variation of conduction band [60]. Consequently, the variation of QCSS of C1-HH1 is dominated by HH valence band in both latticed matched AlGaAs/GaAs QW and InGaAsP/InP QW systems. On the contrary, the modification of interband transition energies is dominated by the conduction band.

Polarization insensitive ON-state is an unbias state in which $\alpha_{TE} = \alpha_{TM}$ and polarization OFF-state is a bias state in which $\alpha_{TE} = \alpha_{TM}$. In order to obtain these ON and OFF states with a polarization insensitive electro-absorptive change (i.e. $\Delta\alpha_{TE} = \Delta\alpha_{TM}$), λ_{op} and the strength of the reverse biased field are two mutual related factors that have to be determined. The chronological steps to determine the ON- and OFF- states by interdiffusion and the electro-absorption modulation are discussed as follows.

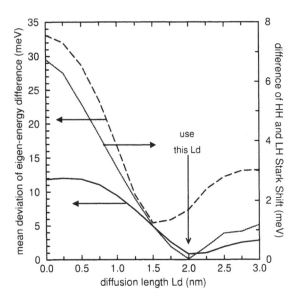

Fig. 21. Mean deviation of energy difference of adjacent bound state (ΔE_{ij}) of HH (solid line) and electron (long dash line), and difference between QCSS of HH and LH transition (dot line) of the $Al_{0.32}Ga_{0.68}As$/GaAs DFQW. [59]

The first step is to obtain a range of λ_{op}, within which the modulator can provide polarization insensitive ON-state. This range can be determined from the unbias ($F = 0$) TE and TM absorption spectra as shown in Figure 22. Due to the broadening of exciton absorption, the same amount of TE and TM absorption coefficient can only be found in the absorption tail at least 10 nm away from the exciton edges (HH for TE and LH for TM). The photon wavelength at the absorption tail with the same TE and TM absorption coefficient starts from ~ 0.858 μm to long wavelengths. This is the range of λ_{op} that can be selected as polarization insensitive ON-state.

The second step is to determine a range of bias fields, with which the polarization insensitive OFF-state can operate within the pre-determined range of wavelengths with the same amount of unbias TE and TM absorption (obtained from step 1). For the case shown in Figure 22, when $F \geq$ 100 kV/cm, the cross point of biased TE and TM absorption (the polarization insensitive OFF-state) can shift (due to Stark effect) into the wavelength range. The wavelength of this cross point is the λ_{op} with polarization insensitive ON- and OFF- states. Since an applied field $F \geq$ 100 kV/cm can be used to produce the ON- and OFF- states, more than one λ_{op} can be obtained for polarization insensitive modulation.

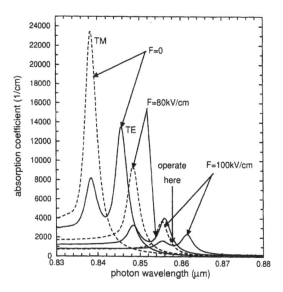

Fig. 22. The absorption spectra of the TE and TM polarization modes of the $Al_{0.32}Ga_{0.68}As$/GaAs DFQW. [59]

In the final step, an appropriate λ_{op} and bias field is determined. When the applied electrical field increases, the absorption coefficient reduces because the overlap integral of electron and heavy wavefunction as well as electron and light hole wavefunction reduces. In the case of Figure 22, when applied field increases from $F = 100$ kV/cm onwards, although polarization insensitive ON- and OFF- states can be obtained, the absorption coefficient of the OFF-state and $\Delta\alpha$ (absorption change between the OFF-state and the ON-state) reduces. In view of this, the tradeoff between α_{loss} (α at the OFF-state) and $\Delta\alpha$ has to be considered. When an applied field increases, λ_{loss} reduces since λ_{op} is moving far from the unbiased exciton edges (high loss), however, $\Delta\alpha$ (and thus CR) also reduces. In order to have a large $\Delta\alpha$ and, at the same time, to maintain polarization insensitive ON- and OFF-states, $F = 100$ kV/cm is used as the bias field and λ_{op} is now set at 0.858 μm for $Al_{0.32}Ga_{0.68}As$/GaAs DFQW with $L_z = 14$ nm and $L_d = 2$ nm. Since λ_{op} is far from the unbiased HH exciton absorption edge (0.846 μm), α_{loss} is low with a value of 258 cm^{-1}. It should be noted that since λ_{op} increases with applied field, these two parameters are mutual related. To compare the result of the modeled DFQW with the experimental result [58], the PQW measured to be $\Delta\alpha \cong 1000$ cm^{-1} against 1162 cm^{-1} for the DFQW at λ_{op} of 0.858 μm. This implies that the calculated DFQWs can perform as well as the PQW.

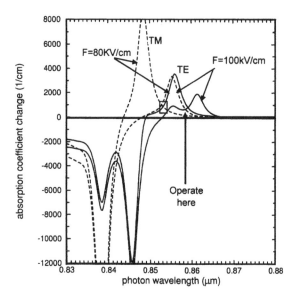

Fig. 23. The electro-absorption spectra of the TE and TM polarization modes of the $Al_{0.32}Ga_{0.68}As/GaAs$ DFQW. [59]

A word of caution from Figure 23. The $\Delta\alpha$ spectra of the DFQW, apart from the fact that λ_{op} (0.858 μm) at $F = 100$ kV/cm can provide polarization insensitivity electro-absorptive modulation, at a lower applied field such as $F = 80$ kV/cm, the polarization insensitivity can also be obtained at $\sim$0.853 μm (see inserted box in Figure 23) where $\Delta\alpha_{TE} = \Delta\alpha_{TM}$. However, as shown in Figure 23, the ON-state $\alpha_{TE} \neq \alpha_{TM}$ means that an extra modulator is required to pre-adjust the TE and TM polarized signal in order to obtain the same amount of modulated intensity. Although there is an advantage in using a smaller power consumption (power consumption is due to load resistance) of $F = 80$ kV/cm than that of $F = 100$ k/cm, the extra device makes the modulation system more complex which is not desirable in practice. As a consequence, DFQW at $F = 80$ kV/cm is not useful for polarization insensitive modulation.

It is known that, within realistic experimental limits, a deviation of few monolayers is possible in QW fabrication to the expected well thickness. For an example of a deviation of ± 5 Å in L_z, the polarization insensitive $\Delta\alpha$ will degrade [52]. In order to develop an parabolic-like DFQW with larger absorption modulation, the as-grown L_z, Al concentration in barrier and extent of interdiffusion have to be selected carefully.

For the InGaAsP/InP QW structure, in order to achieve the polarization insensitivity, an adequate amount of shear strain which can at least counterbalance the splitting of HH1 and LH1 states due to quantum confinement is required [61] so that the HH1 and LH1 exciton peaks can merge together. It is reported that, by using Group V interdiffusion, InGaAs(P)/InP QWs can produce polarization insensitive electro-absorptive modulation by generating enough shear strain [62]. The longer the annealing time, the stronger the induced shear strain. In 1996, polarization insensitive InGaAs/InGaAsP/InP amplifier was experimentally demonstrated by using QW intermixing [63]. This further consolidated that the tensile strain generated by interdiffusion can be used to achieve polarization insensitivity.

A two-phase interdiffusion model [64,65] has been used in the theoretical study [62] that only the Group V sublattices of QWs take part in the interdiffusion mechanism. The results have shown that the two-phase Group V interdiffusion can not only create enough tensile strain in the well, as shown in Figure 24, to merge the C1-HH1 and C1-LH1 transition energies for the polarization insensitivity, as shown by the cross point in Figure 25, but also maintain an abrupt potential profile with diffused L_z equal to that of as-grown QWs. This feature of constant L_z with interdiffusion would improve the potential for developing a polarization insensitive modulator; a graded profile modifies the effective L_z and the broadening factors of exciton and bound state, hence making it difficult to produce polarization insensitivity.

The effects of the P content in the well layer of the as-grown attire-matched InGaAs(P)/InP QW has also been addressed. The increase of the P content creates two disadvantages for the interdiffusion to produce polarization insensitivity. The first one is that the splitting between HH and LH increases, implying that larger tensile strains are required to bring the LH and HH states back together. Thus a longer annealing time is required to generate the stronger tensile strain. However, if a low initial as-grown tensile strain is introduced in an as-grown rectangular QW structure, only a small tensile strain is required to be produced from interdiffusion. The second disadvantage is that the increase of as-grown P content in the well will increase the transition energy of the as-grown lattice matched QW structure, thereby reducing the adjustable range of transition energy by interdiffusion. This reduces the tuning ability of the DFQW to reach the target wavelength of 1.55 μm. Consequently, a material system with a small as-grown tensile strain in the InGaAs/InP QW without any as-grown P in the well should be used.

The DFQW of lattice matched $In_{0.53}Ga_{0.47}As$/InP structures with as-grown L_z of 12 nm has been theoretically demonstrated to produce polarization insensitive electro-absorptive modulation, as shown in Figure 26. The maximum absorption change of ~ 1000 cm^{-1} (CR of ~ 16 dB for 50 μm modulation length) can be obtained at photon wavelength of 1.55 μm as shown in Figure 26. In this structure, the maximum strain at interfaces is

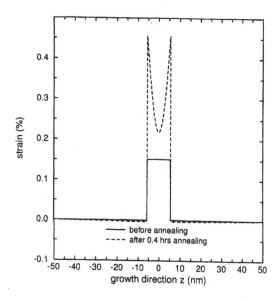

Fig. 24. The strain profile of InGaAs/InP QW with L_z of 11nm and In content of 0.51. (a) square QW (solid line) and (b) DFQW after annealing time of 0.4 hours (dash line). Tensile strain is positive in value and compressive strain in negative in value. [62]

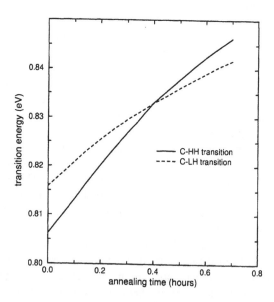

Fig. 25. The change of C-HH (solid line) and C-LH (dash line) transition energies of DFQW (with In content of 0.51, $L_z = 11$ nm) with increasing annealing time. [62]

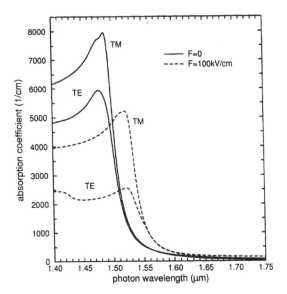

Fig. 26. The TE and TM absorption coefficient spectra of DFQW with In content of 0.53, L_z of 12 nm, and annealing of 1 hour. (a) applied field F = 0 (solid line), (b) F = 100 kV/cm (dash line). [62]

0.32% and the average strain is 0.27%. The lattice matched condition of the starting QW structure is particularly attractive since it only requires an easy (high-yield) fabrication process with a simple post-processing thermal annealing to achieve polarization insensitivity.

It is worth nothing that, apart from the electro-absorptive modulators discussed in section 3, several other waveguide type DFQW devices such as phase modulator [66,12], demultiplexer [67], and prototype blockaded reservior and quantum well electron transfer structures (BRAQWETS) [68] have been produced with promising performance, although they are not discussed in detail in this chapter.

4   FABRY PEROT TRANSVERSE TYPE DFQW MODULATOR

4.1   Device Structure

A transverse modulator is a device that allows light to propagate in and out through the whole p-i-n devices structure and perpendicular to the plane of QW. We can use this modulator as transmission or reflection type modulators depending on whether the modulated light propagates out from the bottom or the top side of the device. The development of this device can bring (a) less complexity for the coupling of light in and out of the modulation range as compared to the waveguide (longitudinal) type of modulators. (b) integratability with vertical-cavity surface emitting lasers [69]. The surface-normal modulators have much shorter optical interaction lengths (typically about a few micrometer) while that of the waveguide type can extend up to about 100 μm. So the total optical modulation amount of the transverse type modulator is lower than that of the waveguide type. Normally, the ON-OFF or contrast ratio of a transverse modulator is about 10:1 [70].

One of the ways to enhance the modulation is to increase the periods of the QW structure but the applied bias voltage also has to be increased simultaneously. Moreover, the thicker the fabricated layers, the higher the non-homogenous field in the QW structure. Generally, a high voltage of ~ 10 V is required. However, the drive voltage for such modulators will be limited to a few volts when they incorporate in parallel arrays with Si-based LSI electronics to form high-bandwidth optical interconnects [71].

In order to (a) reduce the operating voltage swing and thus minimize the amount of electrical input power, (b) increase the effective path of physical interaction for an expected signal modulations, and (c) has sharp resonance characteristics which are very sensitive to the change of refractive index [72], two quarter-wavelength identical Bragg reflectors are applied on the top and bottom of the QW cavity structure. These two identical Bragg reflectors are a stack of alternating layers with different indexes which are sometimes named as grating mirrors [19,73,74]. This is a so-called symmetric Fabry-Perot FP etalon. For the case of making the two grating mirrors with different reflectance, an asymmetric FP structure will be generated. This asymmetric FP configuration has one more merit than the symmetric FP one that the reflectance of the modulator at the OFF-state FP mode will be further reduced and thus perform higher CR [75].

We can model the multi-layer FP modulator using the common transfer matrix method that the whole structure is described as

$$(M_{FP}) = \begin{pmatrix} M_{T11} & M_{T12} \\ M_{T21} & M_{T22} \end{pmatrix} \begin{pmatrix} e^{\frac{1}{2}\alpha_c l_c \sqrt{-1}(\delta_c l_c)} & 0 \\ 0 & e^{\frac{1}{2}\alpha_c l_c \sqrt{-1}(\delta_c l_c)} \end{pmatrix} \begin{pmatrix} M_{B11} & M_{B12} \\ M_{B21} & M_{B22} \end{pmatrix},$$

(15)

where M_{FP} is a 2×2 transfer matrix of the whole device structure. $M_{T_{ij}}$ and $M_{B_{ij}}$ are the matrix elements of the top and bottom Bragg reflector respectively. From this equation, the reflectance R_{tot} of the modulator can be determined as shown in equation (9).

Basically, the FP electro-absorptive modulator can act as both a modulator and a photodetector in smart-pixed systems [76]. The FP-modulator can be applied (a) to optical interconnections [77], where it acts as the transmitting and receiving element of a free-space link, (b) to 2-D switching networks,

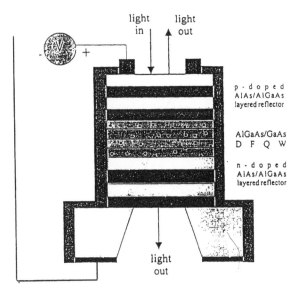

light
in

light
out

p - d o p e d
AlAs/AlGaAs
layered reflector

AlGaAs/GaAs
D F Q W

n - d o p e d
AlAs/AlGaAs
layered reflector

light
out

Fig. 27. Schematic diagram of the layer sequence of the FP reflection layer. The active cavity is $Al_{0.3}Ga_{0.7}As$/GaAs DFQWs structure. The structure is sandwiched with two reflectors, produced by two alternative layers $Al_{0.3}Ga_{0.7}As$ and AlAs the top is p-doped while the another one growth on n^+-GaAs is n-doped. [28]

an array of modulators fabricated to perform switching such as the cross-bar switch [78], (c) to self-electro-optic effect device (SEED), the modulator integrates with a resistor (R-SEED) and capacitor (C-SEED), or two identical FP modulators integrate together (S-SEED) that perform optical bistable [1].

Although there are not so many reports describing the DFQW modulator, the asymmetric FP reflection DFQW modulator has been experimentally demonstrated by using IFVD [79]. In theory, it has been shown that apart from providing an λ_{op} tuning range, this type of DFQW modulator can improve the modulation properties [28]. A schematic diagram of a DFQW FP reflection modulator is shown in Figure 27.

4.2   Characteristics of the Modulator

The first asymmetric Fabry Perot DFQW reflection modulator was reported in 1991 [79]. The active cavity of the modulator is AlGaAs/GaAs DFQW prepared by IFVD annealed at elevated temperatures. The main principles to develop the Fabry Perot DFQW modulator are: First, to fabricate the whole p-i-n device structure by standard technology such as MOCVD. The FP mode of the top and bottom reflectors is designed at a wavelength (in eV) longer than

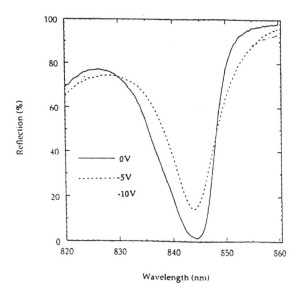

Fig. 28. This shows the reflection spectra, for different biases, from a sample that has been annealed at 930°C for 120s. Normally-off operation is clearly visible. [79]

that of the fundamental transition energy of the as-grown QW in the active cavity of the modulator; and second, by the impurity free vacancy diffusion, where the QW diffuses and the transition energy blue shifts. With a suitable annealing temperature, the transition energy can be adjusted to match the pre-designed operation wavelength (FP mode) of the two reflectors.

With different annealing temperatures, the modulator can be operated in normally-on and normally-off states [79]. The modulation properties of the DFQW modulator in the two states are shown as follows: For normal-off device, the adjusted range of the λ_{op} is a blue shift of 31 meV. The absorption of the exciton at the resonant wavelength results in a reflectivity of $\cong 1\%$ at 0 V. When a voltage swing with a value of -5 V is applied, the reflectivity at the λ_{op} rises up as shown in Figure 28. The device exhibits a contrast ratio of > 10 dB. From the report, the same modulator structure can also be developed as normally-on device by a longer annealing time. The reflectivity of operation is shown in Figure 29. With a reverse voltage increased from 0 to -8 V, the reflectivity drops from 79% to 48%. The OFF-state reflectivity cannot reduce to near zero for good modulation performance. This is mainly due to the structure of the reflectors which is pre-adjusted for a normally-off modulator but not for this normally-on one. In order to optimize the modulation properties of a normally-on DFQW reflection modulator,

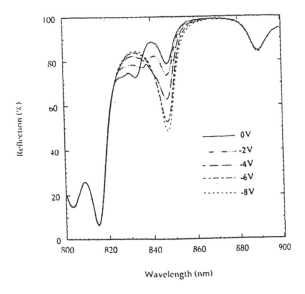

Fig. 29. This shows the reflection spectra of a sample that has been annealed at 930°C for 180s. The device is functioned as normally-on operation. [79]

a theoretical report has shown that the reflectance of the top and bottom Bragg reflector should be tailor-designed to match the impedance condition, i.e. the reflectance of the top Bragg reflector equal to that of the bottom Bragg reflector combined with the QWs active cavity, at the λ_{op} of the DFQW in the active cavity [28]. The optimized modeling result is shown in Figure 30. The CR at $\lambda_{op} = 0.7713$ μm is as high as 41 dB and the OFF-state reflectance of is almost zero.

From a theoretical report, other features such as the modification of the optical properties that interdiffusion can bring to the modulator have been investigated. Moreover, a 100 Å/100 Å $Al_{0.3}Ga_{0.7}As/GaAs$ DFQW structure has been optimized for the FP device [28]. In principle, there still remains a challenge to both symmetric and asymmetric FP structures in terms of precise control of the active QW layer thickness and composition. For this reason, it becomes difficult to operate the modulator precisely at the desired wavelength with good modulator performance. Several ways to modify λ_{op} either permanently or non-permanently have been reported [18,80,81]. The other way is to use a thermally interdiffused QW structure for the active region of the FP cavity. This permits λ_{op} and the optical properties of the modulator to be finely adjusted in a controllable manner by annealing the as-grown QW [28]. It works as follows. Interdiffusion of the QW structure will increase the

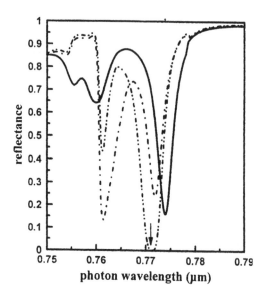

Fig. 30. The reflectance spectra of $Al_{0.3}Ga_{0.7}As$/GaAs 100 Å/100 Å DFQW with $L_d = 4$ nm at applied field F=0 (solid line), F=50 kV/cm (dashed-dot line), and 80 kV/cm (dashed-double dot line). The arrow shown the operation wavelength. [28]

transition energy (and thus increases λ_{op}) from both heavy and light holes in the valence band to electron states in the conduction band, and modify the confinement of these carriers in the QW. As a consequence, the joint density of states changes and the optical parameters of the QW material vary. This modification of λ_{op} in the FP-modulator provides an alternative option apart from the permanent and the non-permanent method.

Three advantageous features of DFQW which can improve the performance of the Fabry-Perot modulator are also indicated in the theoretical paper.

(a) the absorption coefficient change of 100 Å/100 Å$Al_{0.3}Ga_{0.7}As$/GaAs DFQW (with $L_d \leq 20$ Å, weak interdiffusion), under different applied fields, generally increases over that of $L_d = 0$ case (rectangle QW) due to the improvement of the quantum confinement as discussed in section A of 3.2.1. Since the change of reflectance (the higher the reflectance change, the larger the modulation depth) increases with the change of absorption coefficient, the modulation performance of the DFQW device improves.

(b) In cases of more extensive interdiffusion, such as $L_d \geq 30$ Å, electrons have a higher tunneling rate under bias as shown in Figure 8, which implies that a smaller applied voltage is required and thus reduce the power consumption consumed in the load resistance and capacitance of the devices' electrical circuit.

(c) As the fundamental transition energy increases with interdiffusion, λ_{op} can therefore be tuned accordingly as shown in Figure 13. Basing on these factor, DFQW FP modulator structures for different L_d's which are modified from a rectangular QW FP modulator can provide useful modulation characteristics.

5   INTEGRATION OF MODULATORS TO OTHER PHOTONIC DEVICES

One of the most important applications of intermixing technology is to integrate passive and active waveguide devices such as lasers, modulators, detectors and waveguide in the same wafer. The first monolithic integration of modulator to other photonic components by using an intermixing technology has been demonstrated since 1989 [66], in which a phase modulator is integrated to a laser in the AlGaAs/GaAs QW structure by using IFVD. The laser section is protected by SiN when the diffusion of the QW modulator section is accelerated under the cap layer of SiO_2. The exciton peak of the modulator is designed to blue shift 50–100 Å from the operation wavelength of the laser, so that under QCSE, the exciton peak of the modulator red shifts and the modulator can thus operate with the laser.

In order to operate at the photon wavelength of 1.55 μm, a strained InGaAsP DFQW electroabsorption modulator has been monolithically integrated with a distributed feedback (DFB) laser [34]. The intermixing in the modulator section is accelerated by SiO:P. After RTA, the photoluminescence (PL) peak of the modulator and laser locate at 1.46 μm and 1.54 μm respectively as shown in Figure 31. The laser illustrated single frequency operation at 1.55 μm with the feature of low threshold currents and the transmission of the modulator is 14 dB under reverse bias of 6 V. The insertion losses are low with value of 0.2 dB/100 μm. These good device performances indicate that intermixing is an attractive method to produce laser-modulator PIC.

Recently, another intermixing technique of pulsed laser disordering has also been proposed to integrate active and passive photonic components [82] in InGaAsP QW material. In this report, PAID is demonstrated to be an effective disordering technique to various individual components. However, the spatial selectively of the technique is limited by thermal conduction which leads to a lateral heat flow in integrating photonic components. Another technique of pulsed laser disordering is then proposed for the integration. Although no clear result of an integrated circuit is obtained by using this

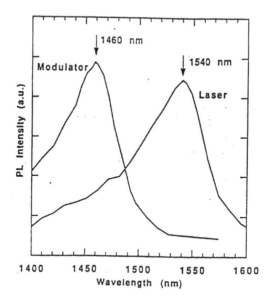

Fig. 31. Room temperature photoluminescence spectra of the laser and modulator sections after annealing. [34]

technique (this means that more work has to be done in the future), their results show that the technique can provide a spatial resolution with value better 25 μm for realizing photonic integrated circuits.

References

1. A.L. Lentine, D.A.B. Miller, L.M.F. Chirovsky, and L.A. D'Asaro, *IEEE J. Quantum Electron.*, **27**, 2431, 1991.
2. G. Bastard, J.A. Brum, and R. Ferreira, *Solid State Physics — Advances in Research and Applications*, **44**, H. Fhrencrieich and D. Turnbull, Eds. New York: Academic Press, 229, 1991.
3. D.S. Chemla, T.C. Damen, D.A.B. Miller, A.C. Gossard, and W. Wiegmann, *Appl. Phys. Lett.*, **42**, 864, 1983.
4. D.A.B. Miller, D.S. Chemla, T.C. Damen, A.C. Gossard, and W. Wiegmann, T. H. Wood, and C.A. Burrus, *Phys. Rev. B*, **32**, 1043, 1985.
5. D.A.B. Miller, M.D. Feuer, T.Y. Chang, S.C. Shunk, J.E. Henry, D.H. Burrows, and D.S. Chemla, *IEEE Photon. Technol. Lett.*, **1**, 62, 1989.
6. T.H. Wood, and C.A. Burrus, D.A.B. Miller, D.S. Chemla, and T.C. Damen, A.C. Gossard and W. Wiegmann, *Appl. Phys. Lett.*, **44**, 16, 1984.
7. T.H. Wood, and C.A. Burrus, R.S. Tucker, J.S. Weiner, D.A.B. Miller, D.S. Chemla, and T.C. Damen, A.C. Gossard and W. Wiegmann, *Electron. Lett.*, **21**, 693, 1995.

8. D.C. Hutchings, M. Sheik-Bahae, D.J. Hagan, E.W. Van Stryland, *Opt. Quantum Electron.*, **24**, 1, 1992.

9. H.K. Tsang, J.B.D. Soole, H.P. LeBlanc, Bhat, and M.A. Koza, I.H. White, *Appl. Phys. Lett.*, **57**, 2285, 1990.

10. P. Zouganeli, J. Stevens, D. Atkinson, and G. Parry, *IEEE J. Quantum Electron.*, **31**, 927, 1995.

11. N. Yoshimoto, H. Kawano, U.Hasumi, H. Takeuchi, S.Kondo, Y. Noguchi, *IEEE Photon. Technol. Lett.*, **6**, 208, 1994

12. W.C.H. Choy, B. L. Weiss, and E. H. Li, *IEEE J. of Quantum Electron.*, **34**, 84, 1998.

13. S. Weisser, E.C. Larkins, K. Czotscher, W. Benz, J. Daleiden, J. Fleissner, J. Maier, J. D. Ralston, B. Romero, A. Schonfelder and J. Rosenzweig, paper SCL 1.1, *IEEE Lasers and Electro-Optics Society Annual Meeting*, San Joes, CA, 1995.

14. T. Ido, S. Tanaka, M. Suzuki, and H. Inoue, IOOC'95, PD1-1., Hong Kong, 1995.

15. F. Koyama and K. Igo, *J. Lightwave Technol.*, **6**, 87, 1988.

16. G.A. Evan, D.P. Bour, N.W. Carlson, R. Amantea, J.M. Hammer, H. Lee, M. Lurie, R.C. Lai, F. Pilka, R.E. Farkas, J.B. Kirk, S.K. Liew, W.F. Reichert, C.A. Wang, H.K. Choi, J.M. Walpole, K.K. butle, W.F. Ferguson, Jr., K.D. Freez, and M. Felisky, *IEEE J. Quantum Electron.*, **27**, 1594, 1991.

17. T. Okoshi, *IEEE Trans. Micarowave Theory Tech.*, **MMT-30**, 1138, 1982.

18. R.C. Alferness, C.H. Joyner, L.L. Buhl, and S.R. Jorotky, *IEEE J. Quantum Eelctron.*, **QE-19**, p.1339, 1983 and also C.M. Gee, D. Thurmond, and H.W. Yen , *Appl. Phys. Lett.*, **43**, 998, 1983.

19. R.H.Yan, R.J. Simes, and L.A. Coldren, *IEEE Photon. Tech. Lett*, **1**, 273, 1989.

20. T.H. Wood, *J. Lightwave Technol.*, **6**, 743, 1988.

21. W.D. Laidig, N. Holonyak, Jr., M.D. Camras, K. Hess, J.J. Coleman, P.D. Dapkus, and J. Hardeen, *Appl. Phys, Lett.*, **38**, 776, 1981.

22. D.G Deppe, L.J. Guido, N. Holonyak, Jr., C. Hsieh, R.D. Burnham, R.L. Thornton, T.L. Paoli, *Appl. Phys. Lett.*, **49**, 510, 1986.

23. C.J. Mchlean, J.H. March, and R.M. De La Rue, A.C. Bryce, B. Garrett, and R.W. Glew, *Electron. Lett.*, **28**, 1117, 1992.

24. Yong Kim, Shu Yuan, R. Leon, C. Jagadish, M. Gal, M.B. Johnston, M.R. Philips, M.A. Stevens Kalceff, J. Zou, and D.J.H. Cockayne, *J. Appl. Phys.*, **80**, 5014, 1996.

25. J.D. Ralston, W.J. Schaff, D.P. Bour, and L.F. Eastman, *Appl. Phys. Lett.*, **54**, 534, 1989.

26. G.K. Kothiyal and P. Bhattacharya, *J. Appl. Phys.*, **3**, 2760, 1988.

27. Shu Yuan, Yong Kim, R. Leon, C. Jagadish, T. Burke, M. Gal, D.J.H. Cockayne, R.M. Cohen, *Appl. Phys. Lett.*, **70**, 1269, 1997.

28. W.C.H. Choy, and E.H. Li, *IEEE J. of Quantum Electron.*, **33**, 382, 1997.

29. W.C.H. Choy, E.H. Li, and B.L. Weiss, *J. Appl. Phys.*, **83**, 858, 1998.

30. E.H. Li and W.C.H. Choy, *IEEE Photon. Technol. Lett.*, **7**(8), 881, 1995.

31. R.H.Yan, R.J. Simes, and L.A. Coldren., *IEEE J. of Quantum Electron.*, **25**, 2272, 1989.

32. M. Ghisoni, P.J. Stevens, G. Parry, J.S. Roberts, *Opt. Quantum Electron.*, **23**, S915, 1991.

33. G. Lullo, A. McKee, C.J. McLean, A.C. Bryce, C. Button, and J.H. Marsh, *Electron. Lett.*, **30**, 1623, 1994.

34. A. Ramdane, Krauz, E.V.K. Rao, A. Hamoudi, A. Ougazzaden, D. Robein, A. Gloukhian, and M. Carrÿ82, *IEEE Photon. Technol. Lett.*, **7**, 1016, 1995.

35. Y. Chen, J.E. Zucker, B. Tell, N.J. Sauer, and T.Y, Chang, *Electron. Lett.*, **29**, 87, 1993.

36. D.G. Deppe, L.J. Guido, N. Holonyak Jr., K.C. Hsieh, R.D. Burnham, R.L. Thornton and T.L. Paoli, *Appl. Phys. Lett.*, **49**, 510, 1986.

37. J.D. Ralston, S. O'Brien, G.W. Wicks,, and L.F. Eastman, *Appl. Phys. Lett.*, **52**, 1551, 1998.

38. C.J. Mchlean, A. Mckee, J.H. March and R.M. De La Rue, *Electron. Lett.*, **29**, 1657, 1993.

39. E.V.K. Rao, A. Hamoudi, Ph. Krauz. M. Juhel, and H. Thibierge, *Appl. Phys. Lett.*, **66**, 472, 1995.

40. It should be noted that the square of this length represents half of the variance of the interdiffusion distribution in a linear flow situation (i.e. $L_d^2 = 1/2$ variance, and $L_d = 1/2$ standard deviation). The stage of interdiffusion starts with a small L_d where QW compositional profile remains almost rectangular with a graded interface, As L_d increases further, the interdiffusion of wells becomes extensive, which results in complete interdiffusion, as L_d approaches infinity as averaged bulk material between the wells and barriers results.

41. F. Devaux, D. Bigan, A. Ougazzaden, B. Pierre, F. Huet, M. Carre, and A. Carenco, *IEEE Photon. Technol. Lett.*, **4**, 720, 1992.

42. W.C.H. Choy and E.H. Li, *IEEE J. Quantum Electron.*, **34**, July 1998.

43. E.H. Li and W.C.H. Choy, *J. of Appl. Phys.*, **82**, 3861, 1997.

44. J. Singh, S. Hong, P.K. Bhatacharya, R. Sahai, C. Lastufka, and H.R. Sobel, *J. of Lightwave Tech.*, **6**, 818, 1988.

45. T. Hayakawa, K. Takahashi, M. Kondo, T. Suyama, S. Yamamoto, and T. Hijikata, *Phys. Rev. Lett.*, **60**, 349, 1988.

46. A.M. Fox, D.A.B. Miller, G. Livescu, J.E. Cuningham, and W.Y. Jan, *IEEE J. of Quantum Electron.*, **27**, 2281, 1991.

47. T.C. Hasenberg, S.D. Kothler, D. Yap, A. Kost, and E.M. Garmine, *IEEE Photon Technol. Lett.*, **6**, 1210, 1994.

48. Z. Yang, B.L. Weiss, and E.H. Li, *Superlattice Microstruct.*, **17**, 177, 1995.

49. N. Debbar, S. Hong, J. Singh, Bhattacharya, and R. Sahai, *J. Appl. Phys.*, **65**, 383, 1989.

50. D.A.B. Miller, *Appl. Phy. Lett.*, **54**, 202, 1989.

51. P.J. Hughes, E.H. Li, and B.L. Weiss, *J. Vac. Sci. Techno. B.*, **13**, 2276, 1995.

52. B. Pezeshki, S.M. Lord, T.B. Boykin, and J.S. Harris, Jr., *Appl. Phys. Lett.*, **60**, 2779, 1992.

53. E.H. Li, K.S. Chan, B.L. Weiss, and J. Micallef, *Appl. Phys. Lett.*, **63**, 533, 1993.

54. K. Nakashima, Y. Kawaguchi, Y. Kawamura, Y. Imamura, and H. Asahi, *Appl. Phys. Lett.*, **52**, 1383, 1988.

55. C. Francis, F.H. Julien, J.Y. Emery, R. Simes, and L. Goldstein, *J. Appl. Phys.*, **75**, 3607, 1994.

56. J. Micallef, E.H. Li, and B.L. Weiss, *J. Appl. Phys.*, **73**, 7524, 1993.

57. R.C. Miller, A.C. Gossard, D.A. Kleinman, and O. Munteamu, *Phys. Rev. B*, **29**, 3740–3743, 1984.

58. T. Ishikawa, S. Nishimura, and K. Tada, *Jpn. J. of Appl. Phys.*, **29**, 1466 1990.

59. W.C.H. Choy and E.H. Li, *Appl. Optics*, **37**, 1674–1681, 1998.

60. W.C.H. Choy, Hughes, B.L. Weiss and E.H. Li, *Mat. Res. Soc. Symp. Proc.*, **450**, 425 1997.

61. J.C. Zucker, K.L. Jones, T.H. Chiu, B. Tell and K.B. Goebeler, *IEEE J. Lightwave Technol.*, **10**, 1926, 1992.

62. W.C.H. Choy, E.H. Li, and J. Micallef, *IEEE J. of Quantum Electron.*, **33**, 1316, 1997.

63. J.J. He, S. Charbonnneau, J. Poole, G.C. Aers, Y. Feng, E.S. Koteles, R.D. Goldberg, and I.V. Mitchell, *Appl. Phys, Lett.*, **69**, 562, 1996.

64. K. Mukai, M. Sugawara, and S. Yamazaki, *Phys. Rev. B*, **50**, 2273, 1994.

65. T. Fujii, M. Sugawara, S. Yazaki and K. Nakajima, *J. Cryst. Growth*, **105**, 348, 1990.

66. H. Ribot, K.W. Lee, R.J. Simes, P.H. Yan, and L.A. Coldren, *Appl. Phys. Lett.*, **55**, 672, 1989.

67. T. Miyazawa, T. Kagawa, H. Iwamura, O. Mikami, and M. Naganuma, *Appl. Phys, Lett.*, **55**, 828, 1989.

68. J.E. Zucker, M.D. Divino, T.Y. Chang, and N.J. Sauer, *IEEE Photon. Technol. Lett.*, **6**, 1105, 1994.

69. G.A. Evan, D.P. Bour, N.W. Carlson, R. Amantea, J.M. Hammer, H. Lee, M. Lurie, R.C. Lai, P.F. Pilka, R.E. Farkas, J.B. Kirk, S.K. Liew, W.F. Reichert, C.A. Wang, H.K. Choi, J.N. Walpole, J.K. Butler, W.F. Ferguson, Jr., R.K. D.Freez. and M.Felisky, *IEEE J. of Quantum electron.*, **27**, 1594, 1991.

70. H. Shen, M. Wraback, *Appl. Phys. Lett.*, **62**, 2908, 1993.

71. M. Whitehesd, G. Parry, Wheatly, *IEE proceedings*, **136, Pt. J**, 52, 1989

72. A. Tomita, Y. Kohga, and A. Suzuki, *Appl. Phys, Lett.*, **55**, 1817, 1989.

73. J.P. van der Ziel and M. Ilegems, *Appl. Optics*, **14**, 2627, 1975.

74. P.L. Gouley and T.J. Drummond, *Appl. Phys. Lett.*, **49**, 489, 1986.

75. K.K. Law, R.H. Yan. L.A. Coldren, and J.L. Merg, *Appl. Phys. Lett.*, **57**, 1345, 1990.
76. C.J.G. Kirkby, R.M. Ash, A.J. Moseley, and A.C. Carter, *Electron. Lett.*, **27**, 2373, 1991.
77. K.K. Law, M. Whitehead, J.L. Merz, and L.A. Coldren, *Electron. Lett.*, **27**, 1863, 1991.
78. N. Barnes, Healey, M.A.Z. Rejman-Greene, E.G. Scott, and R.B. Webb, *Electron. Lett.*, **26**, 1126, 1990.
79. M. Ghisoni, G. Parry, M. Pate, G. Hill, and J. Roberts, *Jpn. J. Appl. Phys.*, **30**, L1018, 1991.
80. J. Bleuse, G. Bastrad, and Voisin, *Phys. Rev. Lett.*, **60**, 220, 1988.
81. B. Joseph, K.W. Goossen, J.M. Kuo, F. Kopf, D.A.B. Miller and D.S. Chemla, *Appl. Phys. Lett.*, **55**, 340, 1989.
82. A. McKee, C.J. McLean, G. Luilo, A.C. Bryce, R.M. De La Rue, J.H. March, C.C. Button, *IEEE J. quantum Electron.*, **33**, 45, 1997.

CHAPTER 16

Analysis and Design of Semiconductor Lasers Using Diffused Quantum Wells Structure

S.F. YU AND C.W. LO

Department of Electrical and Electronic Engineering, The University of Hong Kong, Pokfulam Road, Hong Kong

1 INTRODUCTION

Semiconductor lasers with high output power, single-mode operation and low threshold current are essential for the applications in long-haul optical fiber communication systems as the coherent light sources. In the past decade, intensive studies have been concentrated on the design and fabrication of high-performance semiconductor lasers such as Fabry-Perot (FP) lasers [1–4] with high output power, distributed feedback (DFB) lasers [5][6] with single-mode operation and vertical-cavity surface-emitting lasers (VCSEL's) [7,8] with extremely low threshold current. However, some of their fabrication procedures are relatively complicated [9,10]. In order to reduce the production cost and enhance the yield rate, a relatively simple fabrication technique, the compositional-induced disordering of quantum wells (QW's), is proposed to fabricate semiconductor lasers [11].

Compositional-induced disordering of QW's is a simple, controllable and reproducible technique to change (i) the band-gap energy and (ii) the refractive index of the as-grown QW's such that the lateral confinement of optical field can be selectively defined at different locations of the QW wafer. Therefore, the re-growth of passive confinement layers for the ridge-waveguide [12] and buried heterostructure [13] devices is not required and the fabrication procedures of semiconductor lasers can be further simplified. Recently, several types of QW lasers using compositional-induced disordering process have been fabricated, including stripe-geometry [14], ridge-waveguide [12], and external-cavity lasers [15,16]. There are two possible methods to realize diffused QW's structure: (i) impurity-induced disordering (IID) and (ii) impurity-free vacancy diffusion (IFVD).

1.1 Facet Emitting Lasers

The lateral optical confinement of facet emitting lasers (i.e. FP and DFB lasers) can be realized by the disordering of QW's active layer. The main purpose of the disordering process is to change the refractive-index profile of the QW's active layer for the light confinement within the waveguide region. In the following paragraphs, the use of IID and IFVD to fabricate facet emitting lasers will be discussed.

1.1.1. Impurity-Induced Disordering of QW's

IID can be achieved by a direct implantation of ionized, energetic atoms or molecules into the as-growth QW's material. The kinetic energy and the implantation dose of the injected impurities can be varied from 1k to 10MeV and from 10^{10} to more than 10^{16} ions/cm^2, respectively. The implantation process will then be followed by rapid thermal annealing (RTA). The annealing process will last for several seconds within the temperature range from 900 to 1130°C. The use of implanted impurities is to induce free carriers for the generation of equal amount of vacancies during the annealing process. The role of RTA in the IID process is to (i) enhance the interdiffusion rate of the implanted impurities and to (ii) remove any lattice damage created by the implanted impurities. It must be noted that the penetration depth and the degree of damage to the lattice structure are increased with the implantation energy of the impurities. In addition, the interdiffusion rate is directly proportional to the amount of implanted dose. Moreover, the selective implantation of impurities at different locations to form the disordered region of QW's structure can be realized with the control of interdiffusion rate along lateral and longitudinal directions (see Figure 1).

In the fabrication of semiconductor lasers, different types of implanted ions can be used. They are simply divided into two major categories as follows:

Electrically active species — n- and p-type ions

Silicon can be used as n-type impurity in the IID process to fabricate semiconductor lasers. It is shown that excellent lateral-current confinement can be achieved in ridge-waveguide InGaAs/GaAs QW's laser by silicon induced disordering [17]. Other examples such as, (i) strip-geometry AlGaAs/GaAs QW's heterostructure laser on n-type substrates [18], (ii) index-guided buried heterostructure AlGaAs/GaAs multiple QW's laser for a high-power single-mode operation [2–4], and (iii) the integrated external-cavity AlGaAs/GaAs QW's lasers with low cavity loss, are fabricated by IID process with silicon as an n-type impurity [15,16,19].

Zinc is the common p-type impurity used in the IID process. It is reported that the aluminum-gallium interdiffusion rate can be enhanced at the heterointerfaces of an AlGaAs/GaAs QW's heterostructure by zinc diffusion such that uniform compositional disordering of AlGaAs can be created at low temperature [20]. Using this technique, zinc diffusion is utilized for the enhancement of optical and electrical confinements in a long-wavelength InGaAs/AlGaInAs multiple QW's laser [21].

A major problem associated with the electrically active ions is that they can be ionized at room temperature. Therefore, the generation of free carriers

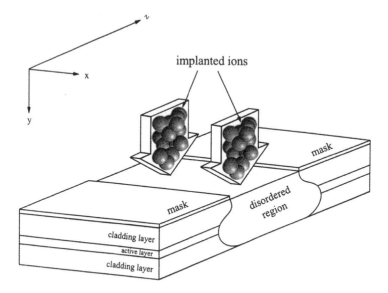

Fig. 1. Schematic of a semiconductor laser with a step diffused structure by impurity induced disordering of QW's.

increases the optical absorption loss inside the diffused QW's material and deteriorates the performance of the semiconductor lasers.

Non-electrically active species — neutral type and constituent ions

Oxygen, boron and fluorine are the typical neutral impurities used in IID process. These electronically inactive implanted ions can reduce the free-carrier losses and enhance electrical resistivity for the disordered regions. The use of neutral impurities can reduce (i) the leakage current and (ii) threshold current density in the semiconductor lasers.

For example, stripe-geometry AlGaAs/GaAs graded-barrier QW's semi-conductor laser can be realized by the implantation of high energy oxygen ions [14]. It is found that using low dose concentration ($10^{14} - 10^{16}$ cm^{-2}) of oxygen ions can eliminate leakage current by the formation of semi-insulating region while implanting high dose concentration ($>10^{17}$ cm^{-2}) can induce refractive index change. It must be noted that for the devices with narrow stripe-width (<5 μm), the lateral distribution of oxygen ions can increase the resistivity of the stripe region and thereby increase its threshold current.

The absorption edge of QW's material can be shifted by tens of meV away from its gain peak by the implantation of boron ions and long annealing time. This characteristic of boron is highly desirable for the fabrication of single-frequency distributed Bragg reflector (DBR) lasers with low optical losses and leakage current [11]. Moreover, a double QW's metal-clad ridge-waveguide laser can also be fabricated using fluorine implantation to form a passive cavity followed by RTA process [12]. With the usage of disordered passive waveguide, the propagation loss is significantly reduced by $4.5 \pm 2 \, \text{cm}^{-1}$.

Constituent implanted ions are already existed in the crystal lattice, so no doping effects are expected. Several semiconductor lasers are fabricated by using this type of implanted ions including (i) gallium implanted and annealed gain-coupled DFB InGaAs/InGaAlAs multiple QW's lasers for single-mode operation [22], (ii) high energy arsenic implanted in InGaAs/GaAs and (iii) phosphorus implanted in InGaAs/InP followed by annealing process for achieving spatially selective tuning of semiconductor laser structures [23,24].

1.1.2   Impurity-Free Vacancy Diffusion of QW's

IFVD is a simple way to control the degree of disordering in QW's material [25]. In order to achieve IFVD, a dielectric capping layer (e.g. SiO_2) is disposed on the III-V compound (e.g. GaAs) and annealed at a temperature of $900^o C$ or higher. As group III atoms (e.g. Ga) have a very high diffusion coefficient in dielectric capping layer at high temperature, this leads to the out-diffusion of group III atoms (e.g. Ga). The vacancies that are created in the III-V compound (e.g. GaAs) diffuse rapidly through the structure. When the vacancies cross the III-V compound QW's interface (e.g. AlGaAs/GaAs), QW intermixing occurs. The shape of QW's is slightly distorted by IFVD such that the bandgap of the QW's is shifted to a higher energy side and its refractive index is reduced [26]. This method is very attractive because no impurities are introduced into the process but only vacancies.

The IFVD technique can be used to fabricate multiple wavelength laser arrays by laterally-selective disordering. The control of IFVD depends upon the variation of the thickness and porosity of the dielectric capping layer over the QW's structure [27]. However, it is difficult to control the thickness of the dielectric capping layer due to the requirement of extra processing steps which makes the preparation procedure rather complicated. In addition, the thickness of dielectric capping layer is not reproducible in practice. In order to solve the above problem, a modified technique which is based on IFVD, called the selective intermixing in selected areas is proposed [28]. It uses a small pattern of strontium fluoride (SrF_2) as a band-gap control mask, followed by the deposition of a dielectric capping layer over the QW's structure. The SrF_2 masked patterns are smaller than or comparable with the diffusion length of the point defects to allow a uniform disordering at the QW's depth by overlapping of the vacancy diffusion fronts. Figure 2 shows

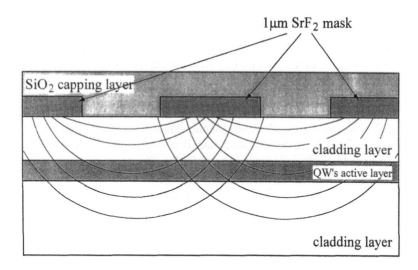

Fig. 2. Schematic diagram of vacancy diffusion profile with SrF₂ mask [29].

the schematic diagram of vacancy diffusion profile of SrF$_2$ mask [29]. As a result, a spatial control of the band-gap shift can be achieved by using a single RTA step only. The degree of disordering depends upon the contact area with the SiO$_2$ layer. This technique is a simple, one-step process and is reproducible. The application of the selective intermixing technique has been used on the fabrication of band-gap tuned lasers on a single chip.

1.2.   Surface Emitting Lasers

In gain-guided VCSEL's, the transverse confinement of injection current and optical field inside the Bragg reflector can be realized by ion implantation and IID. This is because the electrical resistivity is enhanced by the implanted ions but the refractive index is reduced by IID. However, the IID process has not been applied to the QW's active layer and this is the major difference when compared with the application of IID process to fabricate facet emitting lasers, see pervious sections. In the following paragraphs, the fabrication process of VCSEL's by using ion implantation and IID will be discussed.

1.2.1   Ion Implantation into Bragg Reflector

Current-confined structure of gain-guided VCSEL's can be obtained by proton implantation into the p-DBR stack mirror [30]. Figure 3 shows

(a) Shallow implantation   (b) Deeper implantation

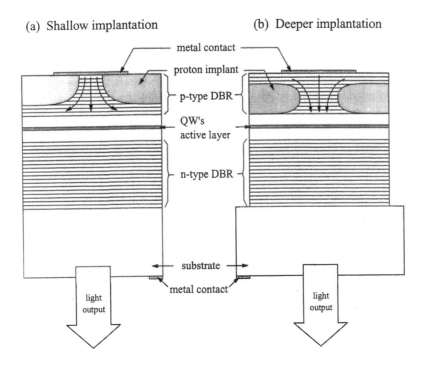

Fig. 3. Schematic of proton-implanted gain-guided VCSEL's with (a) shallow and (b) deep implantation [30].

the schematic diagram of gain-guided VCSEL's with shallow (i.e. low implantation energy) and deep (i.e. high implantation energy) penetration of ions. For a shallow penetration, the VCSEL's have higher series resistance and threshold current due to the significant current spreading. For a deep penetration, current spreading and threshold current are reduced. The advantages of using proton-implanted structure are (i) the simple fabrication process and (ii) the stable single-mode operation at high power. However, the broadening of beam divergence of output light may be resulted due to the lack of optical confinement [31].

In order to overcome the broadening of beam divergence, IID with zinc impurity implanted into the Bragg reflector has been proposed to form a spatial-mode filter [32]. The schematic diagram of a VCSEL using zinc diffusion technique is shown in Figure 4. The use of disordered p-type DBR mirror stack can enhance the lateral optical confinement and minimize the optical loss arising from diffraction. It is found that low threshold current and single transverse-mode operation can be achieved even in the devices of large diameter (>11 μm).

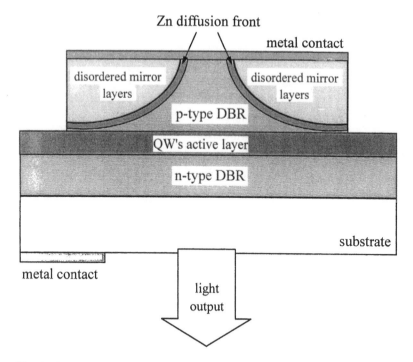

Fig. 4.   Schematic of a VCSEL with zinc diffusion and disordered mode filter [33].

1.2.2. Ion Implantation into Active Layer

Stable single-mode operation in VCSEL's can be further improved by selective ion implantation into the QW's active layers. The active layer is partially destroyed to form a high resistive region for the confinement of injection current. Figure 5 shows the schematic diagram of helium-implanted and zinc-diffused VCSEL [33]. The buried-insulating layer around the active region is formed by high energy implantation of helium ions. Furthermore, zinc IID process is implemented into the periphery of the p-DBR stack mirror. The purpose of zinc diffusion is to (i) form a low resistance path for the injection current and to (ii) improve the transverse confinement of optical field inside the stack mirror. It is found that the device can maintain single-mode operation even at high injection level.

2   UNDESIRED PROPERTIES OF SEMICONDUCTOR LASERS

Fabry-Perot, distributed feedback and vertical-cavity surface-emitting semi-conductor lasers are the most common devices used in the optical fiber

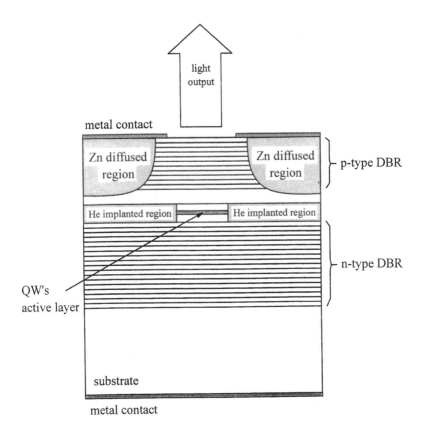

Fig. 5. Schematic of a VCSEL with helium implantation and zinc diffusion [33].

communication systems and the optoelectronics integrated circuits. The performance of these optical systems depends on the quality of the semiconductor lasers such as the purity of spatial and frequency spectra. Therefore, the control of the side modes in semiconductor lasers is essential to provide high-quality coherent light sources for the optical systems.

Resonant modes of FP and DFB lasers are supported in transverse, lateral and longitudinal directions of a rectangular cavity as shown in Figure 6(a). The condition for single-transverse-mode operation can be easily achieved by using a thin active layer. For conventional facet emitting lasers, the active layer is too thin (about 0.2 μm or less) to support higher-order transverse modes. Hence, single-transverse-mode operation is maintained. However, the higher-order lateral modes can be excited under high injection condition. This is due to the reduction of modal gain arising from spatial hole burning (SHB)

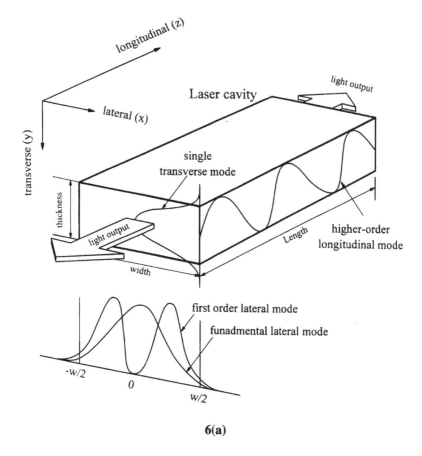

6(a)

of carrier concentration (i.e. a hole is burnt at the center of the waveguide due to the stimulated recombination of carriers). This can be avoided by using a narrow stripe-width design. In fact, the control of longitudinal modes is the most important issue in facet emitting lasers and will be discussed in Section 2.1 and 2.2.

In VCSEL's, the azimuthal symmetry of the cylinder cavity supports resonant modes in transverse and longitudinal directions, as shown in Figure 6(b). The typical length of the devices is less than its operating wavelength. Hence, the higher-order longitudinal modes will not be supported. In fact, higher-order transverse modes may dominate especially for the devices with large core diameter (>5 μm) under high current injection. Multiple transverse-mode operation in VCSEL's is affected by SHB of carrier concentration and thermal lensing inside the active layer. The issue of multiple transverse-mode operation in VCSEL's will be discussed in Section 2.3.

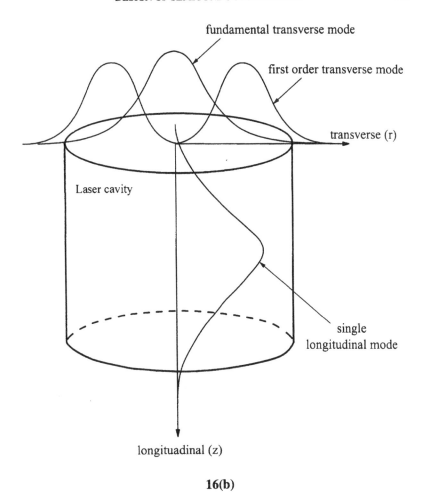

fundamental transverse mode

first order transverse mode

transverse (r)

Laser cavity

single
longitudinal mode

longituadinal (z)

16(b)

Fig. 6. Optical field profiles inside : (a) the facet emitting laser and (b) the surface emitting laser.

2.1 Lasing Characteristics of Fabry-Perot Semiconductor Lasers

The excitation of longitudinal modes in FP lasers is formed by the oscillation of all optical fields (which satisfy the round-trip amplitude and phase conditions) inside the laser cavity. In FP lasers, the modal gain of the longitudinal modes is mainly determined by the optical gain spectrum. However, the gain difference between the adjacent longitudinal modes is very small (<1 cm^{-1}). Therefore, side modes can reach their threshold simultaneously with the main mode and lead to a multiple longitudinal-mode operation.

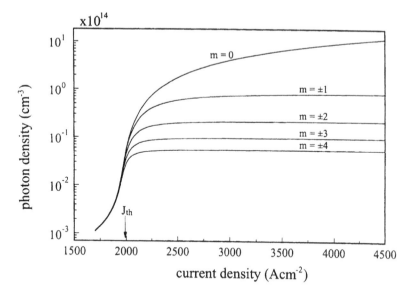

Fig. 7. The calculated photon density of different longitudinal modes various with injection current density [34].

Figure 7 shows the numerical simulation of photon densities of a FP laser as a function of injection current density. It is assumed that the length of the laser cavity is 250 μm and the other parameters used in the calculation are given in reference [34]. The threshold current density of the device is indicated by an arrow at the bottom of the figure. As we can see, for the current density increases above the threshold, the photon density of main mode ($m = 0$) grows faster than that of side modes. The corresponding output spectra at three different injection levels are shown in Figure 8. It is found that the side modes are strongly suppressed at high injection current because of the highest modal gain in main mode among all the longitudinal modes. Figure 9(a) shows the evolution of the resonant modes with P_0 representing the output power of main mode. It is assumed that the device is modulated by a step current of magnitude 1.5 times its threshold. The corresponding transient (i.e. at the first overshoot) and steady-state optical spectra of the laser are also shown in Figure 9(b). It is observed that the side modes are strongly excited at the time of first overshoot but hardly suppressed at steady state. During this time interval, the carrier and photon densities exhibit damped oscillation before reaching its steady state. The oscillation frequency is referred to the relaxation oscillation frequency which is defined as the reciprocal of the oscillation period of the lasers. It is noted that the relaxation oscillation frequency of all

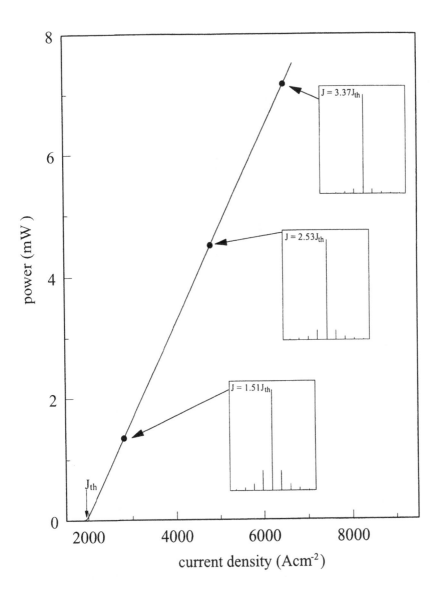

Fig. 8. Output power versus injection current characteristic with inserts of power spectra under different bias current [34].

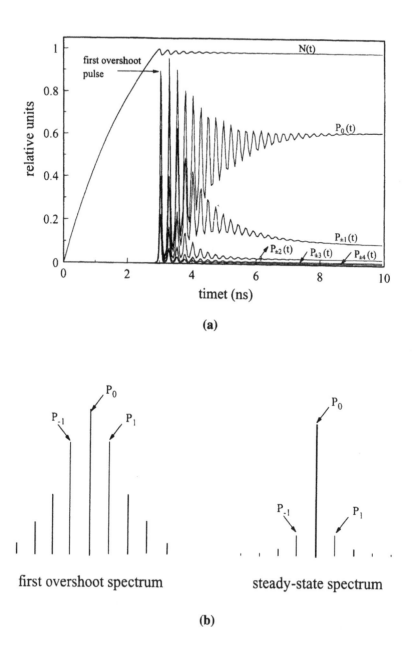

Fig. 9. (a) Switch-on transient response of carrier density and photon densities with different longitudinal modes under the step current modulation. (b) The spectra of the laser output at first peak overshoot and steady state [34].

longitudinal modes is identical and insensitive to the number of modes. It is expected that the spectral width is broadened due to the excitation of the side modes and that the modulation bandwidth of the optical systems is limited.

2.2 Lasing Characteristics of $\lambda/4$ Phase-Shifted Distributed Feedback Semiconductor Lasers

Single-longitudinal-mode operation can be achieved in FP lasers with the introduction of Bragg grating. This type of lasers is denoted as DFB lasers. This is because the wavelength selectivity in laser cavity is achieved by the grating structure. However, band-edge modes can be excited in DFB lasers due to the symmetrical geometry of the grating [35]. Alternatively, the problem of two-mode degeneracy can be overcome by introducing a $\lambda/4$ phase-shifted [36,37]. However, the side modes can still be excited in $\lambda/4$ phase-shifted DFB lasers with large coupling-length product κL due to the longitudinal SHB of carrier concentration.

$\lambda/4$ phase-shifted DFB laser with perfect AR coated on both facets is investigated. The structural and material parameters of the DFB laser can be found in reference [38]. It is assumed that the laser is initially biased at threshold and modulated with a current step of magnitude twice its threshold. Figure 10 shows the switch-on transient response of device with κL equals 1.25. The transient signal exhibits single-mode operation without mode beating after 0.75 ns. The corresponding steady-state spectrum is also shown in Figure 11. It is observed that single-mode operation is maintained at steady state. However, a completely different transient response can be obtained for the device with large κL. Figure 12 shows the transient response of device with $\kappa L = 2.8$. It is observed that the strong mode beating is excited after 1.5 ns. The corresponding steady-state spectrum is given in Figure 13 which indicates the excitation of side mode. The occurrence of side mode is caused by a non-uniform (concave-up) distribution of the refractive index, see Figure 14(a) which is a consequence of the longitudinal SHB of carrier concentration [39], see Figure 14(b). In order to maintain stable single-mode operation in DFB lasers, longitudinal SHB of carrier concentration should be avoided.

2.3 Lasing Characteristics of Vertical-Cavity Surface-Emitting Lasers

Multiple transverse-mode operation of VCSEL's can be attributed to transverse SHB of carrier concentration [40–42] as well as thermal lensing. SHB arises from the stimulated recombination of transverse field inside the active layer. On the other hand, due to the high thermal resistivity of the laser cavity, heat generated inside the active layer can be very high [43,44] (i.e. 25–30°C higher than the substrate temperature [45]). This leads to a substantial increase in refractive index and which is denoted as thermal lensing effect.

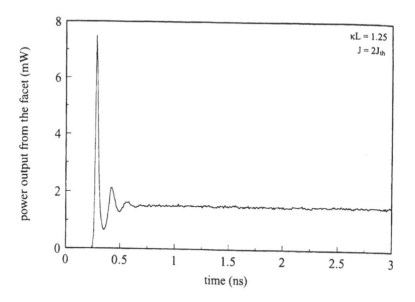

Fig. 10. Transient response of output power from AR coated λ/4-shifted DFB laser with $\kappa L = 1.25$ under the step current modulation.

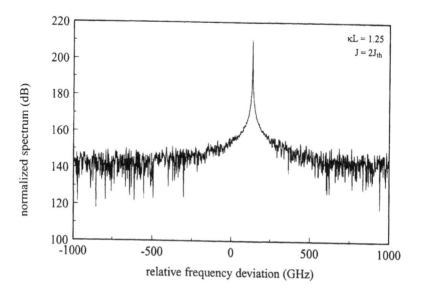

Fig. 11. The corresponding steady-state spectrum of Figure 10.

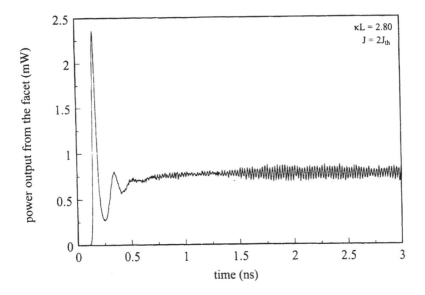

Fig. 12. Transient response of output power from AR coated λ/4-shifted DFB laser with $\kappa L = 2.8$ under the step current modulation.

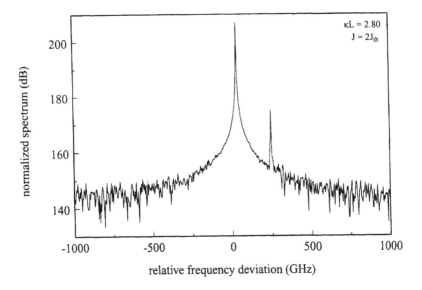

Fig. 13. The corresponding steady-state spectrum of Figure 12.

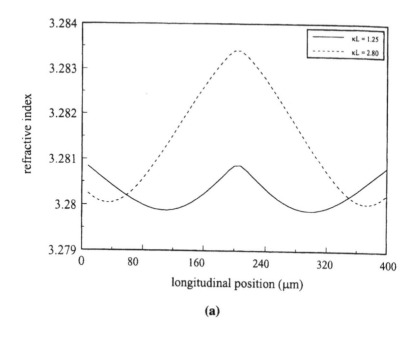

(a)

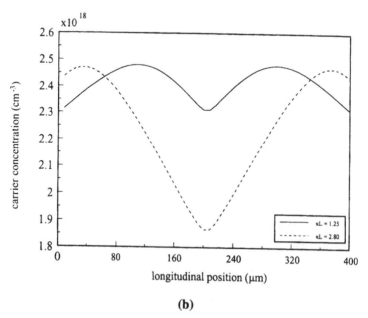

(b)

Fig. 14. Longitudinal spatial distribution of (a) the refractive-index profile and (b) the carrier concentration at steady state.

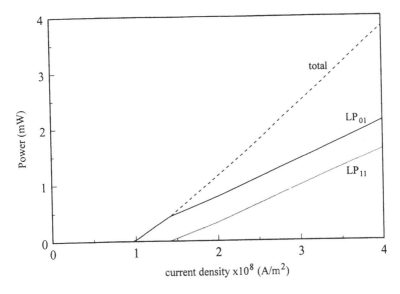

Fig. 15. Calculated light-current characteristics of VCSEL with LP_{01} mode (solid line) and LP_{11} mode (dotted line). [41].

In the following sections, the influence of carrier SHB and thermal lensing on the transverse-mode characteristics of VCSEL's will be studied.

2.3.1   Influence of Carrier Spatial Hole Burning

Figure 15 shows the light-current characteristics of a VCSEL with core radius equals 3 μm. The structural and material parameters used in the calculation can be found in reference [41]. It is shown that the output power of fundamental $(LP_{01})^\dagger$ and first-order (LP_{11}) transverse modes increase almost linearly above their respective threshold. A 'kink' is observed in the light-current curve when the LP_{11} mode starts to lase. The influence of carrier SHB leads to the excitation of LP_{11} mode because the modal gain of LP_{01} mode is reduced.

The evolution of the transverse-mode behavior is analyzed by initially biasing the device at threshold and modulating it with a current step of magnitude 3 times its threshold. Figure 16(a) shows the switch-on transient response of LP_{01} and LP_{11} modes. It is observed that the VCSEL operating at multiple transverse-mode with LP_{01} mode has slightly higher power at steady

---

†Transverse modes can be expressed in terms of linearly-polarized $(LP_{\ell p})$ modes, where the indexes ℓ and p denote the azimuthal and radial order of modes, respectively.

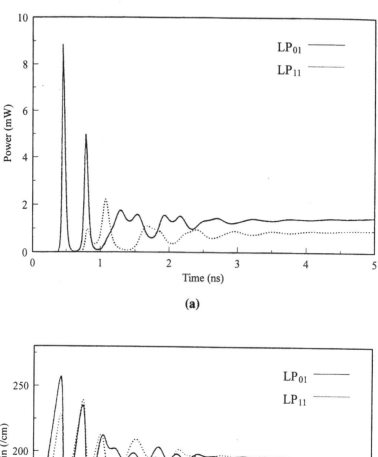

Fig. 16. Transient response of (a) output power and (b) modal gain of the LP_{01} mode (solid line) and LP_{11} mode (dotted line) for VCSEL [41].

$\times 10^6$ Carrier density (um^{-3})

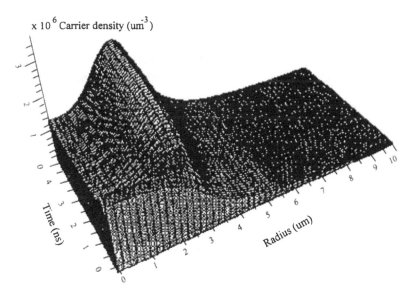

Fig. 17. Time evolution of carrier concentration profile of a VCSEL [41].

state. The competition between LP_{01} and LP_{11} modes can be explained by the interaction between the distribution of transverse modes and the carrier concentration profile inside the active layer. When the modal fields are spatially separated, it is found that the modes may share the available gain. Hence, a multiple transverse-mode operation is obtained. This is confirmed by the corresponding profile of modal gain as shown in Figure 16(b). The evolution of the transverse distribution of carrier concentration is also shown in Figure 17. As we can see, SHB arises from the switch-on of LP_{01} mode. However, the existence of SHB reduces the modal gain of LP_{01} mode due to the reduction of magnitude in spatial overlap between the profile of carrier concentration and LP_{01} mode. Therefore, the excitation of LP_{11} mode is achieved with the presence of SHB. For large-area devices, severe SHB is expected and the tendency of a multiple transverse-mode operation is enhanced [46]. In the above investigation, the influence of thermal effects is ignored in the calculation.

2.3.2 Influence of Thermal Lensing Effect

Figure 18(a) shows the near-field intensity distributions of LP_{01} and LP_{11} modes in VCSEL's with and without the consideration of thermal lensing. The VCSEL's used to investigate have core radii equal 5 μm, see reference [47].

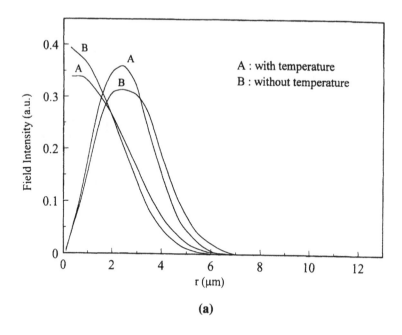

(a)

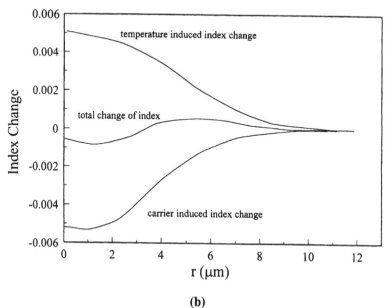

(b)

Fig. 18. (a) Near-field intensity distribution of the LP_{01} and LP_{11} modes with (denoted as 'A') and without (denoted as 'B') taking the temperature into account. (b) The refractive-index changes induced by carriers and heat in the VCSEL above lasing threshold situation [47].

It is found that the intensity of LP_{01} mode decreases while the LP_{11} mode increases due to self-heating effect. The influence of the temperature distribution becomes significant when compared with the mode intensities of the VCSEL without taking the temperature into account. Figure 18(b) shows the half width of the refractive-index profile. As we can see, the refractive-index profile induced by heating is larger than that induced by carriers (i.e. the former is 8 μm and the latter is 6.9 μm). It is expected that the temperature profile with a wider full-width half-maximum is favorable to emit the LP_{11} mode. Therefore, in order to achieve stable single-transverse-mode operation in VCSEL's, careful control of temperature distribution is required.

3   OPTICAL PROPERTIES OF DIFFUSED QUANTUM WELLS — DEVICE ENGINEER'S VIEWPOINT

The variation of optical gain and background refractive index of QW's material as a function of injection carrier concentration is very important to determine the performance of the semiconductor lasers. The compositional-induced disordering alters the band structure of the as-grown QW's such that the optical gain and background refractive index spectra are affected. This is expected as the effective band-gap energy increases, the optical gain peak shifts to the shorter-wavelength region and its magnitude reduces.

Figure 19 shows the influence of interdiffusion of zinc impurities on the optical gain and background refractive index spectra (TE polarization) of $Al_{0.3}Ga_{0.7}As/GaAs$ as-grown QW's at room temperature (i.e. 300 K). It is assumed that the well width and barrier thickness of as-grown QW's are equal to 100 Å and 280 Å, respectively. The effect of interdiffusion is characterized by a diffusion length, L_d where $L_d = 0$ Å represents the as-grown QW's. In the analysis, the Lorentzian broadening factor with half-width half-maximum of 5 meV is used for the calculation of TE mode gain spectra. Other parameters used in this calculation can be found in reference [48]. The detailed modeling of optical gain and background refractive index spectra can be found in the previous chapters. In Figure 19, the injection carrier concentration, N is set to 3×10^{18} cm^{-3}. It is observed that as L_d increases from 0 Å to 35 Å, the gain peak shifts to the shorter-wavelength side (from 0.85 μm to 0.77 μm). The magnitude of the gain peak, G_p, increases slightly to a maximum value, and then decreases beyond $L_d = 10$ Å until L_d reaches 35 Å, where no gain appears. In fact, at $L_d = 30$ Å, the gain peak reduces to 1/5 of the as-grown QW's and the explanation can be found in reference [49]. For L_d greater than 50 Å, no gain peak occurs and the magnitudes become almost constant over the range of wavelength. This unique feature of diffused QW's structure can be utilized to design a novel structure of

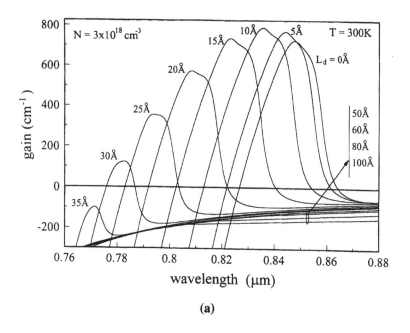

(a)

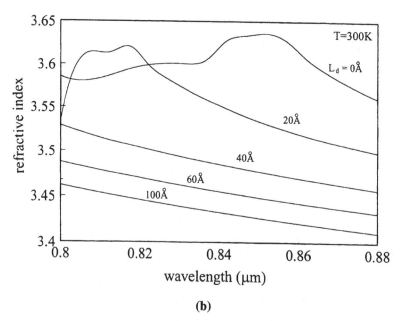

(b)

Fig. 19. Calculated (a) optical gain and (b) background refractive-index spectra of $Al_{0.3}Ga_{0.7}As$/GaAs quantum-well at carrier concentration, $N = 3 \times 10^{18}$ cm^{-3} with various levels of L_d.

high-performance semiconductor lasers and will be discussed in the next chapter.

The carrier concentration at transparency, magnitude and gain-peak wavelength are usually extracted from the optical gain spectra for the analysis of semiconductor lasers. Firstly, we consider the relationship between optical gain peak and injection carrier concentration. Using the approach in reference [50], the gain peak, G_p, and injection carrier concentration, N, can be fitted by a logarithmic relation with different magnitudes of L_d as shown below:

$$G_p = a_N(1 + a_1 T + a_2 T^2) \ln \left[\frac{N}{N_o(1 + b_1 T)} \right], \qquad (1)$$

where T ($\leq$ 300 K) is the temperature, the parameters a_N (cm^{-1}) and N_o (cm^{-3}) are the gain coefficient and carrier concentration at transparency, respectively. In equation (1), a_N and N_o are assumed to vary with L_d and their dependence on L_d are given in Table 1. Figure 20 shows the variation of gain coefficient, a_N, and transparency carrier concentration, N_o, with L_d (at 300 K). The temperature dependence of G_p is approximated by the coefficients $a_1 = 3.0815 \times 10^{-3}$K^{-1}, $a_2 = 5.3170 \times 10^{-6}$K^{-2}, and $b_1 = 6.0170 \times 10^{-3}$K^{-1}.

Figure 21 shows the variation of gain spectra with N for the as-grown QW's at 300K. As we can see, the gain-peak wavelength, λ_p, is varied rigorously for N lower than 3×10^{18} cm^{-3} but nearly clamped at a fixed value for large injection values. In semiconductor lasers, high injection of carrier concentration ($> 3 \times 10^{18}$ cm^{-3}) is required to obtain enough optical gain for lasing operation due to the large optical loss inside the laser cavity (>30 cm^{-1}). Therefore, the dependence of λ_p on N is ignored in our consideration. It will be more appropriate to consider the average value of gain-peak wavelength, $\overline{\lambda_p}$, for N greater than 3×10^{18} cm^{-3} but less than 10×10^{18} cm^{-3}. Figure 22 shows the variation of $\overline{\lambda_p}$ with L_d at 300 K. It is observed that decreases with L_d varying from 10 Å to 100 Å. At this range of L_d, the relation between $\overline{\lambda_p}$ and L_d at 300 K can be approximated by:

$$\overline{\lambda_p} = \overline{\lambda_o} - \lambda_N \log \left(\frac{L_d}{L_o} \right), \qquad (2)$$

where $\overline{\lambda_o}$ (≈ 0.9352 μm) and λ_N (≈ 0.04818 μm) are the fitting parameters and L_o ($= 1$ Å) is a normalizing number.

The refractive-index change induced by carrier, Δn, which varies with the background refractive-index profile of QW's active region can be calculated from the change of optical gain, $\Delta G(\omega) = G(\omega) - G_o(\omega)$, through the Kramers-Kronig dispersion relation [51]:

$$\Delta n(\omega) = \frac{\pi}{c} PV \int_0^\infty \frac{\Delta G(\omega')}{\omega'^2 - \omega^2} d\omega', \qquad (3)$$

TABLE 1

TE mode optical gain parameters of the $Al_{0.3}Ga_{0.7}As/GaAs$ diffused quantum well at gain-peak wavelength.

L_d (Å)	$\overline{\lambda}_p$ (μm)	a_N (cm^{-1})	$N_o \times 10^{18}$ (cm^{-3})	m	$N_r \times 10^{18}$ (cm^{-3})
0	0.8462	1799.7755	1.9917	−0.02636	2.0445
3	0.8443	1787.2177	1.9756	−0.02588	2.0080
5	0.8462	1798.1860	1.9499	−0.02529	1.9872
7	0.8400	1768.7200	1.9195	−0.02457	1.9452
10	0.8347	1742.0925	1.8979	−0.02368	1.9360
13	0.8281	1715.4632	1.9031	−0.02309	1.9285
15	0.8226	1690.0462	1.9306	−0.02282	1.9776
17	0.8165	1659.3795	1.9771	−0.02257	2.0005
20	0.8063	1656.4868	2.1345	−0.02345	2.1914
23	0.7971	1618.3227	2.2889	−0.02371	2.3626
25	0.7913	1597.0120	2.4125	−0.02412	2.4816
30	0.7781	1540.6246	2.7887	−0.02494	2.8845
35	0.7664	1519.2041	3.2924	−0.02720	3.4370
40	0.7572	1501.3819	3.8650	−0.02866	3.9478
45	0.7494	1511.2080	4.5418	−0.03149	4.6454
50	0.7428	1532.6064	5.2774	−0.03436	5.3823
55	0.7364	1641.3285	6.0169	−0.04352	6.4311
60	0.7318	1665.1642	6.8269	−0.04554	7.0765
65	0.7275	1738.6366	7.6483	−0.05072	7.9316
70	0.7238	1811.7305	8.4476	−0.05574	8.7699
75	0.7215	1822.3990	9.2495	−0.05959	9.5587
80	0.7177	1994.2443	10.0222	−0.06578	10.3892
85	0.7157	2011.4179	10.7258	−0.06751	10.9439
90	0.7143	2023.6172	11.4268	−0.07124	11.6623
95	0.7119	2159.0696	12.0870	−0.07635	12.3632
100	0.7105	2183.3428	12.6977	−0.07820	12.8887

where $G_o(\omega)$ is the optical gain profile at transparency. The symbol PV stands for the Cauchy principle value. It can be shown that Δn, at a particular wavelength, λ, and L_d, can be approximated by:

$$\Delta n(\lambda) = m(1 + d_1 T + d_2 T^2) \ln \left[\frac{N}{N_r(1 + e_1 T)} \right] \qquad (4)$$

where the two fitting parameters, m and N_r (cm^{-3}), are assumed to vary with L_d only and their dependence with L_d are given in Table 1. Figure 23 shows the variation of m and N_r with L_d at $\overline{\lambda}_p$ (at 300 K). It is noted that as L_d increases, m decreases whereas N_r increases. The temperature dependence of Δn is approximated by the coefficients $d_1 = 2.0791 \times 10^{-4}$K^{-1}, $d_2 = 1.3658 \times 10^{-7}$K^{-2} and $e_1 = 3.62 \times 10^{-2}$K^{-1}.

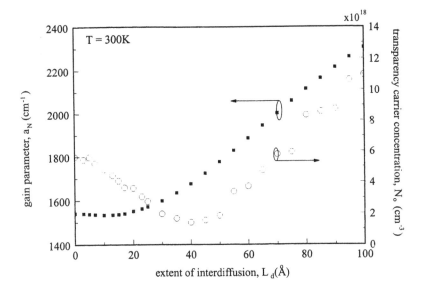

Fig. 20.   The variation of transparency carrier concentration and gain parameter against L_d at wavelength of gain peak.

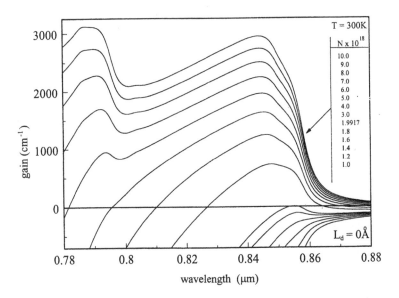

Fig. 21.   Calculated optical gain spectra of $Al_{0.3}Ga_{0.7}As/GaAs$ quantum-well at interdiffusion length, $L_d = 0$ Å with various levels of N.

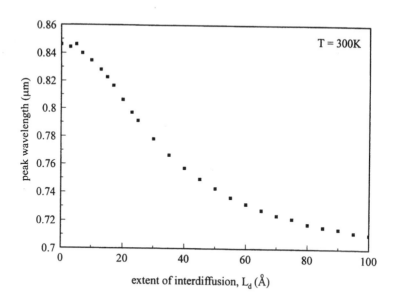

Fig. 22. The variation of peak wavelength against L_d.

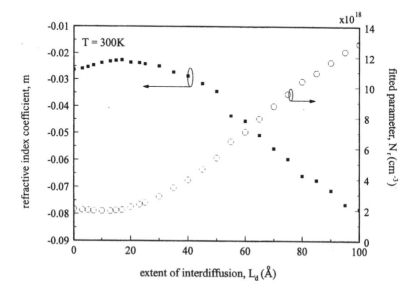

Fig. 23. The variation of fitted parameters, m and N_r against L_d at wavelength of gain peak.

In some applications, the operating wavelength of $Al_{0.3}Ga_{0.7}As/GaAs$ semiconductor lasers may set to around 0.85mm. Hence, the fitting parameters given in equations (1), (2) & (4) have to be calculated again. The variation of optical gain at this wavelength is quite different from that at optical gain-peak wavelength. Firstly, the logarithmic relation between gain and injection carrier concentration is only satisfied for a small range of L_d (0 Å $\leq L_d \leq 20$ Å). However, for $L_d > 20$ Å, the TE optical gain, G, at 0.85 μm can be fitted by a linear approximation and is given as follows:

$$G = a_g(N - N_g), \tag{5}$$

where a_g (cm^2) and N$_g$ (cm^{-3}) are the two fitting parameters and are assumed to vary with $L_d > 20$ Å only. The value of these parameters are given in Table 2. The temperature dependence on G when $L_d > 20$ Å is ignored in the calculation because the change of optical loss with the operating wavelength is almost negligible (i.e. shift of gain spectrum due to temperature effect) as shown in Figure 19(a).

Similarly, for L_d larger than 20 Å, Δn cannot be calculated through the Kramers-Kronig dispersion relationship due to the fact that the range of injection carrier concentration is not large enough to achieve optical gain. Therefore, Δn is not determined at this range of L_d and the effects of carrier-induced refractive-index change is ignored in our consideration. In Table 2, the corresponding fitting parameters for L_d varying between 0 and 100 Åare given at the operating wavelength of 0.85 μm. It must be noted that the refractive-index change due to the variation of temperature is also ignored in the calculation for the same reason discussed before.

It is observed that the error occurs for the logarithmic fitting of optical gain for the range of L_d varies between 0 and 20 Å is less than 5% at 0.85 μm. However, the error may be larger than 10% for L_d larger than 20 Å. Therefore, linear fitting of optical gain is used for L_d larger than 20 Å in order to keep the error down to 10%, see equation (5). Furthermore, Dn is only fitted for L_d less than or equal to 20 Å because optical gain is not achieved in the range of carrier injection level for L_d larger than 20 Å, see Table 2.

4   ENHANCEMENT OF SINGLE-MODE OPERATION IN SEMICONDUCTOR LASERS BY USING DIFFUSED QW'S STRUCTURE

Simple fabrication procedure and low production cost are the major advantages of FP lasers over the other devices. However, the side-mode suppression ratio (SMSR) in FP lasers is poor especially for direct electrical modulation [52] and it is highly desired to improve the SMSR of FP lasers.

TABLE 2

TE mode optical gain parameters of the $Al_{0.3}Ga_{0.7}As$/GaAs diffused quantum well at wavelength, $\lambda = 0.85\ \mu m$.

L_d (Å)	a_N (cm^{-1})	$N_o \times 10^{18}$ (cm^{-3})	m	$N_r \times 10^{18}$ (cm^{-3})	$a_g \times 10^{-18}$ (cm^2)	$N_g \times 10^{18}$ (cm^{-3})[†]
0	1591.6434	1.9399	−0.02830	2.0557		
3	1500.8038	1.9333	−0.02790	2.0464		
5	1384.6995	1.8587	−0.02783	1.9771		
7	1154.3764	1.8074	−0.02780	1.9260		
10	432.1735	2.2196	−0.02644	2.4109		
13	181.8557	3.6493	−0.02327	3.7733		
15	128.7685	5.1113	−0.02264	5.2314		
17	100.2318	7.0739	−0.02199	7.1131		
20	65.5869	11.7549	−0.02539	11.7883		
23					9.3088	13.0104
25					8.2996	14.7520
30					6.6167	18.8419
35					5.6701	22.3176
40					5.0912	25.2615
45					4.6867	27.7333
50					4.3819	29.8423
55					4.3145	31.1034
60					4.1967	32.3840
65					4.1525	33.3089
70					4.7535	31.7092
75					4.8307	32.1574
80					7.0395	27.3708
85					6.9883	28.2895
90					7.0668	28.6972
95					7.0030	29.5391
100					7.1861	29.6957

[†] a_g and N_g are used to approximate the optical gain for $L_d = 23$ Å to 100 Å.

Therefore, a periodic diffused QW's structure is proposed to improve the side-mode discrimination of FP semiconductor lasers. A periodic variation of refractive index and gain is created in the extent of interdiffusion along the longitudinal direction of the QW's active region which acts as a spatial-mode filter to against side modes. This diffused QW's FP laser is similar to a complex-coupled DFB laser with a higher-order grating.

The excitation of band-edge modes in $\lambda/4$ phase-shifted DFB lasers is due to the non-uniform distribution of refractive index which arises from the longitudinal SHB of carrier concentration. In order to enhance stable single-mode operation in the $\lambda/4$ phase-shifted DFB, we can (i) modify the longitudinal refractive-index profile or (ii) remove the $\lambda/4$ phase-shifted. In fact, it is possible to vary the refractive-index profile by disordering of QW's active layer. This can be achieved by modifying the interdiffusion length, L_d, of the diffused QW's [53]. The active region of this modified laser structure consists of non-uniform (step change in the extent of interdiffusion) diffused QW's structure along the longitudinal direction of the device. This will create a non-uniform stepped refractive-index profile in the longitudinal direction and can compensate the SHB effect.

Furthermore, the $\lambda/4$ phase-shifted can also be replaced by a phase-adjustment region (PAR) with a step diffused QW's structure. The use of the PAR is to provide an equal amount of phase-shifted for single-longitudinal-mode oscillation. Since the DFB laser has a uniform-grating structure, the longitudinal SHB is less severe than the conventional $\lambda/4$-shifted DFB laser [39]. In addition, the step diffused QW's profile compensates for any variation of refractive index arises from temperature effects [54] such that single-longitudinal-mode operation can be maintained at high power.

The excitation of multiple transverse modes in VCSEL's is caused by the increase in refractive index, arises from carrier spatial hole burning and thermal lensing, inside the core region of the active layer [47]. Therefore, the self-focusing increases the effective optical gain of higher-order transverse modes [40]. It is suggested using passive anti-guiding structure for VCSEL's to enhance single-mode operation [55]. However, organometallic chemical vapor deposition and molecular beam epitaxy are required for the re-growth of passive anti-guiding structure which is much more complicated than gain-guided structure [56]. Therefore, the possibility of using diffused QW's structure in VCSEL's, to create a non-uniform stepped refractive-index profile to against multiple transverse-modes operation, is proposed. The purpose of the step-diffused QW's structure is to compensate for the influence of self-focusing due to SHB and thermal lensing.

In the following paragraphs, the possibility of using (i) periodic diffused QW's structure in FP lasers, (ii) step diffused QW's structure in DFB lasers and (iii) non-uniform diffused QW's structure in VCSEL's are proposed and analyzed in details.

4.1 Fabry-Perot Lasers with Periodic Diffused QW's Structure

Figure 24 shows the schematic diagram of a FP laser with periodic diffused QW's structure. The laser is composed of six layers; n^+-GaAs substrate, n-GaAs buffer layer, n-AlGaAs guiding layer, followed by an active QW's layer, and finally the p-AlGaAs ($\cong 0.2 \ \mu m$) and p^+-GaAs ($<0.3 \ \mu m$) layers which form the cladder and contact. The active layer re-grows on the top of the guiding layer which consists of four $Al_{0.3}Ga_{0.7}As$/GaAs single QW's with well and barrier thickness of 100 Å and 280 Å, respectively. The periodic

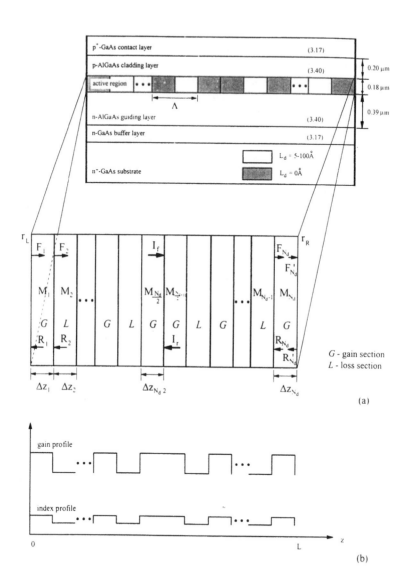

Fig. 24. (a) Schematic of a FP laser with periodic diffused QW's structure. (b) The periodic variation of gain and built-in refractive-index profile along the diffused QW's active region.

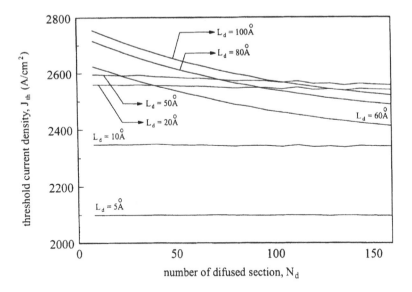

Fig. 25. The threshold current density against total number of diffused sections with the gain region is defined by diffusion length, $L_d = 0$ Å and L_d in loss region is set to 5 Å, 10 Å, 20 Å, 50 Å, 60 Å, 80 Å, and 100 Å, respectively.

variation of gain and refractive index (along the longitudinal direction of the active region) is obtained by periodic interdiffusion into the active region. The as-grown QW's section ($L_d = 0$ Å) serves as a gain region while the diffused section ($L_d = 5 - 100$ Å) serves as a loss region with large differences of refractive index as well as optical gain. The longitudinal length of each diffused section, Δz_k (for $k = 1, 2, \ldots, N_d$), is equal to a multiple of $\lambda_o/4$, where N_d is the total number of diffused sections and λ_o is the operating wavelength. The total length of the device is set to 400 μm with N_d varying between 8 and 160, which can be done by alternating the period of the diffusion grating. The left and right facet reflectivities of the laser cavity are both assumed to be 0.55. Detailed models and parameters used in the calculations can be found in the reference [57].

Figure 25 shows a plot of the threshold current density, J_{th}, of the periodic diffused QW's FP laser against N_d with L_d as a variable parameter. It is found that for large L_d (>50 Å), J_{th} is inversely proportional to N_d. In addition, a maximum J_{th} is located at L_d's combination which approximately equals 0|100 Å for $N_d < 80$ and equals 0|50 Å for $N_d > 100$. This is because the optical distributed feedback is affected by the design of the periodic diffused QW's structure. It is noted that optical gain at 0.85 μm can only be obtained by external carrier injection for diffused QW's with $L_d < 15$ Å, see Figure 19(a).

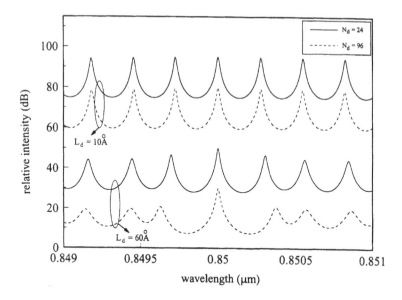

Fig. 26. Threshold emission spectra of diffused QW's FP lasers for (i) $L_d = 10$ Å and $N_d = 24$, (ii) $L_d = 10$ Å and $N_d = 96$, (iii) $L_d = 60$ Å and $N_d = 24$ and (iv) $L_d = 60$ Å and $N_d = 96$.

For $L_d > 60$ Å, optical feedback is enhanced due to the large difference in gain and refractive index between grating sections, [see Figure 19]. Therefore, low threshold current density can be obtained at $L_d < 10$ Å or $L_d > 60$ Å and it is also a function of N_d. It must be noted that the threshold current density of FP lasers without periodic diffused QW's structure is equal to 2123 A/cm^2. Figure 26 shows the corresponding threshold amplified spontaneous spectra for L_d's combination equals 0|10 and 0|60 Å (with N_d equal to 24 and 96). As shown in the figure, band-gap mode is dominant in the spectra especially when L_d's combination is equal to 0|60 Å and $N_d = 96$. This is expected as the filtering properties of the periodic diffused QW's structure are more efficient for large magnitude of L_d and N_d. In the design of devices with diffused QW's structure, the value of L_d should not be greater than 100 Å; otherwise the electrical and optical properties of QW's will be removed.

Figure 27(a) shows the amplified spontaneous spectra for device with L_d's combination equals 0|60 Å and $N_d = 96$. As we can see, when the current density increases from $1.0J_{th}$ to $1.3J_{th}$, the band-gap mode dominates over the other longitudinal modes. Further increase in current density excites band-edge mode (with longer wavelength) due to the influence of longitudinal SHB. Figure 27(b) shows the corresponding longitudinal distribution of carrier concentration at different injection levels. At the injection level

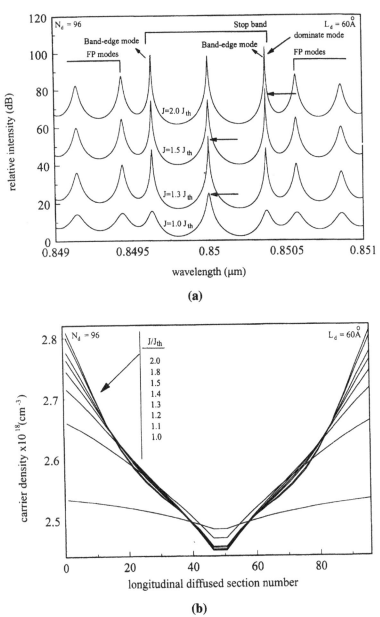

(a)

(b)

Fig. 27. (a) The spontaneous emission power spectra of diffused QW's FP lasers. The device has 96 gain/loss diffused sections and L_d's combination is equal to 0ÿ8Fl60ÿ8F. The injection current density is set to $1.0J_{th}$, $1.3J_{th}$, $1.5J_{th}$ and $2.0J_{th}$. (b) The carrier concentration profile of the lasers with injection current density set to $1.0J_{th}$, $1.1J_{th}$, $1.2J_{th}$, $1.3J_{th}$, $1.4J_{th}$, $1.5J_{th}$, $1.8J_{th}$ and $2.0J_{th}$.

equals $1.5J_{th}$, the longitudinal carrier distribution changes rapidly due to the excitation of band-edge mode but it is stabilized with further increase in the injection current.

The period of the diffusion grating used in the above calculation is in the order of 5 μm. It is expected that the DFB modes are repeated in the optical spectrum. In order to avoid the influence of the higher-order DFB modes, the band-gap mode should be selected in coherent with the optical gain peak of the active region ($L_d = 0$ Å) such that other DFB modes away from the gain-peak wavelength are suppressed. In the above analysis, it is assumed that the wavelength of the band-gap mode (i.e. around 0.85 μm) is coherent with the optical gain peak of the as-grown QW's.

4.2   Distributed Feedback Lasers with Step-Diffused QW's Structure

In the following paragraphs, two types of DFB lasers, (i) $\lambda/4$ phase-shifted DFB lasers and (ii) uniform-grating DFB lasers, using step-diffused QW's structure to enhance single-mode operation are discussed.

4.2.1 $\lambda/4$ Phase-Shifted DFB Lasers with Step-Diffused QW's Structure

The $\lambda/4$ phase-shifted DFB laser structure under investigation is shown in Figure 28. The structure is similar to the conventional FP lasers except that the grating is defined in the guiding layer. It is assumed that the active region has $L_d = 0$ Å except near the $\lambda/4$ phase-shifted region where a single step of L_d ($\cong 10$ Å) is introduced for a longitudinal length of $L_s = 200$ μm, and where the total longitudinal length of the laser cavity, L, is equal to 400 μm. Detailed models and parameters used in the calculations can be found in the reference [58]. It is assumed that the operational wavelength is 0.843 μm which is the effective gain peak wavelength of diffused QW's in the active region.

Figure 29(a) shows the variation of SMSR with normalized (by the threshold) current density, J/J_{th}, for the $\lambda/4$ phase-shifted devices with ($L = 2.8$ and 3.2. It is assumed that the facets are AR coated and current is uniformly injected along the laser cavity. The maximum output power for single-longitudinal-mode operation (defined by a drop of SMSR from 40 to 10 =30dB) of the uniform devices (i.e. $L_d = 0$ Å) with $\kappa L = 2.8$ and 3.2 are equal to 14 and 5 mW, respectively. However, it is clearly seen that the introduction of the built-in refractive-index step profile increases the SMSR, as these devices maintain a single-longitudinal-mode operation up to at least 50 mW. Figure 29(b) shows the refractive-index profile for both the uniform and step-diffused devices with $\kappa L = 3.2$ at high power. The effects of SHB on the refractive-index profile are effectively minimized by the step change of L_d, and the coupling strength between the longitudinal

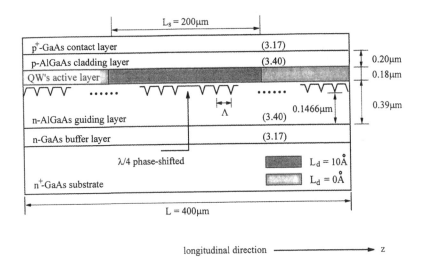

Fig. 28. Schematic diagram of a $\lambda/4$-shifted DFB laser. It is assumed that the $\lambda/4$ phase-shifted is located at the center of the device. L_s is the longitudinal length for the step change ($L_d = 10$ Å) region.

optical intensity profile of the short wavelength band-edge mode and the SHB of carriers is also reduced. The influence of longitudinal heat distribution on the refractive-index profile is also analyzed. It is shown that the variation of temperature is less than 2°C and therefore can be ignored in our analysis [59].

For $\lambda/4$ phase-shifted DFB lasers under large signal modulation, SHB is not taken place during the first overshoot of the output power and single-longitudinal-mode operation is maintained. However, the built-in refractive-index step may reduce the gain requirement of the band-edge modes as well as the single-longitudinal-mode operation. Therefore, it is necessary to consider the influence of built-in step refractive-index profile on the spectral purity of the turn-on transient signal. Figure 30(a) shows the output power spectra at the first overshoot of the step-diffused device ($\kappa L = 2.8$, $L_d = 0$ Å$|10$ Å) for two modulated currents, J$=2.0$J$_{th}$ and 4.5J$_{th}$. The $\kappa L = 2.8$ device is biased initially at the threshold and then modulated by a step current. It is observed that a side mode is excited during the turn-on transient interval for larger current densities and hence it should impose a limitation to the modulation depth requirement. SMSR of the first overshoot power spectra are also shown in Figure 30(b) as a function of the normalized current density for the step-diffused devices with ($L = 2.8$ and 3.2. The results show that no side mode is excited for the case of ($L = 3.2$

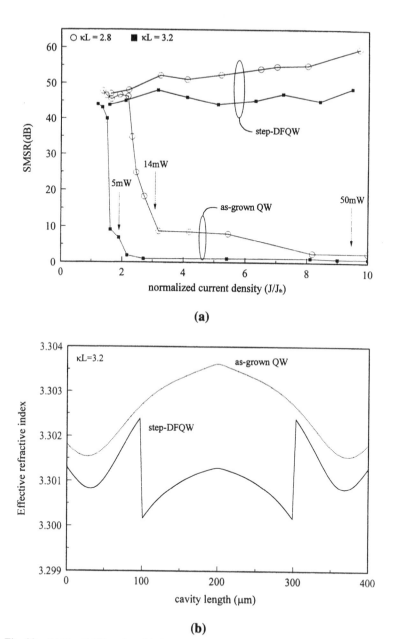

Fig. 29. (a) The SMSR varies with the normalized external injection current, J/J_{th}, for lasers with $\kappa L = 2.8$ (◯) and 3.2 (●). The solid and dotted lines represent cases for the step-diffused devices ($L_d = 0\,\text{Å}|10\,\text{Å}$) and the uniform ($L_d = 0\,\text{Å}$) devices, respectively. (b) The longitudinal refractive index profile of the lasers with notation as in (a). The devices are biased at 8 times of its threshold value.

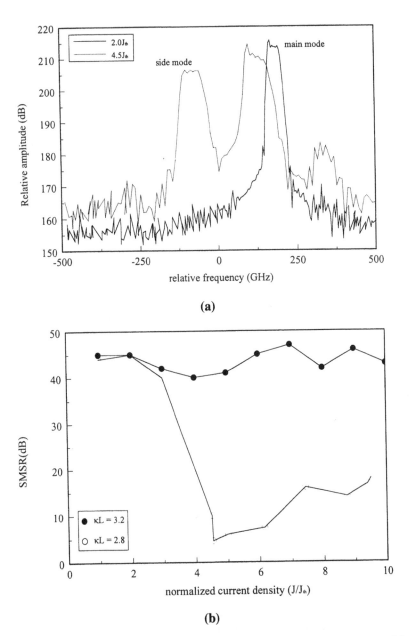

(a)

(b)

Fig. 30. (a) Output power spectra at the first overshoot of the step-diffused device ($\kappa L = 2.8$, $L_d = 0$ Å|10 Å). The device is modulated by a step current with magnitude, $J = 2.0J_{th}$ and $4.5J_{th}$. (b) SMSR of the first overshoot power spectrum varies with the normalized magnitude of the step current, J/J_{th}, for he step-diffused device with $\kappa L = 2.8$ ($\bigcirc$) and 3.2 ($\bullet$).

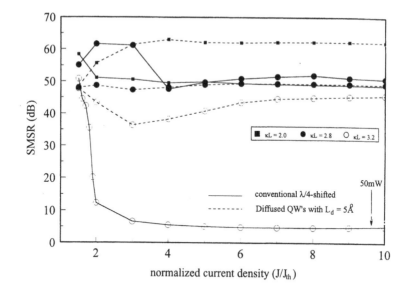

Fig. 31. The side-mode suppression ratio various with normalized current, J/J_{th}, for lasers with $\kappa L = 2.0$ (•), $\kappa L = 2.8$ (•) and $\kappa L = 3.2$ (•). The solid and dash lines represent the cases for the conventional $\lambda/4$ phase-shifted DFB laser and step-diffused device ($L_d = 5$ Å).

and, for ($L = 2.8$, domination of side mode prevails (SMSR < 30 dB) at larger current densities.

4.2.2 Uniform-Grating DFB Lasers with Step-Diffused QW's Structure

The laser structure used in the analysis is similar to Figure 28 except the $\lambda/4$ phase-shifted in the grating structure is removed. A diffusion step is introduced in the center of the active region to form a PAR with optical gain and refractive index slightly less than that of the as-grown region. The product of propagation coefficient difference and effective length of the PAR is set to $\pm(\frac{1}{2} + n)\pi$ (where n is a positive integer [39,60]) to produce a $\lambda/4$ phase-shifted structure. The advantage of using interdiffusion technique to form a phase-adjusted region over (i) phase-adjusted waveguide [39,60] or (ii) corrugation-pitch modulation [37] is that only simple fabrication procedures are required (i.e. diffused QW's structure can be easily obtained by impurities implantation and thermal annealing).

Figure 31 compares the variation of SMSR with normalized injected current density, J/J_{th} (where J_{th} is the threshold current density) at steady state for lasers with ($L_d = 5$ Å) and without (conventional discrete $\lambda/4$ phase-shifted

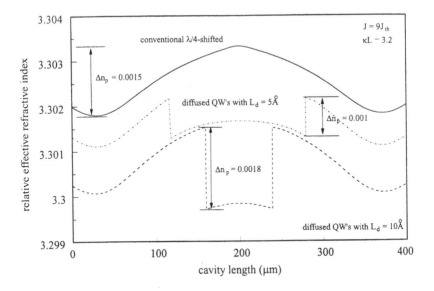

Fig. 32. Longitudinal refractive index profile of the conventional $\lambda/4$ phase-shifted DFB laser (solid line), diffused QW's laser with $L_d = 5$ Å (dash-dotted line) and diffused QW's laser with $L_d = 10$ Å (dash line). The devices are biased at $J = 9J_{th}$ with κL equal to 3.2.

DFB laser) step diffusion profile. Multi-mode operation is observed for conventional discrete $\lambda/4$ phase-shifted DFB laser with $\kappa L \geq 3.2$. However, a stable single-longitudinal-mode is maintained for laser with step diffusion profile. This is because the built-in step refractive-index profile opposes the carrier-induced refractive-index change inside the active region. The maximum output power of the lasers is larger than 50mW at $J/J_{th} = 10$. For the device with ($L_d = 10$ Å), similar behavior is observed and will not be repeated.

The relative effective refractive-index profiles of devices with $\kappa L = 3.2$ and biased at $J = 9J_{th}$ are shown in Figure 32. As we can see, the conventional $\lambda/4$ phase-shifted DFB laser exhibits non-uniform (concave-up) distribution of refractive index with peak to peak value, (n_p, equals 0.0015. However, the uniformity of the effective refractive index is maintained for $L_d = 5$ Å with ($n_p = 0.001$. On the other hand, the built-in refractive-index step for $L_d = 10$ Å is larger than required to overcome the SHB effects and Δn_p is found to be equal to 0.0018. Therefore, only a small range of L_d (5 Å $\leq L_d < 10$ Å) will satisfy the requirement to minimize the influence of SHB effects.

The spectrum purity of the turn-on transient signal determines the maximum modulation bandwidth of the lasers. In step-diffused DFB lasers,

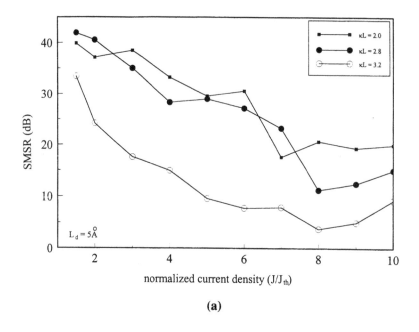

(a)

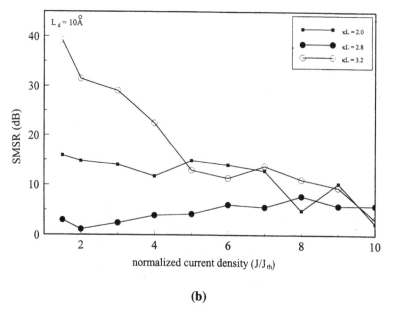

(b)

Fig. 33. The side-mode suppression ratio of the first overshoot power spectrum various with normalized current, J/J_{th}, for device with diffusion length (a) $L_d = 5$ Å and (b) $L_d = 10$ Å. The symbols (•), (•) and (○) represent $\kappa L = 2.0$, 2.8 and 3.2, respectively.

a built-in step refractive-index profile is introduced (by interdiffusion) which may affect the spectrum purity of the lasers as SHB is negligible during the turn-on time interval. Figure 33 shows the influence of built-in refractive-index step on the SMSR at the first overshoot of lasers with $L_d = 5$ and 10 Å. As we can see, (i) SMSR is reduced with the increase of injection current and (ii) devices with low κL exhibit better SMSR. The reduction of SMSR can be attributed to the built-in index profile because no SHB is taken place during the first overshoot of the output power. The built-in refractive-index step reduces the gain requirement of the band-edge mode. Therefore, excitation of band-edge mode is observed in both cases and the case $L_d = 10$ Å is more pronounced than $L_d = 5$ Å.

4.3   Vertical-Cavity Surface-Emitting Lasers with Non-uniform Diffused QW's Structure

The VCSEL with diffused QW's structure under investigation is shown in Figure 34. It is assumed that the laser has a circular metal contact of diameter 10(m on the epitaxial side (p-side) for current injection. An active layer is sandwiched between two undoped spacer layers and two Bragg reflectors. Each undoped spacer layer has half-wavelength thick (~ 0.1 μm). The Bragg reflectors are formed by alternating layers of AlAs and AlGaAs with quarter-wavelength thick of dielectric layers on both the n- and p-side, respectively. The total number of layers is equal to 36 (~ 2.1 μm) on the n-side and 90 on the p-side (~ 5.4 μm). The active layer consists of three $Al_{0.3}Ga_{0.7}As$/GaAs quantum-wells with total thickness of half-wavelength (~ 0.1 μm). The step diffused QW's structure can be realized by applying compositional disordering into the QW's active layers. The selective injection of impurities across the p-Bragg reflector can be done by using a circular hollow mask to shelter from the implantation. It must be noted that the thickness of p-Bragg reflector and the impurities used in implantation should be carefully selected, otherwise, penetration of impurities into the active layer may not be possible.

The numerical laser model and parameters used to investigate the modal behavior of VCSEL with non-uniform diffused QW's structure can be found in reference [62]. Figure 35 shows the L-I curves of VCSEL's (a) with and (b) without a step diffused QW's structure. In the investigation, the radius of the diffusion area is equal to 3.8 μm. It is observed that a 'kink' is occurred in the L-I curves with the excitation of first-order mode (LP_{11}). With the selective defined step diffused QW's structure introduced into the active region, the 'kink' is shifted upward and a stable fundamental-mode (LP_{01}) operation can be maintained at a high power level. Figure 36 compares the refractive-index profile of the lasers, with and without diffused QW's structure, operating at 1.3 mW output power. The refractive-index profile at

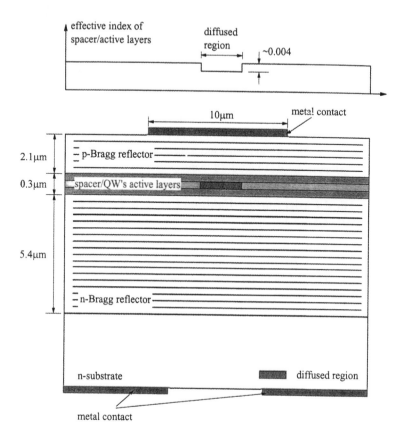

Fig. 34.   Schematic diagram of a VCSEL with diffused QW's structure.

threshold for gain-guided laser (i.e. without a step diffused QW's structure) is also shown in the figure. For the case without a step-diffused QW's structure, the refractive index of the active layer increases near the center of the core region due to the self-focusing effect. This results in the optical field with transverse modes moves towards the center region and LP_{01} mode is excited. On the other hand, the increase in refractive index is counteracted by the step diffused QW's structure and a single-transverse-mode operation is maintained. Figure 37 shows the profile of LP_{01} and LP_{11} modes with and without a step-diffused QW's structure. For diffused QW's laser, a 'kink' is appeared in the LP_{01} and LP_{11} modes. However, for gain-guided laser, no 'kink' is observed. The purpose of deforming the profile of the transverse modes is to minimize the influence of self-focusing. Figure 38

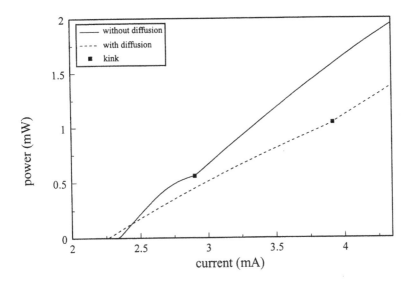

Fig. 35. Light-current characteristics of VCSEL's without (solid line) and with (dotted line) a step-diffused QW's structure.

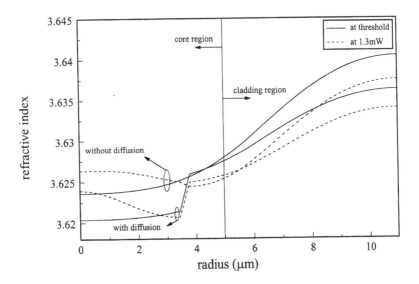

Fig. 36. Transverse distribution of refractive index of VCSEL's with and without diffused QW's structure at threshold (solid line) and output power of 1.3mW (dotted line).

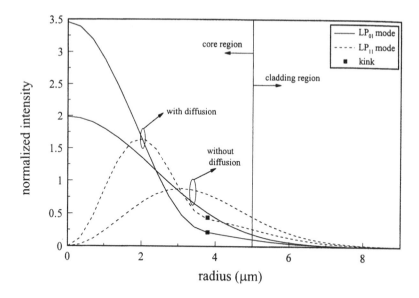

Fig. 37. Profile of LP$_{01}$ (solid line) and LP$_{11}$ (dotted line) modes for VCSEL's with and without diffused QW's structure.

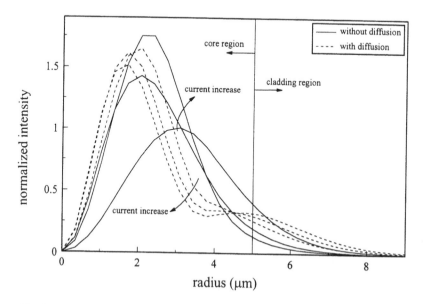

Fig. 38. Variation of LP$_{11}$ mode with the injection current increase from I_{th} to $1.9I_{th}$, where I_{th} is the threshold current. The VCSEL's without (solid line) and with (dotted line) a step-diffused QW's structure are chosen for investigation.

shows the variation of LP_{11} mode with the injection current. It is observed that self-focusing has less influence on LP_{11} mode of diffused QW's laser. The volume of LP_{11} mode inside the cladding region is not varied by an increasing injection current such that a stable operation of LP_{01} mode can be achieved. However, the beam-width of LP_{11} mode is reduced significantly by a self-focusing effect of gain-guided laser.

5   CONCLUSION

Novel semiconductor lasers using diffused QW's structure are proposed and analyzed in this chapter. It is found that stable single-mode operation of semiconductor lasers can be improved significantly by using diffused QW's structure. This is due to the unique electrical and optical properties of QW's material under compositional-induced disordering. In addition, the design criteria of diffused QW's lasers is also discussed and the results given in this chapter can be utilized as a guideline to realize high-performance semiconductor lasers using diffused QW's structure.

In Section 1, various techniques for compositional-induced disordering of QW's have been reviewed. Semiconductor lasers using diffused QW's structure are fabricated with significant improvement in the confinement of injection current and light. This is due to the reduction of refractive index and electrical conductivity of the diffused QW's. In practice, (i) impurity-induced disordering and (ii) impurity-free vacancies diffusion are the two possible ways to realized diffused QW's. The former method consists of ion implantation and rapid thermal annealing processes for the diffusion of atoms. However, the scattering and free carrier absorption losses of the diffused QW's material are increased due to the ionization of ions. In order to cope with this problem, the latter method, impurity-free vacancies diffusion is utilized to obtain the interdiffusion without the presence of ion implantation process.

The spectral characteristics of conventional semiconductor lasers are discussed in Section 2. It is noted that the performance of the high-speed long-haul optical fiber communication systems is limited by the multiple longitudinal-mode operation of facet-emitted lasers (i.e. FP and DFB lasers). Multiple longitudinal-mode operation in FP lasers is attributed to the wide linewidth of the gain spectrum. The excitation of side modes in DFB lasers is due to the SHB of carrier concentration. On the other hand, multiple transverse-mode operation is observed in VCSEL's especially for devices with large area and under high current injection. The origin of side-mode excitation can be attributed to SHB of carrier concentration and thermal lensing. Under these circumstances, novel structure of high-performance semiconductor lasers is required for a stable single-mode operation. The optical properties (i.e. optical gain and background refractive index) of diffused $Al_{0.3}Ga_{0.7}As/GaAs$ QW's have been analyzed in Section 3. It is shown that the gain-peak wavelength, optical gain peak and the corresponding refractive-index change can be approximated by a logarithmic function with the carrier concentration and the extent of interdiffusion as the variables. However, at the wavelength equals 0.85(m and $L_d > 20$ Å, the optical gain can only be approximated by a linear function of carrier concentration but the corresponding change of refractive-index change is negligible. The fitting parameters given in this chapter can be utilized to estimate the characteristics of semiconductor lasers with diffused QW's structure.

In Section 4, the improvement of semiconductor lasers by using the diffused QW's structure has been analyzed. A novel structure of FP semiconductor lasers by interdiffusion of QW's active region has been proposed and analyzed in Section 4.1. A single-longitudinal-mode operation is achieved in periodic diffused QW's FP lasers at and above threshold. It is shown that the diffused QW's structure enhances SMSR of FP lasers. This is due to the filtering effects arise from the difference of refractive index and optical gain between the diffused sections. It is revealed that device with L_d's combination equals 0|60 Å exhibits better spectrum purity than device with L_d's combination equals 0|10 Å even the threshold current densities of both devices are close together. This is because the periodic diffused QW's structure demonstrates better optical filtering efficiency with large L_d which compensates for the increase of optical loss arises from the interdiffusion effects. Although the threshold current density of the proposed FP lasers with diffused QW's structure is deteriorated by 16% (compares with FP lasers without interdiffusion structure), the SMSR is enhanced by more than 10 dB at threshold.

The proposed DFB semiconductor lasers with step diffused QW's structure defined in the active region have been analyzed in Section 4.2. The devices under investigation have grating structure with and without $\lambda/4$ phase-shifted region. It is observed that the former laser structure exhibits a stable single-longitudinal-mode operation with large coupling-length product. This is because the diffusion step counteracts the influence of carrier SHB inside the QW's active layers. On the other hand, phase-adjustment region can be applied into the DFB lasers with uniform grating structure through the disordering of QW's active layer. It is shown that side-mode suppression ratio of the devices can be enhanced even for devices with large κL. The merit of these laser structures is that the longitudinal SHB is less severe than the conventional $\lambda/4$-shifted DFB lasers.

In Section 4.3, the possibility of using a step diffused QW's structure in VCSELs has been discussed. It is observed that the maximum output power for a stable single-mode operation can be enhanced. The step diffused QW's structure creates a non-uniform refractive-index profile inside the active region in which the influence of carrier SHB and thermal lensing can be minimized. The success of the above proposal depends upon whether the diffused QW's semiconductor lasers can be realized in practice by using the existing fabrication techniques such that the production cost and wastage in devices' fabrication can be further reduced. In fact, the possibility to realize diffused QW's structure depends upon the following criteria:

- The proposed semiconductor lasers have typical dimensions which require simple processing technique and are compatible with the existing fabrication techniques.
- The penetration depth of impurities or vacancies is in the same order of magnitude to the depth of the QW's active layer below the p-cladding layer of the lasers such that the diffusion process can be carried out.
- The interdiffusion length, L_d, of the diffused QW's active region is determined by the implantation energy and thermal annealing time. With careful control of annealing temperature and time, L_d, down to 5 Å, can be obtained without any difficulty.
- The formation accuracy of diffusion pattern can be realized by the combined techniques of Electron Beam Lithography and implantation enhanced disordering to achieve nanometer structure which is far more precise than our requirement.

The other advantages for adopting diffused QW's structure are: (i) the tunability of the operating wavelength by modifying the diffusion profile and (ii) the use of interdiffusion technique can avoid the complex fabrication process of semiconductor lasers.

References

1. S.Y. Hu, S.W. Corzine, K.K. Law, D.B. Young, A.C. Gossard, L.A. Coldren, and J.L. Merz, *J. Appl. Phys.*, **76**, 4479 (1994).
2. R.L. Thornton, R.D. Burnham, T.L. Paoli, N. Holonyak, Jr., and D.G. Deppe, *Appl. Phys. Lett.*, **47**, 1239 (1985).
3. R.L. Thornton, R.D. Burnham, T.L. Paoli, N. Holonyak, Jr., and D.G. Deppe, *Appl. Phys. Lett.*, **49**, 133 (1986).
4. R.L. Thornton, R.D. Burnham, and T.L. Paoli, *Appl. Phys. Lett.*, **48**, 7 (1986).
5. K. Sekartedjo, N. Eda, K. Furuya, et al., *Electron. Lett.*, **20**, 80 (1984).
6. V. Paschoas, T. Sphicopoulos, D. Syvridis, and C. Caroubalos, *IEEE J. Quantum Electron.*, **QE-30**, 660 (1994).
7. K.L. Lear, K.D. Choquette, R.P. Schneider, Jr., S.P. Kilcoyne, and K.M. Geib, *Electron. Lett.*, **31**, 208 (1995).
8. Y. Hayashi, T. Mukaihara, N. Hatori, N. Ohnoki, A. Matsutani, F. Koyama, and K. Iga, *Electron. Lett.*, **31**, 560 (1995).
9. T. Tsukuda, *J. Appl. Phys.*, **45**, 4899 (1974).
10. W.T. Tasng, R.A. Logan, and J.A. Ditzenberger, *Electron. Lett.*, **18**, 845 (1982).
11. J.H. Marsh, S.I. Hansen, A.C. Bryce, and R.M. De La Rue, *Opt. and Quantum Electron.*, **23**, S941 (1991).

12. S.R. Andrew, J.H. Marsh, M.C. Holland, and A.H. Kean, *IEEE Photon. Technol.*, **4**, 426 (1992).

13. G.A. Vawter, D.R. Myers, M.T. Brennan, and B.E. Hammons, *Appl. Phys. Lett.*, **56**, 1945 (1990).

14. R.P. Bryan, J.J. Coleman, L.M. Miller, M.E. Givens, R.S. Averback, and J.L. Klatt, *Appl. Phys. Lett.*, **55**, 94 (1989).

15. J. Werner, T.P. Lee, E. Kapon, E. Colas, N.G. Stoffel, S.A. Schwarz, L.C. Scheartz, and N.C. Andreadakis, *Appl. Phys. Lett.*, **57**, 810 (1990).

16. P. Cusumano, J.H. Marsh, M.J. Rose, and J.S. Roberts, *IEEE Photon. Technol.*, **9**, 282 (1997).

17. S.Y. Hu, M.G. Peters, D.B. Young, A.C. Gossard, and L.A. Coldren, *IEEE Photon. Technol.*, **7**, 712 (1995).

18. K. Meehan, P. Gavrilovic, and N. Holonyak, Jr., R.D. Burnham, and R.L. Thornton, *Appl. Phys. Lett.*, **46**, 75 (1985).

19. J. Werner, E. Kapon, N.G. Stoffel, E. Colas, S.A. Schwarz, C.L. Schwartz, and N. Andreadakis, *Appl. Phys. Lett.*, **55**, 540 (1989).

20. W.D. Laidig, N. Holonyak, Jr., and M.D. Camras, *Appl. Phys. Lett.*, **38**, 776 (1981).

21. K. Goto, F. Uesugi, S. Takahashi, T. Takiguchi, E. Omura, and Y. Mihashi, *Jpn. J. Appl. Phys.*, **33**, 5774 (1994).

22. C. Kaden, H.P. Gauggel, V. Hofsäss, A. Hase, and H. Schweizer, *Appl. Phys, Lett.*, **65**, 3170 (1994).

23. S. Charbonneau, P.J. Poole, P.G. Piva, M. Buchanan, R.D. Goldberg, and I.V. Mitchell, *Nucl. Instru. and Meth. In Phys. Res. B*, **106**, 457 (1995).

24. S. Charbonneau, P.J. Poole, P.G. Piva, G.C. Aers, E.S. Koteles, M. Fallahi, J.J. He, J.P. McCaffrey, M. Buchanan, and M. Dion, *J. Appl. Phys.*, **78**, 3697 (1995).

25. J. Marsh, *Semicond. Sci. Technol.*, **8**, 1136 (1993).

26. J.D. Ralston, S. O'Brien, G.W. Wick, and L.F. Eastman, *Appl. Phys. Lett.*, **52**, 1511 (1988).

27. S. Bürkner, M. Maier, E.C. Larkins, W. Rothemund, E.P. O'Reilly, and J.D. Ralston, *J. Electron. Mater.*, **24**, 805 (1995).

28. S.G. Ayling, J. Beauvais, and J.H. Marsh, *Electron. Lett.*, **28**, 2240 (1992).

29. B.S. Ooi, G. Ayling, A.C. Bryce, and J.H. Marsh, *IEEE Photon. Technol. Lett.*, **7**, 944 (1995).

30. C.J. Chang-Hasnain, J.P. Harbison, G. Hasnain, A.C. Von Lehmen, L.T. Florez, and N.G. Stoffel, *IEEE J. Quantuam Electron.*, **QE-27**, 1402 (1991).

31. K.L. Lear, R.P. Schneider, Jr., K.D. Choquette, and S.P. Kilcoyne, *IEEE Photon. Technol. Lett.*, **8**, 740 (1996).

32. P.D. Floyd, M.G. Peters, L.A. Coldern, and J.L. Merz, *IEEE Photon. Technol. Lett.*, **7**, 1388 (1995).

33. Y.J. Yang, T.G. Dziura, T. Bardin, S.C. Wang, and R. Fernandez, *Electron. Lett.*, **28**, 274 (1992).

34. D. Marcuse, and T.P. Lee, *IEEE J. Quantum Electron.*, **QE-19**, 1397 (1983).

35. H. Kogelnik, and C.V. Shank, *J. Appl. Phys.*, **43**, 2327 (1972).

36. H.A. Haus, and C.V. Shank, *IEEE J. Quantum Electron.*, **QE-12**, 532 (1976).

37. J.E.A. Whiteaway, G.H.B. Thompson, A.J. Collar, and C.J. Armistead, *IEEE J. Quantum Electron.*, **QE-25**, 1761 (1989).

38. L.M. Zhang, S.F. Yu, M.C. Nowell, D.D. Marcenac, J.E. Carroll, and R.G.S. Plumb, *IEEE J. Quantum Electron.*, **QE-30**, 1389 (1994).

39. H. Soda, Y. Kotaki, H. Sudo, H. Ishikawa, S. Yamakoshi, and H. Imai, *IEEE J. Quantum Electron.*, **QE-23**, 804 (1987).

40. G.C. Wilson, D.M. Kuchta, J.D. Walker, and J.S. Smith, *Appl. Phys. Lett.*, **64**, 542 (1994).

41. A. Valle, J. Sarma, and K.A. Shore, *IEEE J. Quantum Electron.*, **QE-31**, 1423 (1995).

42. D. Vakhshoori, J.D. Wynn, G.J. Zydzik, R.E. Leibenguth, M.T. Asom, K. Kojima, and R.A. Morgan, *Appl. Phys. Lett.*, **62**, 1448 (1993).

43. R. Baets, *IEE Proc. Pt. J*, **135**, 233 (1988).

44. W. Nakwaski, *IEE Proc. Pt. J*, **137**, 129 (1990).

45. K. Tai, R.J. Fischer, C.W. Seabury, N.A. Olsson, T.C.D. Huo, Y. Ota, and A.Y. Cho, *Appl. Phys. Lett.*, **55**, 2473 (1989).

46. A. Valle, J. Sarma, and K.A. Shore, *Opt. Commun.*, **115**, 297 (1995).
47. Y.G. Zhao, and J.G. McInerney, *IEEE J. Quantum Electron.*, **QE-32**, 1950 (1996).
48. E.H. Li and K.S. Chan, *Electron. Lett.*, **29**, 1233 (1993).
49. E.H. Li, B.L. Weiss, and K.S. Chan, *Phys. Rev. B.*, **46**, 15181 (1992).
50. T.A. DeTemple and C.M. Herzinger, *IEEE J. Quantum Electron.*, **QE-29**, 1246 (1993).
51. C.H. Herny, R.A. Logan, and K.A. Bertness, *J. Appl. Phys.*, **52**, 4457 (1981).
52. G.P. Agrawal, and N.K. Dutta, *"Long-wavelength Semiconductor Lasers"*, Van Nostrand Reinhold Company, New York (1986).
53. E.H. Li, B.L. Weiss, K.S. Chan, and J. Micallef, *Appl. Phys. Lett.*, **62**, 550 (1992).
54. C.C. Lee, and D.H. Chien, *IEEE J. Lightwave Technol.*, **14**, 1847 (1996).
55. Y.A. Wu, G.S. Li, R.F. Nabiev, K.D. Choquette, C. Caneau, and C.J. Chang-Hasnain, *IEEE J. of Selected Topics in Quantum Electron.*, **1**, 629 (1995).
56. U. Fiedler, B. Moller, G. Reiner, D. Wiedenmann, and K.J. Ebeling, *IEEE Photon. Technol. Lett.*, **7**, 1116 (1995).
57. S.F. Yu, C.W. Lo, and E.H. Li, *IEEE J. Quantum Electron.*, **QE-33**, 999 (1997).
58. S.F. Yu, and E.H. Li, *IEEE Photon. Technol. Lett.*, **8**, 482 (1996).
59. W.B. Joyce, and R.W. Dixon, *J. Appl. Phys.*, **46**, 855 (1975).
60. J. Kinoshita, K. Ohtsuka, H. Agatsuma, A. Tanaka, T. Matsuyama, A. Makuta, and H. Kobayashi, *IEEE J. Quantum Electron.*, **QE-27**, 1759 (1991).
61. R.F. Kazarinov, and C.H. Henry, *IEEE J. Quantum Electron.*, **QE-21**, 144 (1985).
62. S.F. Yu, and C.W. Lo, *IEE Proc.-Optoelectron.*, **143**, 189 (1996).

Index

In preparation

**The Optics of Semiconductor Quantum Wires and Dots:
Fabrication, Characterization, Theory and Application**
Edited by Garnett W. Bryant

Vertical-Cavity Surface-Emitting Lasers and Their Applications
Edited by Julian Cheng and Niloy K. Dutta

**Characterization of Reduced Dimensional Semiconductor
Microstructures**
Edited by Fred H. Pollak

II–VI Semiconductor Materials and Their Applications
Edited by Maria C. Tamargo